T0238117

Lecture Notes in Computer Science 3557

Commenced Publication in 1973
Founding and Former Series Editors:
Gerhard Goos, Juris Hartmanis, and Jan van Leeuwen

Henri Gilbert Helena Handschuh (Eds.)

Fast Software Encryption

12th International Workshop, FSE 2005
Paris, France, February 21-23, 2005
Revised Selected Papers

Volume Editors

Henri Gilbert
France Telecom, 92794 Issy les Moulineaux, France
E-mail: henri.gilbert@francetelecom.com

Helena Handschuh
Gemplus SA, Issy-les-Moulineaux, France
E-mail: Helena.Handschuh@gemplus.com

Library of Congress Control Number: 2005928340

CR Subject Classification (1998): E.3, F.2.1, E.4, G.2, G.4

ISSN 0302-9743
ISBN-10 3-540-26541-4 Springer Berlin Heidelberg New York
ISBN-13 978-3-540-26541-2 Springer Berlin Heidelberg New York

Springer is a part of Springer Science+Business Media

springeronline.com

Printed in Germany

Typesetting: Camera-ready by author, data conversion by Scientific Publishing Services, Chennai, India
Printed on acid-free paper SPIN: 11502760 06/3142 5 4 3 2 1 0

Preface

The Fast Software Encryption 2005 Workshop was the twelfth in a series of annual workshops on symmetric cryptography, sponsored for the fourth year by the International Association for Cryptologic Research (IACR). The workshop concentrated on all aspects of fast primitives for symmetric cryptology, including the design, cryptanalysis and implementation of block and stream ciphers as well as hash functions and message authentication codes. The first FSE workshop was held in Cambridge in 1993, followed by Leuven in 1994, Cambridge in 1996, Haifa in 1997, Paris in 1998, Rome in 1999, New York in 2000, Yokohama in 2001, Leuven in 2002, Lund in 2003, and New Delhi in 2004.

This year, a total of 96 submissions were received. After an extensive review by the Program Committee, 30 submissions were accepted. Two of these submissions were merged into a single paper, yielding a total of 29 papers accepted for presentation at the workshop. Also, we were very fortunate to have in the program an invited talk by Xuejia Lai on "Attacks and Protection of Hash Functions" and a very entertaining rump session that Bart Preneel kindly accepted to chair. These proceedings contain the revised versions of the accepted papers; the revised versions were not subsequently checked for correctness.

We are very grateful to the Program Committee members and to the external reviewers for their hard work. Each paper was refereed by at least three reviewers, and at least five reviewers in the case of papers (co-)authored by Program Committee members; eventually, an impressive total of 334 reviews was produced. Special thanks are also due to the members of the Local Organizing Committee, Côme Berbain, Olivier Billet (who designed the FSE 2005 Web pages and assembled the preproceedings), Julien Brouchier (who managed the submission and Webreview servers), Stanislas Francfort, Aline Gouget, Françoise Levy, Pierre Loidreau, and Pascal Paillier (who managed on-site registration), for their generous efforts and strong support.

Many thanks to Kevin McCurley for handling the registration server, to Patrick Arditti, Virginie Berger and Claudine Campolunghi for providing assistance with the registration process, and to the research group COSIC of the K.U.Leuven for kindly providing their Webreview software.

Last but not least, we would like to thank the conference sponsors France Telecom, Gemplus, and Nokia for their financial support, DGA and ENSTA for hosting the conference on their premises, and all submitters and workshop participants who made this year's workshop such an enjoyable event.

April 2005 Henri Gilbert and Helena Handschuh

FSE 2005

February 21–23, 2005, Paris, France

Sponsored by
the International Association for Cryptologic Research (IACR)

Program and General Chairs

Henri Gilbert .... France Telecom, France
Helena Handschuh .... Gemplus, France

Program Committee

Kazumaro Aoki .... NTT, Japan
Steve Babbage .... Vodafone, UK
Eli Biham .... Technion, Israel
Anne Canteaut .... INRIA, France
Don Coppersmith .... IBM Research, USA
Joan Daemen .... STMicroelectronics, Belgium
Thomas Johansson .... Lund University, Sweden
Antoine Joux .... DGA and Université de Versailles, France
Xuejia Lai .... Shanghai Jiaotong University, China
Stefan Lucks .... Universität Mannheim, Germany
Mitsuru Matsui .... Mitsubishi Electric, Japan
Willi Meier .... FH Aargau, Switzerland
Kaisa Nyberg .... Nokia, Finland
Bart Preneel .... K.U.Leuven, Belgium
Matt Robshaw .... Royal Holloway, University of London, UK
Palash Sarkar .... Indian Statistical Institute, India
Serge Vaudenay .... EPFL, Switzerland
Moti Yung .... Columbia University, USA

Local Organizing Committee

Côme Berbain, Olivier Billet, Julien Brouchier, Stanislas Francfort, Henri Gilbert, Aline Gouget, Helena Handschuh, Françoise Levy, Pierre Loidreau, Pascal Paillier

Industry Sponsors

France Télécom
Gemplus SA
Nokia

External Referees

Frederik Armknecht
Daniel Augot
Gildas Avoine
Thomas Baignères
Elad Barkan
An Braeken
Claude Carlet
Pascale Charpin
Sanjit Chatterjee
Rafi Chen
Debra L. Cook
Christophe De Cannière
Orr Dunkelman
Matthieu Finiasz
Pierre-Alain Fouque
Soichi Furuya
Louis Granboulan
Tor Helleseth
Shoichi Hirose
Tetsu Iwata
Pascal Junod
Charanjit Jutla
Grigory Kabatyanskiy
Jonathan Katz
Alexander Kholosha
Yuichi Komano
Matthias Krause
Ulrich Kühn
Simon Künzli
Shreekanth Laksmeshwar
Joseph Lano
Cédric Lauradoux
Yi Lu
Marine Minier
Håvard Molland
Jean Monnerat
Shiho Moriai
Frédéric Muller
Sean Murphy
Philippe Oechslin
Kenji Ohkuma
Katsuyuki Okeya
Elisabeth Oswald
Souradyuti Paul
Gilles Piret
Zulfikar Ramzan
Vincent Rijmen
Akashi Satoh
Takeshi Shimoyama
Taizo Shirai
François-Xavier Standaert
Dirk Stegemann
Henk van Tilborg
Hiroki Ueda
Kan Yasuda
Erik Zenner

Table of Contents

New Designs

Stream Ciphers I

Boolean Functions

Block Ciphers I

Stream Ciphers II

Hash Functions

Modes of Operation

Stream Ciphers III

Block Ciphers II

Implementations

A New MAC Construction ALRED and a Specific Instance ALPHA-MAC

Joan Daemen[1] and Vincent Rijmen[2,3,⋆]

[1] STMicroelectronics Belgium
joan.daemen@st.com
[2] IAIK, Graz University of Technology,
vincent.rijmen@iaik.tugraz.at
[3] Cryptomathic A/S

Abstract. We present a new way to construct a MAC function based on a block cipher. We apply this construction to AES resulting in a MAC function that is a factor 2.5 more efficient than CBC-MAC with AES, while providing a comparable claimed security level.

1 Introduction

Message Authentication Codes (MAC functions) are symmetric primitives, used to ensure authenticity of messages. They take as input a secret key and the message, and produce as output a short tag.

Basically, there are three approaches for designing MACs. The first approach is to design a new primitive from scratch, as for instance MAA [15] and, more recently, UMAC [8]. This approach allows to optimize the security-performance trade-off. The second approach is to define a new *mode of operation* for existing primitives. In this category, we firstly have numerous variants based on the CBC encryption mode for block ciphers, e.g. CBC-MAC [5], OMAC [16], TMAC [22], XCBC [9], and RMAC [17]. Secondly, there are the designs based on an unkeyed hash function: NMAC, HMAC [7, 4]. Finally, one can design new MACs using components from existing primitives, e.g. MDx-MAC [24] and Two-Track MAC [10].

In this paper, we propose a new MAC design method which belongs in the third category, cf. Section 3. We also present a concrete construction in Section 5. Before going there, we start with a discussion of security requirements for MACs and we present a new proposal for MAC security claims in Section 2. We discuss internal collisions for the new model in Section 4, and for the concrete construction in Section 6. Section 7 contains more details on extinguishing differentials, a special case of internal collisions. We briefly discuss performance in Section 8 and conclude in Section 9.

⋆ This researcher was supported financially by the A-SIT, Austria.

H. Gilbert and H. Handschuh (Eds.): FSE 2005, LNCS 3557, pp. 1–17, 2005.

2 Iterative MAC Functions and Security Claims

A MAC function maps a key-message pair to a tag. The basic property of a MAC function is that it provides an unpredictable mapping between messages and the tag for someone who does not know, or only partially knows the key. Usually, one defines a number of design objectives that a cryptographic primitive of a given type must satisfy in order to be considered as secure. For MAC functions, we find the following design objectives in [23–Table 9.2]:

- *Key non-recovery*: the expected complexity of any key recovery attack is of the order of 2^{ℓ_k} MAC function executions.
- *Computation resistance*: there is no forgery attack with probability of success above $\max(2^{-\ell_k}, 2^{-\ell_m})$.

Here ℓ_k is the key length and ℓ_m the tag length. By forgery one means the generation of a message-tag pair (m, t) using only information on pairs (m_i, t_i) with $m \neq m_i$ for all i.

2.1 Iterative MAC Functions

Most practical MAC functions are *iterative*. An iterative MAC function operates on a working variable, called the *state*. The message is split up in a sequence of message blocks and after a (possibly keyed) initialization the message blocks are sequentially injected into the state by a (possibly keyed) iteration function. Then a (possibly keyed) final transformation may be applied to the state resulting in the tag.

Iterative MAC functions can be implemented in hardware or software with limited amount of working memory, irrespective of the length of the input messages. They have the disadvantage that different messages may be found that lead to the same value of the state before the final transformation. This is called an *internal collision* [26].

2.2 Internal Collisions

Internal collisions can be used to perform forgery. Assume we have two messages m_1 and m_2 that result in an internal collision. Then for any string m_3 the two messages $m_1 \| m_3$ and $m_2 \| m_3$ have the same tag value. So given the tag of any message $m_1 \| m_3$, one can forge the tag of the message $m_2 \| m_3$. Internal collisions can often be used to speed up key recovery as well [25]. If the number of bits in the state is n, finding an internal collision takes at most $2^n + 1$ known pairs. If the state transformation can be modeled by a random transformation, one can expect to find a collision with about $2^{n/2}$ known pairs due to the birthday paradox. One may consider to have a final transformation that is not reversible to make the detection of internal collisions infeasible. However, as described in Appendix A, this is impossible.

The presence of internal collisions makes that even the best iterative MAC function cannot fulfill the design objectives given above: if the key is used to generate tags over a very large number of messages, an internal collision is likely to occur and forgery is easy.

For many MAC schemes based on CBC-MAC with the DES as underlying block cipher, internal collisions can be used to retrieve the key: the ISO 9797 [5] schemes are broken in [11, 18]. More sophisticated variants like Retail MAC [1] and MacDES [19] are broken in [25], respectively [12, 13].

One approach to avoid the upper limit due to the birthday paradox in iterative MAC functions is *diversification*. The MAC function has next to the message and the key a third input parameter that serves to diversify the MAC computation to make the detection of internal collisions impossible. For the proofs of security that accompany these schemes, the implementation of a tag generating device must impose that its value is non-repeating or random. This method has several important drawbacks. First of all, not only the tag must be sent along with the message, but also this third parameter, typically doubling the overhead. In case of a random value, this puts the burden on the developer of the tag generating device to implement a cryptographic random generator, which is a non-trivial task. Moreover the workload of generating the random value should be taken into account in the performance evaluation of the primitive. In case of a non-repeating value the MAC function becomes stateful, i.e., the tag generation device must keep track of a counter or multiple counters and guarantee that these counters cannot be reset. But in some cases even the randomization mechanism itself introduces subtle flaws. The best known example of a randomized MAC is RMAC [17] cryptanalyzed in [21].

Another way to avoid internal collisions is to impose an upper bound on the number of tags that may be generated with a given key. If this upper bound is large enough it does not impose restrictions in actual applications. This is the approach we have adopted in this paper.

2.3 A Proposal for New Security Claims

We formulate a set of three orthogonal security claims that an iterative MAC function should satisfy to be called secure.

Claim 1 *The probability of success of any forgery attack not involving key recovery or internal collisions is* $2^{-\ell_m}$.

Claim 2 *There are no key recovery attacks faster than exhaustive key search, i.e. with an expected complexity less than* $2^{\ell_k - 1}$ *MAC function executions.*

We model the effect of internal collisions by a third dimension parameter, the *capacity* ℓ_c. The capacity is the size of the internal memory of a random map with the same probability for internal collisions as the MAC function.

Claim 3 *The probability that an internal collision occurs in a set of* A *((adaptively) chosen message, tag) pairs, with* $A < 2^{\ell_c/2}$, *is not above* $1 - \exp(-A^2/2^{\ell_c+1})$.

Note that for $A < 1/4 \times 2^{\ell_c/2}$ we have $1 - \exp(-A^2/2^{\ell_c+1}) \approx A^2/2^{\ell_c+1}$. In the best case the capacity ℓ_c is equal to the number of bits of the state. It is up to the designer to fix the value of the capacity ℓ_c used in the security claim.

3 The ALRED Construction

We describe here a way to construct a MAC function based on an iterated block cipher. The key length of the resulting MAC function is equal to that of the underlying block cipher, the length of the message must be a multiple of ℓ_w bits, where ℓ_w is a characteristic of a component in the MAC function. In our presentation, we use the term *message word* to indicate ℓ_w-bit message blocks and call ℓ_w the *word length.*

3.1 Definition

The ALRED construction consists of a number of steps:

1. Initialization:
   (a) Initialize the state with the all-zero block.
   (b) Apply the *block cipher* to the state.
2. Chaining: for each message word perform an iteration:
   (a) Map the bits of the message word to an *injection input* that has the same dimensions as a sequence of r round keys of the block cipher. We call this mapping the *injection layout.*
   (b) Apply a sequence of r *block cipher round functions* to the state, with the injection input taking the place of the round keys.
3. Final transformation:
   (a) Apply the *block cipher* to the state.
   (b) Truncation: the tag is the first ℓ_m bits of the state.

Let the message words be denoted by x_i, the state after iteration i by y_i, the key by k and the tag by z. Let f denote the iteration function, which consists of the combination of the injection layout and the sequence of r block cipher round functions. Then we can write:

$$y_0 = \mathrm{Enc}_k(0) \quad (1)$$
$$y_i = f(y_{i-1}, x_i), \quad i = 1, 2, \ldots, q \quad (2)$$
$$z = \mathrm{Trunc}(\mathrm{Enc}_k(y_q)) \quad (3)$$

The construction is illustrated in Figure 1 for the case $r = 1$. With this approach, the design of the MAC function is limited to the choice of the block cipher, the number of rounds per iteration r, the injection layout and ℓ_m. The goal is to choose these such that the resulting MAC function fulfills the security claims for iterated MAC functions for some value of ℓ_m and ℓ_c near the block length.

3.2 Motivation

Prior to the chaining phase, the state is initialized to zero and it is transformed by applying the block cipher, resulting in a state value unknown to the attacker. In the chaining phase every iteration injects ℓ_w message bits into the state with an

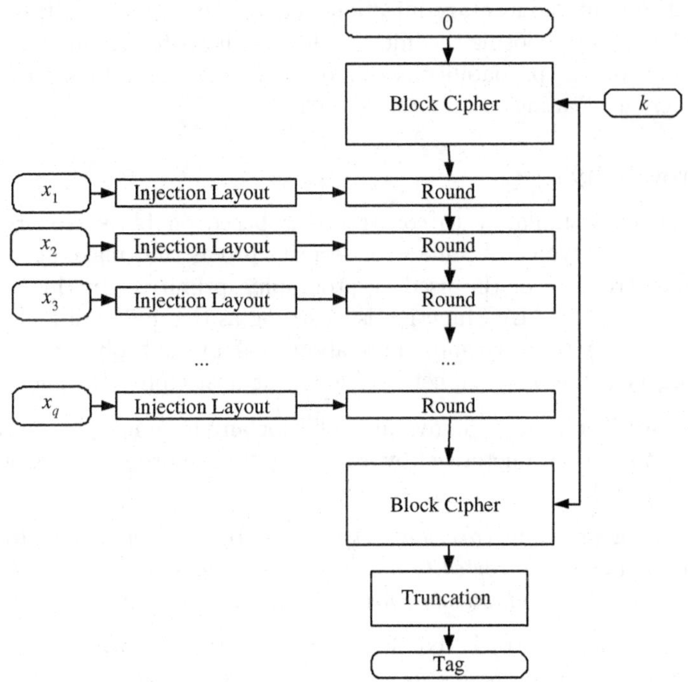

Fig. 1. Scheme of the ALRED construction with $r = 1$

unkeyed iteration function. Without the block cipher application in the initialization, generating an internal collision would be similar to finding a collision in an unkeyed hash function which can be conducted without known message tag pairs. The initial block cipher application makes the difference propagation through the chaining phase, with its nonlinear iteration function, depend on the key.

The iteration function consists of r block cipher rounds where message word bits are mapped onto the round key inputs. The computational efficiency of ALRED depends on the word length. Where in CBC based constructions for long messages there is one block cipher execution per block, ALRED takes merely r rounds per word. Clearly, the performance of ALRED becomes interesting if the word length divided by r is larger than the block length divided by the number of rounds of the block cipher.

Decreasing the message word length with respect to the round key length makes the MAC function less efficient, but also reduces the degrees of freedom available to an attacker to generate internal collisions (see Section 6.1 for an example). Another way to reduce these degrees of freedom is to have the message words first undergo a message schedule, and apply the result as round keys. This is similar to the key schedule in a block cipher, the permutation of message words between the rounds in MD4 [27] or the message expansion in SHA-1 [2]. However, such a message schedule also introduces need for additional memory and additional workload per iteration. Therefore, and for reasons of simplicity, we decided to limit the message injection to a simple layout. Limiting the word length and carefully choosing the injection layout allows to demonstrate powerful upper bounds on the probability of sets of known or chosen messages with any chosen difference leading to an internal collision.

3.3 Provability

ALRED has some similarity to constructions based on block ciphers in CBC mode. The modes typically come with security proofs that make abstraction of the internal structure of the used cryptographic primitive. In this section we prove that an ALRED MAC function is as strong as the underlying block cipher with respect to key recovery and, in the absence of internal collisions, is resistant against forgery if the block cipher is resistant against ciphertext guessing.

Observation: The proofs we give are valid for any chaining phase that transforms y_0 into y_{final} parameterized by a message. In the proofs we denote this by $y_{\text{final}} = F_{\text{cf}}(y_0, m)$.

Theorem 1. *Every key recovery attack on* ALRED, *requiring* t *(adaptively) chosen messages, can be converted to a key recovery attack on the underlying block cipher, requiring* $t+1$ *adaptively chosen plaintexts.*

Proof: Let $\mathcal{A}$ be an attack requiring the t tag values corresponding to the t (adaptively) chosen messages m_j, yielding the key. Then, the attack on the underlying block cipher works as follows.

1. Request $c_0 = \text{Enc}(k, 0)$, where '0' denotes the all-zero plaintext block.
2. For $j = 1$ to t, compute $p_j = F_{\text{cf}}(c_0, m_j)$.
3. For $j = 1$ to t, request $c_j = \text{Enc}(k, p_j)$.
4. Input the tag values $\text{Trunc}(c_j)$ to $\mathcal{A}$ and obtain the key.

□

Theorem 2. *Every forgery attack on* ALRED *not involving internal collisions, requiring* t *(adaptively) chosen messages, can be converted to a ciphertext guessing attack on the underlying block cipher, requiring* $t+1$ *adaptively chosen plaintexts.*

Proof: Let $\mathcal{B}$ be an attack, not involving internal collisions, requiring the t tag values corresponding to the t (adaptively) chosen messages m_j yielding a forged tag for the message m. Then, the ciphertext guessing attack on the underlying block cipher works as follows.

1. Request $c_0 = \text{Enc}(k, 0)$, where '0' denotes the all-zero plaintext block.
2. For $j = 1$ to t, compute $p_j = F_{\text{cf}}(c_0, m_j)$.
3. For $j = 1$ to t, request $c_j = \text{Enc}(k, p_j)$.
4. Input the tag values $\text{Trunc}(c_j)$ to $\mathcal{B}$ and obtain the tag for the message m.
5. Compute $p = F_{\text{cf}}(c_0, m)$.
6. If there is a j for which $p = p_j$, then $\mathcal{B}$ has generated an internal collision, which conflicts with the requirement on $\mathcal{B}$. Otherwise, input the tag values $\text{Trunc}(c_j)$ to $\mathcal{B}$ and obtain the tag, yielding the truncated ciphertext of p.

□

3.4 On the Choice of the Cipher

One may use any block cipher in the ALRED construction. The block length imposes an upper limit to the capacity ℓ_c relevant in the number of tags that may be generated with the same key before an internal collision occurs. When using ciphers with a block length of 64 bits as (Triple) DES and IDEA, the number of tags generated with the same keys should be well below 2^{32}.

The use of the round function for building the iteration function restricts the ALRED construction somewhat. Ciphers that are well suited in this respect are (Triple) DES, IDEA, Blowfish, Square, RC6, Twofish and AES. An ALRED construction based on Serpent, with its eight different rounds, would typically have $r = 8$, with the iteration function consisting of the eight Serpent rounds. MARS with its non-uniform round structure is less suited for ALRED. The choice of the injection layout is a design exercise specific for the round function of the chosen cipher. Note that whatever the choice of the underlying cipher, the strength of the ALRED construction with respect to key search is that of the underlying cipher.

4 On Internal Collisions in ALRED

In general, any pair of message sequences, possibly of different length, that leads to the same value of the internal state is an internal collision. We see two approaches to exploit knowledge of the iteration function to generate internal collisions. The first is to generate pairs of messages of equal length that have a difference (with respect to some group operation at the choice of the attacker) that may result in a zero difference in the state after the difference has been injected. We call this *extinguishing differentials.* The second is to insert in a message a sequence of words that do not impact the state. We call this *fixed points.*

4.1 Extinguishing Differentials

Finding high-probability extinguishing differentials is similar to differential cryptanalysis of block ciphers. In differential cryptanalysis the trick is to find an input difference that leads to an output difference with high probability. For an iterative MAC function, the trick is to find an extinguishing differential with high probability. Resistance against differential cryptanalysis is often cited as one of the main criteria for the design of the round function of block ciphers. Typically the round function is designed in such a way that upper bounds can be demonstrated on the probability of differentials over a given number of rounds. One may design MAC functions in a similar way: design the iteration function such that upper bounds can be demonstrated on the probability of extinguishing differentials. In ALRED the only part of the iteration function to be designed is the injection layout. So the criterion for the choice of the injection layout is the upper bound on the probability of extinguishing differentials.

4.2 Fixed Points

Given a message word value x_i, one may compute the number of state values that are invariant under the iteration function $y_i = f(y_{i-1}, x_i)$, called *fixed points*. If the number of fixed points is w, the probability that inserting the message word x_i in a message will not impact its tag is $w \times 2^{-n}$ with n the block length.

We can try to find the message word value $x_{\max}$ with the highest number of fixed points. If this maximum is $w_{\max}$, inserting $x_{\max}$ in a message and assuming that the resulting message has the same tag, is a forgery attack with success probability $w_{\max} \times 2^{-n}$. Since Claim 3 requires that this probability be smaller than $1 - \exp(-2^2/2^{\ell_c+1}) = 1 - \exp(-(2^{1-\ell_c})) \approx 2^{1-\ell_c}$, this imposes a limit to the capacity: $\ell_c < n + 1 - \log_2 w_{\max}$.

If the iteration function can be modeled as a random permutation, the number of fixed points has a Poisson distribution with $\lambda = 1$. The expected value of $w_{\max}$ depends on the number of different iteration functions with a given message word, i.e. the word length ℓ_w. For example, the expected $w_{\max}$ values for 16, 32, 64 and 128 bits are 8, 12, 20 and 33 respectively. However, if $r = 1$, the iteration function is just a round function of a block cipher and it may not behave as a random function in this respect. If the round function allows it, one may determine the number of fixed points for a number of message word values to determine whether the Poisson distribution applies.

One may consider the number of fixed points under a sequence of g rounds. In the random model, the expected value of $w_{\max}$ over all possible sequences of g message words now is determined by the total number of messagebits injected in the g rounds. For most round functions determining the number of fixed points given the message word values is hard even for $g = 2$. However, for multiple iterations it is very likely that the random model will hold. The value of $w_{\max}$ grows with g but actually finding message word sequences with a number of fixed points of the order $w_{\max}$ becomes quickly infeasible as g grows. If we consider a sequence of iterations taking 500 message bits (for example 10 iterations taking each 50 message bits), the expected value of $w_{\max}$ is 128. In conclusion, if analysis of the iteration function confirms that the number of fixed points has a Poisson distribution, taking $\ell_c \leq n - 8$ provides a sufficient security margin with respect to forging using fixed points.

5 Alpha-MAC

Alpha-MAC is an Alred construction with AES [3] as underlying block cipher. As AES, Alpha-MAC supports keys of 16, 24 and 32 bytes. Its iteration function consists of a single round, its word length is 4 bytes and the injection layout places these bytes in 4 byte positions of the AES state. We have chosen AES mainly because we expect AES to be widely available thanks to its status as a standard. Additionally, AES is efficient in hardware and software and it has withstood intense public scrutiny very well since its publication as Rijndael [14].

5.1 Specification

The number of rounds per iteration r is 1 and the word length ℓ_w is equal to 32 bits. The AES round function takes as argument a 16-byte round key, represented in a 4×4 array. The injection layout positions the 4 bytes of the message word $[q_1, q_2, q_3, q_4]$ in 4 positions of this array, resulting in the following injection input:

$$\begin{bmatrix} q_1 & 0 & q_2 & 0 \\ 0 & 0 & 0 & 0 \\ q_3 & 0 & q_4 & 0 \\ 0 & 0 & 0 & 0 \end{bmatrix} \quad (4)$$

The length of the tag ℓ_m may have any value below or equal to 128. ALPHA-MAC should satisfy the security claims for iterative MAC functions for the three key lengths of AES with $\ell_m \leq 128$ and $\ell_c = 120$. Appendix B gives and equivalent description of ALPHA-MAC.

5.2 Message Padding

ALPHA-MAC is only defined for messages with a length that is a multiple of 32 bits. One may extend ALPHA-MAC to message of any length by preprocessing the message with a reversible padding scheme. We propose to use the following padding scheme: append a single 1 followed by the minimum number of 0 bits such that the length of the result is a multiple of 32. This corresponds to padding method 2 in [5].

6 Internal Collisions and Injection Layout of ALPHA-MAC

With respect to fixed points, we implemented a program that determines the number of fixed points for the ALPHA-MAC iteration function. It turns out that the number of fixed points behaves as a Poisson distribution with $\lambda = 1$. The choice of the ALPHA-MAC injection was however mainly guided by the analysis of extinguishing differentials, as we explain in the following subsections.

6.1 A Simple Attack on a Variant of ALPHA-MAC

Let us consider a simple extinguishing differential for a variant of ALPHA-MAC with an injection layout mapping a 16 byte message block to a 16-byte round key. Assume the difference in the first message word has a single active byte with value a in position i, j (row i, column j).

- AddRoundKey (AK) injects the difference in the state giving a single active byte with value a in the state in position i, j.
- SubBytes (SB) converts this to a single active byte in position i, j with value b. Given a, b has 127 possible values. 126 of these values have probability 2^{-7} and one has probability 2^{-6}.
- ShiftRows (SR) moves the active byte with value b to position i, ℓ with $\ell = j - i \bmod 4$.

– MixColumns (MC) converts the single active byte to an active column, column ℓ. The value of the single active byte completely determines the values of bytes of the active column.

Hence a message difference with a single active byte may lead to 127 state differences before the injection of the second message word. Of these, one has probability 2^{-6} and 126 have probability 2^{-7}. Assume now that the second message word has a difference given by the active column that has probability 2^{-6}. Clearly, the probability of the resulting extinguishing differential is 2^{-6} and the expected number of message pairs that must be queried to obtain an internal collision using it is 2^6.

We can reduce the number of required messages to query by applying a set of n messages that have pairwise differences of the type described above. About half of the $n(n-1)/2$ differences are extinguishing differentials with a probability of 2^{-7} and due to the birthday paradox a set of only 20 messages are likely to generate an internal collision.

When achieving an internal collision, the fact that the active S-box converts difference a to b gives 6 or 7 bits of information on the absolute value of the state. Applying this attack for all byte positions allows the reconstruction of the state at the beginning of the iteration phase for the given key. When this is known, generating internal collisions is easy.

In the described internal collision attack, the attacker is not hampered by the injection layout. He has full liberty in positioning the differences in the injection inputs. We see that the described difference leads to an internal collision if there is a match in a single S-box: the S-box must map the difference a to difference b. This is an extinguishing characteristic with one active S-box.

For the injection layout of ALPHA-MAC this attack is not possible as it requires in the second injection input a difference pattern with four active bytes in the same column. However, attacks can be mounted with other message difference patterns. The goal of the injection layout is exactly to impose that there are no extinguishing differentials with high probability and hence that there are no extinguishing characteristics with a small number of active S-boxes. Together with the implementation complexity, this has been the main criterion for the selection of the injection layout. We will treat this in the following sections.

6.2 Choosing the Injection Layout

In order to select the injection layout we have written a program that determines the minimum number of active S-boxes over all extinguishing truncated characteristics for a given injection layout. Truncated characteristics are *clusters* of ordinary characteristics [20]. All characteristics in a cluster have intermediate state differences with active bytes in the same positions. The probability of a truncated characteristic is the sum of the probabilities of all its characteristics. Similar to ordinary characteristics, the probability of a truncated characteristic can be expressed in terms of active S-boxes, but only a subset of the S-boxes with non-zero input difference are counted as active.

Table 1. Number of injection layout equivalence classes

Word length (in bytes)	1	2	3	4	5
Total number of layout classes	3	21	77	252	567
with minimum extinguishing cost equal to 16	3	21	68	87	0

Our program iteratively builds a tree with the 2^{16} possible state difference patterns (only distinguishing between 'zero' and 'non-zero' values) as nodes. The root is the all-zero pattern and each edge has an S-box cost and message difference pattern. The program builds the tree such that the minimum extinguishing cost of a pattern is the sum of the S-box costs of the edges leading to the root. It also includes the all-zero pattern as a leaf in the tree, and hence the minimum extinguishing cost is that of this leaf. Note that for any injection layout an upper bound for the minimum extinguishing cost is 16 as it is always possible to guess all bits of the state at a given time.

In total there are 2^{16} different injection layouts. With respect to this propagation analysis, they are partitioned in 8548 equivalence classes:

- Thanks to the horizontal symmetry in the AES round function injection layouts that can be mapped one to the other by means of a horizontal shift $(i, j) \mapsto (i, j + a \bmod 4)$ are equivalent.
- Injection layouts that can be mapped one to the other by means of a mirroring around $(0, 0)$, i.e. $(i, j) \mapsto (-i \bmod 4, -j \bmod 4)$, are equivalent. This is thanks to the fact that SB and SR are invariant under this transformation and the branch number of MC, the only aspect of MC relevant in this propagation analysis, is not modified.

The results for word lengths 5 and below are summarized in Table 1.

As ALPHA-MAC requires one round function computation per message word, the performance is proportional to message word length. Note that the minimum extinguishing cost of an injection layout is upper bounded by those of all injection layouts that can be formed by removing one or more of its bytes. We see that the maximum word length for which there are injection layouts with a minimum extinguishing cost of 16, is 4 bytes. In the choice of the injection layout from the 87 candidates we have taken into account the number of operations it takes to apply the message word to the state. As in 32-bit implementations of AES the columns of state and round keys are coded as 32-bit words, the number of active columns in the injection layout is best minimized. Among the 87 layouts, 40 have four active columns, 42 have three and 5 have two. We chose the ALPHA-MAC injection layout from the latter.

7 On Extinguishing Differentials in ALPHA-MAC

We start with a result on the minimum length of an extinguishing differential.

Theorem 3. *An extinguishing differential in* ALPHA-MAC *spans at least 5 message words.*

Proof: The proof is divided into 3 steps.
Step 1: It can easily be verified that for the AES round transformation, there are no two different round keys that result in a common round input being mapped to a common round output. Hence, extinguishing differentials must span at least two message words.
Step 2: There are also no extinguishing differentials spanning only 2 message words. This can be shown as follows.

Let x_i be the first message word with non-zero difference. Hence y_{i-1} has no differences. The state y_i can have non-zero differences in the positions $(0,0)$, $(0,2)$, $(2,0)$ and $(2,2)$ and nowhere else. The application of SR and SB doesn't change this. Since MC has branch number 5, its application will result in a state with at least 3 non-zero bytes in the first or the third column, or both. The choice of the injection layout ensures that these differences can't be extinguished in the next AK step.
Step 3: We show that there are no extinguishing differentials that span 3 or 4 message words by means of an 'impossible differential.' The impossible differential is illustrated in Figure 2.

Let again x_i be the first message word with non-zero difference. The state y_{i+1} has zero differences in the second and the fourth column. At least one of the remaining columns contains a non-zero difference, because there are no extinguishing differentials of length 2.

Assume now that y_{i+3} has zero difference. This is only possible if before the application of the step AK in iteration $i+3$, the second and the fourth column contain no zeroes. Propagating this condition backwards, we obtain that y_{i+2} must have zero differences in the positions $(0,1)$, $(0,3)$, $(1,0)$, $(1,2)$, $(2,1)$, $(2,3)$, $(3,0)$ and $(3,1)$.

Since AK doesn't act on any of these bytes, the same condition must hold on the state before the application of AK in iteration $i+2$. But then the MC step in iteration $i+2$ has an input with at least 2 zeroes in each column, and an output with at least 2 zeroes in each column, and a least one column with at least one non-zero byte. This is impossible because the branch number of MC is 5. □

We have the following corollary.

Corollary 1. *Given y_{t-1}, the state value before iteration t, the map*

$$s : (x_t, x_{t+1}, x_{t+2}, x_{t+3}) \rightarrow y_{t+3}$$

from the sequence of four message words $(x_t, x_{t+1}, x_{t+2}, x_{t+3})$ to the state value before iteration $t+4$ is a bijection.

Proof: This follows directly from the fact that the number of possible message word sequences of length four and the number of state values are equal and that starting from a given state, any pair of such message sequences will lead to different state values. □

The ALPHA-MAC injection layout is one of the few 4-byte injection layouts with this property. Note that $s^{-1}(y_{t+3})$ can easily be computed, for any given

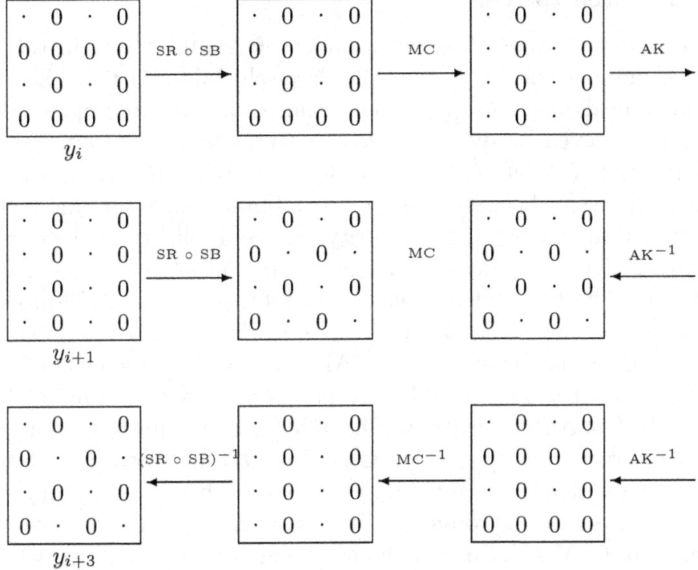

Fig. 2. The impossible differential used in the proof of Theorem 3

y_{t-1}. It follows that if the value of the state leaks, it becomes trivial to construct forgeries forgeries based on internal collisions. However, we see no other methods for obtaining the value of the state than key recovery or the generation of internal collisions.

In the assumption that our program described in Section 6.2 is correct, its result is a proof for the fact that the minimum extinguishing cost is 16. An analytical proof for this minimum cost can be constructed, but is left out here because of the space restrictions. The minimum extinguishing cost imposes an upper bound to the probability of a truncated characteristic of $(2^{-6})^{16} = 2^{-96}$. A closer analysis reveals that in almost all cases an active S-box contributes a factor 2^{-8} rather than 2^{-6}. An active S-box contributes a factor 2^{-6} only if it was 'activated' by the previous application of AK, hence, if it was passive before AK.

We have written a variant of our program taking these aspects into account, resulting in an upper bound for the probability of truncated characteristics of 2^{-126}. For a single extinguishing differential there can be multiple truncated characteristics leading to an extinguishing probability that is the sum of those of the characteristics. In our security claims we have taken a margin by claiming $\ell_c = 120$.

8 Performance

We estimate the relative performance difference between AES and ALPHA-MAC. We compare this estimate with some benchmark performance figures.

8.1 Compared to AES

In this section we express the performance of Alpha-MAC in terms of AES operations, more particularly, the AES key schedule and the AES encryption operation. This allows to use AES benchmarks for software implementations on any platform or even hardware implementations to get a pretty good idea on the performance of Alpha-MAC. We illustrate the comparison for the case of a Pentium processor, because it is easy to obtain figures for this platform. We note however that in most of the security-critical applications the cryptographic services are delivered by dedicated security modules: smart cards, HSMs, set-top boxes, ... These modules typically use 8-bit processors, 486 processors and the new ones may have AES accelerators. Clearly, in this respect Alpha-MAC takes advantage from the efficiency of AES on a wide range of platforms.

One iteration of Alpha-MAC corresponds roughly to 1 round of AES, hence roughly 1/10 of an AES encryption. The differences are due to the following facts. Firstly, the iteration of Alpha-MAC replaces the addition of a 16-byte round key by the injection layout and the addition of the 4 bytes. Some implementations of the AES recompute the round keys for every encryption. This overhead is not present in Alpha-MAC. Finally, the last round of AES is not equal to the first 9 rounds. Using this rough approximation, we can state that MACing a message requires:

setup: 1 AES key schedule and 1 AES encryption,
message processing: 0.1 AES encryptions per 4-byte message word,
finalization: 1 AES encryption to compute the tag.

Hence, the performance of the Alpha-MAC message processing can be estimated at $0.25 \times 0.1^{-1} = 2.5$ times the performance of AES encryption, with a fixed overhead of 1 encryption for the final tag computation. The setup overhead can be written off over many tag computations.

8.2 On the Pentium III

A 32-bit optimized implementation of the AES round transformation implements MC, SR and SB together by means of 16 masking operations, 12 shifts, 12 XOR operations, and 16 table lookups. The implementation of AK requires 4 XOR operations. The iteration function of Alpha-MAC replaces the 4 XORs by 2 masks, 2 XORs and one shift for the combination of the injection layout and AK. If we estimate that all operations have the same cost, then the cost of the iteration function equals $61/60 \approx 1.02$ times the cost of the AES round transformation.

Table 2 is based on the performance figures given by the NESSIE consortium [6]. The performance of Alpha-MAC was estimated using the NESSIE figures for AES. We conclude that the performance of Alpha-MAC is quite good. It outperforms HMAC/SHA-1 and CBC-MAC/AES for all message lengths. Alpha-MAC is slower than Umac-32 but its setup time is a factor 50 shorter.

Table 2. Performance on the Pentium III/Linux platform, as defined by NESSIE

Primitive Name	message processing (cycles/byte)	setup (cycles)	setup + finalization (cycles)
HMAC/MD5	7.3	804	2634
HMAC/SHA-1	15	1346	4697
CBC-MAC/AES	26	616	2056
Umac-32	2.9	54K	55K
ALPHA-MAC (estimate)	10.6	1032	1448

9 Conclusions

We have proposed a set of three security claims for iterated MAC functions, addressing the issue of internal collisions. We presented a new construction method for block cipher based MAC functions. We proved that, in the absence of internal collisions, the security of the construction can be reduced to the security of the underlying block cipher.

Secondly, we proposed ALPHA-MAC, an efficient MAC function constructed from AES with the method presented in the first part. We explained our design decisions and provided the results of our preliminary security analysis. The performance of ALPHA-MAC turns out to be quite competitive.

References

1. ANSI X9.19, *Financial institution retail message authentication,* American Bankers Association, 1986.
2. *Federal Information Processing Standard 180-2, Secure Hash Standard*, National Institute of Standards and Technology, U.S. Department of Commerce, August 2002.
3. *Federal Information Processing Standard 197, Advanced Encryption Standard (AES)*, National Institute of Standards and Technology, U.S. Department of Commerce, November 2001.
4. *Federal Information Processing Standard 198, The Keyed-Hash Message Authentication Code (HMAC)*, National Institute of Standards and Technology, U.S. Department of Commerce, March 2002.
5. *ISO/IEC 9797-1, Information technology - Security Techniques - Message Authentication Codes (MACs) - Part 1: Mechanisms using a block cipher,* ISO 1999.
6. *Performance of optimized implementations of the NESSIE primitives, version 2.0*, The NESSIE Consortium, 2003, https://www.cosic.esat.kuleuven.ac.be/nessie/deliverables/D21-v2.pdf.
7. Mihir Bellare, Ran Canetti, Hugo Krawczyk, "Keying hash functions for message authentication," *Advances in Cryptology - Crypto 96, LNCS 1109,* N. Koblitz, Ed., Springer-Verlag, 1996, pp. 1–15.
8. John Black, Shai Halevi, Hugo Krawczyk, Ted Krovetz, Phillip Rogaway, "UMAC: Fast and Secure Message Authentication," *Advances in Cryptology – Crypto '99, LNCS 1666,* M.J. Wiener, Ed., Springer-Verlag, 1999, pp. 216–233.

9. John Black and Phillip Rogaway, "CBC MACs for Arbitrary-Length Messages: The Three-Key Constructions," *Advances in Cryptology - CRYPTO '00, LNCS 1880,* M. Bellare, Ed., Springer-Verlag, 2000, pp. 197–215.
10. Bert den Boer, Bart Van Rompay, Bart Preneel and Joos Vandewalle., "New (Two-Track-)MAC based on the two trails of RIPEMD — Efficient, especially on short messages and for frequent key-changes," *Selected Areas in Cryptography — SAC 2001, LNCS 2259,* S. Vaudenay and Amr M. Youssef, Eds., Springer-Verlag, pp. 314–324.
11. K. Brincat and C. J. Mitchell, "New CBC-MAC forgery attacks," *Information Security and Privacy, ACISP 2001, LNCS 2119,* V. Varadharajan and Y. Mu, Eds., Springer-Verlag, 2001, pp. 3–14.
12. Don Coppersmith and Chris J. Mitchell, "Attacks on MacDES MAC Algorithm," *Electronics Letters, Vol. 35,* 1999, pp. 1626–1627
13. Don Coppersmith, Lars R. Knudsen and Chris J. Mitchell, "Key recovery and forgery attacks on the MacDES MAC algorithm," *Advances in Cryptology - CRYPTO'2000 LNCS 1880,* M. Bellare, Ed., Springer Verlag, 2000, pp. 184–196.
14. Joan Daemen, Vincent Rijmen, "AES Proposal: Rijndael," *AES Round 1 Technical Evaluation CD-1: Documentation,* National Institute of Standards and Technology, Aug 1998.
15. Donald W. Davies, "A message authenticator algorithm suitable for a mainframe computer," *Advances in Cryptology – Proceedings of Crypto '84, LNCS 196,* G. R. Blakley and D. Chaum, Eds., Springer-Verlag, 1985, pp. 393-400.
16. Tetsu Iwata and Kaoru Kurosawa, "OMAC: One-key CBC MAC," *Fast Software Encryption 2003, LNCS 2887,* T. Johansson, Ed., Springer-Verlag, 2003, pp. 129–153.
17. E. Jaulmes, A. Joux and F. Valette, "On the security of randomized CBC-MAC beyond the birthday paradox limit: A new construction," *Fast Software Encryption 2002, LNCS 2365,* J. Daemen and V. Rijmen, Eds., Springer-Verlag, 2002, pp. 237-251.
18. Antoine Joux, Guillaume Poupard and Jacques Stern, "New Attacks against Standardized MACs," *Fast Software Encryption 2003, LNCS 2887,* T. Johansson, Ed., Springer-Verlag, 2003, pp. 170–181.
19. Lars R. Knudsen and Bart Preneel, "MacDES: a new MAC algorithm based on DES," *Electronics Letters, Vol. 34, No. 9,* 1998, pp. 871–873.
20. Lars R. Knudsen, "Truncated and higher order differentials," Fast Software Encryption '94, LNCS 1008, B. Preneel, Ed., Springer-Verlag, 1995, pp. 196–211.
21. Lars R. Knudsen and Chris J. Mitchell, "Partial key recovery attack against RMAC," *Journal of Cryptology, to appear.*
22. Kaoru Kurosawa and Tetsu Iwata, "TMAC: Two-Key CBC MAC," *Topics in Cryptology: CT-RSA 2003, LNCS 2612,* M. Joye, Ed., Springer-Verlag, 2003, pp. 265–273.
23. Alfred J. Menezes, Paul C. van Oorschot and Scott A. Vanstone, *Handbook of Applied Cryptography,* CRC Press, 1997.
24. Bart Preneel and Paul C. van Oorschot, "MDx-MAC and building fast MACs from hash functions," *Advances in Cryptology, Proceedings Crypto'95, LNCS 963,* D. Coppersmith, Ed., Springer-Verlag, 1995, pp. 1–14.
25. Bart Preneel and Paul C. van Oorschot, "A key recovery attack on the ANSI X9.19 retail MAC," *Electronics Letters, Vol. 32,* 1996, pp. 1568-1569.
26. Bart Preneel and Paul C. van Oorschot, "On the security of iterated Message Authentication Codes," *IEEE Trans. on Information Theory, Vol. IT45, No. 1,* 1999, pp. 188-199.
27. Ron Rivest, "The MD4 message digest algorithm," *Network Working Group Request for Comments:1186,* 1990.

A Detecting Internal Collisions

If the final transformation is reversible, any pair of messages with the same tag form an internal collision. Otherwise, for two messages with the same tag, the collision could have taken place in the final transformation itself. If the ratio between the size of the state and the tag length is v, one can query the MAC function for message tuples $\{m_i \| a_1, m_i \| a_2, \ldots, m_i \| a_{\lceil v \rceil}\}$, with the a_j a set of $\lceil v \rceil$ different strings. If we have two messages m_i and m_j for which all components of the corresponding tuples have matching tags, m_i and m_j very probably form an internal collision. With respect to a tag with the same length as the size of the state, having a shorter tag only multiplies the required number of MAC function queries by v.

B Another Way to Write ALPHA-MAC

The standard [3] explains how to construct an equivalent description for the inverse cipher of AES. We have a similar effect here. Firstly, in the definition of f, the order of the steps SR and SB plays no role. Therefore a sequence of applications of f can also be written as follows:

$$\cdots \circ f \circ f \circ \cdots = \cdots \circ \mathrm{AK}[LI(x_{i+1})] \circ \mathrm{MC} \circ \mathrm{SB} \circ \mathrm{SR} \circ \mathrm{AK}[LI(x_i)] \circ \mathrm{MC} \circ \mathrm{SB} \circ \mathrm{SR} \circ \cdots$$

Secondly, the order of the steps SR and AK can be changed, if the injection layout is adapted accordingly:

$$\mathrm{SR} \circ \mathrm{AK}[LI(x_i)] = \mathrm{AK}[\mathrm{SR}(LI(x_i))] \circ \mathrm{SR} = \mathrm{AK}[LI'(x_i)] \circ \mathrm{SR}.$$

We obtain the following:

$$\cdots \circ f \circ f \circ \cdots = \cdots \circ \mathrm{MC} \circ \mathrm{SB} \circ \mathrm{AK}[LI'(x_i)] \circ \mathrm{SR} \circ \mathrm{MC} \circ \mathrm{SB} \circ \mathrm{AK}[LI'(x_{i-1})] \circ \mathrm{SR} \circ \cdots$$

Concluding, when we ignore the boundary effects at the beginning and the end of the message, ALPHA-MAC can also be described using an alternative iteration function f' and an alternative injection layout function LI', given by:

$$f'(y_{i-1}, x_i) = (\mathrm{AK}[LI'(x_i)] \circ \mathrm{SR} \circ \mathrm{MC} \circ \mathrm{SB})\,(y_{i-1})$$

$$LI'(m) = \begin{bmatrix} q_0 & 0 & q_1 & 0 \\ 0 & 0 & 0 & 0 \\ q_3 & 0 & q_2 & 0 \\ 0 & 0 & 0 & 0 \end{bmatrix}.$$

The alternative injection layout is equivalent to the original injection layout applied to message words with the rightmost two bytes swapped.

New Applications of T-Functions in Block Ciphers and Hash Functions

Alexander Klimov and Adi Shamir

Computer Science department,
The Weizmann Institute of Science,
Rehovot 76100, Israel
{ask, shamir}@wisdom.weizmann.ac.il

Abstract. A *T-function* is a mapping from n-bit words to n-bit words in which for each $0 \le i < n$, bit i of any output word can depend only on bits $0, 1, \ldots, i$ of any input word. All the boolean operations and most of the numeric operations in modern processors are T-functions, and all their compositions are also T-functions. Our earlier papers on the subject dealt with "crazy" T-functions which are invertible mappings (including Latin squares and multipermutations) or single cycle permutations (which can be used as state update functions in stream ciphers). In this paper we use the theory of T-functions to construct new types of primitives, such as MDS mappings (which can be used as the diffusion layers in substitution/permutation block ciphers), and self-synchronizing hash functions (which can be used in self-synchronizing stream ciphers or in "fuzzy" string matching applications).

1 Introduction

There are two basic approaches to the design of secret key cryptographic schemes, which we can call "tame" and "wild". In the tame approach we try to use only simple primitives (such as linear mappings or LFSRs) with well understood behavior, and try to prove mathematical theorems about their cryptographic properties. Unfortunately, the clean mathematical structure of such schemes can also help the cryptanalyst in his attempt to find an attack which is faster than exhaustive search. In the wild approach we use crazy compositions of operations (which mix a variety of domains in a non-linear and non-algebraic way), hoping that neither the designer nor the attacker will be able to analyze the mathematical behavior of the scheme. The first approach is typically preferred in textbooks and toy schemes, but real world designs often use the second approach.

In several papers published in the last few years [4, 5, 6] we tried to bridge this gap by considering "semi-wild" constructions, which look like crazy combinations of boolean and arithmetic operations, but have many analyzable mathematical properties. In particular, in these papers we defined the class of T-functions which contains arbitrary compositions of *plus, minus, times, and, or,* and *xor* operations on n-bit words, and showed that it is easy to analyze their invertibility

H. Gilbert and H. Handschuh (Eds.): FSE 2005, LNCS 3557, pp. 18–31, 2005.

and cycle structure for arbitrary word sizes. This led to the efficient construction of multipermutations and stream ciphers. In this paper we explore additional applications of the theory of T-functions, which do not depend on their invertibility or cycle structure. In particular, we develop new classes of MDS mappings for block ciphers and hash functions, self-synchronizing stream ciphers, and self-synchronizing hash functions which can be used in "fuzzy" string matching to compare strings with a relatively large edit distance.

2 Basic Definitions

Let us first introduce our notation. We denote the set $\{0,1\}$ by $\mathbb{B}$. We denote by $[x]_i$ bit number i of word x (with $[x]_0$ being the least significant bit). We use the same symbol x to denote the n-bit vector $([x]_{n-1}, \ldots, [x]_0) \in \mathbb{B}^n$ and an integer modulo 2^n, with the usual conversion rule: $x \longleftrightarrow \sum_{i=0}^{n-1} 2^i [x]_i$.

A collection of m n-bit numbers is described either as a column vector of values or as an $m \times n$ bit matrix x. We start numbering its rows and columns from zero, and refer to its i-th row $x_{i-1,\star}$ as x_{i-1} and to its j-th column $x_{\star,j-1}$ as $[x]_{j-1}$.

The basic operations we allow in our mappings are the following *primitive operations*: "+", "−", "×" (*addition*, *subtraction*, and *multiplication* modulo 2^n), "∨", "∧", and "⊕" (bitwise *or*, *and*, and *xor* on n-bit words). Note that left shift is allowed (since it is equivalent to multiplication by a power of two), but right shift and circular rotations are not included in this definition, even though they are available as basic machine instructions in most microprocessors. It does not mean that we exclude them from further consideration, we just want to use them in a more restricted way.

Definition 1 (T-function). *A function f from $\mathbb{B}^{m\times n}$ to $\mathbb{B}^{l\times n}$ is called a* T-function *if the k-th column of the output $[f(x)]_{k-1}$ depends only on the first k columns of the input $[x]_{k-1}, \ldots, [x]_0$:*

$$\begin{aligned}
[f(x)]_0 &= f_0([x]_0), \\
[f(x)]_1 &= f_1([x]_0, [x]_1), \\
[f(x)]_2 &= f_2([x]_0, [x]_1, [x]_2), \\
&\vdots \\
[f(x)]_{n-1} &= f_{n-1}([x]_0, \ldots, [x]_{n-2}, [x]_{n-1}).
\end{aligned} \tag{1}$$

The name is due to the Triangular form of (1). It turns out that T-functions are very common since any combination of constants, variables, and primitive operations is a T-function.

Definition 2. *A T-function is called a* parameter *(and denoted by a Greek letter such as α) if each* bit-slice function *f_i does not depend on $[x]_i$.*

If T-functions can be viewed as triangular matrices, then parameters can be viewed as triangular matrices with zeroes on the diagonal (note that these functions are typically non-linear, and thus the matrix form is a visualization metaphor rather than an actual definition). The name "parameter" usually refers to some unspecified constant in an expression, and in this context we use it to denote that in many applications it suffices to analyze the dependence of a bit-slice of the output $[f(x)]_i$ only on the current bit-slice of the input $[x]_i$, and to consider the effect of all the previous bit-slices of the input (e.g., in the form of addition or multiplication carries) as unspecified values.

Given an arbitrary expression with primitive operations, we can recursively apply the following rules to produce a simple representation of its bit-slice mappings using such unspecified parameters. Note that in this representation we only have to distinguish between the least significant bit-slice and all the other bit-slices, regardless of the word length n:

Theorem 1. *For $i > 0$ the following equalities hold*

$$\begin{array}{ll} [x \times y]_0 = [x]_0 \wedge [y]_0, & [x \times y]_i = [x]_i \alpha_{[y]_0} \oplus \alpha_{[x]_0} [y]_i \oplus \alpha_{xy}, \\ [x \pm y]_0 = [x]_0 \oplus [y]_0, & [x \pm y]_i = [x]_i \oplus [y]_i \oplus \alpha_{x \pm y}, \\ [x \oplus y]_0 = [x]_0 \oplus [y]_0, & [x \oplus y]_i = [x]_i \oplus [y]_i, \\ [x \overset{\wedge}{\vee} y]_0 = [x]_0 \overset{\wedge}{\vee} [y]_0, & [x \overset{\wedge}{\vee} y]_i = [x]_i \overset{\wedge}{\vee} [y]_i, \end{array} \tag{2}$$

where the unspecified parameters α's denote the dependence of the subscripted operation on previous bit-slices.

Consider, for example, the following mapping: $x \to x + (x^2 \vee 5)$.

$$[x + (x^2 \vee 5)]_0 = [x]_0 \oplus [x^2 \vee 5]_0 = [x]_0 \oplus ([x^2]_0 \vee [5]_0) = [x]_0 \oplus 1$$

and, for $i > 0$,

$$\begin{aligned} [x + (x^2 \vee 5)]_i &= [x]_i \oplus [x^2 \vee 5]_i \oplus \alpha_{x+(x^2 \vee 5)} = [x]_i \oplus ([x^2]_i \vee [5]_i) \oplus \alpha_{x+(x^2 \vee 5)} \\ &= [x]_i \oplus ([x]_i \alpha_{[x]_0} \oplus \alpha_{[x]_0} [x]_i \oplus \alpha_{x^2}) \vee [5]_i \oplus \alpha_{x+(x^2 \vee 5)} \\ &= [x]_i \oplus \alpha_{x^2} \vee [5]_i \oplus \alpha_{x+(x^2 \vee 5)} = [x]_i \oplus \alpha. \end{aligned}$$

This mapping is clearly invertible since we can uniquely recover the consecutive bit-slices of the input (from LSB to MSB) from the given bit-slices of the output. A summary of the simplest recursive constructions of parameters can be found in Figure 1 at the end of the paper.

3 A New Class of MDS Mappings

In this section we consider the efficient construction of new types of MDS mappings, which are a fundamental building block in the construction of many block ciphers. Unlike all the previous constructions, our mappings are non-linear and non-algebraic, and thus they can provide enhanced protection against differential and linear attacks.

Let X be a finite set and ϕ be an invertible mapping on m-tuples of values from X ($\phi : X^m \to X^m$). Let $y = \phi(x)$ and $y' = \phi(x')$, where $x = (x_0, \ldots, x_{m-1})^T$, $y = (y_0, \ldots, y_{m-1})^T$, and, similarly, $x' = (x'_0, \ldots, x'_{m-1})^T$, $y' = (y'_0, \ldots, y'_{m-1})^T$, and $x \neq x'$. Let d_x be the number of i's such that $x_i \neq x'_i$, and, similarly, let d_y be the number of differences between y and y'. Let $D_\phi = min_{x,x'}(d_x + d_y)$. Since $d_y \leq m$ and d_x can be equal to 1 it follows that $D_\phi \leq m + 1$ for arbitrary ϕ.

Definition 3. *A mapping ϕ is called* Maximum Distance Separable (MDS) *if* $D_\phi = m + 1$.[1]

If we use ϕ as a diffusion layer in a Substitution Permutation[2] encryption Network (SPN),[3] then every differential [1] or linear [7] characteristic has at least D_ϕ active S-boxes in each pair of consecutive layers of the network. Using this property we can demonstrate resistance to differential and linear cryptanalysis, because in combination with the probability bounds on a single S-box it provides an upper bound on the probability of any differential or linear characteristic. Consequently, MDS mappings are used in many modern block cipher designs including AES [3].

Common constructions of MDS mappings use linear algebra over the finite field $GF(2^n)$. This makes the analysis easier, but has the undesirable side effect that a linear diffusion layer by itself is "transparent" (i.e., has transition probability of 1) to differential and linear attacks. If we could use a "non-transparent" MDS diffusion layer we would simultaneously achieve two goals by spending the same computational effort—forcing many S-boxes to be active and further reducing the upper bound on the probability of characteristics in each diffusion layer.

One way to construct a linear MDS mapping over a finite field is to use the following method. Let

$$\mathcal{W}(a_0, \ldots, a_{m-1}) = \begin{pmatrix} 1 & a_0 & a_0^2 & \ldots & a_0^{m-1} \\ 1 & a_1 & a_1^2 & \ldots & a_1^{m-1} \\ \vdots & \vdots & \vdots & \ddots & \vdots \\ 1 & a_{m-1} & a_{m-1}^2 & \ldots & a_{m-1}^{m-1} \end{pmatrix}.$$

It is known that if all the a_i are distinct then this matrix is non-singular. Consider the following mapping

$$x \to y = \mathcal{W}(a_0, \ldots, a_{m-1})\mathcal{W}^{-1}(a_m, \ldots, a_{2m-1})x.$$

[1] In our definition ϕ can be an arbitrary mapping, even though the name MDS usually relates to *linear* mappings or error correcting codes. The alternative definition which counts the number of non-zero entries in a single input/output pair is applicable only to linear codes.

[2] Note that the name "permutation" here is due to the historical tradition since modern designs use for diffusion not a bit permutation (as, e.g., in DES) but a general linear or affine transformation (as, e.g., in AES).

[3] Alternatively ϕ can be used as a diffusion layer in a Feistel construction. Note that in this case ϕ need not to be calculated backwards even during decryption.

Let us show that if all the a_i's are distinct then this mapping is MDS. Let

$$p = \mathcal{W}^{-1}(a_m, \ldots, a_{2m-1})x.$$

If we consider p as the vector of the coefficients of a polynomial then

$$\begin{aligned} x_i &= p_0 + p_1 a_{m+i} + p_2 a_{m+i}^2 + \cdots + p_{m-1} a_{m+i}^{m-1} = p(a_{i+m}), \\ y_i &= p_0 + p_1 a_i \quad + p_2 a_i^2 \quad + \ldots + p_{m-1} a_i^{m-1} = p(a_i). \end{aligned}$$

The number of common values c of two distinct polynomials (p and p', defined by the two sets of input/output values) of degree $m-1$ is at most $m-1$ and thus the number of unequal pairs of primed and non-primed values among all the inputs and outputs satisfies

$$d = d_x + d_y = 2m - c \geq m + 1.$$

Consider an example in $\mathbb{F}_{2^3}$ (modulo $\mathsf{b} = 1011_2 = t^3 + t + 1$):

$$\mathcal{W}(1,2,3) \times \mathcal{W}^{-1}(4,5,6) = \begin{pmatrix} 1\,1\,1 \\ 1\,2\,4 \\ 1\,3\,5 \end{pmatrix} \times \begin{pmatrix} 4\,3\,6 \\ 4\,7\,3 \\ 5\,6\,3 \end{pmatrix} = \begin{pmatrix} 5\,2\,6 \\ 5\,3\,7 \\ 4\,2\,7 \end{pmatrix}. \tag{3}$$

Notice that this mapping uses multiplication in a finite field. We prefer to use arithmetic modulo 2^n, which is much more efficient in software implementations on modern microprocessors, and would also like to mix arithmetic and boolean operations in order to make cryptanalysis harder. The general T-function methodology in such cases can be summarized as follows:

1. Find a skeleton bitwise mapping from 1-bit inputs to 1-bit outputs which has the desired property (e.g., invertibility).
2. Extend the mapping in a natural way to n-bit words.
3. Add some parameters in order to obtain a larger class of mappings with the same bit-slice properties, and to provide some inter–bit-slice mixing.
4. Change some $\oplus$ operations to *plus* or *minus*, using the fact that they have the same bit-slice mappings (up to the exact definition of some parameters).

Let us apply this T-function methodology to the construction of an MDS mapping with $m = 3$ input words. First of all, we have to represent x_i as a bit vector $(x_{i,u}, x_{i,v}, x_{i,w})$ and represent (3) as a mapping of bits:

$$\begin{aligned}
y_{0,u} &= (x_{0,w}) \oplus (x_{1,v}) \oplus (x_{2,v} \oplus x_{2,u} \oplus x_{2,w}), \\
y_{0,v} &= (x_{0,u}) \oplus (x_{1,u} \oplus x_{1,w}) \oplus (x_{2,w} \oplus x_{2,v}), \\
y_{0,w} &= (x_{0,w} \oplus x_{0,v}) \oplus (x_{1,u}) \oplus (x_{2,u} \oplus x_{2,v}), \\
y_{1,u} &= (x_{0,w}) \oplus (x_{1,u} \oplus x_{1,v}) \oplus (x_{2,v} \oplus x_{2,w}), \\
y_{1,v} &= (x_{0,u}) \oplus (x_{1,v} \oplus x_{1,u} \oplus x_{1,w}) \oplus (x_{2,w}), \\
y_{1,w} &= (x_{0,w} \oplus x_{0,v}) \oplus (x_{1,w} \oplus x_{1,u}) \oplus (x_{2,w} \oplus x_{2,u} \oplus x_{2,v}), \\
y_{2,u} &= (x_{0,u} \oplus x_{0,w}) \oplus (x_{1,v}) \oplus (x_{2,v} \oplus x_{2,w}), \\
y_{2,v} &= (x_{0,u} \oplus x_{0,v}) \oplus (x_{1,u} \oplus x_{1,w}) \oplus (x_{2,w}), \\
y_{2,w} &= (x_{0,v}) \oplus (x_{1,u}) \oplus (x_{2,w} \oplus x_{2,u} \oplus x_{2,v}).
\end{aligned}$$

Note that multiplication in our field works as follows:

$$\begin{aligned}(x_{i,u}, x_{i,v}, x_{i,w}) \times 1 &= (x_{i,u}, \quad x_{i,v}, \quad x_{i,w}),\\ (x_{i,u}, x_{i,v}, x_{i,w}) \times 2 &= (x_{i,v}, \quad x_{i,w} \oplus x_{i,u}, x_{i,u}),\\ (x_{i,u}, x_{i,v}, x_{i,w}) \times 4 &= (x_{i,w} \oplus x_{i,u}, x_{i,u} \oplus x_{i,v}, x_{i,v}).\end{aligned}$$

So, for example, to get the topmost-leftmost block we calculate

$$5x_0 = 1x_0 \oplus 4x_0 = \begin{pmatrix} x_{0,u} \oplus (x_{0,w} \oplus x_{0,u}) \\ x_{0,v} \oplus (x_{0,u} \oplus x_{0,v}) \\ x_{0,w} \oplus \quad x_{0,v} \end{pmatrix} = \begin{pmatrix} x_{0,w} \\ x_{0,u} \\ x_{0,w} \oplus x_{0,v} \end{pmatrix}.$$

Let us now consider each $x_{i,\cdot}$ and $y_{i,\cdot}$ not as a single-bit variable but as a whole n-bit word, so each x_i and y_i is now of length $3n$. Suppose that for $(x_0, \ldots, x_m)^T$ and $(x'_0, \ldots, x'_m)^T$ the number of differences $d_x = D > 0$, that is there are D values of i such that $x_i \neq x'_i$. It follows that in each bit-slice $d_{[x]_j} \leq D$ and so, since at least one bit-slice was changed and the bit-slice mapping

$$\begin{pmatrix} [x_0]_j \\ \vdots \\ [x_{m-1}]_j \end{pmatrix} \rightarrow \begin{pmatrix} [y_0]_j \\ \vdots \\ [y_{m-1}]_j \end{pmatrix}$$

is MDS, it follows that $d_{[y]_j} \geq m + 1 - D$, and thus $d_y \geq m + 1 - D$, that is $d_x + d_y \geq m + 1$, and thus the whole mapping $x \rightarrow y$ is also MDS.

Our next goal is to introduce arbitrary parameters in order to define a much larger class of mappings. Note that if

$$\begin{pmatrix} x_0 \\ \vdots \\ x_{m-1} \end{pmatrix} \rightarrow \begin{pmatrix} y_0 \\ \vdots \\ y_{m-1} \end{pmatrix} \text{ is MDS then } \begin{pmatrix} x_0 \\ \vdots \\ x_{m-1} \end{pmatrix} \rightarrow \begin{pmatrix} \phi_0(y_0) \\ \vdots \\ \phi_{m-1}(y_{m-1}) \end{pmatrix},$$

where the ϕ_i's are any invertible mappings, is also MDS. Since $\phi : x \rightarrow x \oplus \alpha$ is an invertible mapping it follows that the introduction of *additive* parameters preserves the MDS property of bit-slices. Consequently, we can replace some "$\oplus$"s with "+"s or "−"s, and add arbitrary parameters in order to obtain the following "crazier" mapping which is also provably MDS:

$$\begin{aligned}
y_{0,u} &= x_{0,w} - (x_{1,v} \oplus x_{2,v}) + (x_{2,u} \oplus x_{2,w}) && \oplus\, 2x_{0,u}x_{1,v},\\
y_{0,v} &= (x_{0,u} + x_{1,u} - (x_{1,w} \oplus x_{2,w})) \oplus x_{2,v} && \oplus\, 2x_{0,v}x_{1,w},\\
y_{0,w} &= x_{0,w} \oplus (x_{0,v} + x_{1,u}) \oplus x_{2,u} \oplus x_{2,v} && \oplus\, 2x_{0,w}x_{2,u},\\
y_{1,u} &= x_{0,w} + (x_{1,u} \oplus x_{1,v} \oplus x_{2,v}) + x_{2,w} && \oplus\, 2x_{1,u}x_{2,v},\\
y_{1,v} &= x_{0,u} \oplus (x_{1,v} + x_{1,u} - x_{1,w}) \oplus x_{2,w} && \oplus\, 2x_{1,v}x_{2,w},\\
y_{1,w} &= (x_{0,w} - x_{0,v} - x_{1,w}) \oplus x_{1,u} \oplus (x_{2,w} - (x_{2,u} \oplus x_{2,v})) && \oplus\, 2x_{1,w}x_{0,u},\\
y_{2,u} &= x_{0,u} \oplus (x_{0,w} + x_{1,v} + x_{2,v}) \oplus x_{2,w} && \oplus\, 2x_{2,u}x_{0,v},\\
y_{2,v} &= x_{0,u} - x_{0,v} + (x_{1,u} \oplus x_{1,w} \oplus x_{2,w}) && \oplus\, 2x_{2,v}x_{0,w},\\
y_{2,w} &= (x_{0,v} + x_{1,u} \oplus x_{2,w}) - (x_{2,u} \oplus x_{2,v}) && \oplus\, 2x_{2,w}x_{1,u}.
\end{aligned}$$

This mapping allows us to intermix 3 S-boxes of $3n$ bits each. It is possible to construct a similar mapping which allows us to intermix m S-boxes of ln bits each as long as $2m \leq 2^l$, since in this case $\mathbb{F}_{2^l}$ contains $2m$ different elements. In the above example $m = l = 3$, and so the block size is $mln = 576$ bits for $n = 64$, and the size of each S-box is $ln = 192$ bits. Although in some applications (e.g., hash functions or stream ciphers) this is not a limitation, in others (e.g., block ciphers) such long blocks can be a problem. Note that for embedded low-end devices $n = 8$ and so the above example is too small ($mln = 72$ and $ln = 24$), but if we use larger parameters, such as $l = 4$ and $m \leq 8$ in a 128-bit block cipher, we can intermix, for example, the outputs of four 32-bit S-boxes by an MDS mapping.

4 Simpler Mappings Which Are Almost MDS

The constructions in the previous section were somewhat complicated and did not have ideal parameter sizes (even if we take into account a slight improvement described in the appendix). The source of the problem was that a non-trivial linear mapping cannot be MDS modulo 2 and thus it is provably impossible to have $l = 1$ and $m > 1$. Fortunately, we can use much simpler functions which are almost MDS, and which are almost as useful as actual MDS functions in cryptographic applications.

Define a mapping as an *almost MDS mapping* if $d_x + d_y \geq m$. Such a diffusion layer guarantees that at least m (instead of $m + 1$) S-boxes are active in any pair of consecutive layers in a substitution permutation network, and thus in many block cipher designs they provide almost the same upper bound on the probability of the cipher's differential and linear characteristics.

Let us construct an almost MDS mapping with conveniently sized parameters. As usual we start with a skeleton of the bit-slices. In this case we use the simple skeleton:

$$\begin{pmatrix} y_0 \\ y_1 \\ y_2 \\ y_3 \end{pmatrix} = \begin{pmatrix} 0\,1\,1\,1 \\ 1\,0\,1\,1 \\ 1\,1\,0\,1 \\ 1\,1\,1\,0 \end{pmatrix} \times \begin{pmatrix} x_0 \\ x_1 \\ x_2 \\ x_3 \end{pmatrix} + \begin{pmatrix} \alpha_0 \\ \alpha_1 \\ \alpha_2 \\ \alpha_3 \end{pmatrix},$$

where the α_i are arbitrary constants. It is easy to check that $d_x + d_y \geq 4$. Using this skeleton and replacing the constants α_i by simple parameters we can construct, for example, the following non-linear almost MDS mapping:

$$\begin{aligned} y_0 &= x_1 + (x_2 \oplus x_3)(2x_0 + 1), \\ y_1 &= x_2 + (x_3 \oplus x_0)(2x_1 + 1), \\ y_2 &= x_3 + (x_0 \oplus x_1)(2x_2 + 1), \\ y_3 &= x_0 + (x_1 \oplus x_2)(2x_3 + 1). \end{aligned}$$

This mapping can diffuse the outputs of four S-boxes of arbitrary sizes (e.g., 32-bit to 32-bit computed S-boxes in a 128-bit block cipher) in a way that guarantees that at least 4 S-boxes are active in any pair of consecutive layers. A similar construction can be used to diffuse a higher number of smaller (computed or stored) S-boxes in other designs.

An interesting application of this mapping is to enhance the security of SHA-1 against the recently announced collision attacks. The first step in SHA-1 is to linearly expand the 16 input words into 80 output words. The linearity of this process makes it easy to search for low Hamming weight differential patterns in the output words without committing to actual input values. We propose to replace the linear expansion by the following process: Arrange the 16 input words in a 4×4 array, and alternately apply our 4-word nonlinear mapping to its rows and columns. This is similar to the AES encryption process, but without keys and S-boxes, and using 32-bit words rather than bytes as array elements. Each application produces a new batch of 16 output words, and thus two row mixings and two column mixings suffice to expand the 16 input words into 80 output words. To enhance the mixing of the weak LSBs in T-functions, we propose to cyclically rotate each generated word by a variable number of bits. The nonlinearity of the mapping makes it difficult to predict the evolution of differential patterns, the MDS property provides lower bounds on their Hamming weights, and thus the modified SHA-1 offers greatly enhanced protection against the new attacks.

5 Self-synchronizing Functions

In this section we propose several novel applications of T-functions which are based on the observation that the iterated application of a parameter slowly "forgets" its distant history in the same way that a triangular matrix with zeroes on the diagonal converges to the zero matrix when raised to a sufficiently high power. Let us start with the following definition:

Definition 4 (SSF). *Let* $\{c^{(i)}\}_{i=0,\ldots}$ *and* $\{\hat{c}^{(i)}\}_{i=0,\ldots}$ *be two input sequences, let* $s^{(0)}$ *and* $\hat{s}^{(0)}$ *be two initial states, and let* K *be a common key. Assume that the function* $\mathcal{U}$ *is used to update the state based on the current input and the key:* $s^{(i+1)} = \mathcal{U}(s^{(i)}, c^{(i)}, K)$ *and* $\hat{s}^{(i+1)} = \mathcal{U}(\hat{s}^{(i)}, \hat{c}^{(i)}, K)$*. The function* $\mathcal{U}$ *is called a* self-synchronizing function (SSF) *if equality of any* k *consecutive inputs implies the equality of the next state, where* k *is some integer:*

$$c^{(i)} = \hat{c}^{(i)}, \quad \ldots, \quad c^{(i+k-1)} = \hat{c}^{(i+k-1)} \qquad \Longrightarrow \qquad s^{(i+k)} = \hat{s}^{(i+k)}.$$

Let us now show why the T-function methodology is ideally suited to the construction of SSFs:

Theorem 2. *Let the T-function* $\mathcal{U}(s, c, K)$ *be a parameter with respect to the state* s *and an arbitrary function of the input* c *and the key* K*:*

$$[\mathcal{U}(s, c, K)]_i = f_i([s]_{0,\ldots,i-1}, [c]_{0,\ldots,n-1}, K),$$

then $\mathcal{U}$ *is an SSF.*

Proof. To prove the theorem it is enough to note that bit number i of $s^{(t)}$ can depend only on

$$\left(\left[s^{(t-1)}\right]_{0,\ldots,i-1}, c^{(t-1)}, K\right),$$

that, in turn, can depend only on

$$\left(\left[s^{(t-2)}\right]_{0,\ldots,i-2}, c^{(t-1)}, c^{(t-2)}, K\right),$$

... that, in turn, can depend only on

$$\left(\left[s^{(t-i)}\right]_{0}, c^{(t-1)}, \ldots, c^{(t-i)}, K\right),$$

and, finally, this can depend only on

$$\left(c^{(t-1)}, \ldots, c^{(t-i-1)}, K\right).$$

So if all the calculations are done modulo 2^n, $s^{(t)}$ can depend on $c^{(t-1)}, \ldots, c^{(t-n)}$, but cannot depend on any earlier input.

By using the parameters described in Figure 1 it is easy to construct a large variety of SSFs, for example, $\mathcal{U}(s, c, K) = 2s \oplus cK$ or $\mathcal{U}(s, c, K) = ((s \oplus K)^2 \vee 1) + c$. In different applications it may be important to have different k values (representing the number of steps needed to resynchronize). Our constructions seem to be limited to $k = n$ steps (using n^2 input bits), where n is the word-size of the processor (usually, $n = 32$ or $n = 64$). However, it is easy to enlarge or shrink the size of the *effective region* by adjusting the size of c used in each iteration. For example, on a 64-bit processor the above construction has an effective region of size 2^{12} bits, using one byte of c in each iteration we can reduce it down to 2^9, or enlarge it up to 2^{15} if we use eight 64-bit words at a time. Alternatively, to avoid performance penalties, we can use $\mathcal{U}(s, c, K)$ which is a *multiple parameter* with respect to s:

$$[\mathcal{U}(s, c, K)]_i = f_i([s]_{0,\ldots,i-p}, [c]_{0,\ldots,n-1}),$$

where p is some integer. For such function $s^{(t)}$ depends only on $c^{(t-1)}, \ldots, c^{(t-\frac{n}{p})}$. For example, $\mathcal{U}(s, c, K) = ((s \bullet 8) \oplus c) \times (c \vee 1)$, where $a \bullet b$ denotes left shift of a by b bit positions, depends only on eight ($\frac{64}{8} = 8$) previous c's.

SSFs have many applications including cryptography, fast methods for bringing remote files into sync, finding duplications on the web, et cetera. Let us now describe those applications in more detail.

Self-synchronizing stream ciphers allow parties to continue their communication even if they temporarily lose synchronization: after processing k additional ciphertexts the receiver automatically resynchronizes it state with the sender. The standard way to achieve this is to create a state which is the concatenation of the last k ciphertext symbols, and to compute the next pseudo random value as the keyed hash of this state. However, in stream cipher construction speed is extremely important, and thus the active maintenance of such a concatenation (adding the newest input, deleting the oldest input, and shifting the other inputs) wastes precious cycles. In addition, the opponent always knows and can sometimes manipulate this state, and thus the hash function has to be relatively strong (and thus relatively slow) in order to withstand cryptanalysis. We propose to combine the state maintenance and the hash operation (and thus eliminate the computational overhead of the state maintenance) by applying a mapping which is a parameter with respect to the state and an arbitrary function with respect to the ciphertext and secret key. This keeps the current state secret, and allows us to use a potentially weaker hash function to produce the next output. More formally, let $\{p^{(i)}\}_{i=0,\ldots}$ denote the plaintext, K denote the secret key, $\{s^{(i)}\}_{i=0,\ldots}$ denote the internal state, and $\mathcal{I}$ denote the initialization function. The state is initially set to $s^{(0)} = \mathcal{I}(K)$, and then it evolves over time by an SSF update operation $\mathcal{U}$: $s^{(i+1)} = \mathcal{U}(s^{(i)}, c^{(i)}, K)$, where $\{c^{(i)}\}_{i=0,\ldots}$ denotes the ciphertext which is produced using an output function $\mathcal{O}$: $c^{(i)} = p^{(i)} \oplus \mathcal{O}(s^{(i)}, K)$.

The actual construction of a *secure* self-synchronizing stream cipher requires great care. Unlike PRNG, where the known-plaintext attack is usually the only one to be considered, there are many reasonable attacks on a self-synchronizing stream cipher:

- known plaintext attack,
- chosen plaintext attack,
- chosen ciphertext attack, and probably even
- related key attack.

To avoid some of these attacks, it is recommended to use a *nonce* in the initialization process to make sure that the opponent cannot restart the stream cipher in the same state.

In this paper we propose a general methodology for the construction of cryptographic primitives rather than fully specified schemes, but let us give one concrete example in order to demonstrate our ideas and encourage further research on the new approach. Let the state s consist of three 64-bit words: $s = (s_0, s_1, s_2)^T$. At each iteration, we would like to output a 64-bit pseudo random value which can be xored with the next plaintext to produce the next ciphertext. Since in the T-function–based constructions the LSBs are usually weaker than the MSBs, the proposed output function swaps the high and low halves:

$$\mathcal{O}(s_0, s_1, s_2) = ((s_0 \oplus s_2 \oplus K_{\mathcal{O}}) \bullet\ 32) \times ((s_1 \oplus K'_{\mathcal{O}}) \bullet\ 32) \vee 1),$$

where $a \bullet\ b$ denotes circular left shift of a by b bit positions. The state is updated by the following function:

$$\begin{pmatrix} s_0 \\ s_1 \\ s_2 \end{pmatrix} \to \begin{pmatrix} (((s'_1 \oplus s'_2) \vee 1) \oplus K_0)^2 \\ (((s'_0 \oplus s'_2) \vee 1) \oplus K_1)^2 \\ (((s'_0 \oplus s'_1) \vee 1) \oplus K_2)^2 \end{pmatrix},$$

where $s'_0 = s_0 + c$, $s'_1 = s_1 - (c \bullet\ 21)$, and $s'_2 = s_2 \oplus (c \bullet\ 21)$. The best attack we are aware of against this particular example requires $O(2^{96})$ time.

Let us now consider some non-cryptographic applications of self-synchronizing functions. Suppose that we want to update a file on one machine (receiver) to be identical to a similar file on another machine (sender) and we assume that the two machines are connected by a low-bandwidth high-latency bidirectional communications link. The simplest solution is to break the file into blocks, and to send across the hashed value of each block in order to identify (and then correct) mismatches. However, if one of the two files has a single missing (or added) character, then all the blocks from that point onwards will have different hash values due to framing errors. The *rsync* algorithm [8] allows two parties to find and recover from such framing errors by asking one party to send the hash values of all the non-overlapping k-symbol blocks, and asking the other party to compare them to the locally computed hash values of all the possible k-symbol blocks (at any offset, not just at locations which are multiples of k). To get the fastest possible speed in such a comparison, we can again eliminate the block maintenance overhead by using a single pass SSF computation which repeatedly updates its internal state with the next character and compares the result to the hash values received from the other party. Such an incremental hash computation can overcome framing errors by automatically forgetting its distant history.

Self-synchronizing functions are also useful in "fuzzy" string matching applications, in which we would like to determine if two documents are sufficiently similar, even though they can have a relatively large edit distance (of changed, added, deleted or rearranged words). Computing the edit distance between two documents is very expensive, and finding a pair of similar documents in a large collection is even harder. To overcome this difficulty, Broder et al. [2] introduced the following notion of *resemblance* of two documents A and B:

$$r(A,B) = \frac{|S(A) \cap S(B)|}{|S(A) \cup S(B)|},$$

where $S(\cdot)$ denotes the set of all the overlapping word k-grams (called *shingles*) of a document, and $|\cdot|$ denotes the size of a set. It was suggested [2] that a good estimate of this resemblance can be obtained by computing the set of hash values of all the shingles in each document, reducing each set into a small number of representative values (such as the hash values which are closest to several dozen particular target values), and then computing the similarity expression above for just the representative values. Since each document can be independently summarized, we get a linear rather than a quadratic algorithm for finding similar pairs of documents in a large collection. In web applications, this makes it possible to analyze the structure of the web, to reduce the size of its cached copies, to find popular documents, or to identify copyright violations.

Notice that the notion of the adversary in the non-cryptographic applications of SSFs is rather limited, and thus we are not bothered by some of the inherent limitations of any such similarity checking procedure for web documents. For example, there are many ways to trick a web crawler into "thinking" that your arbitrary document is similar to a totally different document, or that very similar documents are quite different. The techniques range from checking the IP address of the tester (returning to the crawler a different page than to other users), to creating web pages in such a way that after the execution of `JavaScript` it displays the text on the screen in a completely different way than the raw text which is "seen" by a crawler. Thus it seems that we should not over-design the application to withstand sophisticated attacks against it, and just make sure that the hashed values are random looking and reasonably unpredictable when we use a secret key K in the initialization and state update functions.

References

1. E. Biham and A. Shamir, "Differential Cryptanalysis of DES-like Cryptosystems," CRYPTO 1990.
2. A. Broder, S. Glassman, M. Manasse, and G. Zweig, "Syntactic Clustering of the Web." Available from `http://decweb.ethz.ch/WWW6/Technical/Paper205/Paper205.html`
3. J. Daemen, V. Rijmen, "AES Proposal: Rijndael," version 2, 1999.
4. A. Klimov and A. Shamir, "A New Class of Invertible Mappings," Workshop on Cryptographic Hardware and Embedded Systems (CHES), 2002.
5. A. Klimov and A. Shamir, "Cryptographic Applications of T-functions," Selected Areas in Cryptography (SAC), 2003.
6. A. Klimov and A. Shamir, "New Cryptographic Primitives Based on Multiword T-functions," Fast Software Encryption Workshop (FSE), 2004.
7. M. Matsui, "Linear Cryptanalysis Method for DES Cipher," EUROCRYPT 1993.
8. A. Tridgell and P. Mackerras, "The rsync algorithm." Available from `http://rsync.samba.org/tech_report/`

Smaller MDS Mappings

In Section 3 we considered MDS mappings which allow us to intermix m S-boxes of ln bits each as long as $2m \le 2^l$. In order to study if this inequality is an essential condition or just an artifact of that skeleton construction method let us first consider the following question: is it possible to construct an MDS T-function such that $l = 1$ and $m > 1$. The following reasoning shows that it is impossible. Suppose that we constructed such an MDS T-function θ. Let

$$\begin{aligned} x &= (0, \quad 0, \quad 0, \dots 0)^T, \\ x' &= (2^{n-1}, 0, \quad 0, \dots 0)^T, \\ x'' &= (0, \quad 2^{n-1}, 0, \dots 0)^T, \end{aligned}$$

and θ maps them into y, y', and y'' respectively. Since $x = x' = x'' \pmod{2^{n-1}}$ and θ is a T-function it follows that $y = y' = y'' \pmod{2^{n-1}}$. So the only difference between y, y', and y'' is in the most significant bit. Our mapping θ is an MDS and $d_{x,x'} = 1$ so $d_{y,y'} \ge m$ and thus the most significant bit in y'_i is the inverse of y_i:

$$\forall i, [y'_i]_{m-1} = \overline{[y_i]_{m-1}}.$$

For the same reason

$$\forall i, [y_i'']_{m-1} = \overline{[y_i]_{m-1}},$$

and so $y' = y''$ which is a contradiction because any MDS mapping has to be invertible.

Although the algorithm which uses finite field arithmetic does not allow us to construct a mapping intermixing three S-boxes such that each one of them consists of fewer than three words ($m = 3$ and $l < 3$), it is possible to construct such a mapping using a different algorithm. Since we already know that the case of $m = 3$ and $l = 1$ is impossible let us try to construct an MDS mapping with $m = 3$ and $l = 2$. To do it we need a skeleton for bit-slices that is an MDS mapping

$$\psi : \mathbb{B}^2 \times \mathbb{B}^2 \times \mathbb{B}^2 \to \mathbb{B}^2 \times \mathbb{B}^2 \times \mathbb{B}^2.$$

Using a computer search we found the following mapping:[4]

```
00⋆ → 000 111 222 333   01⋆ → 123 032 301 210   02⋆ → 231 320 013 102   03⋆ → 312 203 130 021
10⋆ → 132 023 310 201   11⋆ → 011 100 233 322   12⋆ → 303 212 121 030   13⋆ → 220 331 002 113
20⋆ → 213 302 031 120   21⋆ → 330 221 112 003   22⋆ → 022 133 200 311   23⋆ → 101 010 323 232
30⋆ → 321 230 103 012   31⋆ → 202 313 020 131   32⋆ → 110 001 332 223   33⋆ → 033 122 211 300
```

Interestingly, this mapping is linear:

$$\begin{pmatrix} x_{0,u} \\ x_{0,v} \\ \\ x_{1,u} \\ x_{1,v} \\ \\ x_{2,u} \\ x_{2,v} \end{pmatrix} \to \begin{pmatrix} 1\,0 & 1\,0 & 1\,0 \\ 0\,1 & 0\,1 & 0\,1 \\ \\ 0\,1 & 1\,1 & 1\,0 \\ 1\,1 & 1\,0 & 0\,1 \\ \\ 1\,1 & 0\,1 & 1\,0 \\ 1\,0 & 1\,1 & 0\,1 \end{pmatrix} \begin{pmatrix} x_{0,u} \\ x_{0,v} \\ \\ x_{1,u} \\ x_{1,v} \\ \\ x_{2,u} \\ x_{2,v} \end{pmatrix}.$$

Using it as a skeleton we can construct, for example, the following "crazy" mapping:

$$\begin{pmatrix} x_{0,u} \\ x_{0,v} \\ \\ x_{1,u} \\ x_{1,v} \\ \\ x_{2,u} \\ x_{2,v} \end{pmatrix} \to \begin{pmatrix} x_{0,u} + (x_{1,u} \oplus x_{2,u})(2x_{0,v}x_{1,v} + 1) \\ x_{0,v} - x_{1,v} + x_{2,v} \oplus ((x_{0,u} \oplus x_{1,u})^2 \vee 1) \\ \\ (x_{0,v} \oplus x_{1,u})(2x_{0,v}x_{2,v} - 1) - x_{1,v} + x_{2,u} \\ (x_{0,u} + x_{0,v}) \oplus (x_{1,u} - x_{2,v})((x_{1,v} \oplus x_{2,u})^2 \vee 1) \\ \\ (x_{0,u} - x_{0,v})(2x_{1,u}x_{2,v} + 1) + x_{1,v} \oplus x_{2,u} \\ (x_{0,u} \oplus x_{1,u}) - (x_{1,v} \oplus x_{2,v}) + ((x_{0,v} \oplus x_{2,u})^2 \vee 1) \end{pmatrix}.$$

---

[4] This notation means that

$$\begin{array}{ccc} (x_0 = 0, x_1 = 0, x_2 = 0) & \mapsto & (y_0 = 0, y_1 = 0, y_2 = 0), \\ & \vdots & \\ (x_0 = 3, x_1 = 3, x_2 = 3) & \mapsto & (y_0 = 3, y_1 = 0, y_2 = 0). \end{array}$$

Formal representation	Examples
$\mathcal{P}$:	
$\mathcal{C}$	3
$\mathcal{E} \times \mathcal{E}$	$(x+5)^2$
$\mathcal{P}_0$	$2x$
$(\mathcal{P} \pm \mathcal{E}) \oplus \mathcal{E}$	$(5+x) \oplus x$
$\mathcal{P}' \circ \mathcal{P}''$	$x^2 + 2x$
$\mathcal{P}_0$:	
$\mathcal{C}_0$	2
$\mathcal{C}_0 \times \mathcal{E}$	$2(x^3 \wedge x)$
$\mathcal{P} \wedge \mathcal{C}_0$	$x^2 \wedge 1 \ldots 10_2$
$\mathcal{P}_0 \times \mathcal{P}$	$(x^2 \wedge 1 \ldots 10_2)((5+x) \oplus x)$
$\mathcal{P}_0' \circ \mathcal{P}_0''$	$2x - (x^2 \wedge 1 \ldots 10_2)$

$\mathcal{C}$	constant
$\mathcal{P}$	parameter
$\mathcal{E}$	expression
$\mathcal{C}_0$	constant with 0 in the least significant bit
$\mathcal{P}_0$	parameter with 0 in the least significant bit
$\circ$	primitive operation

This table summarize techniques most commonly used to construct parameters. Note that in a single line the same symbol denotes the same expression (e.g., $\mathcal{E} \times \mathcal{E}$ denotes squaring). Keep in mind that expressions obtained by means of this table are not necessary parameters in the least significant bit-slice (clearly, $\mathcal{P}_0$ are parameters everywhere).

Fig. 1. Common parameter construction techniques

The Poly1305-AES Message-Authentication Code

Daniel J. Bernstein*

Department of Mathematics,
Statistics, and Computer Science (M/C 249),
The University of Illinois at Chicago,
Chicago, IL 60607–7045
djb@cr.yp.to

Abstract. Poly1305-AES is a state-of-the-art message-authentication code suitable for a wide variety of applications. Poly1305-AES computes a 16-byte authenticator of a variable-length message, using a 16-byte AES key, a 16-byte additional key, and a 16-byte nonce. The security of Poly1305-AES is very close to the security of AES; the security gap is at most $14D\lceil L/16\rceil/2^{106}$ if messages have at most L bytes, the attacker sees at most 2^{64} authenticated messages, and the attacker attempts D forgeries. Poly1305-AES can be computed at extremely high speed: for example, fewer than $3.1\ell + 780$ Athlon cycles for an ℓ-byte message. This speed is achieved *without* precomputation; consequently, 1000 keys can be handled simultaneously without cache misses. Special-purpose hardware can compute Poly1305-AES at even higher speed. Poly1305-AES is parallelizable, incremental, and not subject to any intellectual-property claims.

1 Introduction

This paper introduces and analyzes Poly1305-AES, a state-of-the-art secret-key message-authentication code suitable for a wide variety of applications.

Poly1305-AES computes a 16-byte authenticator $\text{Poly1305}_r(m, \text{AES}_k(n))$ of a variable-length message m, using a 16-byte AES key k, a 16-byte additional key r, and a 16-byte nonce n. Section 2 of this paper presents the complete definition of Poly1305-AES.

Poly1305-AES has several useful features:

- **Guaranteed security if AES is secure.** The security gap is small, even for long-term keys; the only way for an attacker to break Poly1305-AES is to break AES. Assume, for example, that messages are packets up to 1024 bytes; that the attacker sees 2^{64} messages authenticated under a Poly1305-AES key; that the attacker attempts a whopping 2^{75} forgeries; and that the

* The author was supported by the National Science Foundation under grant CCR–9983950, and by the Alfred P. Sloan Foundation. Date of this document: 2005.03.29. Permanent ID of this document: 0018d9551b5546d97c340e0dd8cb5750. This version is final and may be freely cited.

H. Gilbert and H. Handschuh (Eds.): FSE 2005, LNCS 3557, pp. 32–49, 2005.

attacker cannot break AES with probability above δ. Then, with probability at least $0.999999 - \delta$, *all* of the 2^{75} forgeries are rejected.

- **Cipher replaceability.** If anything does go wrong with AES, users can switch from Poly1305-AES to Poly1305-AnotherFunction, with an identical security guarantee. All the effort invested in the non-AES part of Poly1305-AES can be reused; the non-AES part of Poly1305-AES cannot be broken.
- **Extremely high speed.** My published Poly1305-AES software takes just 3843 Athlon cycles, 5361 Pentium III cycles, 5464 Pentium 4 cycles, 4611 Pentium M cycles, 8464 PowerPC 7410 cycles, 5905 PowerPC RS64 IV cycles, 5118 UltraSPARC II cycles, or 5601 UltraSPARC III cycles to verify an authenticator on a 1024-byte message. Poly1305-AES offers *consistent* high speed, not just high speed for one CPU.
- **Low per-message overhead.** The same software takes just 1232 Pentium 4 cycles, 1264 PowerPC 7410 cycles, or 1077 UltraSPARC III cycles to verify an authenticator on a 64-byte message. Poly1305-AES offers *consistent* high speed, not just high speed for long messages. Most competing functions have much larger overhead for each message; they are designed for long messages, without regard to short-packet performance.
- **Key agility.** Poly1305-AES offers *consistent* high speed, not just high speed for single-key benchmarks. The timings in this paper do *not* rely on any pre-expansion of the 32-byte Poly1305-AES key (k, r); Poly1305-AES can fit thousands of simultaneous keys into cache, and remains fast even when keys are out of cache. This was my primary design goal for Poly1305-AES. Almost all competing functions use a large table for each key; as the number of keys grows, those functions miss the cache and slow down dramatically.
- **Parallelizability and incrementality.** The circuit depth of Poly1305-AES is quite small, even for long messages. Consequently, Poly1305-AES can take advantage of additional hardware to reduce the latency for long messages. For essentially the same reason, Poly1305-AES can be recomputed at low cost for a small modification of a long message.
- **No intellectual-property claims.** I am not aware of any patents or patent applications relevant to Poly1305-AES.

Section 3 of this paper analyzes the security of Poly1305-AES. Section 4 discusses the software achieving the speeds stated above. Section 5 discusses the speed of Poly1305-AES in other contexts.

Genealogy

Gilbert, MacWilliams, and Sloane in [15] introduced the idea of provably secure authentication. The Gilbert-MacWilliams-Sloane system is fast, but it requires keys longer than L bytes to handle L-byte messages, and it requires a completely new key for each message.

Wegman and Carter in [32] pointed out that the key length could be merely $64 \lg L$ for the first message plus 16 bytes for each additional message. At about the same time, in a slightly different context, Karp and Rabin achieved a key length of 32 bytes for the first message; see [19] and [26]. The system in [19] is fast once keys are generated, but key generation is slow.

The idea of using a cipher such as AES to expand a short key into a long key is now considered obvious. Brassard in [12] published the idea in the Wegman-Carter context; I don't know whether the idea was considered obvious back then.

Polynomial-evaluation MACs—MACs that treat each message as a univariate polynomial over a finite field and then evaluate that polynomial at the key—were introduced in three papers independently: [14] by den Boer; [31–Section 3] by Taylor; [9–Section 4] by Bierbrauer, Johansson, Kabatianskii, and Smeets. Polynomial-evaluation MACs combine several attractive features: short keys, fast key generation, and fast message authentication. Several subsequent papers reported implementations of polynomial-evaluation MACs over binary fields: [28] by Shoup; [4] by Afanassiev, Gehrmann, and Smeets, reinventing Kaminski's division algorithm in [18]; [22] by Nevelsteen and Preneel.

Polynomial-evaluation MACs over prime fields can exploit the multipliers built into many current CPUs, achieving substantially better performance than polynomial-evaluation MACs over binary fields. This idea was first published in my paper [5] in April 1999, and explained in detail in [7]. Another MAC, avoiding binary fields for the same reason, was published independently by Black, Halevi, Krawczyk, Krovetz, and Rogaway in [11] in August 1999.

I used 32-bit polynomial coefficients modulo $2^{127} - 1$ ("hash127") in [5] and [7]. The short coefficients don't allow great performance (for short messages) without precomputation, so I casually precomputed a few kilobytes of data for each key; this is a disaster for applications handling many keys simultaneously, but I didn't think beyond a single key. Similarly, [11] ("UMAC") uses large keys.

Krovetz and Rogaway in [21] suggested 64-bit coefficients modulo $2^{64} - 59$, with an escape mechanism for coefficients between $2^{64} - 59$ and $2^{64} - 1$. They did not claim competitive performance: their software, run twice to achieve a reasonable 100-bit security level, was more than three times slower than hash127 (and more than six times slower for messages with all bits set). Krovetz and Rogaway did point out, however, that their software did not require large tables.

In `http://cr.yp.to/talks.html#2002.06.15`, posted July 2002, I pointed out that 128-bit coefficients over the slightly *larger* prime field $\mathbf{Z}/(2^{130} - 5)$ allow excellent performance without precomputation. This paper explains Poly1305-AES in much more detail.

Kohno, Viega, and Whiting subsequently suggested 96-bit coefficients modulo $2^{127} - 1$ ("CWC HASH"). They published some non-competitive timings for CWC HASH and then gave up on the idea. A careful implementation of CWC HASH without precomputation would be quite fast, although still not as fast as Poly1305-AES.

2 Specification

This section defines the Poly1305-AES function. The Poly1305-AES formula is a straightforward polynomial evaluation modulo $2^{130} - 5$; most of the detail is in key format and message padding.

Messages

Poly1305-AES authenticates messages. A **message** is any sequence of bytes $m[0], m[1], \ldots, m[\ell-1]$; a **byte** is any element of $\{0, 1, \ldots, 255\}$. The length ℓ can be any nonnegative integer, and can vary from one message to another.

Keys

Poly1305-AES authenticates messages using a 32-byte secret key shared by the message sender and the message receiver. The key has two parts: first, a 16-byte AES key k; second, a 16-byte string $r[0], r[1], \ldots, r[15]$. The second part of the key represents a 128-bit integer r in unsigned little-endian form: i.e., $r = r[0] + 2^8 r[1] + \ldots + 2^{120} r[15]$.

Certain bits of r are required to be 0: $r[3], r[7], r[11], r[15]$ are required to have their top four bits clear (i.e., to be in $\{0, 1, \ldots, 15\}$), and $r[4], r[8], r[12]$ are required to have their bottom two bits clear (i.e., to be in $\{0, 4, 8, \ldots, 252\}$). Thus there are 2^{106} possibilities for r. In other words, r is required to have the form $r_0 + r_1 + r_2 + r_3$ where $r_0 \in \{0, 1, 2, 3, \ldots, 2^{28} - 1\}$, $r_1/2^{32} \in \{0, 4, 8, 12, \ldots, 2^{28} - 4\}$, $r_2/2^{64} \in \{0, 4, 8, 12, \ldots, 2^{28} - 4\}$, and $r_3/2^{96} \in \{0, 4, 8, 12, \ldots, 2^{28} - 4\}$.

Nonces

Poly1305-AES requires each message to be accompanied by a 16-byte **nonce**, i.e., a unique message number. Poly1305-AES feeds each nonce n through AES_k to obtain the 16-byte string $\mathrm{AES}_k(n)$.

There is nothing special about AES here. One can replace AES with an arbitrary keyed function from an arbitrary set of nonces to 16-byte strings. This paper focuses on AES for concreteness.

Conversion and Padding

Let $m[0], m[1], \ldots, m[\ell-1]$ be a message. Write $q = \lceil \ell/16 \rceil$. Define integers $c_1, c_2, \ldots, c_q \in \{1, 2, 3, \ldots, 2^{129}\}$ as follows: if $1 \le i \le \lfloor \ell/16 \rfloor$ then

$$c_i = m[16i-16] + 2^8 m[16i-15] + 2^{16} m[16i-14] + \cdots + 2^{120} m[16i-1] + 2^{128};$$

if ℓ is not a multiple of 16 then

$$c_q = m[16q-16] + 2^8 m[16q-15] + \cdots + 2^{8(\ell \bmod 16)-8} m[\ell-1] + 2^{8(\ell \bmod 16)}.$$

In other words: Pad each 16-byte chunk of a message to 17 bytes by appending a 1. If the message has a final chunk between 1 and 15 bytes, append 1 to the chunk, and then zero-pad the chunk to 17 bytes. Either way, treat the resulting 17-byte chunk as an unsigned little-endian integer.

Authenticators

$\mathrm{Poly1305}_r(m, \mathrm{AES}_k(n))$, the Poly1305-AES authenticator of a message m with nonce n under secret key (k, r), is defined as the 16-byte unsigned little-endian representation of

$$(((c_1 r^q + c_2 r^{q-1} + \cdots + c_q r^1) \bmod 2^{130} - 5) + \mathrm{AES}_k(n)) \bmod 2^{128}.$$

Here the 16-byte string $\mathrm{AES}_k(n)$ is treated as an unsigned little-endian integer, and $c_1, c_2, \ldots, c_q$ are the integers defined above. See Appendix B for examples.

Sample Code

The following C++ code reads k from k[0], k[1], ..., k[15], reads r from r[0], r[1], ..., r[15], reads $\mathrm{AES}_k(n)$ from s[0], s[1], ..., s[15], reads m from m[0], m[1], ..., m[l-1], and places $\mathrm{Poly1305}_r(m, \mathrm{AES}_k(n))$ into out[0], out[1], ..., out[15]:

```
#include <gmpxx.h>

void poly1305_gmpxx(unsigned char *out,
  const unsigned char *r,
  const unsigned char *s,
  const unsigned char *m,unsigned int l)
{
  unsigned int j;
  mpz_class rbar = 0;
  for (j = 0;j < 16;++j)
    rbar += ((mpz_class) r[j]) << (8 * j);
  mpz_class h = 0;
  mpz_class p = (((mpz_class) 1) << 130) - 5;
  while (l > 0) {
    mpz_class c = 0;
    for (j = 0;(j < 16) && (j < l);++j)
      c += ((mpz_class) m[j]) << (8 * j);
    c += ((mpz_class) 1) << (8 * j);
    m += j; l -= j;
    h = ((h + c) * rbar) % p;
  }
  for (j = 0;j < 16;++j)
    h += ((mpz_class) s[j]) << (8 * j);
  for (j = 0;j < 16;++j) {
    mpz_class c = h % 256;
    h >>= 8;
    out[j] = c.get_ui();
  }
}
```

See [16] for the underlying integer-arithmetic library, gmpxx.

This code is not meant as a high-speed implementation; it does not have even the simplest speedups; it should be expected to provide intolerable performance. It is simply a secondary statement of the definition of Poly1305-AES.

Design Decisions

I considered various primes above 2^{128}. I chose $2^{130} - 5$ because its sparse form makes divisions particularly easy in both software and hardware. My encoding of messages as polynomials takes advantage of the gap between 2^{128} and $2^{130} - 5$.

There are several reasons that Poly1305-AES uses nonces. First, comparable protocols without nonces have security bounds that look like $C(C + D)L/2^{106}$ rather than $DL/2^{106}$—here C is the number of messages authenticated by the sender, D is the number of forgery attempts, and L is the maximum message length—and thus cannot be used with confidence for large C. Second, nonces allow the invocation of AES to be carried out in parallel with most of the other operations in Poly1305-AES, reducing latency in many contexts. Third, most protocols have nonces anyway, for a variety of reasons: nonces are required for secure encryption, for example, and nonces allow trivial rejection of replayed messages.

I constrained r to simplify and accelerate implementations of Poly1305-AES in various contexts. A wider range of r—e.g., all 128-bit integers—would allow a quantitatively better security bound, but the current bound $DL/2^{106}$ will be perfectly satisfactory for the foreseeable future, whereas slower authenticator computations would not be perfectly satisfactory.

I chose little-endian instead of big-endian to improve overall performance. Little-endian saves time on the most popular CPUs (the Pentium and Athlon) while making no difference on most other CPUs (the PowerPC, for example, and the UltraSPARC).

The definition of Poly1305-AES could easily be extended from byte strings to bit strings, but there is no apparent benefit of doing so.

3 Security

This section discusses the security of Poly1305-AES.

Responsibilities of the User

Any protocol that uses Poly1305-AES must ensure unpredictability of the secret key (k, r). This section assumes that secret keys are chosen from the uniform distribution: i.e., probability 2^{-234} for each of the 2^{234} possible pairs (k, r).

Any protocol that uses Poly1305-AES must ensure that the secret key is, in fact, kept secret. This section assumes that all operations are independent of (k, r), except for the computation of authenticators by the sender and receiver. (There are safe ways to reuse k for encryption, but those ways are not analyzed in this paper.)

The sender must *never* use the same nonce for two different messages. The simplest way to achieve this is for the sender to use an increasing sequence of nonces in, e.g., reverse-lexicographic order of 16-byte strings. (Problem: If a key is stored on disk, while increasing nonce values are stored in memory, what happens when the power goes out? Solution: Store a safe nonce value—a new nonce larger than any nonce used—on disk alongside the key.) Any protocol that uses Poly1305-AES must specify a mechanism of nonce generation and maintenance that prevents duplicates.

Security Guarantee

Poly1305-AES guarantees that the only way for the attacker to find an (n, m, a) such that $a = \text{Poly1305}_r(m, \text{AES}_k(n))$, other than the authenticated messages (n, m, a) sent by the sender, is to break AES. If the attacker cannot break AES, and the receiver discards all (n, m, a) such that $a \neq \text{Poly1305}_r(m, \text{AES}_k(n))$, then the receiver will see only messages authenticated by the sender.

This guarantee is not limited to "meaningful" messages m. It is true even if the attacker can see all the authenticated messages sent by the sender. It is true even if the attacker can see whether the receiver accepts a forgery. It is true even if the attacker can influence the sender's choice of messages and unique nonces. (But it is not true if the nonce-uniqueness rule is violated.)

Here is a quantitative form of the guarantee. Assume that the attacker sees at most C authenticated messages and attempts at most D forgeries. Assume that the attacker has probability at most δ of distinguishing AES_k from a uniform random permutation after $C + D$ queries. Assume that all messages have length at most L. Then, with probability at least

$$1 - \delta - \frac{(1 - C/2^{128})^{-(C+1)/2} 8D\lceil L/16 \rceil}{2^{106}},$$

all of the attacker's forgeries are discarded. In particular, if $C \leq 2^{64}$, then the attacker's chance of success is at most $\delta + 1.649 \cdot 8D\lceil L/16 \rceil / 2^{106} < \delta + 14D\lceil L/16 \rceil / 2^{106}$.

The most important design goal of AES was for δ to be small. There is, however, no hope of *proving* that δ is small. Perhaps AES will be broken someday. If that happens, users should switch to Poly1305-AnotherFunction. Poly1305-AnotherFunction provides the same security guarantee relative to the security of AnotherFunction.

Proof of the Security Guarantee

For each message m, write $\underline{m}$ for the polynomial $c_1 x^q + c_2 x^{q-1} + \cdots + c_q x^1$, where $q, c_1, c_2, \ldots, c_q$ are defined as in Section 2. Define $H_r(m)$ as the 16-byte unsigned little-endian representation of $(\underline{m}(r) \bmod 2^{130} - 5) \bmod 2^{128}$; note that H_r and k are independent. Define a group operation $+$ on 16-byte strings as addition modulo 2^{128}, where each 16-byte string is viewed as the unsigned little-endian representation of an integer in $\{0, 1, 2, \ldots, 2^{128} - 1\}$. Then the authenticator $\text{Poly1305}_r(m, \text{AES}_k(n))$ is equal to $H_r(m) + \text{AES}_k(n)$.

The crucial property of H_r is that it has small differential probabilities: if g is a 16-byte string, and m, m' are distinct messages of length at most L, then $H_r(m) = H_r(m') + g$ with probability at most $8\lceil L/16 \rceil / 2^{106}$. See below.

Theorem 5.4 of [8] now guarantees that $H_r(m) + \text{AES}_k(n)$ is secure if AES is secure: specifically, that the attacker's success chance against $H_r(m) + \text{AES}_k(n)$ is at most $\delta + D(1 - C/2^{128})^{-(C+1)/2} 8\lceil L/16 \rceil / 2^{106}$.

The rest of this section is devoted to proving that H_r has small differential probabilities.

Theorem 3.1. $2^{130} - 5$ *is prime.*

Proof. Define $p_1 = (2^{130} - 6)/1517314646$ and $p_2 = (p_1 - 1)/222890620702$. Observe that 37003 and 221101 are prime divisors of $p_2 - 1$; $(37003 \cdot 221101)^2 > p_2$; $2^{p_2-1} - 1$ is divisible by p_2; $2^{(p_2-1)/37003} - 1$ and $2^{(p_2-1)/221101} - 1$ are coprime to p_2; $p_2^2 > p_1$; $2^{p_1-1} - 1$ is divisible by p_1; $2^{(p_1-1)/p_2} - 1$ is coprime to p_1; $p_1^2 > 2^{130} - 5$; $2^{2^{130}-6} - 1$ is divisible by $2^{130} - 5$; and $2^{(2^{130}-6)/p_1} - 1$ is coprime to $2^{130} - 5$. Hence p_2, p_1, and $2^{130} - 5$ are prime by Pocklington's theorem. □

Theorem 3.2. *Let m and m' be messages. Let u be an integer. If the polynomial $\underline{m'} - \underline{m} - u$ is zero modulo $2^{130} - 5$ then $m = m'$.*

Proof. Define $c_1, c_2, \ldots, c_q$ as above, and define $c'_1, c'_2, \ldots, c'_{q'}$ for m' similarly.

If $q > q'$ then the coefficient of x^q in $\underline{m'} - \underline{m}$ is $0 - c_1$. By construction c_1 is in $\{1, 2, 3, \ldots, 2^{129}\}$, so it is nonzero modulo $2^{130} - 5$; contradiction. Thus $q \leq q'$. Similarly $q \geq q'$. Hence $q = q'$.

If $i \in \{1, 2, \ldots, q\}$ then $c_i - c'_i$ is the coefficient of x^{q+1-i} in $\underline{m'} - \underline{m} - u$, which by hypothesis is divisible by $2^{130} - 5$. But $c_i - c'_i$ is between -2^{129} and 2^{129} by construction. Hence $c_i = c'_i$. In particular, $c_q = c'_q$.

Define ℓ as the number of bytes in m. Recall that $q = \lceil \ell/16 \rceil$; thus ℓ is between $16q - 15$ and $16q$. The exact value of ℓ is determined by q and c_q: it is $16q$ if $2^{128} \leq c_q$, $16q - 1$ if $2^{120} \leq c_q < 2^{121}$, $16q - 2$ if $2^{112} \leq c_q < 2^{113}$, ..., $16q - 15$ if $2^8 \leq c_q < 2^9$. Hence m' also has ℓ bytes.

Now consider any $j \in \{0, 1, \ldots, \ell - 1\}$. Write $i = \lfloor j/16 \rfloor + 1$; then $16i - 16 \leq j \leq 16i - 1$, and $1 \leq i \leq \lceil \ell/16 \rceil = q$, so $m[j] = \lfloor c_i/2^{8(j-16i+16)} \rfloor \bmod 256 = \lfloor c'_i/2^{8(j-16i+16)} \rfloor \bmod 256 = m'[j]$. Hence $m = m'$. □

Theorem 3.3. *Let m, m' be distinct messages, each having at most L bytes. Let g be a 16-byte string. Let R be a subset of $\{0, 1, \ldots, 2^{130} - 6\}$. Then there are at most $8\lceil L/16 \rceil$ integers $r \in R$ such that $H_r(m) = H_r(m') + g$.*

Consequently, if $\#R = 2^{106}$, and if r is a uniform random element of R, then $H_r(m) = H_r(m') + g$ with probability at most $8\lceil L/16 \rceil / 2^{106}$.

Proof. Define U as the set of integers in $[-2^{130} + 6, 2^{130} - 6]$ congruent to g modulo 2^{128}. Note that $\#U \leq 8$.

If $H_r(m) = H_r(m') + g$ then $(\underline{m'}(r) \bmod 2^{130} - 5) - (\underline{m}(r) \bmod 2^{130} - 5) \equiv g \pmod{2^{128}}$ so $(\underline{m'}(r) \bmod 2^{130} - 5) - (\underline{m}(r) \bmod 2^{130} - 5) = u$ for some $u \in U$. Hence r is a root of the polynomial $\underline{m'} - \underline{m} - u$ modulo the prime $2^{130} - 5$. This polynomial is nonzero by Theorem 3.2, and has degree at most $\lceil L/16 \rceil$, so it has at most $\lceil L/16 \rceil$ roots modulo $2^{130} - 5$. Sum over all $u \in U$: there are most $8\lceil L/16 \rceil$ possibilities for r. □

4 A Floating-Point Implementation

This section explains how to compute $\text{Poly1305}_r(m, \text{AES}_k(n))$, given (k, r, n, m), at very high speeds on common general-purpose CPUs.

These techniques are used by my `poly1305aes` software library to achieve the Poly1305-AES speeds reported in Section 1. See Appendix A for further speed information. The software itself is available from `http://cr.yp.to/mac.html`.

The current version of `poly1305aes` includes separate implementations of Poly1305-AES for the Athlon, the Pentium, the PowerPC, and the UltraSPARC; it also includes a backup C implementation to handle other CPUs. This section focuses on the Athlon for concreteness.

Outline

The overall strategy to compute $\text{Poly1305}_r(m, \text{AES}_k(n))$ is as follows. Start by setting an accumulator h to 0. For each chunk c of the message m, first set $h \leftarrow h + c$, and then set $h \leftarrow rh$. Periodically reduce h modulo $2^{130} - 5$, not necessarily to the smallest remainder but to something small enough to continue the computation. After all input chunks c are processed, fully reduce h modulo $2^{130} - 5$, and add $\text{AES}_k(n)$.

Large-Integer Arithmetic in Floating-Point Registers

Represent each of h, c, r as a sum of floating-point numbers, as in [7]. Specifically:

- As in Section 2, write r as $r_0 + r_1 + r_2 + r_3$ where $r_0 \in \{0, 1, 2, \ldots, 2^{28} - 1\}$, $r_1/2^{32} \in \{0, 4, 8, \ldots, 2^{28} - 4\}$, $r_2/2^{64} \in \{0, 4, 8, \ldots, 2^{28} - 4\}$, and $r_3/2^{96} \in \{0, 4, 8, \ldots, 2^{28} - 4\}$. Store each of r_0, r_1, r_2, r_3, $5 \cdot 2^{-130} r_1$, $5 \cdot 2^{-130} r_2$, $5 \cdot 2^{-130} r_3$ in memory in 8-byte floating-point format.
- Write each message chunk c as $d_0 + d_1 + d_2 + d_3$ where $d_0, d_1/2^{32}, d_2/2^{64} \in \{0, 1, 2, 3, \ldots, 2^{32} - 1\}$ and $d_3/2^{96} \in \{0, 1, 2, 3, \ldots, 2^{34} - 1\}$.
- Write h as $h_0 + h_1 + h_2 + h_3$ where h_i is a multiple of 2^{32i} in the range specified below. Store each h_i in one of the Athlon's floating-point registers.

Warning: The FreeBSD operating system starts each program by instructing the CPU to round all floating-point mantissas to 53 bits, rather than using the CPU's natural 64-bit precision. Make sure to disable this instruction. Under `gcc`, for example, the code `asm volatile("fldcw %0"::"m"(0x137f))` specifies full 64-bit mantissas.

To set $h \leftarrow h + c$, set $h_0 \leftarrow h_0 + d_0$, $h_1 \leftarrow h_1 + d_1$, $h_2 \leftarrow h_2 + d_2$, $h_3 \leftarrow h_3 + d_3$. Before these additions, $h_0, h_1/2^{32}, h_2/2^{64}, h_3/2^{96}$ are required to be integers in $[-(63/128) \cdot 2^{64}, (63/128) \cdot 2^{64}]$. After these additions, $h_0, h_1/2^{32}, h_2/2^{64}, h_3/2^{96}$ are integers in $[-(127/256) \cdot 2^{64}, (127/256) \cdot 2^{64}]$.

Before multiplying h by r, reduce the range of each h_i by performing four parallel carries as follows:

- Define $\alpha_0 = 2^{95} + 2^{94}$, $\alpha_1 = 2^{127} + 2^{126}$, $\alpha_2 = 2^{159} + 2^{158}$, and $\alpha_3 = 2^{193} + 2^{192}$.
- Compute $y_i = \text{fp}_{64}(h_i + \alpha_i) - \alpha_i$ and $x_i = h_i - y_i$. Here $\text{fp}_{64}(h_i + \alpha_i)$ means the 64-bit-mantissa floating-point number closest to $h_i + \alpha_i$, with ties broken in the usual way; see [3]. Then $y_0/2^{32}, y_1/2^{64}, y_2/2^{96}, y_3/2^{130}$ are integers.
- Set $h_0 \leftarrow x_0 + 5 \cdot 2^{-130} y_3$, $h_1 \leftarrow x_1 + y_0$, $h_2 \leftarrow x_2 + y_1$, and $h_3 \leftarrow x_3 + y_2$.

This substitution changes h by $(2^{130} - 5)2^{-130}y_3$, so it does not change h mod $2^{130} - 5$. There are 17 floating-point operations here: 8 additions, 8 subtractions, and 1 multiplication by the constant $5 \cdot 2^{-130}$.

Ranges: x_0, $x_1/2^{32}$, and $x_2/2^{64}$ are in $[-(1/2) \cdot 2^{32}, (1/2) \cdot 2^{32}]$; $x_3/2^{96}$ is in $[-2 \cdot 2^{32}, 2 \cdot 2^{32}]$; $y_0/2^{32}$, $y_1/2^{64}$, $y_2/2^{96}$, and $y_3/2^{128}$ are in $[-(127/256) \cdot 2^{32}, (127/256) \cdot 2^{32}]$; h_0 is in $[-(1147/1024) \cdot 2^{32}, (1147/1024) \cdot 2^{32}]$; $h_1/2^{32}$ is in $[-(255/256) \cdot 2^{32}, (255/256) \cdot 2^{32}]$; $h_2/2^{64}$ is in $[-(255/256) \cdot 2^{32}, (255/256) \cdot 2^{32}]$; $h_3/2^{96}$ is in $[-(639/256) \cdot 2^{32}, (639/256) \cdot 2^{32}]$.

To multiply h by r modulo $2^{130} - 5$, replace (h_0, h_1, h_2, h_3) with

$$\begin{array}{llll}(r_0 h_0 + 5 \cdot 2^{-130} r_1 h_3 + 5 \cdot 2^{-130} r_2 h_2 + 5 \cdot 2^{-130} r_3 h_1, \\ r_0 h_1 + & r_1 h_0 + 5 \cdot 2^{-130} r_2 h_3 + 5 \cdot 2^{-130} r_3 h_2, \\ r_0 h_2 + & r_1 h_1 + & r_2 h_0 + 5 \cdot 2^{-130} r_3 h_3, \\ r_0 h_3 + & r_1 h_2 + & r_2 h_1 + & r_3 h_0).\end{array}$$

Recall that $2^{-34}r_1$, $2^{-66}r_2$, and $2^{-98}r_3$ are integers, so $2^{-130}r_1h_3$, $2^{-130}r_2h_2$, and $2^{-130}r_3h_1$ are integers; similarly, $2^{-130}r_2h_3$ and $2^{-130}r_3h_2$ are multiples of 2^{32}, and $2^{-130}r_3h_3$ is a multiple of 2^{64}. There are 28 floating-point operations here: 16 multiplications and 12 additions.

Ranges: $h_0, h_1/2^{32}, h_2/2^{64}, h_3/2^{96}$ are now integers of absolute value at most $2^{28}(1147/1024 + 2 \cdot (5/4)255/256 + (5/4)639/256)2^{32} < (63/128)2^{64}$, ready for the next iteration of the inner loop.

Note that the carries can be omitted on the first loop: d_0 is an integer in $[0, 2^{32}]$; $d_1/2^{32}$ is an integer in $[0, 2^{32}]$; $d_2/2^{64}$ is an integer in $[0, 2^{32}]$; $d_3/2^{96}$ is an integer in $[0, 3 \cdot 2^{32}]$; and $2^{28}(1 + (5/4) + (5/4) + (5/4)3)2^{32} < (63/128)2^{64}$.

Output Conversion

After the last message chunk is processed, carry one last time, to put h_0, h_1, h_2, h_3 into the small ranges listed above.

Add $2^{130} - 2^{97}$ to h_3; add $2^{97} - 2^{65}$ to h_2; add $2^{65} - 2^{33}$ to h_1; and add $2^{33} - 5$ to h_0. This makes each h_i positive, and puts $h = h_0 + h_1 + h_2 + h_3$ into the range $\{0, 1, \ldots, 2(2^{130} - 5) - 1\}$.

Perform a few integer add-with-carry operations to convert the accumulator into a series of 32-bit words in the usual form. Subtract $2^{130} - 5$, and keep the result if it is nonnegative, being careful to use constant-time operations so that no information is leaked through timing.

Finally, add $\text{AES}_k(n)$. There are two reasons to pay close attention to the AES computation:

- It is extremely difficult to write high-speed *constant-time* AES software. Typical AES software leaks key bytes to the simplest conceivable timing attack. See [6]. My new AES implementations go to extensive effort to reduce the AES timing variability.
- The time to compute $\text{AES}_k(n)$ from (k, n) is more than half of the time to compute $\text{Poly1305}_r(m, \text{AES}_k(n))$ for short messages, and remains quite

noticeable for longer messages. My new AES implementations are, as far as I know, the fastest available software for computing $\text{AES}_k(n)$ from (k, n). Of course, *if* there is room in cache, then one can save some time by instead computing $\text{AES}_k(n)$ from (K, n), where K is a pre-expanded version of k.

Details of the AES computation are not discussed in this paper but are discussed in the `poly1305aes` documentation.

Instruction Selection and Scheduling

Consider an integer (such as d_0) between 0 and $2^{32} - 1$, stored in the usual way as four bytes. How does one load the integer into a floating-point register, when the Athlon does not have a load-four-byte-unsigned-integer instruction? Here are three possibilities:

- Concatenate the four bytes with $(0, 0, 0, 0)$, and use the Athlon's load-eight-byte-signed-integer instruction. Unfortunately, the four-byte store forces the eight-byte load to wait for dozens of cycles.
- Concatenate the bytes with $(0, 0, 56, 67)$, producing an eight-byte floating-point number. Load that number, and subtract $2^{52}+2^{51}$ to obtain the desired integer. This well-known trick has the virtue of also allowing the integer to be scaled by (e.g.) 2^{32}: replace 67 with 69 and $2^{52} + 2^{51}$ with $2^{84} + 2^{83}$. Unfortunately, as above, the four-byte store forces the eight-byte load to wait for dozens of cycles.
- Subtract 2^{31} from the integer, use the Athlon's load-four-byte-signed-integer instruction, and add 2^{31} to the result. This has smaller latency, but puts more pressure on the floating-point unit.

Top performance requires making the right choice.

(A variant of Poly1305-AES using signed 32-bit integers would save time on the Athlon. On the other hand, it would lose time on typical 64-bit CPUs.)

This is merely one example of several low-level issues that can drastically affect speed: instruction selection, instruction scheduling, register assignment, instruction fetching, etc. A "fast" implementation of Poly1305-AES, with just a few typical low-level mistakes, will use twice as many cycles per byte as the software described here.

Other Modern CPUs

The same floating-point operations also run at high speed on the Pentium 1, Pentium MMX, Pentium Pro, Pentium II, Pentium III, Pentium 4, Pentium M, Celeron, Duron, et al.

The UltraSPARC, PowerPC, et al. support fast arithmetic on floating-point numbers with 53-bit, rather than 64-bit, mantissas. The simplest way to achieve good performance on these chips is to break a 32-bit number into two 16-bit pieces before multiplying it by another 32-bit number.

As in the case of the Athlon, careful attention to low-level CPU details is necessary for top performance.

5 Other Implementation Strategies

Some people, upon hearing that there is a tricky way to use the Athlon's floating-point unit to compute a function quickly, leap to the unjustified conclusion that the same function cannot be computed quickly except on an Athlon. Consider, for example, the incorrect statement "hash-127 needs good hardware support for a fast implementation" in [17–footnote 3].

This section outlines three non-floating-point methods to compute Poly1305-AES, and indicates contexts where the methods are useful.

Integer Registers

The 130-bit accumulator in Poly1305-AES can be spread among several integer registers rather than several floating-point registers.

This is good for low-end CPUs that do not support floating-point operations but that still have reasonably fast integer multipliers. It is also good for some high-end CPUs, such as the Athlon 64, that offer faster multiplication through integer registers than through floating-point registers.

Tables

One can make a table of the integers $r, 2r, 4r, 8r, \ldots, 2^{129}r$ modulo $2^{130} - 5$, and then multiply any 130-bit integer by r by adding, on average, about 65 elements of the table.

One can reduce the amount of work by using both additions and subtractions, by increasing the table size, and by choosing table entries more carefully. For example, one can include $3r, 24r, 192r, \ldots$ in the table, and then multiply any 130-bit integer by r by adding and subtracting, on average, about 38 elements of the table. This is a special case of an algorithm often credited to Brickell, Gordon, McCurley, Wilson, Lim, and Lee, but actually introduced much earlier by Pippenger in [23].

One can also balance the table size against the effort in reduction modulo $2^{130} - 5$. Consider, for example, the table $r, 2r, 3r, 4r, \ldots, 255r$.

Table lookups are often the best approach for tiny CPUs that do not have any fast multiplication operations. Of course, their key agility is poor, and they are susceptible to timing attacks if they are not implemented very carefully.

Special-Purpose Circuits

An 1800MHz AMD Duron, costing under \$50, can feed 4 gigabits per second of 1500-byte messages through Poly1305-AES with the software discussed in Section 4. Hardware implementations of Poly1305-AES can strip away a great deal of unnecessary cost: the multiplier is only part of the cost of the Duron; furthermore, some of the multiplications are by sparse constants; furthermore, only about 20% of the multiplier area is doing any useful work, since each input is much smaller than 64 bits; furthermore, almost all carries can be deferred until the end of the Poly1305-AES computation, rather than being performed after each multiplication; furthermore, hardware implementations need not, and should not, imitate traditional software structures—one can directly build a fast multiplier modulo $2^{130}-5$, taking advantage of more sophisticated multiplication algorithms than those used in the Duron. Evidently Poly1305-AES can handle next-generation Ethernet speeds at reasonable cost.

References

1. —, *17th annual symposium on foundations of computer science*, IEEE Computer Society, Long Beach, California, 1976. MR 56:1766.
2. —, *20th annual symposium on foundations of computer science*, IEEE Computer Society, New York, 1979. MR 82a:68004.
3. —, *IEEE standard for binary floating-point arithmetic*, Standard 754–1985, Institute of Electrical and Electronics Engineers, New York, 1985.
4. Valentine Afanassiev, Christian Gehrmann, Ben Smeets, *Fast message authentication using efficient polynomial evaluation*, in [10] (1997), 190–204. URL: `http://cr.yp.to/bib/entries.html#1997/afanassiev`.
5. Daniel J. Bernstein, *Guaranteed message authentication faster than MD5 (abstract)* (1999). URL: `http://cr.yp.to/papers.html#hash127-abs`.
6. Daniel J. Bernstein, *Cache-timing attacks on AES* (2004). URL: `http://cr.yp.to/papers.html#cachetiming`. ID `cd9faae9bd5308c440df50fc26a517b4`.
7. Daniel J. Bernstein, *Floating-point arithmetic and message authentication* (2004). URL: `http://cr.yp.to/papers.html#hash127`. ID `dabadd3095644704c5cbe9690ea3738e`.
8. Daniel J. Bernstein, *Stronger security bounds for Wegman-Carter-Shoup authenticators*, Proceedings of Eurocrypt 2005, to appear (2005). URL: `http://cr.yp.to/papers.html#securitywcs`. ID `2d603727f69542f30f7da2832240c1ad`.
9. Jürgen Bierbrauer, Thomas Johansson, Gregory Kabatianskii, Ben Smeets, *On families of hash functions via geometric codes and concatenation*, in [30] (1994), 331–342. URL: `http://cr.yp.to/bib/entries.html#1994/bierbrauer`.
10. Eli Biham (editor), *Fast Software Encryption '97*, Lecture Notes in Computer Science, 1267, Springer-Verlag, Berlin, 1997. ISBN 3–540–63247–6.
11. John Black, Shai Halevi, Hugo Krawczyk, Ted Krovetz, Phillip Rogaway, *UMAC: fast and secure message authentication*, in [34] (1999), 216–233. URL: `http://www.cs.ucdavis.edu/~rogaway/umac/`.
12. Gilles Brassard, *On computationally secure authentication tags requiring short secret shared keys*, in [13] (1983), 79–86. URL: `http://cr.yp.to/bib/entries.html#1983/brassard`.
13. David Chaum, Ronald L. Rivest, Alan T. Sherman (editors), *Advances in cryptology: proceedings of Crypto 82*, Plenum Press, New York, 1983. ISBN 0–306–41366–3. MR 84j:94004.
14. Bert den Boer, *A simple and key-economical unconditional authentication scheme*, Journal of Computer Security **2** (1993), 65–71. ISSN 0926–227X. URL: `http://cr.yp.to/bib/entries.html#1993/denboer`.

15. Edgar N. Gilbert, F. Jessie MacWilliams, Neil J. A. Sloane, *Codes which detect deception*, Bell System Technical Journal **53** (1974), 405–424. ISSN 0005–8580. MR 55:5306. URL: `http://cr.yp.to/bib/entries.html#1974/gilbert`.
16. Torbjorn Granlund (editor), *GMP 4.1.2: GNU multiple precision arithmetic library* (2004). URL: `http://www.swox.com/gmp/`.
17. Shai Halevi, Phil Rogaway, *A tweakable enciphering mode* (2003). URL: `http://www.research.ibm.com/people/s/shaih/pubs/hr03.html`.
18. Michael Kaminski, *A linear time algorithm for residue computation and a fast algorithm for division with a sparse divisor*, Journal of the ACM **34** (1987), 968–984. ISSN 0004–5411. MR 89f:68033.
19. Richard M. Karp, Michael O. Rabin, *Efficient randomized pattern-matching algorithms*, IBM Journal of Research and Development **31** (1987), 249–260. ISSN 0018–8646. URL: `http://cr.yp.to/bib/entries.html#1987/karp`.
20. Neal Koblitz (editor), *Advances in cryptology—CRYPTO '96*, Lecture Notes in Computer Science, 1109, Springer-Verlag, Berlin, 1996.
21. Ted Krovetz, Phillip Rogaway, *Fast universal hashing with small keys and no preprocessing: the PolyR construction* (2000). URL: `http://www.cs.ucdavis.edu/~rogaway/papers/poly.htm`.
22. Wim Nevelsteen, Bart Preneel, *Software performance of universal hash functions*, in [29] (1999), 24–41.
23. Nicholas Pippenger, *On the evaluation of powers and related problems (preliminary version)*, in [1] (1976), 258–263; newer version split into [24] and [25]. MR 58:3682. URL: `http://cr.yp.to/bib/entries.html#1976/pippenger`.
24. Nicholas Pippenger, *The minimum number of edges in graphs with prescribed paths*, Mathematical Systems Theory **12** (1979), 325–346; see also older version [23]. ISSN 0025–5661. MR 81e:05079. URL: `http://cr.yp.to/bib/entries.html#1979/pippenger`.
25. Nicholas Pippenger, *On the evaluation of powers and monomials*, SIAM Journal on Computing **9** (1980), 230–250; see also older version [23]. ISSN 0097–5397. MR 82c:10064. URL: `http://cr.yp.to/bib/entries.html#1980/pippenger`.
26. Michael O. Rabin, *Fingerprinting by random polynomials*, Harvard Aiken Computational Laboratory TR-15-81 (1981). URL: `http://cr.yp.to/bib/entries.html#1981/rabin`.
27. Victor Shoup, *On fast and provably secure message authentication based on universal hashing*, in [20] (1996), 313–328; see also newer version [28].
28. Victor Shoup, *On fast and provably secure message authentication based on universal hashing* (1996); see also older version [27]. URL: `http://www.shoup.net/papers`.
29. Jacques Stern (editor), *Advances in cryptology: EUROCRYPT '99*, Lecture Notes in Computer Science, 1592, Springer-Verlag, Berlin, 1999. ISBN 3–540–65889–0. MR 2000i:94001.
30. Douglas R. Stinson (editor), *Advances in cryptology—CRYPTO '93: 13th annual international cryptology conference, Santa Barbara, California, USA, August 22–26, 1993, proceedings*, Lecture Notes in Computer Science, 773, Springer-Verlag, Berlin, 1994. ISBN 3–540–57766–1, 0–387–57766–1. MR 95b:94002.
31. Richard Taylor, *An integrity check value algorithm for stream ciphers*, in [30] (1994), 40–48. URL: `http://cr.yp.to/bib/entries.html#1994/taylor`.

32. Mark N. Wegman, J. Lawrence Carter, *New classes and applications of hash functions*, in [2] (1979), 175–182; see also newer version [33]. URL: `http://cr.yp.to/bib/entries.html#1979/wegman`.
33. Mark N. Wegman, J. Lawrence Carter, *New hash functions and their use in authentication and set equality*, Journal of Computer and System Sciences **22** (1981), 265–279; see also older version [32]. ISSN 0022–0000. MR 82i:68017. URL: `http://cr.yp.to/bib/entries.html#1981/wegman`.
34. Michael Wiener (editor), *Advances in cryptology—CRYPTO '99*, Lecture Notes in Computer Science, 1666, Springer-Verlag, Berlin, 1999. ISBN 3–5540–66347–9. MR 2000h:94003.

A Appendix: Speed Graphs

These graphs show the time to verify an authenticator in various situations. The horizontal axis on the graphs is message length, from 0 bytes to 4096 bytes. The vertical axis on the graphs is time, from 0 CPU cycles to 24576 CPU cycles; time includes function-call overhead, timing overhead, etc. The bottom-left-to-top-right diagonal is 6 CPU cycles per byte. Color scheme:

- Non-reddish (black, green, dark blue, light blue): Keys are in cache.
- Reddish (red, yellow, purple, gray): Keys are not in cache. Loading the keys from DRAM takes extra time.
- Non-greenish (black, red, dark blue, purple): Messages, authenticators, and nonces are in cache.
- Greenish (green, yellow, light blue, gray): Messages, authenticators, and nonces are not in cache. Loading the data from DRAM takes extra time, typically growing with the message length.
- Non-blueish (black, red, green, yellow): Keys, message, authenticators, and nonces are aligned.
- Blueish (dark blue, purple, light blue, gray): Keys, message, authenticators, and nonces are unaligned. This hurts some CPUs.

The graphs include code in cache and code out of cache, with no color change. The out-of-cache case costs between 10000 and 30000 cycles, depending on the CPU; it is often faintly visible as a cloud above the in-cache case.

Lengths divisible by 16 are slightly faster than lengths not divisible by 16. The best case in (almost) every graph is length divisible by 16, everything in cache, everything aligned; this case is visible as 256 black dots at the bottom of the graph.

In black-and-white printouts, the keys-not-in-cache case is a slightly higher line at the same slope; the data-not-in-cache case is a line at a considerably higher slope; the unaligned case is a line at a slightly higher slope.

See `http://cr.yp.to/mac/speed.html` for much more speed information.

AMD Athlon, 900MHz:

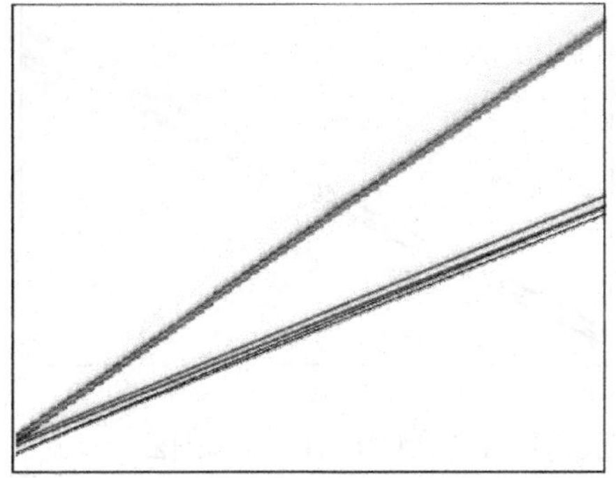

IBM PowerPC RS64 IV, 668MHz:

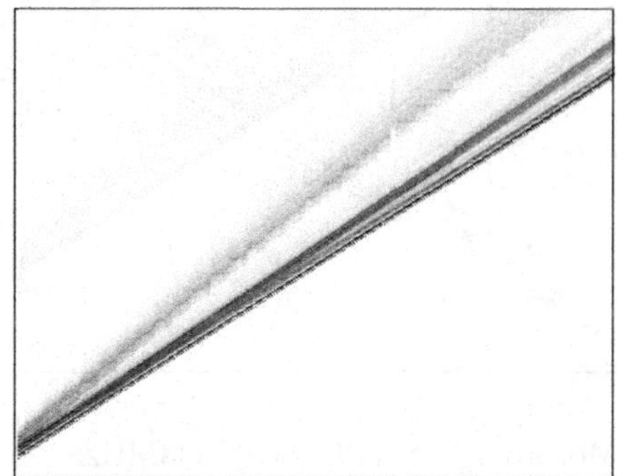

Intel Pentium III, 500MHz:

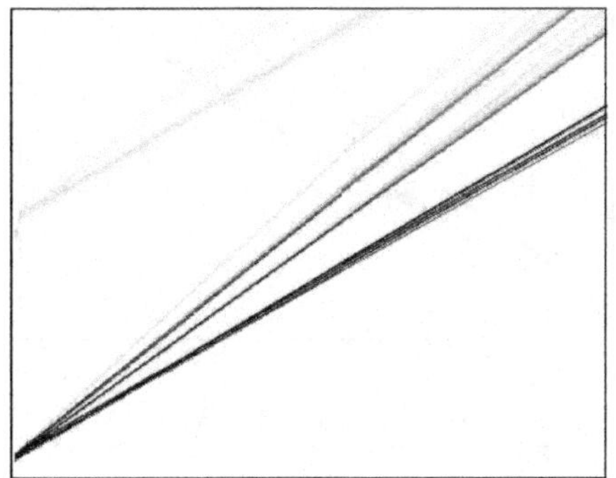

Intel Pentium III, 850MHz:

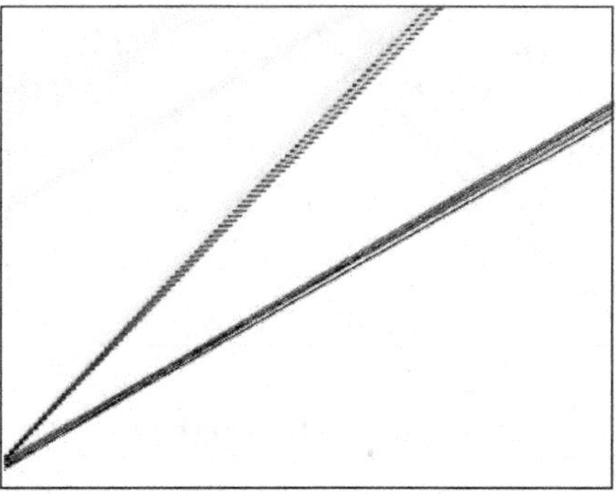

Intel Pentium III, 1000MHz:

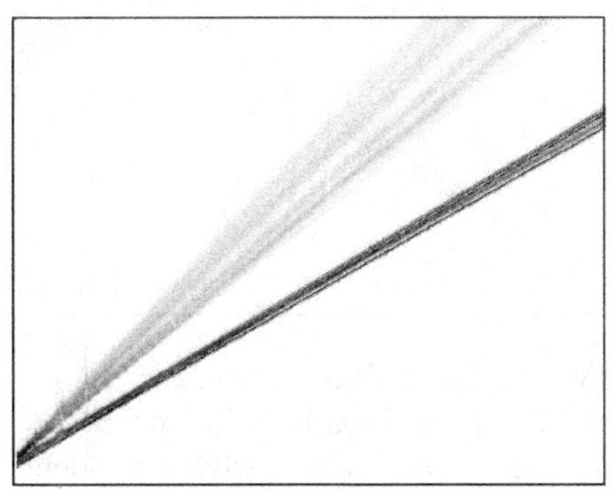

Intel Pentium 4, 1900MHz:

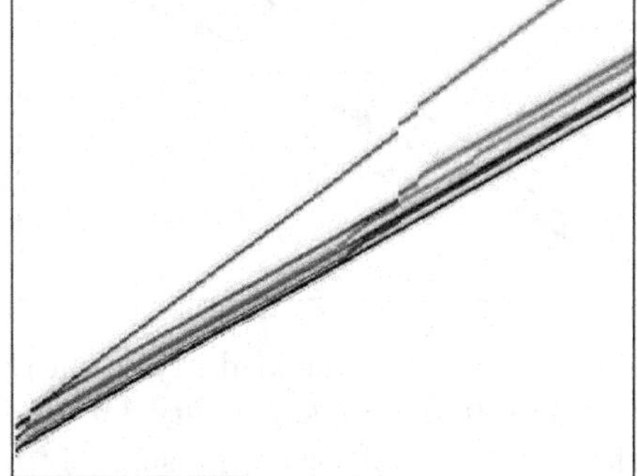

Intel Pentium 4, 3400MHz:

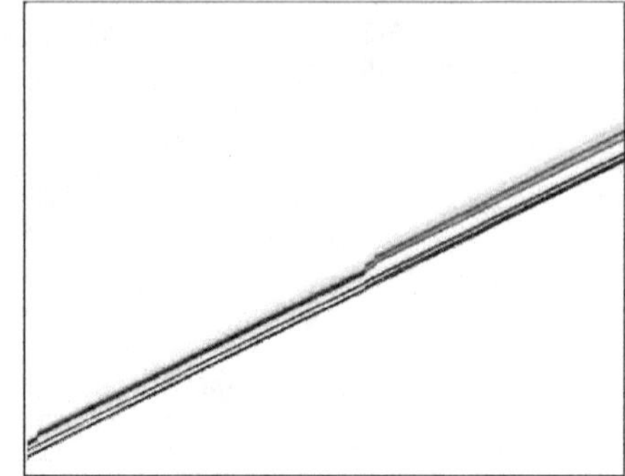

Motorola PowerPC 7410, 533MHz:

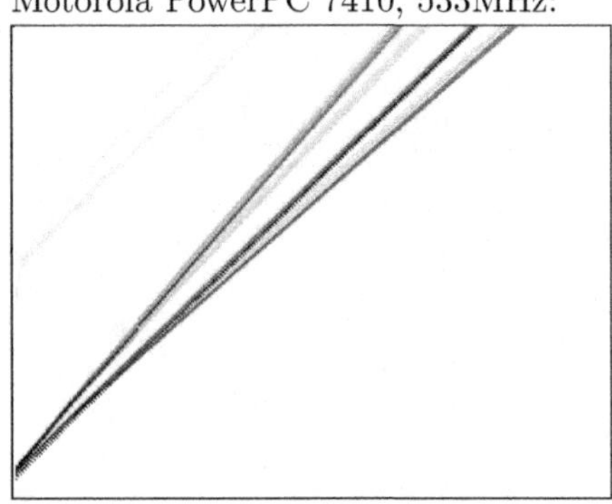

Sun UltraSPARC II, 296MHz:

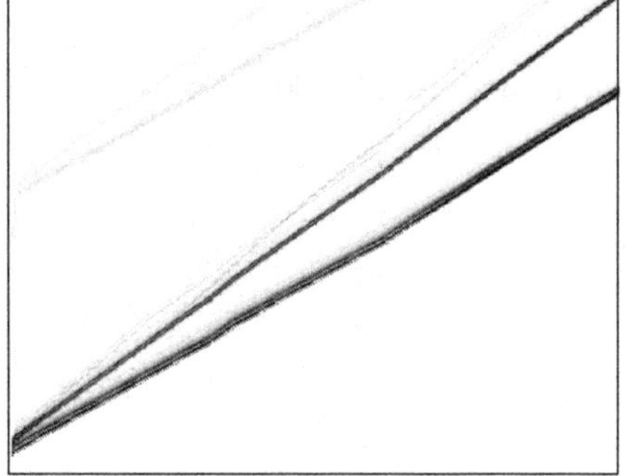

Sun UltraSPARC IIi, 360MHz:

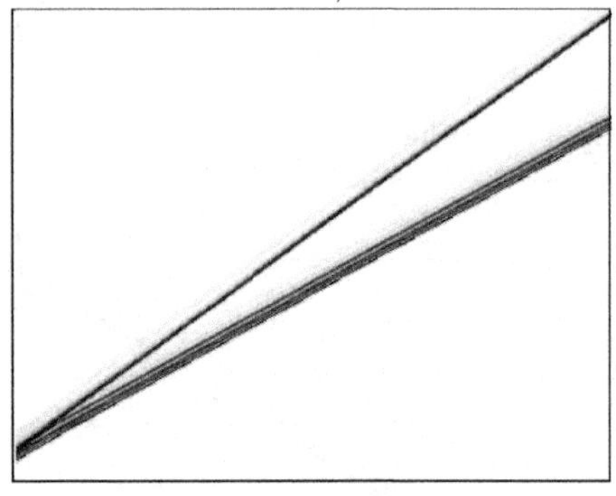

Sun UltraSPARC III, 900MHz:

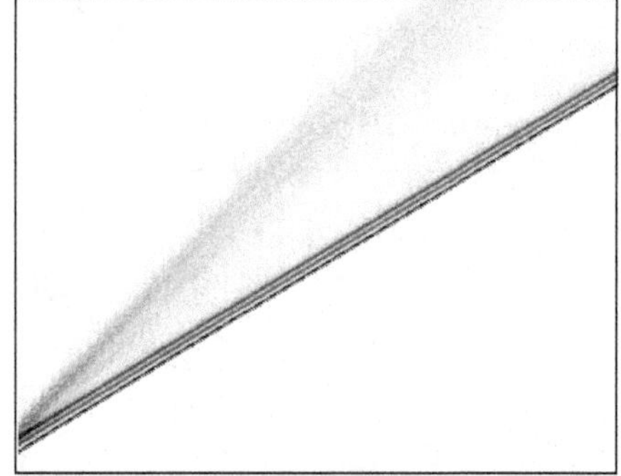

Two notes: 1. The load-keys-from-DRAM penalty (red) is quite small, thanks to Poly1305-AES's key agility. On the PowerPC 7410, keys in cache are *slower* than keys out of cache, presumably because of a cache-associativity accident that slightly more sophisticated code will be able to avoid.

2. The load-data-from-DRAM penalty (green) is generally quite noticeable. I have not yet experimented with prefetch instructions. But the penalty is small on the Pentium 4 and almost invisible on the Pentium M; the Pentium M does a good job of figuring out for itself which data to prefetch.

B Appendix: Examples

The following table, with all integers on the right displayed in hexadecimal, illustrates authenticator computations for strings of length 2, 0, 32, and 63. The notation $\underline{m}(r)$ means $c_1 r^q + c_2 r^{q-1} + \cdots + c_q r^1$. A much more extensive test suite appears in `http://cr.yp.to/mac/test.html`.

m	f3 f6
c_1	00000000000000000000000000001f6f3
r	85 1f c4 0c 34 67 ac 0b e0 5c c2 04 04 f3 f7 00
$\underline{m}(r) \bmod 2^{130} - 5$	321e58e25a69d7f8f27060770b3f8bb9c
k	ec 07 4c 83 55 80 74 17 01 42 5b 62 32 35 ad d6
n	fb 44 73 50 c4 e8 68 c5 2a c3 27 5c f9 d4 32 7e
$\mathrm{AES}_k(n)$	58 0b 3b 0f 94 47 bb 1e 69 d0 95 b5 92 8b 6d bc
$\mathrm{Poly1305}_r(m, \mathrm{AES}_k(n))$	f4 c6 33 c3 04 4f c1 45 f8 4f 33 5c b8 19 53 de
m	
r	a0 f3 08 00 00 f4 64 00 d0 c7 e9 07 6c 83 44 03
$\underline{m}(r) \bmod 2^{130} - 5$	000000000000000000000000000000000
k	75 de aa 25 c0 9f 20 8e 1d c4 ce 6b 5c ad 3f bf
n	61 ee 09 21 8d 29 b0 aa ed 7e 15 4a 2c 55 09 cc
$\mathrm{AES}_k(n)$	dd 3f ab 22 51 f1 1a c7 59 f0 88 71 29 cc 2e e7
$\mathrm{Poly1305}_r(m, \mathrm{AES}_k(n))$	dd 3f ab 22 51 f1 1a c7 59 f0 88 71 29 cc 2e e7
m	66 3c ea 19 0f fb 83 d8 95 93 f3 f4 76 b6 bc 24
	d7 e6 79 10 7e a2 6a db 8c af 66 52 d0 65 61 36
c_1	124bcb676f4f39395d883fb0f19ea3c66
c_2	1366165d05266af8cdb6aa27e1079e6d7
r	48 44 3d 0b b0 d2 11 09 c8 9a 10 0b 5c e2 c2 08
$\underline{m}(r) \bmod 2^{130} - 5$	1cfb6f98add6a0ea7c631de020225cc8b
k	6a cb 5f 61 a7 17 6d d3 20 c5 c1 eb 2e dc dc 74
n	ae 21 2a 55 39 97 29 59 5d ea 45 8b c6 21 ff 0e
$\mathrm{AES}_k(n)$	83 14 9c 69 b5 61 dd 88 29 8a 17 98 b1 07 16 ef
$\mathrm{Poly1305}_r(m, \mathrm{AES}_k(n))$	0e e1 c1 6b b7 3f 0f 4f d1 98 81 75 3c 01 cd be
m	ab 08 12 72 4a 7f 1e 34 27 42 cb ed 37 4d 94 d1
	36 c6 b8 79 5d 45 b3 81 98 30 f2 c0 44 91 fa f0
	99 0c 62 e4 8b 80 18 b2 c3 e4 a0 fa 31 34 cb 67
	fa 83 e1 58 c9 94 d9 61 c4 cb 21 09 5c 1b f9
c_1	1d1944d37edcb4227341e7f4a721208ab
c_2	1f0fa9144c0f2309881b3455d79b8c636
c_3	167cb3431faa0e4c3b218808be4620c99
c_4	001f91b5c0921cbc461d994c958e183fa
r	12 97 6a 08 c4 42 6d 0c e8 a8 24 07 c4 f4 82 07
$\underline{m}(r) \bmod 2^{130} - 5$	0c3c4f37c464bbd44306c9f8502ea5bd1
k	e1 a5 66 8a 4d 5b 66 a5 f6 8c c5 42 4e d5 98 2d
n	9a e8 31 e7 43 97 8d 3a 23 52 7c 71 28 14 9e 3a
$\mathrm{AES}_k(n)$	80 f8 c2 0a a7 12 02 d1 e2 91 79 cb cb 55 5a 57
$\mathrm{Poly1305}_r(m, \mathrm{AES}_k(n))$	51 54 ad 0d 2c b2 6e 01 27 4f c5 11 48 49 1f 1b

Narrow T-Functions

Magnus Daum*

CITS Research Group, Ruhr-University Bochum
daum@cits.rub.de

Abstract. *T-functions* were introduced by Klimov and Shamir in a series of papers during the last few years. They are of great interest for cryptography as they may provide some new building blocks which can be used to construct efficient and secure schemes, for example block ciphers, stream ciphers or hash functions.

In the present paper, we define the *narrowness* of a T-function and study how this property affects the strength of a T-function as a cryptographic primitive. We define a new data strucure, called a *solution graph*, that enables solving systems of equations given by T-functions. The efficiency of the algorithms which we propose for solution graphs depends significantly on the narrowness of the involved T-functions. Thus the subclass of T-functions with small narrowness appears to be weak and should be avoided in cryptographic schemes.

Furthermore, we present some extensions to the methods of using solution graphs, which make it possible to apply these algorithms also to more general systems of equations, which may appear, for example, in the cryptanalysis of hash functions.

Keywords: Cryptanalysis, hash functions, solution graph, T-functions, w-narrow.

1 Introduction

Many cryptanalytical problems can be described by a system of equations. A well-known example are the algebraic attacks on block and stream ciphers which use systems of multivariate quadratic equations for describing the ciphers.

However, many cryptographic algorithms use a mixture of different kinds of operations (e.g. bitwise defined functions, modular additions or multiplications and bit shifts or rotations) such that they cannot be described easily by some relatively small or simple system of linear or quadratic equations. As these operations are algebraically rather incompatible, it is hard to solve equations which include different ones algebraically.

* The work described in this paper has been supported in part by the European Commission through the IST Programme under Contract IST-2002-507932 ECRYPT. The information in this document reflects only the author's views, is provided as is and no guarantee or warranty is given that the information is fit for any particular purpose. The user thereof uses the information at its sole risk and liability.

H. Gilbert and H. Handschuh (Eds.): FSE 2005, LNCS 3557, pp. 50–67, 2005.

In a series of papers [6, 7, 8] Klimov and Shamir introduced the notion of *T-functions*, in order to be able to prove theoretical results at least for some of the constructions mentioned above. Roughly spoken, a T-function is a function for which the k-th bit of the output depends only on the first k input bits. Many basic operations available on modern microprocessors are T-functions and this means that many T-functions can be implemented very efficiently. Furthermore many of the operations mentioned above are T-functions or very similar to T-functions.

In this paper we concentrate on a certain subclass of T-functions, which we call *w-narrow T-functions*. In a w-narrow T-function the dependance of the k-th output bit on the first k input bits is even more restricted: The k-th output bit must be computable from only the k-th input bits and some information of a length of w bits computed from the first $k-1$ input bits.

We present a data structure, called a *solution graph*, which allows to efficiently represent the set of solutions of an equation, which can be described by a w-narrow T-function. The smaller w is, the more efficient is this representation. Additionally we present a couple of algorithms which can be used for analysing and solving such systems of equations described by T-functions. These algorithms include enumerating all solutions, computing the number of solutions, choosing random solutions and also combining two or more solution graphs, e.g. to compute the intersection of two sets of solutions or to compute the concatenation of two T-functions.

However, this paper is not only dedicated to the quite young subject of T-functions. The solution graphs together with the presented algorithms, can be used for cryptanalysis in a lot of contexts, for example also in the cryptanalysis of hash functions. In his attacks on the hash functions MD4, MD5 and RIPEMD (see [3, 4, 5]), Dobbertin used, as one key ingredient, an algorithm which can be described as some kind of predecessor of the algorithms used for constructing solution graphs and enumerating all the solutions (see Appendix A). In this paper we also describe some extensions which allow to apply the algorithms also in contexts which are a little more general than systems of equations describable by "pure" T-functions.

We start in Section 2 by defining the narrowness of a T-function and give some basic examples and properties. Then in Section 3 we describe the new data structure, the solution graph, and give an algorithm for constructing solution graphs from systems of equations of T-functions. Section 4 gives further algorithms for solution graphs.

In Section 5 we present some possible extensions to the definition of a solution graph, which allow to apply these algorithms also in more general situations, for example in the cryptanalysis of hash functions.

In Appendix A we describe the ideas and the original algorithm used by Dobbertin in his attacks. Two actual examples of systems coming from the cryptanalysis of hash functions, which have been solved successfully with solution graphs are given in Appendix B.

In this extended abstract the proofs are omitted. They can be found in [1] and the full version [2] of this paper.

2 Notation and Definitions

For the convenience of the reader, we mainly adopt the notation of [9]. Especially, let n be the word size in bits, $\mathbb{B}$ be the set $\{0,1\}$ and let $[x]_i$ denote the i-th bit of the word $x \in \mathbb{B}^n$, where $[x]_0$ is the least significant bit of x. Hence, $x = ([x]_{n-1}, \ldots, [x]_0)$ also stands for the integer $\sum_{i=0}^{n-1} [x]_i 2^i$.

If $x = (x_0, \ldots, x_{m-1})^T \in \mathbb{B}^{m \times n}$ is a column vector of m words of n bits, then $[x]_i$ stands for the column vector $([x_0]_i, \ldots, [x_{m-1}]_i)^T$ of the i-th bits of those words.

By $x \ll s$ we will denote a left shift by s positions and by $x \lll r$ we denote a left rotation (a cyclic shift) by r positions.

Let us first recall the definition of a T-function from [9]:

Definition 1 (T-Function). *A function $f : \mathbb{B}^{m \times n} \to \mathbb{B}^{l \times n}$ is called a* T-function *if the k-th column of the output $[f(x)]_{k-1}$ depends only on the first k columns of the input $[x]_{k-1}, \ldots, [x]_0$:*

$$\begin{pmatrix} [x]_0 \\ [x]_1 \\ [x]_2 \\ \vdots \\ [x]_{n-1} \end{pmatrix}^T \mapsto \begin{pmatrix} f_0([x]_0) \\ f_1([x]_0, [x]_1) \\ f_2([x]_0, [x]_1, [x]_2) \\ \vdots \\ f_{n-1}([x]_0, [x]_1, \ldots, [x]_{n-1}) \end{pmatrix}^T \tag{1}$$

There are many examples for T-functions. All bitwise defined functions, e.g. a Boolean operation like $(x, y) \mapsto x \wedge y$ or the majority function $(x, y, z) \mapsto (x \wedge y) \vee (x \wedge z) \vee (y \wedge z)$, are T-functions, because the k-th output bit depends only on the k-th input bits. But also other common functions, like addition or multiplication of integers (modulo 2^n) are T-functions, as can be easily seen from the schoolbook methods. For example, when executing an addition, to compute the k-th bit of the sum, the only necessary information (besides the k-th bits of the addends) is the carrybit coming from computing the $(k-1)$-th bit.

This is also a good example for some other more special property that many T-functions have: You need much less information than "allowed" by the definition of a T-function: In order to compute the k-th output column $[f(x)]_{k-1}$ you need only the k-th input column $[x]_{k-1}$ and very little information about the first $k-1$ columns $[x]_{k-2}, \ldots, [x]_0$, for example some value $\alpha_k([x]_{k-2}, \ldots, [x]_0) \in \mathbb{B}^w$ of w bits width. This leads to our definition of a *w-narrow T-function*:

Definition 2 (w-narrow).
A T-function f is called w-narrow *if there are mappings*

$$\alpha_1 : \mathbb{B}^m \to \mathbb{B}^w, \qquad \alpha_k : \mathbb{B}^{m+w} \to \mathbb{B}^w, k = 2, \ldots, n-1 \tag{2}$$

and auxiliary variables

$$a_1 := \alpha_1([x]_0), \qquad a_k := \alpha_k([x]_{k-1}, a_{k-1}), k = 2, \ldots, n-1 \tag{3}$$

such that f can be written as

$$\begin{pmatrix} [x]_0 \\ [x]_1 \\ [x]_2 \\ [x]_3 \\ \vdots \\ [x]_{n-1} \end{pmatrix}^T \mapsto \begin{pmatrix} f_0([x]_0) \\ f_1([x]_1, a_1) \\ f_2([x]_2, a_2) \\ f_3([x]_3, a_3) \\ \vdots \\ f_{n-1}([x]_{n-1}, a_{n-1}) \end{pmatrix}^T \tag{4}$$

The smallest w such that some f is w-narrow is called the narrowness *of f.*

Let us take a look at some examples of w-narrow T-functions.

Example 1.

1. The identity function and all bitwise defined functions are 0-narrow.
2. As described above, addition of two integers modulo 2^n is a 1-narrow T-function, as you only need to remember the carrybit in each step.
3. A left shift by s bits is an s-narrow T-function.
4. Each T-function $f : \mathbb{B}^{m\times n} \to \mathbb{B}^{l\times n}$ is $(m(n-1))$-narrow.

Directly from Definition 2 one can derive the following lemma about the composition of narrow functions:

Lemma 1. *Let $f, g_1, \ldots, g_r$ be T-functions which are w_f-, w_{g_1}-,...,w_{g_r}-narrow respectively. Then the function h defined by*

$$h(x) := f(g_1(x), \ldots, g_r(x))$$

is $(w_f + w_{g_1} + \ldots + w_{g_r})$-narrow.

Note that this lemma (as the notion of w-narrow itself) gives only an upper bound on the narrowness of a function: For example, the addition of 4 integers can be composed of three (1-narrow) 2-integer-additions. Thus by Lemma 1 it is 3-narrow. But it is also 2-narrow, because the carry value to remember can never become greater than 3 (which can be represented in $\mathbb{B}^2$) when adding 4 bits and a maximum (earlier) carry of 3.

3 Solution Graphs for Narrow T-Functions

In this section we will describe a data structure which allows to represent the set of solutions of a system of equations of T-functions.

Common approaches for finding solutions of such equations are doing an exhaustive or randomized search or using some more sophisticated algorithms as the one used by Dobbertin in his attacks on the hash functions MD4, MD5 and RIPEMD in [3, 4, 5]. This algorithm, which gave us the idea of introducing the data structure presented here, is described in Appendix A.

In general, the trees build in Dobbertin's algorithm and thus its complexity, needed for building them, may become quite large, in the worst case up to the complexity of an exhaustive search. But this can be improved a lot in many cases, or, to be more precise, in the case of T-functions which are w-narrow for some small w, as we will show in the sequel.

Let us first note, that it suffices to consider only the problem of solving one equation

$$f(x) = 0, \tag{5}$$

where $f : \mathbb{B}^{m\times n} \to \mathbb{B}^n$ is some T-function:
If we had an equation described by two T-functions $g(x) = h(x)$ we could simply define $\hat{g}(x) := g(x) \oplus h(x)$ and consider the equation $\hat{g}(x) = 0$ instead. If we had a system of several such equations $\hat{g}_1(x) = 0, \ldots, \hat{g}_r(x) = 0$ (or a function $\hat{g} : \mathbb{B}^{m\times n} \to \mathbb{B}^{l\times n}$ with component functions $\hat{g}_1, \ldots, \hat{g}_r$) we could simply define $f(x) := \bigvee_{i=1}^{r} \hat{g}_i(x)$ and consider only the equation $f(x) = 0$.
As both operations, $\oplus$ and $\vee$, are 0-narrow, due to Lemma 1, the narrowness of f is at most the sum of the narrownesses of the involved functions.

If f in (5) is a w-narrow T-function for some "small" w, a solution graph, as given in the following definition, can be efficiently constructed and allows many algorithms which are useful for cryptanalysing such functions.

Definition 3 (Solution Graph). *A directed graph $\mathcal{G}$ is called a* solution graph *for an equation $f(x) = 0$ where $f : \mathbb{B}^{m\times n} \to \mathbb{B}^n$, if the following properties hold:*

1. *The vertices of $\mathcal{G}$ can be arranged in $n+1$ layers such that each edge goes from a vertex in layer l to some vertex in layer $l+1$ for some $l \in \{0, \ldots, n-1\}$.*
2. *There is only one vertex in layer 0, called the* root*.*
3. *There is only one vertex in layer n, called the* sink*.*
4. *The edges are labelled with values from $\mathbb{B}^m$ such that the labels for all edges starting in one vertex are pairwise distinct.*
5. *There is a 1-to-1 correspondence between paths from the root to the sink in $\mathcal{G}$ and solutions of the equation $f(x) = 0$:*
   For each solution x there exists a path from the root to the sink such that the k-th edge on this path is labelled with $[x]_{k-1}$ and vice versa.

The maximum number of vertices in one layer of a solution graph $\mathcal{G}$ is called the width *of $\mathcal{G}$.*

In the following we will describe how to efficiently construct a solution graph which represents the complete set of solutions of (5). Therefore let f be w-narrow with some auxiliary functions $\alpha_1, \ldots, \alpha_{n-1}$ as in Definition 2. To identify the vertices during the construction we label them with two numbers (l, a) each, where $l \in \{0, \ldots, n\}$ is the number of the layer and $a \in \mathbb{B}^w$ corresponds to a possible output of one of the auxiliary functions α_i. This labelling is only required for the construction and can be deleted afterwards.

Then the solution graph can be constructed by the following algorithm:

Algorithm 1 (Construction of a Solution Graph).

1. Start with one vertex labelled with $(0, *)$
2. For each possible value for $[x]_0$, for which it holds that $f_0([x]_0) = 0$:
   Add an edge
   $$(0, *) \longrightarrow (1, \alpha_1([x]_0))$$
   and label this edge with the value of $[x]_0$.
3. For each layer l, $l \in \{1, \ldots, n-2\}$, and each vertex (l, a_l) in layer l:
   For each possible value for $[x]_l$ for which $f_l([x]_l, a_l) = 0$:
   Add some edge
   $$(l, a_l) \longrightarrow (l+1, \alpha_{l+1}([x]_l, a_l))$$
   and label this edge with the value of $[x]_l$.
4. For each vertex $(n-1, a)$ in layer $n-1$ and each possible value for $[x]_{n-1}$ for which $f_{n-1}([x]_{n-1}, a) = 0$:
   Add an edge
   $$(n-1, a) \longrightarrow (n, *)$$
   and label it with the value of $[x]_{n-1}$.

Toy examples of the results of this construction can be found in Figure 1. Compared with the trees in Figure 5 and 6, resulting from Dobbertin's algorithm, this shows that these solution graphs are much more efficient.

From the description of Algorithm 1 the following properties can be easily deduced:

Theorem 1. *Let $f : \mathbb{B}^{m \times n} \to \mathbb{B}^n$ be a w-narrow T-function and $\mathcal{G}$ the graph for $f(x) = 0$ constructed by Algorithm 1. Then $\mathcal{G}$*

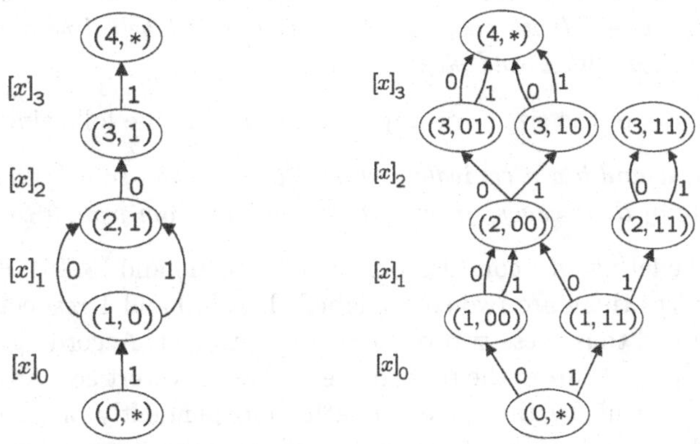

Fig. 1. Solution graphs for the equations $((x \vee 0010_2) + 0110_2) \oplus 0001_2 = 0$ (on the left) and $((0100_2 \oplus (x + 0101_2)) - (0100_2) \oplus x) \oplus 1101_2 = 0$ (on the right) with $n = 4$

– *is a solution graph for* $f(x) = 0$,
– *has width at most* 2^w*, i.e.* $\mathcal{G}$ *has* $v \leq (n-1)2^w + 2$ *vertices and* $e \leq (v-1)2^m$ *edges.*

Proof. For the proof, see [1] or the full version [2] of this paper.

This theorem gives an upper bound on the size of the constructed solution graph, which depends significantly on the narrowness of the examined function f. This shows that, as long as f is w-narrow for some small w, such a solution graph can be constructed quite efficiently.

4 Algorithms for Solution Graphs

The design of a solution graph, as presented in Section 3 is very similar to that of binary decision diagrams (BDDs). Thus it is not surprising, that many ideas of algorithms for BDDs can be adopted to construct efficient algorithms for solution graphs. For an introduction to the subject of BDDs, see for example [10].

The complexity of these algorithms naturally depends mainly on the size of the involved solution graphs. Thus, we will first describe how to reduce this size.

4.1 Reducing the Size

We describe this using the example of the solution graph on the right hand side of Figure 1: There are no edges starting in $(3, 11)$ and thus there is no path from the root to the sink which crosses this vertex. This means, due to Definition 3, this vertex is of no use for representing any solution, and therefore it can be deleted. After this deletion the same applies for $(2, 11)$ and thus this vertex can also be deleted.

For further reduction of the size let us define what we mean by *equivalent* vertices:

Definition 4. *Two vertices* a *and* b *in a solution graph are called* equivalent, *if for each edge* $a \rightarrow c$ *(with some arbitrary vertex* c*) labelled with* x *there is an edge* $b \rightarrow c$ *labelled with* x *and vice versa.*

For the reduction of the size, it is important to notice the following lemma:

Lemma 2. *If* a *and* b *are equivalent, then there are the same paths (according to the labelling of their edges) from* a *to the sink as from from* b *to the sink.*

For example let us now consider the vertices $(3, 01)$ and $(3, 10)$. From each of these two vertices there are two edges, labelled with 0 and 1 respectively, which point to $(4, *)$ and thus these two vertices are equivalent. According to Lemma 2 this means that a path from the root to one of those two vertices can be extended to a path to the sink by the same subpaths, independently of whether it goes through $(3, 01)$ or $(3, 10)$. Due to the defining property of a solution graph, this means, that we can merge these two equivalent vertices into one, reducing the size once more. The resulting solution graph is presented in Figure 2. In this

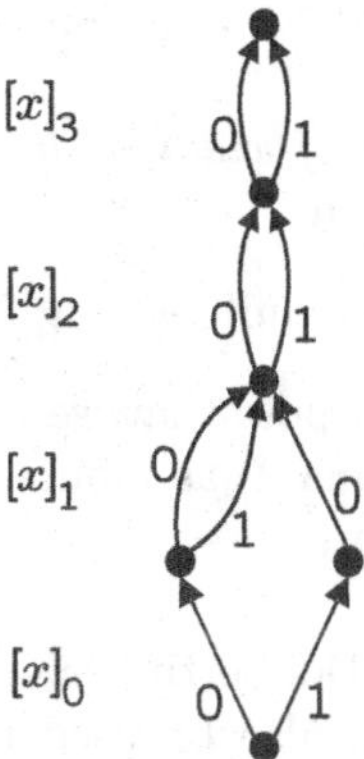

Fig. 2. Solution graph for the equation $((0100_2 \oplus (x + 0101_2)) - (0100_2 \oplus x)) \oplus 1101_2 = 0$ (compare Figure 1) after reducing its size

figure the labels of the vertices are omitted as they are only required for the construction algorithm.

Of course, merging two equivalent vertices, and also the deletion of vertices as described above, may again cause two vertices to become equivalent, which have not been equivalent before. But this concerns only vertices in the layer below the layer in which two vertices were merged. Thus for the reduction algorithm it is important to work from top (layer $n-1$) to bottom (layer 1):

Algorithm 2 (Reduction of the Size).

1. Delete each vertex (together with corresponding edges) for which there is no path from the root to this vertex or no path from this vertex to the sink.
2. For each layer l starting from $n-1$ down to 1 merge all pairs of vertices in layer l which are equivalent.

To avoid having to check all possible pairs of vertices in one layer for equivalence separately to find the equivalent vertices (which would result in a quadratic complexity), in Algorithm 2 one should first sort the vertices of the active layer according to their set of outgoing edges. Then equivalent vertices can be found in linear time.

Similar to what can be proven for ordered BDDs, for solution graphs reduced by Algorithm 2 it can be shown that they have minimal size:

Theorem 2. *Let $\mathcal{G}$ be a solution graph for some function f and let $\tilde{\mathcal{G}}$ be the output of Algorithm 2 applied to $\mathcal{G}$. Then there is no solution graph for f which has less vertices than $\tilde{\mathcal{G}}$.*

Proof. For the proof, see [1] or the full version [2] of this paper. □

With the help of the following theorem it is possible to compute the narrowness of f, i.e. the smallest value w such that f is w-narrow. Like Theorem 1 gives a bound on the width of a solution graph based on a bound for the narrowness of the considered function, the following theorem provides the other direction:

Theorem 3. *Let* $f : \mathbb{B}^{m\times n} \to \mathbb{B}^n$ *be a T-function and define* $\tilde{f} : \mathbb{B}^{(m+1)\times n} \to \mathbb{B}^n$ *by* $\tilde{f}(x,y) := f(x) \oplus y$.
If $\mathcal{G}$ *is a minimal solution graph of width* W *for the equation* $\tilde{f}(x,y) = 0$*, then* f *is a* $\lceil \log_2 W \rceil$*-narrow T-function.*

Proof. For the proof, see [1] or the full version [2] of this paper. □

In the following we always suppose that we have solution graphs of minimal size (from Algorithm 2 and Lemma 2) as inputs.

4.2 Computing Solutions

Similar to what can be done by Dobbertin's algorithm (see Algorithm 7 in Appendix A), a solution graph can also be used to enumerate all the solutions:

Algorithm 3 (Enumerate Solutions).
Compute all possible paths from the root to the sink by a depth-first search and output the corresponding labelling of the edges.

Of course, the complexity of this algorithm is directly related to the number of solutions. If there are many solutions, it is similar to the complexity of an exhaustive search (as for Algorithm 7), simply because all of them need to be written. But if there are only a few, it is very fast, usually much faster than Algorithm 7.

However, often we are only interested in the number of solutions of an equation which can be computed much more efficiently, namely, with a complexity linear in the size of the solution graph. The following algorithm achieves this by labeling every vertex with the number of possible paths from that vertex to the sink. Then the number computed for the root gives the number of solutions:

Algorithm 4 (Number of Solutions).

1. Label the sink with 1.
2. For each layer l from $n-1$ down to 0:
   Label each vertex A in l with the sum of the labels of all vertices B (in layer $l+1$) for which an edge $A \to B$ exists.
3. Output the label of the root.

An application of this algorithm is illustrated in Figure 3.

After having labelled all vertices by Algorithm 4 it is even possible to choose solutions from the represented set uniformly at random:

Algorithm 5 (Random Solution).
Prerequisite: The vertices have to be labelled as in Algorithm 4.

1. Start at the root.
2. Repeat
   - From the active vertex A (labelled with n_A) randomly choose one outgoing edge such that the probability that you choose $A \to B$ is $\frac{n_B}{n_A}$ where n_B is the label of B.

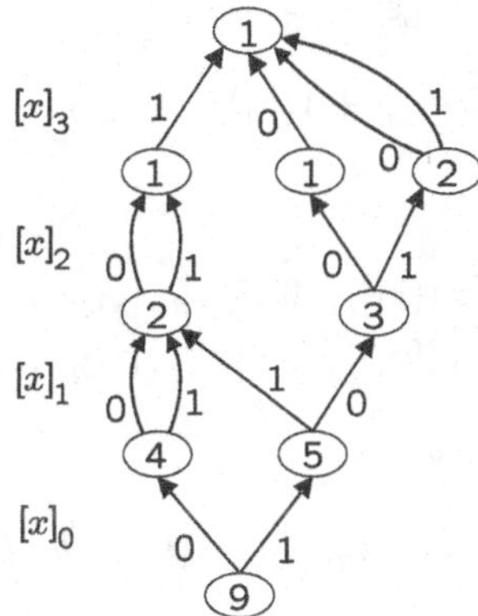

Fig. 3. A solution graph after application of Algorithm 4

   - Remember the label of $A \to B$
   - Make B the active vertex.

   until you reach the sink.
3. Output the solutions corresponding to the remembered labels of the edges on the chosen path.

4.3 Combining Solution Graphs

So far, we only considered the situation in which the whole system of equations is reduced to one equation $f(x) = 0$, as described at the beginning of Section 3, and then a solution graph is constructed from this equation. Sometimes it is more convenient to consider several (systems of) equations separately and then combine their sets of solutions in some way. Therefore let us now consider two equations

$$g(x_1, \ldots, x_r, y_1, \ldots, y_s) = 0 \tag{6}$$
$$h(x_1, \ldots, x_r, z_1, \ldots, z_t) = 0 \tag{7}$$

which include some common variables $x_1, \ldots, x_r$ as well as some distinct variables $y_1, \ldots, y_s$ and $z_1, \ldots, z_t$ respectively. Let $\mathcal{G}_g$ and $\mathcal{G}_h$ be the solution graphs for (6) and (7) respectively.

Then the set of solutions of the form $(x_1, \ldots, x_r, y_1, \ldots, y_s, z_1, \ldots, z_t)$ which fulfill both equations simultaneously can be computed by the following algorithm.

Algorithm 6 (Intersection). Let the vertices in $\mathcal{G}_g$ be labelled with $(l, a_g)_g$ where l is the layer and a_g is some identifier which is unique per layer, and those of $\mathcal{G}_h$ analogously with some $(l, a_h)_h$. Then construct a graph whose vertices will be labelled with (l, a_g, a_h) by the following rules:

1. Start with the root $(0, *_g, *_h)$.
2. For each layer $l \in \{0, \ldots, n-1\}$ and each vertex (l, a_g, a_h) in layer l:

– Consider each pair of edges

$$((l, a_g)_g \to (l+1, b_g)_g, (l, a_h)_h \to (l+1, b_h)_h)$$

labelled with

$$(X_g, Y_g) = ([x_1]_l, \ldots, [x_r]_l, [y_1]_l, \ldots, [y_s]_l)$$

and $(X_h, Z_h) = ([x_1]_l, \ldots, [x_r]_l, [z_1]_l, \ldots, [z_t]_l)$ respectively.

– If $X_g = X_h$, add an edge

$$(l, a_g, a_h) \to (l+1, b_g, b_h)$$

and label it with (X_g, Y_g, Z_h).

The idea of this algorithm is to traverse the two input graphs $\mathcal{G}_g$ and $\mathcal{G}_h$ in parallel and to simulate computing both functions in parallel in the output graph by storing all necessary information in the labels of the output graph. For an illustration of this algorithm, see Figure 4. Also notice that this algorithm can be easily generalized to having more than two input graphs.

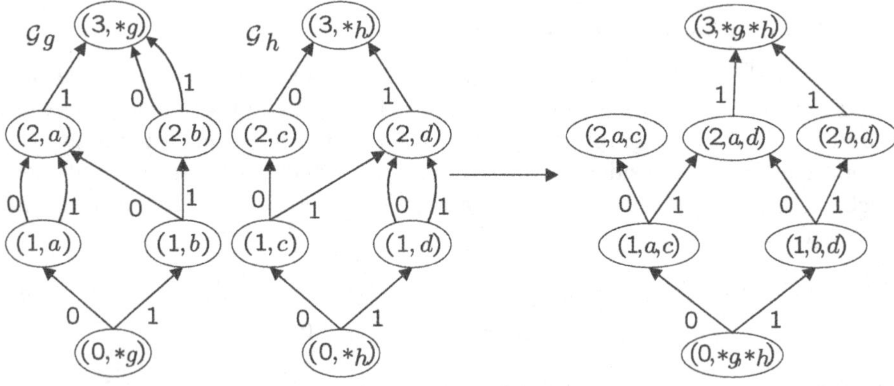

Fig. 4. Intersection of two solution graphs by Algorithm 6

Apart from just computing mere intersections of sets of solutions, Algorithm 6 can also be used to solve equations given by the concatenation of two T-functions:

$$f(g(x)) = y \tag{8}$$

To solve this problem, just introduce some auxiliary variable z and apply Algorithm 6 to the two solution graphs which can be constructed for the equations $f(z) = y$ and $g(x) = z$ respectively.

Combining this idea (applied to the situation $f = g$) with some square-and-multiply technique, allows for some quite efficient construction of a solution graph for an equation of the form $f^i(x) = y$ with some (small) fixed value i. This may be of interest for example for cryptanalysing stream ciphers which are constructed as suggested for example by Klimov in [9], but use T-functions with some small narrowness instead of one of the functions proposed by Klimov which seem to have a large narrowness.

5 Extensions of This Method

In many cryptographical systems the operations used are usually *not* restricted to T-functions. Often such systems also include other basic operations, as, for example, right bit shifts or bit rotations, which are quite similar, but not T-functions according to Definition 1. Hence, systems of equations used in the cryptanalysis of such ciphers usually cannot be solved directly by applying solution graphs as presented in Sections 3 and 4. In this section we give some examples of how such situations can be handled, for example by extending the definition of a solution graph such that it is still applicable.

5.1 Including Right Shifts

Let us first consider a system of equations which includes only T-functions and some right shift expressions $x \gg r$. This can be transformed by substituting every appearance of $x \gg r$ by an auxiliary variable z_r and adding an extra equation

$$z_r \ll r = x \wedge (\underbrace{11\dots1}_{n-r}\underbrace{0\dots0}_{r}) \tag{9}$$

which defines the relationship between x and z_r. Then the resulting system is completely described by T-functions and can be solved with a solution graph.

Here, similarly as when solving (8) some problem occurs: We have to add an extra (auxiliary) variable z, which potentially increases the size of the needed solution graph. This is even worse as the solution graph stores all possible values of z corresponding to solutions for the other variables, even if we are not interested in them at all. This can be dealt with by softening Definition 3 to *generalized solutions graphs:*

5.2 Generalized Solution Graphs

For a generalized solution graph we require every property from Definition 3 with the exception that the labels of edges starting in one vertex are *not* required to be pairwise distinct.

Then we can use similar algorithms as those described above, e.g. for reducing the size or combining two graphs. But usually these algorithms are a little bit more sophisticated: For example, for minimizing the size, it does not suffice to consider equivalent vertices as defined in Definition 4. In a generalized solution graph it is also possible that the sets of *incoming* edges are equal and, clearly, two such vertices with equal sets of incoming edges (which we will also call equivalent in the case of general solution graphs) can also be merged. But this also means that merging two equivalent vertices in layer l may not only cause vertices in layer $l-1$ to become equivalent, but also vertices in layer $l+1$. Thus, in the generalized version of Algorithm 2 we have to go back and forth in the layers to ensure that in the end there are no equivalent vertices left.

This definition of a generalized solution graph allows to "remove" variables without losing the information about their existence. This means, instead of representing the set $\{(x, y) \mid f(x, y) = 0\}$ with a solution graph $\mathcal{G}$, we can represent the set $\{x \mid \exists y : f(x, y) = 0\}$ with a solution graph $\mathcal{G}'$ which is constructed from $\mathcal{G}$ by simply deleting the parts of the labels which correspond to y. Of course, this does not decrease the size of the generalized solution graph directly but (hopefully) it allows further reductions of the size.

5.3 Including Bit Rotations

Let us now take a look at another commonly used function which is not a T-function, a bit rotation by r bits:

$$f(x) := x \lll r \tag{10}$$

If we would fix the r most significant bits of x, for example to some value c, then this function can be described by a bit shift of r positions and a bitwise defined function

$$f(x) := (x \ll r) \vee c \tag{11}$$

which is an r-narrow T-function. Thus, by looping over all 2^r possible values for c an equation involving (10) can also be solved by solution graphs.

If we use generalized solution graphs, it is actually possible to combine all 2^r such solution graphs to one graph, in which again the complete set of solutions is represented: This can be done by simply merging all the roots and all the sinks of the 2^r solution graphs as they are clearly equivalent in the generalized sense.

Two examples of actual systems of equations which were solved by applying solution graphs and the extensions from this section are given in Appendix B.

6 Conclusion

In this paper we defined a subclass of weak T-functions, the *w-narrow T-functions*. We showed that systems of equations involving only w-narrow T-functions (with small w) can be solved efficiently by using *solution graphs* and thus such functions should be avoided in cryptographical schemes.

Let us stress again that this does *not* mean that the concept of using T-functions for constructing cryptosystems is bad. One just has to assure that the used T-functions are not too narrow. For example, it is a good idea to always include multiplications and bit shifts of some medium size in the functions, as those are examples of T-functions which are not very narrow.

Additionally we presented some extensions to our proposal of a solution graph. These extensions allow to use the solution graphs also in other contexts than pure T-functions, for example as a tool in the cryptanalysis of hash functions.

Acknowledgements. This work is part of PhD thesis [1] and I would like to thank my supervisor Hans Dobbertin for support and discussions. Also I would like to thank Tanja Lange for many helpful discussions and comments.

References

1. M. Daum. *Cryptanalysis of Hash Functions of the MD4-Family*. PhD Thesis, Ruhr-University Bochum, in preparation.
2. M. Daum. *Narrow T-functions.* Cryptology ePrint Archive, Report 2005/016. (available under: `http://eprint.iacr.org/2005/016`)
3. H. Dobbertin: *The status of MD5 after a recent attack.* CryptoBytes, 2(2), 1996, pp. 1-6.
4. H. Dobbertin: *RIPEMD with two-round compress function is not collision-free.* Journal of Cryptology 10, 1997, pp. 51-68.
5. H. Dobbertin: *Cryptanalysis of MD4.* Journal of Cryptology 11, 1998, pp. 253-274.
6. A. Klimov and A. Shamir: *A New Class of Invertible Mappings.* Workshop on Cryptographic Hardware and Embedded Systems (CHES), 2002.
7. A. Klimov and A. Shamir: *Cryptographic Applications of T-functions.* Selected Areas in Cryptography (SAC), 2003.
8. A. Klimov and A. Shamir: *New Cryptographic Primitives Based on Multiword T-functions.* FSE 2004, 2004.
9. A. Klimov: *Applications of T-functions in Cryptography.* PhD Thesis, Weizmann Institute of Science, submitted, 2004. (available under: `http://www.wisdom.weizmann.ac.il/~{}ask/`)
10. I. Wegener: *Branching Programs and Binary Decision Diagrams: Theory and Applications.* SIAM Monographs on Discrete Mathematics and Applications, 2000.

A Dobbertin's Original Algorithm from the Attacks on MD4, MD5 and RIPEMD

In this section we describe the algorithm used by Dobbertin in his attacks from [3, 4, 5]. However, we do this using the same terminology as in the other sections of the present paper to maximize the comparability.

Let $\mathcal{S}$ be a system of equations which can be completely described by T-functions and let $\mathcal{S}_k$ denote the system of equations in which only the k least significant bits of each equation are considered. As those k bits only depend on the k least significant bits of all the inputs, we will consider the solutions of $\mathcal{S}_k$ to have only k bits per variable as well.

Then, from the defining property of a T-function, the following theorem easily follows:

Theorem 4. *Every solution of $\mathcal{S}_k$ is an extension of a solution of $\mathcal{S}_{k-1}$.*

This theorem directly leads to the following algorithm for enumerating all the solutions of $\mathcal{S}$.

Algorithm 7.

1. Find all solutions (having only 1 bit per variable) of $\mathcal{S}_1$.
2. For every found solution of some $\mathcal{S}_k, k \in \{1, \ldots, n-1\}$, recursively check which extensions of this solution by 1 bit per variable are solutions of $\mathcal{S}_{k+1}$.
3. Output the found solutions of $\mathcal{S}_n (= \mathcal{S})$.

An actual toy example application of this algorithm – finding the solutions x of the equation $\mathcal{S}$ given by $(x \vee \mathtt{0010}_2) + \mathtt{0110}_2 = \mathtt{0001}_2$ with $n = 4$ – is illustrated in Figure 5: We start at the root of the tree and check whether 0 or 1 are possible values for $[x]_0$, i.e. if they are solutions of $\mathcal{S}_1$ which is given by $([x]_0 \vee 0) + 0 = 1$. Obviously 0 is not a solution of this equation and thus we need not consider any more values for x starting with 0. But 1 is a solution of $\mathcal{S}_1$, thus we have to check whether extensions (i.e. $\mathtt{01}_2$ or $\mathtt{11}_2$) are solutions of $\mathcal{S}_2$: $(x \vee \mathtt{10}_2) + \mathtt{10}_2 = \mathtt{01}_2$. Doing this recursively finally leads to the "tree of solutions", illustrated on the left hand side of Figure 5.

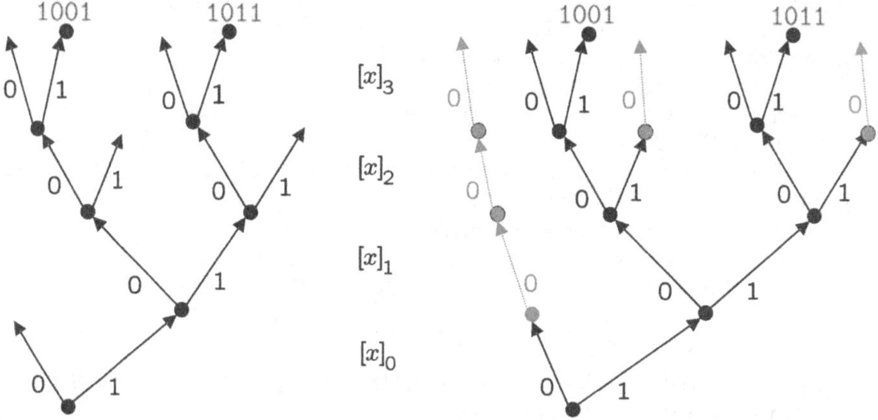

Fig. 5. "Solution tree" for the equation $(x \vee \mathtt{0010}_2) + \mathtt{0110}_2 = \mathtt{0001}_2$ with $n = 4$

If this method is implemented directly as described in Algorithm 7, it has a worst case complexity which is about twice as large as that of an exhaustive search, because the full solution tree of depth n has $2^{n+1} - 1$ vertices. An example of such a "worst case solution tree" is given in Figure 6. To actually achieve a worst case complexity similar to that of an exhaustive search a little modification is necessary to the algorithm: The checking should be done for complete paths (as indicated by the *grey* arrows in the tree on the right hand side in Figure 5), which can also be done in one machine operation, and not bit by bit. This

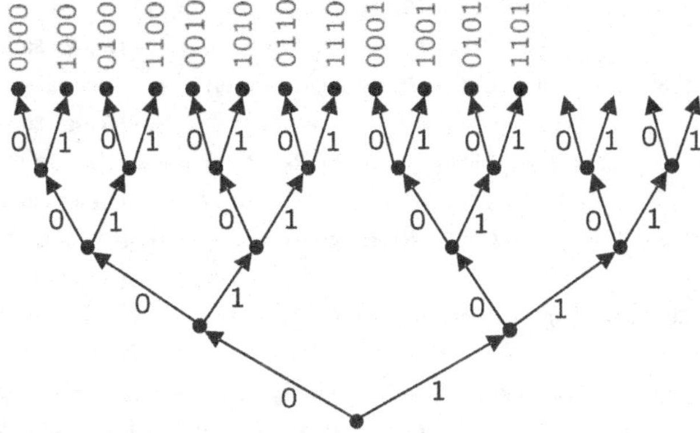

Fig. 6. "Solution tree" for the equation $(0100_2 \oplus (x+0101_2)) - (0100_2 \oplus x) = 1101_2$ with $n = 4$

means, we would start by checking 0000_2 and recognize that this fails already in the least significant bit. In the next step we would check 0001_2 and see that the three least significant bits are okay. This means in the following step we would only change the fourth bit and test 1001_2 which would give us the first solution. All in all we would need only 7 checks for this example as indicated by the grey arrows.

The worst case complexity of this modified algorithm (which is what was actually implemented in Dobbertin's attacks) is clearly 2^n as this is the number of leaves of a full solution tree. However, it is also quite clear, that in the average case, or rather in the case of fewer solutions, this algorithm is much more efficient.

B Examples of Applications

In this section we present two examples of systems of equations which were actually solved by using the techniques presented in this paper. They have both appeared as one small part in an attempt to apply Dobbertin's methods from [3, 4, 5] to SHA-1. In this paper we concentrate on describing how these systems were solved and omit a detailed description of their meanings.

The first system comes from looking for so-called "inner collisions" and includes 14 equations and essentially 22 variables $R_1, \ldots, R_{13}, \varepsilon_3, \ldots, \varepsilon_{11}$:

$$
\begin{aligned}
0 &= \varepsilon_3 + 1\\
0 &= \varepsilon_4 - (\widetilde{R_3} \lll 5 - R_3 \lll 5) + 1\\
\mathrm{Ch}(\widetilde{R_3}, R_2 \lll 30, R_1 \lll 30) - \mathrm{Ch}(R_3, R_2 \lll 30, R_1 \lll 30) &= \varepsilon_5 - (\widetilde{R_4} \lll 5 - R_4 \lll 5) + 1\\
\mathrm{Ch}(\widetilde{R_4}, \widetilde{R_3} \lll 30, R_2 \lll 30) - \mathrm{Ch}(R_4, R_3 \lll 30, R_2 \lll 30) &= \varepsilon_6 - (\widetilde{R_5} \lll 5 - R_5 \lll 5)\\
\mathrm{Ch}(\widetilde{R_5}, \widetilde{R_4} \lll 30, \widetilde{R_3} \lll 30) - \mathrm{Ch}(R_5, R_4 \lll 30, R_3 \lll 30) &= \varepsilon_7 - (\widetilde{R_6} \lll 5 - R_6 \lll 5) + 1
\end{aligned}
$$

$$\begin{aligned}
\mathrm{Ch}(\widetilde{R_6}, \widetilde{R_5} \lll 30, \widetilde{R_4} \lll 30) - \mathrm{Ch}(R_6, R_5 \lll 30, R_4 \lll 30) &= \varepsilon_8 - (\widetilde{R_7} \lll 5 - R_7 \lll 5) \\
&\quad -(\widetilde{R_3} \lll 30 - R_3 \lll 30) + 1 \\
\mathrm{Ch}(\widetilde{R_7}, \widetilde{R_6} \lll 30, \widetilde{R_5} \lll 30) - \mathrm{Ch}(R_7, R_6 \lll 30, R_5 \lll 30) &= \varepsilon_9 - (\widetilde{R_8} \lll 5 - R_8 \lll 5) \\
&\quad -(\widetilde{R_4} \lll 30 - R_4 \lll 30) + 1 \\
\mathrm{Ch}(\widetilde{R_8}, \widetilde{R_7} \lll 30, \widetilde{R_6} \lll 30) - \mathrm{Ch}(R_8, R_7 \lll 30, R_6 \lll 30) &= \varepsilon_{10} - (\widetilde{R_9} \lll 5 - R_9 \lll 5) \\
&\quad -(\widetilde{R_5} \lll 30 - R_5 \lll 30) \\
\mathrm{Ch}(\widetilde{R_9}, \widetilde{R_8} \lll 30, \widetilde{R_7} \lll 30) - \mathrm{Ch}(R_9, R_8 \lll 30, R_7 \lll 30) &= \varepsilon_{11} - (\widetilde{R_{10}} \lll 5 - R_{10} \lll 5) \\
&\quad -(\widetilde{R_6} \lll 30 - R_6 \lll 30) \\
\mathrm{Ch}(\widetilde{R_{10}}, \widetilde{R_9} \lll 30, \widetilde{R_8} \lll 30) - \mathrm{Ch}(R_{10}, R_{19} \lll 30, R_8 \lll 30) &= -(\widetilde{R_{11}} \lll 5 - R_{11} \lll 5) \\
&\quad -(\widetilde{R_7} \lll 30 - R_7 \lll 30) + 1 \\
\mathrm{Ch}(\widetilde{R_{11}}, \widetilde{R_{10}} \lll 30, \widetilde{R_9} \lll 30) - \mathrm{Ch}(R_{11}, R_{10} \lll 30, R_9 \lll 30) &= -(\widetilde{R_8} \lll 30 - R_8 \lll 30) \\
\mathrm{Ch}(R_{12}, \widetilde{R_{11}} \lll 30, \widetilde{R_{10}} \lll 30) - \mathrm{Ch}(R_{12}, R_{11} \lll 30, R_{10} \lll 30) &= -(\widetilde{R_9} \lll 30 - R_9 \lll 30) \\
\mathrm{Ch}(R_{13}, R_{12} \lll 30, \widetilde{R_{11}} \lll 30) - \mathrm{Ch}(R_{13}, R_{12} \lll 30, R_{11} \lll 30) &= -(\widetilde{R_{10}} \lll 30 - R_{10} \lll 30) + 1 \\
0 &= -(\widetilde{R_{11}} \lll 30 - R_{11} \lll 30) + 1
\end{aligned}$$

Here we use $\widetilde{R_i} := R_i + \varepsilon_i$ for a compact notation, the word size is $n = 32$, and the Ch in these equations stands for the bitwise defined choose-function

$$\mathrm{Ch}(x, y, z) = (x \wedge y) \vee (\overline{x} \wedge z).$$

It was not possible to solve this system in full generality, but for the application it sufficed to find some fixed values for $\varepsilon_3, \ldots, \varepsilon_{11}$ such that there are many solutions for the R_i and then to construct a generalized solution graph for the solutions for $R_1, \ldots, R_{13}$.

The choice for good values for some of the ε_i could be done by either theoretical means or by constructing solution graphs for single equations of the system and counting solutions with fixed values for some ε_i.
For example, from the solution graph for the last equation it is possible (as described in Section 5.2) to remove the R_{11} such that we get a solution graph which represents all values for ε_{11} for which an R_{11} exists such that

$$0 = -(\widetilde{R_{11}} \lll 30 - R_{11} \lll 30) + 1.$$

This solution graph shows that only $\varepsilon_{11} \in \{1, 4, 5\}$ is possible. Then by inserting each of these values in the original solution graph (by Algorithm 6) and counting the possible solutions for R_{11} (by Algorithm 4) it can be seen that $\varepsilon_{11} = 4$ is the best choice. Having fixed $\varepsilon_{11} = 4$ also the last but one equation includes only one of the ε_i, namely ε_{10} (implicitly in $\widetilde{R_{10}}$). Then possible solutions for ε_{10} can be derived similarly as before for ε_{11} and doing this repeatedly gave us some good choices for $\varepsilon_{11}, \varepsilon_{10}, \varepsilon_9, \varepsilon_8, \varepsilon_7$ and (using the first two equations) for ε_3 and ε_4.

Finding values ε_5 and ε_6 such that the whole system still remains solvable was quite hard and could be done by repeatedly applying some of the techniques described in this paper, e.g. by combining generalized solution graphs for different of the equations and removing those variables R_i from the graphs which were no longer of any explicit use. This way we found four possible values for ε_5 and ε_6.

After fixing all the ε_i variables in a second step we were then able to construct the generalized solution graph for the complete system of equations with the remaining variables $R_1, \ldots, R_{13}$. It contains about 700 vertices, more than 80000 edges and represents about 2^{205} solutions.

The second examplary system of equations appeared when looking for a so-called "connection" and after some reduction steps it can be written as follows:

$$
\begin{aligned}
C_1 &= R_9 + \mathrm{Ch}(R_{12} \lll 2, R_{11}, R_{10}) \\
C_2 &= (C_3 - R_{10} - R_{11}) \oplus (C_4 + R_9 \lll 2) \\
C_5 &= (C_6 - R_{11}) \oplus (C_7 + R_{10} \lll 2 - (R_9 \lll 7)) \\
C_8 &= (C_9 - R_{12}) \oplus (C_{10} + R_9 \lll 2) \\
&\qquad \oplus (C_{11} + R_{11} \lll 2 - (R_{10} \lll 7) - \mathrm{Ch}(R_9 \lll 2, C_{12}, C_{13}))
\end{aligned}
$$

In these equations the C_i are constants which come from some transformations of the original (quite large) system of equations together with some random choices of values. For this system we are interested in finding at least one solution for $R_9, R_{10}, R_{11}, R_{12}$.

As the first three equations are quite simple and (after eliminating the rotations) also quite narrow, the idea for solving this system was the following: First compute a generalized solution graph for the first three equations which represents all possible solutions for R_9, R_{10}, R_{11} for which at least one corresponding value for R_{12} exists. For this set of solutions we observed numbers of about 2^{11} to 2^{15} solutions. Then we could enumerate all these solutions from this graph and for each such solution we just had to compute the value for R_{12} corresponding to the last equation

$$
\begin{aligned}
R_{12} = C_9 - (C_8 \oplus (C_{10} + R_9 \lll 2) \\
\oplus (C_{11} + R_{11} \lll 2 - (R_{10} \lll 7) - \mathrm{Ch}(R_9 \lll 2, C_{12}, C_{13})))
\end{aligned}
$$

and check whether it also fulfilled the first equation. If we consider the first equation with random but fixed values for R_9, R_{10}, R_{11} we see that either there is no solution or there are many solutions for R_{12}, as only every second bit of R_{12} (on average) has an effect on the result of $\mathrm{Ch}(R_{12} \lll 2, R_{11}, R_{10})$. However, since the values for R_9, R_{10}, R_{11} were chosen from the solution graph of the first three equations there is at least one solution and thus the probabiliy that the value for R_{12} from the last equation also fulfills the first, is quite good.

This way we succeded in solving this system of equations quite efficiently.

A New Class of Single Cycle T-Functions

Jin Hong, Dong Hoon Lee,
Yongjin Yeom, and Daewan Han

National Security Research Institute,
161 Gajeong-dong, Yuseong-gu
Daejeon, 305-350, Korea
{jinhong, dlee, yjyeom, dwh}@etri.re.kr

Abstract. T-function is a relatively new cryptographic building block suitable for streamciphers. It has the potential of becoming a substitute for LFSRs, and those that correspond to maximum length LFSRs are called single cycle T-functions. We present a family of single cycle T-functions, previously unknown. An attempt at building a hardware oriented streamcipher based on this new T-function is given.

Keywords: T-function, single cycle, streamcipher.

1 Introduction

The appearance of algebraic attack on streamciphers[5, 11, 12, 13] has certainly made the designing of streamciphers a more difficult task. At the same time, as presentations[7, 28] and discussions during a recent streamcipher workshop has shown, the demand for streamciphers is declining. But, we have also seen at the same workshop that at the very extremes, there are still genuine needs for streamciphers. One case is when the cipher has to be ultra-fast(Gbps) in software (on relatively good platforms), as in software routers. The other extreme, namely where efficient hardware implemented ciphers for resource constrained environment is needed, could also benefit from a good streamcipher design.

Most of the recent attempts at streamcipher constructions[30, 14, 31, 16, 15, 9, 10, 17, 6, 26, 29] are mainly focused on software, except for those based on LFSRs. In particular, most of them demand a very large memory space to store its internal state. If we turn to traditional designs that use bitwise LFSRs, which could have advantages in hardware, we find that large registers have to be used to counter algebraic attacks. In short, we have a lack of good hardware oriented streamciphers at the moment. This paper is an attempt at filling this gap.

Few years ago, Klimov and Shamir started developing the theory of T-functions[20, 21, 22]. A T-function is a function acting on a collection of memory words, with a weak one-wayness property. It started out as a tool for blockciphers, but now, its possibility as a building block for streamciphers is drawing attention.

An important class of T-functions consists of those that exhibit the *single cycle* property. This is the T-function equivalent of maximum length LFSRs

H. Gilbert and H. Handschuh (Eds.): FSE 2005, LNCS 3557, pp. 68–82, 2005.

and has potential to bring about a very fast streamcipher. Unfortunately only a very small family of single cycle T-functions is known to the crypto community currently.[1]

The main contribution of this work is to uncover a new class of single cycle T-functions. It is a generalization of a small subset of the previously known single cycle T-functions and it does show some good properties which the previous ones did not. We also give an example of how one might build a cipher on top of this new class of single cycle T-functions. Although previous T-functions targeted software implementations, our T-function based streamcipher is designed to be light and is suitable for constrained hardware environment.

The paper is organized as follows. We start by reviewing the basics of T-functions. In Section 3, we look into the existing single cycle T-functions and show that without the multiplicative part, which is not understood at all, it is a very simple object, far from a random function. With this new way of viewing existing T-functions, we give a new class of single cycle T-functions in Section 4. A streamcipher example built on top of the new T-functions is introduced in the following section. The last section concludes the paper.

2 Review of T-Functions

We shall review the basics of multi-word T-functions in this section. Readers may refer to the original papers [20, 21, 22] for a more in-depth treatment.

Let us consider a gathering of m-many of n-bit words, which we denote by x_i ($i = 0, \ldots, m-1$). Our main interest lies in the case $n = 32$ and $m = 4$. As a shorthand for multiple words, we shall often use the corresponding boldface letter. For example, $\mathbf{x} = (x_k)_{k=0}^{m-1}$. The i-th bit of a word x is denoted by $[x]_i$, where we always count starting from 0. Seen as an integer, we have

$$x = \sum_{i=0}^{n-1} [x]_i 2^i. \tag{1}$$

The i-th bits of the m-tuple of words $\mathbf{x}$ are denoted collectively as $[\mathbf{x}]_i$. We sometimes view $[\mathbf{x}]_i$ also as an m-bit integer by setting

$$[\mathbf{x}]_i = \sum_{k=0}^{m-1} [x_k]_i 2^k. \tag{2}$$

In reading the rest of this paper, it helps to view the various notations pictorially as follows.

[1] It seems T-function was already studied in the mathematics community under various different names[3, 4].

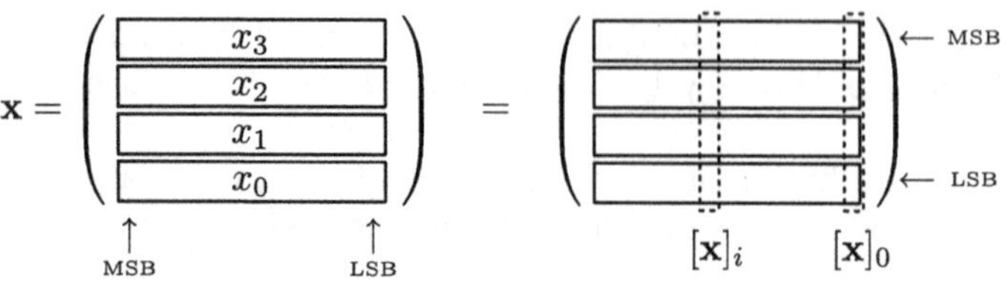

Accordingly, we shall sometimes refer to $[\mathbf{x}]_i$ as the i-th column.

Definition 1. *A (multi-word)* T-function *is a map*

$$\mathbf{T} : (\{0,1\}^n)^m \longrightarrow (\{0,1\}^n)^m, \qquad \mathbf{x} \mapsto \mathbf{T}(\mathbf{x}) = (T_k(\mathbf{x}))_{k=0}^{m-1}$$

sending an m-tuple of n-bit words to another m-tuple of n-bit words, where each resulting n-bit word is denoted as $T_k(\mathbf{x})$, such that for each $0 \leq i < n$, the i-th bits of the resulting words $[\mathbf{T}(\mathbf{x})]_i$ are functions of just the lower input bits $[\mathbf{x}]_0, [\mathbf{x}]_1, \ldots, [\mathbf{x}]_i$.

We shall mainly be dealing with multi-word T-functions, as opposed to single-word T-functions, which is the $m = 1$ case, and hence shall mostly omit writing *multi-word.* Also, unless stated otherwise, we shall always assume a T-function to be acting on m words of n-bit size. The set of words a T-function is acting on is sometimes referred to as *memory* or *register* and the bit values it contains are said to form a *state* of the memory.

Given a T-function $\mathbf{T}$, one may fix an initial state $\mathbf{x}^0$ for the memory and iteratively act $\mathbf{T}$ on it to obtain a sequence defined by

$$\mathbf{x}^{t+1} = \mathbf{T}(\mathbf{x}^t). \tag{3}$$

Such a sequence will always be eventually periodic and if its periodic part passes through all of the 2^{nm} possible states the memory may take, the T-function is said to form a *single cycle.* A single cycle T-function may serve as a good building block for a streamcipher. To state results about single cycle T-functions, we need a few more definitions.

Definition 2. *A (multi-word)* parameter *is a map*

$$\alpha : (\{0,1\}^n)^m \longrightarrow \{0,1\}^n, \qquad \mathbf{x} \mapsto \alpha(\mathbf{x})$$

sending an m-tuple of n-bit words to a single n-bit word such that for each $0 \leq i < n$, the i-th bit of the resulting word $[\alpha(\mathbf{x})]_i$ is a function of just the strictly lower input bits $[\mathbf{x}]_0, [\mathbf{x}]_1, \ldots, [\mathbf{x}]_{i-1}$.

In other words, a parameter is a sort of multi-word to single-word T-function for which the i-th output bit does not depend on the i-th input bits $[\mathbf{x}]_i$. When restricted to just linear functions acting on a single word, T-functions are exactly the upper triangular matrices and parameters correspond to the strictly upper triangular matrices.

Given a parameter α, and fixed $0 \le i < n$, we may consider the bit value

$$B[\alpha, i] = \bigoplus_{\mathbf{x}=(0,\dots,0)}^{(2^i-1,\dots,2^i-1)} [\alpha(\mathbf{x})]_i. \tag{4}$$

Notice that since $[\alpha(\mathbf{x})]_i$ does not depend on any of the high indexed input bits, going any higher in this direct sum would be meaningless. Also, note that $[\alpha(\mathbf{x})]_0$ is constant for all input $\mathbf{x}$ and the sum $B[\alpha, 0]$ is equal to this constant value. If the value $B[\alpha, i] = 0$ for all i, the parameter is said to be *even*. Likewise, if it is 1 for all i, the parameter is *odd*.

Currently, it seems that the only single cycle T-functions known to the crypto community are based on the following theorem from [22].

Theorem 1. *The T-function defined by setting* $\mathbf{T}(\mathbf{x}) = (T_k(\mathbf{x}))_{k=0}^{m-1}$, *with*

$$T_k(\mathbf{x}) = x_k \oplus (\alpha_k(\mathbf{x}) \wedge x_0 \wedge x_1 \wedge \cdots \wedge x_{k-1}) \tag{5}$$

for $k = 0, \dots, m-1$, *exhibits the single cycle property, when each* α_k *is an odd parameter.*

3 Analysis of an Example T-Function

The following example may be found in [22]. We shall try to give a clearer view of the inner workings of this example.

Example 1. Consider the following T-function which acts on four 64-bit words. Fix any odd number C0 and set C1 = 0x12481248, C3 = 0x48124812. Use the notation $a_0 = x_0$ and $a_{i+1} = a_i \wedge x_{i+1}$ for $i = 0, 1, 2$. Then,

$$\alpha = \alpha(\mathbf{x}) = (a_3 + \mathtt{C0}) \oplus a_3 \tag{6}$$

defines an odd parameter. Finally, the mapping

$$\begin{pmatrix} x_3 \\ x_2 \\ x_1 \\ x_0 \end{pmatrix} \mapsto \begin{pmatrix} x_3 \oplus (\alpha \wedge a_2) \oplus (2x_0(x_1 \vee \mathtt{C1})) \\ x_2 \oplus (\alpha \wedge a_1) \oplus (2x_0(x_3 \vee \mathtt{C3})) \\ x_1 \oplus (\alpha \wedge a_0) \oplus (2x_2(x_3 \vee \mathtt{C3})) \\ x_0 \oplus \alpha \qquad\quad \oplus (2x_2(x_1 \vee \mathtt{C1})) \end{pmatrix} \tag{7}$$

gives a single cycle T-function.

This example is not a direct application of Theorem 1, because of the last term in each row that utilizes multiplications. But these last terms are even parameters and it is possible to follow through the proof of Theorem 1 given in [22] with them attached.

Let us have a closer look at this example. Without the even parameter part, it is almost identical to Theorem 1.

$$\begin{pmatrix} x_3 \\ x_2 \\ x_1 \\ x_0 \end{pmatrix} \mapsto \begin{pmatrix} x_3 \oplus (\alpha(\mathbf{x}) \wedge x_0 \wedge x_1 \wedge x_2) \\ x_2 \oplus (\alpha(\mathbf{x}) \wedge x_0 \wedge x_1) \\ x_1 \oplus (\alpha(\mathbf{x}) \wedge x_0) \\ x_0 \oplus \ \alpha(\mathbf{x}) \end{pmatrix}. \tag{8}$$

We will denote this simplified function by $\mathbf{T}$ for the moment. Now, let us look at just the 0-th column. We know $[\alpha(\mathbf{x})]_0 = 1$ for any odd parameter and this can also be checked directly from (6). Hence, as noted in [22], the map (8) restricted to the 0-th column is

$$[\mathbf{T}(\mathbf{x})]_0 = [\mathbf{x}]_0 + 1 \pmod{2^m}. \tag{9}$$

Here, we are using the notation of (2). Let us move onto the higher bits. Since all the odd parameters are set to be the same in (8), the mapping $\mathbf{T}$ may be described by

$$[\mathbf{T}(\mathbf{x})]_i = \begin{cases} [\mathbf{x}]_i & \text{if } [\alpha(\mathbf{x})]_i = 0, \\ [\mathbf{x}]_i + 1 \pmod{2^m} & \text{if } [\alpha(\mathbf{x})]_i = 1. \end{cases} \tag{10}$$

In other words, the role of the odd parameter α is to decide whether or not to apply the map

$$[\mathbf{x}]_i \mapsto [\mathbf{x}]_i + 1 \pmod{2^m} \tag{11}$$

to the i-th bit column.

In the extreme case when $\texttt{C0} = 1$ in the definition (6) for the odd parameter, we have $[\alpha(\mathbf{x})]_i = 1$ if and only if $[x_k]_j = 1$ for all k and all $j < i$. So the mapping (11) is applied to the i-th bit column if and only if all $4i$ of the strictly lower bits are filled with 1. Hence the map (8) literally defines a counter (in hexadecimal numbers). It runs through all the $2^{4 \cdot 64}$ possible values, incrementing the memory by 1 at each application of $\mathbf{T}$.

The mapping (8) is more complex when the constant $\texttt{C0}$ is bigger, but this does not seem to give a fundamental difference. Using a more complex odd parameter would make (8) a lot more random-like. Without this, the real reason for (7) producing a sequence which passes all the statistical tests lies in the even parameter part.

Remark 1. The paper [27] gives an attack on a (very basic) streamcipher base on (7). Essentially, they analyze the multiplicative part, and find a way to apply this technique to (7). From the viewpoint of their attack, the components of (7), excluding the multiplicative part, contribute very little to the security of the system. Arguments of this section show that this is a natural consequence of its inner workings.

4 A New Class of T-Functions

Arguments of the previous section lead us naturally to the following idea. What would happen if we replaced mapping (11) with a more general mapping?

Given an $m \times m$ S-box, $S : \{0,1\}^m \mapsto \{0,1\}^m$, define

$$\mathbf{S} : (\{0,1\}^n)^m \longrightarrow (\{0,1\}^n)^m, \qquad \mathbf{x} \mapsto \mathbf{S}(\mathbf{x})$$

by setting

$$[\mathbf{S}(\mathbf{x})]_i = S([\mathbf{x}]_i).$$

Here, we are using the notation of (2), so that the bold face $\mathbf{S}$ acts on each and every *column* of the registers. We say that an S-box has the *single cycle* property if its cycle decomposition gives a single cycle. That is, starting from any point, if we iteratively act S, we end up going through all possible elements of $\{0,1\}^m$.

Certainly, $\mathbf{S}$ will not define a single cycle T-function, even when S is of single cycle. So let us start by first defining some logical operations on multi-words. Let $\mathbf{x} = (x_k)_{k=0}^{m-1}$ and $\mathbf{y} = (y_k)_{k=0}^{m-1}$, be two multi-words. Define $\mathbf{x} \oplus \mathbf{y}$ by setting

$$\mathbf{x} \oplus \mathbf{y} = (x_k \oplus y_k)_{k=0}^{m-1}.$$

Also, for a (single) word α, define the multi-word

$$\alpha \cdot \mathbf{x} = (\alpha \wedge x_k)_{k=0}^{m-1}.$$

The notation $\sim \alpha$ will denote the bitwise complement of α.

Theorem 2. *Let S be a single cycle S-box and let α be an odd parameter. If S^o is an odd power of S and S^e is an even power of S, the mapping*

$$\mathbf{T}(\mathbf{x}) = \big(\alpha(\mathbf{x}) \cdot \mathbf{S}^{\mathbf{o}}(\mathbf{x})\big) \oplus \big((\sim \alpha(\mathbf{x})) \cdot \mathbf{S}^{\mathbf{e}}(\mathbf{x})\big).$$

defines a single cycle T-function.

Proof. That this is a T-function is easy to check. Notice that due to its definition, any T-function may be restricted to just the lower bits. It suffices to prove that, when restricted to the lower bits $[\mathbf{x}]_0, [\mathbf{x}]_1, \ldots, [\mathbf{x}]_{i-1}$, the period of the above map is $2^{m \cdot i}$. This will be shown by induction.

The mapping $\mathbf{T}$ can be better understood when it is written as

$$[\mathbf{T}(\mathbf{x})]_i = \begin{cases} S^e([\mathbf{x}]_i) & \text{if } [\alpha(\mathbf{x})]_i = 0, \\ S^o([\mathbf{x}]_i) & \text{if } [\alpha(\mathbf{x})]_i = 1. \end{cases} \qquad (12)$$

In particular, since we always have $[\alpha(\mathbf{x})]_0 = 1$, if we restrict $\mathbf{T}$ to the 0-th column of the registers, it acts as just S^o regardless of the input $\mathbf{x}$. Notice that any odd power of S also has the single cycle property. Hence the period of $\mathbf{T}$ is 2^m, when restricted to $[\mathbf{x}]_0$. This gives us the starting point.

So suppose, as an induction hypothesis, that the period of $\mathbf{T}$, restricted to the lower bits $[\mathbf{x}]_0, \ldots, [\mathbf{x}]_{i-1}$, is $2^{m \cdot i}$. The period of $\mathbf{T}$, restricted to the next step

$[\mathbf{x}]_0, \ldots, [\mathbf{x}]_i$, must be a multiple of $2^{m \cdot i}$. Now, with the parameter α being odd, (12) shows that when $\mathbf{T}$ is consecutively applied to the bits $[\mathbf{x}]_0, \ldots, [\mathbf{x}]_i$ exactly $2^{m \cdot i}$ times, S^o and S^e are both applied an odd number of times to $[\mathbf{x}]_i$. In all, this is equivalent to applying S to $[\mathbf{x}]_i$ an odd number of times. Since an odd number is relatively prime to the period 2^m of S, the period of $\mathbf{T}$ restricted to $[\mathbf{x}]_0, \ldots, [\mathbf{x}]_i$ must be $2^{m \cdot (i+1)}$. This completes the induction step and the proof.

Expression (12) shows that this new T-function may be viewed as a twisted product of small S-boxes, each acting on a single column of memory.

The reader may already have noticed that allowing for odd powers of single cycles S-boxes is not really any more general than allowing for just the (single power of) single cycle S-boxes. If we further restrict the above theorem to the case when the even power used is zero, we have the following corollary.

Corollary 1. *Given a single cycle S-box S and an odd parameter α, the following mapping defines a single cycle T-function.*

$$\mathbf{x} \mapsto \mathbf{x} \oplus \big(\alpha(\mathbf{x}) \cdot (\mathbf{x} \oplus \mathbf{S}(\mathbf{x}))\big).$$

5 T-Function Based Streamcipher; TSC-1

In this section, we propose a very bare framework for a streamcipher based on Theorem 2. A distinguishing attack on this example of very low complexity is already known[19] and the authors no longer believe this cipher to be secure, but we include this as a reference for further developments in this direction.

Since the work [27] has shown that disclosing parts of the raw memory state could be fatal, we want to hide the memory while producing output from this T-function. So we shall use the T-function as a substitute for an LFSR in a filter model.

5.1 The Specification

Specification of the cipher will be given by supplying a filter function in addition to fixing various components for the T-function.

Fix $n = 32$ and $m = 4$, that is, we work with four 32-bit words, for a total internal state of 128 bits. Define an odd parameter by setting

$$\alpha(\mathbf{x}) = (p + \mathtt{C}) \oplus p \oplus 2s, \tag{13}$$

where[2]

$$\mathtt{C} = \mathtt{0x12488421}, \quad p = x_0 \wedge x_1 \wedge x_2 \wedge x_3, \quad \text{and} \quad s = x_0 + x_1 + x_2 + x_3.$$

[2] The constant `0x12488421` was chosen so that 1s are denser at the higher bits than at the lower bits. This will help it quickly move away from the all-zero state, should it occur.

All additions are done modulo 2^{32}. This is equal to (6), except that we have added the even parameter $2s$ to allow for stronger inter-column effects. Define a 4×4 S-box S, as given by the following line written in C-style.

```
S[16] = {3,5,9,13,1,6,11,15,4,0,8,14,10,7,2,12};        (14)
```

One may check easily that this is a single cycle S-box. Using this S-box, let us set $S^o = S$ and $S^e = S^2$. We can now define a single cycle T-function $\mathbf{T}$, through the use of Theorem 2.

To actually obtain the keystream, use the filter

$$f(\mathbf{x}) = (x_{0\lll 9} + x_1)_{\lll 15} + (x_{2\lll 7} + x_3) \tag{15}$$

on the memory bits, after each application of $\mathbf{T}$. This will give us a single 32-bit block of keystream per action of the T-function. Here, the symbol $\lll$ denotes left rotation, and the additions should be done modulo 2^{32}. Going back to the notation of (3), the output word produced at time t may be written as $f(\mathbf{x}^t)$.

5.2 Naive Security

Period We already know that the period of the state registers is 2^{128}, as guaranteed by the single cycle property. The output itself also has a period of 2^{128} words.

To see this, first note that the period has to be a divisor of 2^{128}. Now, initialize the register content with the all zero state and consider what the content of the registers would be after 2^{124} iterated applications of the T-function. Since the period of the T-function restricted to the lower 31 columns is 2^{124}, all columns except the most significant column will be zero. Furthermore, when observed every 2^{124} iterations apart, due to description (12) and the definition of an odd parameter, the change of the most significant column follows some fixed odd power of the S-box, which is of cycle length 16. Explicit calculation of the 16 keystream output words for each odd power of the S-box confirms that, in all odd power cases, one has to go through all 16 points before reaching the starting point. Hence the period of the cipher is $16 \cdot 2^{124} = 2^{128}$.

Actually, for a general single cycle T-function, one can always show that at least one bit position in the register will show period equal to the T-function. When any mildly complicated filter is attached to such a T-function, the output keystream has a high chance of inheriting this property and one should be able to show some result on the period of the whole filter generator. For example, the cipher TF-1[23], can be shown to have a period of at least 2^{254}.

Maximum security level Given a single word block of keystream, guessing any three words (96 bits) of memory determines the remaining word uniquely. And it suffices to look at the next three word blocks of keystream to check if the guess is correct. Hence, it is clear that this proposal cannot provide more than 96-bit security.

Bit-flip property In addition to imposing the single cycle property, we had chosen the S-box (14) to satisfy the following conditions.[3]

1. At the application of S, each of the four bits has bit-flip probability of $\frac{1}{2}$.
2. The same if true for S^2, the square of S.

In more exact terms, the first condition states that for each $i = 0, 1, 2, 3$,

$$\#\{\, 0 \le t < 16 \mid \text{the } i\text{-th bit of } t \oplus S(t) \text{ is } 1 \} = 8.$$

Due to this property, regardless of the behavior of the odd parameter α, every bit in the register is guaranteed a $\frac{1}{2}$ bit-flip probability. This is one thing that wasn't satisfied by (8) and which was only naively expected of (7).

Rotations Rotations used in the filter serve two main purposes. The first is to ensure that output from the same S-box, i.e., bits from the same column, do not contribute directly to the same output bit. We want contributions to any single output bit to come from bits that change independently of each other.

The other reason is to remove the possibility of relating a part of the output with a part of the memory that allows some sort of separate handling in view of the action of T-function. In particular, this stops the guess-then-determine attack. Difficulty of correlation attacks can also be understood from this viewpoint. In the last step of a correlation attack, one needs to guess a part of the state bits and compare calculated outputs with the keystream, checking for the occurrence of expected correlation. In our case, any correlation with a single output bit will involve multiple input bits and at least one of them will come near the high ends of the registers. This will force one to guess quite a large part of the registers to be able to apply T-function even once.

Misc We have done most of the tests presented in [2] and have verified that this proposal gives good statistical results. As the S-boxes are nonlinear, the most dangerous attack on streamciphers, the algebraic attack, seems to be out of the question.

With our T-function based filter model, one can view the randomness of keystream as originating from the T-function and as being amplified through the filter. Compared with LFSR based filter models, it seems fair to say that the randomness at the source is better with T-functions. In the Appendix, we present another design that shifts the burden of producing randomness more to the filter.

5.3 Implementation

Let us first consider the cipher's efficiency in hardware. Given a single 4-bit input $t = t_0 + 2t_1 + 4t_2 + 8t_3$, the output $u = u_0 + 2u_1 + 4u_2 + 8u_3$ of $S(t)$ for the

[3] S-box (14) enjoys the added property that S^6, S^{10}, S^{14}, and all odd powers of S also exhibit the $\frac{1}{2}$ bit-flip probability.

S-box (14) may be written as follows. Each line represents a single output bit as a function of four input bits.

$$\begin{aligned} u_3 &= t_1 \oplus (t_3 \wedge t_2 \wedge \bar{t}_0) \\ u_2 &= t_0 \oplus (t_3 \wedge \bar{t}_2 \wedge \bar{t}_1) \\ u_1 &= t_2 \oplus (t_3 \wedge t_1 \wedge t_0) \oplus (\bar{t}_3 \wedge \bar{t}_1 \wedge \bar{t}_0) \\ u_0 &= \bar{t}_3 \oplus (t_2 \wedge \bar{t}_1 \wedge t_0) \end{aligned} \tag{16}$$

Here, the bar denotes bit complement. Similarly, the bits of $S^2(t) = v_0 + 2v_1 + 4v_2 + 8v_3$ may be calculated as follows.

$$\begin{aligned} v_3 &= t_2 \oplus (\bar{t}_3 \wedge \bar{t}_1 \wedge \bar{t}_0) \\ v_2 &= \bar{t}_3 \oplus (t_2 \wedge \bar{t}_1 \wedge t_0) \oplus (\bar{t}_2 \wedge t_1 \wedge \bar{t}_0) \\ v_1 &= t_0 \oplus (\bar{t}_3 \wedge t_2 \wedge t_1) \\ v_0 &= \bar{t}_1 \oplus (t_3 \wedge t_2 \wedge \bar{t}_0) \oplus (\bar{t}_3 \wedge \bar{t}_2 \wedge t_0) \end{aligned} \tag{17}$$

We had deliberately chosen the S-box so that these expressions are simple.[4]

For sake of simplicity, let us assume that the logical operations NOT, AND, and XOR all take the same time in a hardware implementation. Even for a very straightforward implementation of (16) and (17), the critical path for the simultaneous calculation of $S(t)$ and $S^2(t)$ contains only 4 logical operations.

In most hardware implementations, this takes a lot shorter than the time required for a single 32-bit addition, an algebraic operation. Hence the calculation of α, given by (13), whose critical path consists of two 32-bit additions and a single XOR, will take longer than the S-box calculation. In all, the total time cost of the T-function given by Theorem 2 is two 32-bit additions and four logical bit operations.

The filter (15), taking two 32-bit additions, may be run in parallel to the T-function, so our cipher will produce 32-bits of keystream for every clock tick that allows for two 32-bit additions and four logical bit operations.

Thus a straightforward approach will give us a very fast hardware implementation. For example, in an ASIC implementation that uses a 32-bit adder of modest delay time 0.4ns, the cipher will run at somewhere around 32 Gbps. The total cost for such a rough implementation is given in Table 1.

For lack of a good hardware oriented streamcipher, let us try to compare this implementation cost with that of a summation generator on four 256-bit LFSRs. Results of [24] show that this may be broken within time complexity of $2^{77} \sim \binom{4\cdot 256}{3}^{\log_2 7}$. Even this weak summation generator needs 1024 flip-flops, just to get started on the four LFSRs. This will already be larger than what we have in Table 1.

[4] We have rejected S-boxes that contained wholly linear expressions. But even such S-boxes might be used when this cipher is better understood.

Table 1. Implementation const of TSC-1

register	128×(flip-flop)
S-box	32×(11 XOR, 22 AND, 20 NOT)
odd parameter	4×(32-bit addition), 32×(2 XOR, 3 AND)
filter	3×(32-bit addition)
the rest	32×(1 XOR, 8 AND, 1 NOT)

Actually, some tricks may be used to reduce the gates needed for S-box implementation without impacting speed. For example, if we use the expression

$$\begin{aligned} y &:= (\bar{t}_1 \wedge (t_2 \wedge t_0)) \oplus t_3 \\ z &:= \ \bar{t}_1 \vee (t_2 \vee t_0) \\ u_0 &= \bar{y} \\ v_2 &= y \oplus z, \end{aligned}$$

calculation of u_0 and v_2 may be done in 2 XOR, 2 AND, 2 OR, and 2 NOT, whereas, it was carelessly counted as 3 XOR, 6 AND, and 6 NOT in Table 1.

A very small but slower implementation might use just one or two 4×4 S-boxes, and implementations that come somewhere in between are also possible, allowing for a very wide range of implementation choices.

Although we have designed this cipher mainly for hardware, its performance in software is not bad. Using the standard bit-slice technique for S-boxes with the above polynomial expressions (16) and (17), we achieve speeds of up to 1.25 Gbps on a Pentium IV 2.4GHz, Windows XP (SP1) platform using Visual C++ 6.0(SP6). In comparison, the Crypto++ Library[1] cites a speed of 113 MBps for RC4 on a Pentium IV 2.1GHz machine. Scaled up for 2.4GHz, this is only 1.03 Gbps.

6 Conclusion

We have given an analysis of the generic single cycle T-function previously known. With a better understanding of this T-function, we were able to present a new class of single cycle T-functions.

Compared to the old T-function (Theorem 1), our T-function (Theorem 2) certainly gives a better *column mixing*. Also, unlike (8), which is based on the old T-function, the bit-flip probability of the register bits under the action of our new T-function construction can be manipulated, and even made equal to $\frac{1}{2}$, through proper selection of S and α.

On the other hand, unlike previous T-functions, our new T-function does not allow for the addition of an even parameter. This, we admit, is a very disappointing characteristic. But we would like to take the position that the multiplicative even parameter is the less understood part of previous T-function(Example 1), while being at the very core of its randomness. And as we saw in our example ciphers, the reduced randomness of the register contents can be compensated for by an appropriate use of the filter function.

We have also presented an example cipher which shows the possibility of using T-functions to build hardware oriented streamciphers. Our T-function allows for a wide range of implementation choices, so that the final cipher could be either fast or of small footprint in hardware.

Acknowledgments

We would like to thank Alexander Klimov and Adi Shamir for their valuable comments on an earlier version of this paper.

References

1. Crypto++ 5.2.1 benchmarks. Available from `http://www.eskimo.com/~weidai/benchmarks.html`
2. NIST. A statistical test suite for random and psedorandom number generators for cryptographic applications. NIST Special Publication 800-22.
3. V. S. Anashin, Uniformly distributed sequences over p-adic integers. Proceedings of the Intl Conference on Number Theoretic and Algebraic Methods in Computer Science (A. J. van der Poorten, I. Shparlinsky and H. G. Zimmer, eds.), World Scientific, 1995.
4. V. S. Anashin, private communication.
5. F. Armknecht, M. Krause, Algebraic attacks on combiners with memory, *Advances in Cryptology - Crypto 2003*, LNCS 2729, Springer-Verlag, pp.162–175, 2003.
6. M. Boesgaard, M. Besterager, T. Pedersen, J. Christiansen, and O. Scavenius, Rabbit: A new high-performance stream cipher. *FSE 2003*, LNCS 2887, pp.307–329, 2003.
7. S. Babbage, Stream ciphers: What does the industry want? Presented at *State of the Art of Stream Ciphers* workshop, Brugge, 2004.
8. A. Biryukov, A. Shamir. Cryptanalytic time/memory/data tradeoffs for stream ciphers. *Asiacrypt 2000,* LNCS 1976, pp. 1–13, Springer-Verlag, 2000.
9. K. Chen, M. Henricksen, W. Millan, J. Fuller, L. Simpson, E. Dawson, H. Lee, and S. Moon, Dragon: A fast word based stream cipher. To be presented at *ICISC 2004.*
10. A. Clark, E. Dawson, J. Fuller, J. Golić, H-J. Lee, W. Miller, S-J. Moon, and L. Simpson, The LILI-II Keystream Generator. *ACISP 2002.*
11. N. Courtois, Algebraic attacks on combiners with memory and several outputs, E-print archive, 2003/125. To be presented at ICISC 2004.
12. N. Courtois, Higher order correlation attacks, XL algorithm and Cryptanalysis of Toyocrypt, *ICISC 2002*, LNCS 2587, Springer-Verlag, pp.182–199, 2002.
13. N. Courtois, W. Meier, Algebraic attacks on stream ciphers with linear feedback, *Advances in Cryptology - Eurocrypt 2003*, LNCS 2656, Springer-Verlag, pp.345–359, 2003.

14. P. Ekdahl and T. Johansson, A new version of the stream cipher SNOW, *SAC 2002*, pp.47–61, 2002.
15. N. Ferguson, D. Whiting, B. Schneier, J. Kelsey, S. Lucks, and T. Kohno, Helix, fast encryption and authentication in a single cryptographic primitive, *FSE 2003*, 2003.
16. P. Hawkes and G. Rose, Primitive specification and supporting documentation for SOBER-t32, *NESSIE Submission*, 2000.
17. S. Halevi, D. Coppersmith, and C. Jutla, Scream: A software-efficient stream cipher, *FSE2002*, volume 2365 of LNCS, pages 195–209, Springer-Verlag, 2002.
18. J. Hong, D. H. Lee, Y. Yeom, and D. Han, A new class of single cycle T-functions and a stream cipher proposal. *SASC*(State of the Art of Stream Ciphers, Brugge, Belgium, Oct. 2004) workshop record. Available from `http://www.isg.rhul.ac.uk/research/projects/ecrypt/stvl/sasc.html`
19. P. Junod, S. Kuenzlie, and W. Meier, Attacks on TSC. FSE 2005 rump session presentation.
20. A. Klimov and A. Shamir, A new class of invertible mappings. *CHES 2002*, LNCS 2523, Springer-Verlag, pp.470–483, 2003.
21. A. Klimov and A. Shamir, Cryptographic application of T-functions. *SAC 2003*, LNCS 3006, Springer-Verlag, pp.248–261, 2004.
22. A. Klimov and A. Shamir, New cryptographic primitives based on multiword T-functions. *FSE 2004*, LNCS 3017, Springer-Verlag, pp.1–15, 2004.
23. A. Klimov and A. Shamir, The TFi family of stream ciphers. Handout at the *State of the Art of Stream Ciphers* workshop, Brugge, 2004.
24. D. H. Lee, J. Kim, J. Hong, J. W. Han, and D. Moon, Algebraic attacks on summation generators. *FSE 2004*, LNCS 3017, Springer-Verlag, pp.34–48, 2004
25. M. Matsui, Linear cryptanalysis method for DES cipher. *Eurocrypt '93*, LNCS 765, Springer-Verlag, pp.386–397, 1994.
26. D. McGrew and S. Fluhrer, The stream cipher LEVIATHAN. *NESSIE Submission*, 2000.
27. J. Mitra and P. Sarkar, Time-memory trade-off attacks on multiplications and T-functions. *Asiacrypt 2004*, LNCS 3329, Springer-Verlag, pp.468–482, 2004.
28. A. Shamir, Stream ciphers: Dead or alive? Invited talk presented at *State of the Art of Stream Ciphers* workshop, Brugge, 2004 and *Asiacrypt 2004*.
29. K. Sugimoto, T. Chikaraishi, and T. Morizumi, Design criteria and security evaluations on certain stream ciphers. *IEICE Technical Report*, ISEC20000-69, Sept. 2000.
30. D. Watanabe, S. Furuya, H. Yoshida, K. Takaragi, and B. Preneel, A new key stream generator MUGI, *FSE 2002*, pp.179–194, 2002.
31. H. Wu, A new stream cipher HC-256, *FSE 2004*, LNCS 3017, Springer-Verlag, pp.226–244, 2004.

A T-Function Based Streamcipher; TSC-2

A streamcipher based on Corollary 1 will be given in this section.[5] At the time of this writing, this example cipher is known to be insecure[19]. It is included in this paper for reference purposes only.

[5] This example cipher was presented at the SASC workshop. Readers may find more detail, including an example C-code, in [18].

Compared to TSC-1, this cipher example will be lighter in hardware at slightly reduced speed and faster in software. As with TSC-1, we will provide a filter function in addition to fixing various components for the T-function.

A.1 The Specification

Fix $n = 32$ and $m = 4$. Define an odd parameter by setting

$$\alpha(\mathbf{x}) = (p + 1) \oplus p \oplus 2s, \tag{18}$$

where

$$p = x_0 \wedge x_1 \wedge x_2 \wedge x_3 \quad \text{and} \quad s = x_0 + x_1 + x_2 + x_3.$$

Define a 4×4 S-box S as follows.

```
S[16] = {5,2,11,12,13,4,3,14,15,8,1,6,7,10,9,0};
```
(19)

One may check easily that this is a single cycle S-box. We can now define a single cycle T-function $\mathbf{T}$, through the use of Corollary 1.

To actually obtain the keystream, use the filter

$$f(\mathbf{x}) = (x_{0 \lll 11} + x_1)_{\lll 14} + (x_{0 \lll 13} + x_2)_{\lll 22} + (x_{0 \lll 12} + x_3) \tag{20}$$

on the memory bits, after each application of $\mathbf{T}$. This will give us a single 32-bit block of keystream per action of the T-function.

A.2 Naive Security

Most of the arguments of Section 5.2 carry over to TSC-2, word for word. But arguments concerning the bit-flip probability needs to be redone.

With a T-function following the construction of Corollary 1, it is difficult to obtain a $\frac{1}{2}$ bit-flip probability for all the memory bits. If this non-randomly characteristic were to show unfiltered in the final keystream, we could obtain a distinguishing attack. So we took care to make sure one word of memory displayed the $\frac{1}{2}$ bit-flip probability and used it to ensure that the final keystream showed the same characteristic. Let us explain this in more detail.

The S-box (19), in addition to meeting the single cycle property, satisfies the following.

- An even number is sent to an odd number and vice versa.

In other words, the LSB (among the four bits in a single column of memory) is flipped on every application of S. To see the bit-flip probability of $\mathbf{T}$ itself, we should next look at how often S is applied to each column.

Lemma 1. *The odd parameter* (18) *satisfies*

$$\left| \frac{1}{2} - \text{prob}_{\mathbf{x}}([\alpha(\mathbf{x})]_i = 1) \right| = \frac{1}{2^{4i}}$$

for all $i > 0$.

This lemma, which may be proved directly, tells us that except for in the lower few bits, each output bit of α is equal to 1 almost half of the time. Recalling description (12) of $\mathbf{T}$ together with the bit-flip characteristic of S, we conclude that bits of memory x_0 has bit-flip probability close to $\frac{1}{2}$ except at the lower few bits.

Now, the 32-bit addition operation, seen at each bit position, is an XOR of two input bits and a carry bit, so we may apply the Piling-up Lemma[25] to argue that for each $i = 1, 2, 3$, the bits of

$$x_{0\lll k} + x_i$$

will have bit-flip probability very close to $\frac{1}{2}$, except maybe at the points where lower bits of x_0 was used. What discrepancy these bits may show from changing one half of the time disappears, once again through the use of Piling-up Lemma, when these values are rotated relative to each other and added together to form the final output (20). The rotation, while allowing the mixing of lower bits of x_0 with higher bits, also gains independence of the XORed bits needed in applying the Piling-up Lemma.

Using the explicit probability stated by Lemma 1, we have checked that a straightforward distinguishing attack based on the bit-flip probability of (single $\mathbf{T}$ action on) register contents is not possible.

A.3 Implementation

As in TSC-1, the S-box (19) was chosen with its efficient implementation in mind. The mapping $t \mapsto t \oplus S(t)$ allows for an efficient bit slice realization. Also note that because of the even-odd exchange condition, the LSB of $t \oplus S(t)$ will always be 1, leaving only 3 bits to be calculated.

Hardware implementation of TSC-2 will be slightly slower compared to that of TSC-1, because the output filter now exhibits the critical path of three 32-bit additions. But we have halved the count of S-boxes to obtain a lower total implementation cost.

In software, TSC-2 runs on a Pentium-IV 2.4 GHz machine with code compiled using Visual C++ 6.0 (SP6) at speeds over 1.6 Gbps. This is over 1.6 times faster than the speed for RC4 given in [1].

F-FCSR: Design of a New Class of Stream Ciphers

François Arnault and Thierry P. Berger

LACO, Université de Limoges,
123 avenue A. Thomas, 87060 LIMOGES CEDEX, France
{arnault, thierry.berger}@unilim.fr

Abstract. In this paper we present a new class of stream ciphers based on a very simple mechanism. The heart of our method is a Feedback with Carry Shift Registers (FCSR) automaton. This automaton is very similar to the classical LFSR generators, except the fact that it performs operations with carries. Its properties are well mastered: proved period, non-degenerated states, good statistical properties, high non-linearity.

The only problem to use such an automaton directly is the fact that the mathematical structure (2-adic fraction) can be retrieved from few bits of its output using an analog of the Berlekamp-Massey algorithm.

To mask this structure, we propose to use a filter on the cells of the FCSR automaton. Due to the high non-linearity of this automaton, the best filter is simply a linear filter, that is a XOR on some internal states. We call such a generator a Filtered FCSR (F-FCSR) generator.

We propose four versions of our generator: the first uses a static filter with a single output at each iteration of the generator (F-FCSR-SF1). A second with an 8 bit output (F-FCSR-SF8). The third and the fourth are similar, but use a dynamic filter depending on the key (F-FCSR-DF1 and F-FCSR-DF8). We give limitations on the use of the static filter versions, in scope of the time/memory/data tradeoff attack.

These stream ciphers are very fast and efficient, especially for hardware implementations.

Keywords: stream cipher, pseudorandom generator, feedback with carry shift register, 2-adic fractions.

1 Introduction

Linear Feedback Shift Registers (LFSR) are the most popular tool used to design fast pseudorandom generators. Their properties are well known, among them the fact that the structure of an LFSR can be easily recovered from his output by the Berlekamp-Massey algorithm. Many methods have been used to thwart the Berlekamp-Massey attack because the high speed and simplicity of LFSRs are important benefits.

Feedback with Carry Shift Registers (FCSR) were introduced by M. Goresky and A. Klapper in [7]. They are very similar to classical Linear Feedback Shift

H. Gilbert and H. Handschuh (Eds.): FSE 2005, LNCS 3557, pp. 83–97, 2005.

Registers (LFSR) used in many pseudorandom generators. The main difference is the fact that the elementary additions are not additions modulo 2 but with propagation of carries. This generator is almost as simple and as fast as a LFSR generator. The mathematical model for FCSR is the one of rational 2-adic numbers (cf. [9, 10]). This model leads to proved results on period and non degeneration of internal states of the generator. It inherits the good statistical properties of LFSR sequences.

Unfortunately, as for the LFSR case, it is possible to recover the structure of a sequence generated by an FCSR (cf. [8, 2],[1]). To avoid this problem, we propose to use a filter on the cells of the FCSR automaton. Since this automaton has good non linear properties, the filter is simply a linear function, i.e. a XOR on some cells. This method is very efficient for practical implementations.

First we describe the FCSR automaton and recall the properties of its output. For applications, we propose an automaton with a key of 128 bits in the main register.

Then we present the different versions of our generator with a detailed security analysis in each case. For the F-FCSR-SF1 version, we show that the algebraic attack is not possible and we describe some dedicated attacks. For the proposed parameters, this attack is more expensive than the exhaustive one. The main restriction to the use of this version is the fact that the cost of the time/memory/data tradeoffs attack is $O(2^{98})$, which is less than the exhaustive attack.

With the F-FCSR-SF8 version, we explain how our automaton can be filtered in order to obtain an 8-bit output at each iteration. The problem on designing a good filter in that situation is discussed. This leads to some problems on its design. This is why we recommend to use the F-FCSR-DF8 version of our generator to perform a 8-bit output system with high level of security.

In the dynamic filter versions of our generator, we substitute to the static filter a dynamic one, i.e. depending on the secret initialization key. This method increases the cost of the time/memory/data tradeoffs attack. This cost becomes $O(2^{162})$ for a 128-bit key. Moreover this dynamic filter avoids all 2-adic and algebraic attacks. In particular for the 8-bit output version, it avoids some attacks on filter combinations. For practical applications, we propose to use the S-box of Rijndael in order to construct the dynamic filter. This method is very efficient, and generally, this box is already implemented.

In the last section, we explain how it is possible to use our generators as stream ciphers with IV mode of size 64 bits. The 128-bit key is used to initialize the main register, and the initial vector is used to initialize the carries register. For some dedicated applications, we also propose to use a key of 96 bits with an IV of 64 bits.

2 The FCSR Automaton

We first recall the properties of an FCSR automaton used to construct our pseudorandom generators: an FCSR automaton performs the division of two integers following the increasing powers of 2 in their binary decompositions.

This mechanism is directly related to the theory of 2-adic fractions. For more theoretical approach, the reader could refer to [11, 7].

The main results used here are the following:

- Any periodic binary sequence can be expressed as a 2-adic fraction p/q, where q is a negative odd integer and $0 \leq p < |q|$.
- Conversely, if a periodic binary sequence is generated from a 2-adic fraction p/q, then the period of this sequence is known and is exactly the order of 2 modulo q.
- It is easy to choose a prime number q such as the order of 2 is exactly $T = |q|-1$, and therefore the period generated by any initial value $0 < p < |q|$ is exactly T. So, in the rest of this paper, we suppose that q is such that $2^{128} < |q| < 2^{129}$ and that the condition on the order of 2 is always satisfied in order to guarantee a period greater than 2^{128}.
- If p and q are integers of "small size", i.e. 128 bits for p and 129 bits for q, the sequences p/q looks like random sequences of period T in terms of linear complexity (but it remains false for its 2-adic complexity (i.e. the size of q)).

From now, we suppose that the FCSR studied in this section verifies the following conditions: $q < 0 \leq p$, $p < -q$, $p = \sum_{i=0}^{k-1} p_i 2^i$, $q = 1 - 2d$ and $d = \sum_{i=0}^{k-1} d_i 2^i$.

p will be the initial (secret) state of the automaton whereas q will be the equivalent of the "feedback polynomial" of a classical LFSR.

2.1 Modelization of the Automaton

If q is defined as above, the FCSR generator with feedback prime q can be described as a circuit containing two registers:

- The main register M with k binary memories (one for each cell), where k is the bitlength of d, that is $2^{k-1} \leq d < 2^k$.
- The carry register C with ℓ binary memories (one for each cell with a $\boxplus$ at its left) where $\ell+1$ is the Hamming weight of d. Using the binary expansion $\sum_{i=0}^{k-1} d_i 2^i$ of d, we put $I_d = \{i \mid 0 \leq i \leq k-2 \text{ and } d_i = 1\}$. So $\ell = \#I_d$. We also put $d^* = d - 2^{k-1}$.

We will say that the main register contains the integer $m = \sum_{i=0}^{k-1} m_i 2^i$ when it contains the binary values $(m_0, \ldots, m_{k-1})$. The content m of the main register always satisfies $0 \leq m \leq 2^k - 1$. In order to use similar notation for the carry register, we can think of it as a k bit register where the $k-l$ bits of rank **not** in I_d are always 0. The content $c = \sum_{i \in I_d} c_i 2^i$ of the carry register always satisfies $0 \leq c \leq d^*$.

Example 1. *Let $q = -347$, so $d = 174 =$ 0xAE, $k = 8$ and $\ell = 4$. The following diagram shows these two registers:*

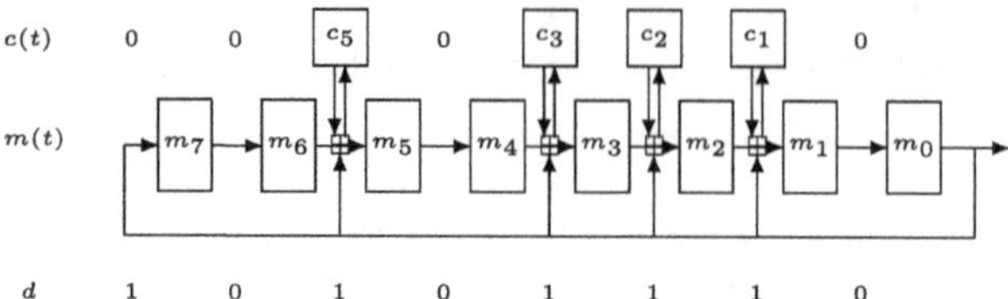

where ⊞ denotes the addition with carry, i.e., it corresponds to the following scheme in hardware:

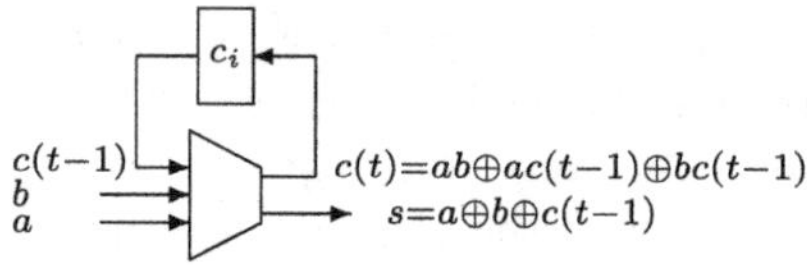

Transition Function. As described above, the FCSR circuit with feedback prime q is an automaton with 2^{k+l} states corresponding to the $k+l$ binary memories of main and carry registers. We say that the FCSR circuit is in state (m, c) if the main and carry registers contain respectively the binary expansion of m and of c.

Suppose that at time t, the FCSR circuit is in state $(m(t), c(t))$ with $m = \sum_{i=0}^{k-1} m_i(t)2^i$ and $c = \sum_{i=0}^{k-1} c_i(t)2^i$. The state $(m(t+1), c(t+1))$ at time $t+1$ is computed using:

- For $0 \le i \le k-2$ and $i \notin I_d$
  $m_i(t+1) := m_{i+1}(t)$
- For $0 \le i \le k-2$ and $i \in I_d$
  $m_i(t+1) := m_{i+1}(t) \oplus c_i(t) \oplus m_0(t)$
  $c_i(t+1) := m_{i+1}(t)c_i(t) \oplus c_i(t)m_0(t) \oplus m_0(t)m_{i+1}(t)$
- For the case $i = k-1$
  $m_{k-1}(t+1) := m_0(t)$.

Note that this transition function is described with (at most) quadratic boolean functions and that for all three cases $m_i(t+1)$ and $c_i(t+1)$ can be expressed with a single formula:

$$m_i(t+1) := m_{i+1}(t) \oplus d_i c_i(t) \oplus d_i m_0(t)$$

$$c_i(t+1) := m_{i+1}(t)c_i(t) \oplus c_i(t)m_0(t) \oplus m_0(t)m_{i+1}(t)$$

if we put $m_k(t) = 0$ and $c_i(t) = 0$ for i not in I_d.

We now study the sequences of values taken by the binary memories of the main register, that is the sequences $M_i = (m_i(t))_{t\in\mathbb{N}}$, for $0 \le i \le k-1$.

The main result is the following theorem:

Theorem 1. *Consider the FCSR automaton with (negative) feedback prime $q = 1-2d$. Let k be the bitlength of d. Then, for all i such that $0 \le i \le k-1$, there exists an integer p_i such that M_i is the 2-adic expansion of p_i/q. More precisely,* *these values p_i can be easily computed from the initial states $m_i(0)$ and $c_i(0)$ using the recursive following formulas:*

$$p_i = \begin{cases} qm_i(0) + 2p_{i+1} & \text{if } d_i = 0 \\ q\big(m_i(0) + 2c_i(0)\big) + 2(p_{i+1} + p_0) & \text{if } d_i = 1. \end{cases}$$

If we consider a prime divisor q such that the period is exactly $T = (|q|-1)/2$, the sequences M_i are distinct shifts of a same sequence (e.g. $1/q$), but each shift amount depends on the initial values of the main register and the carry register, and looks like random shifts on a sequence of period T (remember that, for applications $T \simeq 2^{128}$).

2.2 Hardware and Software Performances of the FCSR

2.2.1 Hardware Realization

As we have just seen before, we could directly implement in hardware the structure of an FCSR using a Galois architecture. Even if the needed number of gates is greater, the speed of such a structure is equivalent to the one of an LFSR.

2.2.2 Software Aspects

The transition function can also be described in pseudocode with the following global presentation expressing integers $m(t), c(t)$ instead of bits $m_i(t), c_i(t)$ more suitable for software implementations.

If $\oplus$ denotes bitwise addition without carries, $\otimes$ denotes bitwise AND, $shift_+$ the shift of one bit on the right, i.e. $shift_+(m) = \lfloor m(t)/2 \rfloor$ and *par* is the parity of a number m (1 if m is odd, 0 if it is even):

$$m(t+1) := shift_+(m(t)) \oplus c(t) \oplus par(m)d$$
$$c(t+1) := shift_+(m(t)) \otimes c(t) \oplus c(t) \otimes par(m)d \oplus par(m)d \otimes shift_+(m(t))$$

And the pseudoalgorithm could be written as:

```
b := par(m)      (boolean)
a := shift+(m)
m := a ⊕ c
c := a ⊗ c
if b = 1 then
      c := c ⊕ (m ⊗ d)
      m := m ⊕ d
end if
```

The number of cycles needed to implement the FCSR in software seems to be twice greater than the one required for an LFSR but as we will see in the following section, due to the very simplicity of our filtering function, the general speed in software of our Filtering FCSR might be more efficient than usual LFSR based generators.

2.2.3 Parameters of the FCSR Automaton for Designing the Stream Ciphers

For a cryptographic use with a security greater than 2^{128}, we recommend the use of a negative retroaction prime $-q$, corresponding to $k = 128$. This implies that $2^{128} < |q| < 2^{129} - 1$.

In order to maximize the period of the generated sequence, the order of 2 modulo q must be maximal i.e. equals to $|q|-1$. Moreover, to avoid some potential particular cases, we propose to choose a prime q such that $(|q| - 1)/2$ is also a prime.

The FCSR retroaction prime must be public. We propose

$$-q = 493877400643443608883820482007839438271 \quad (1)$$
$$= \text{0x1738D53D56FC4BFAD3D0C6D3430ADD893}$$

The binary expansion of $d = (|q| + 1)/2$ is
10111001 11000110 10101001 11101010 10110111 11100010 01011111 11010110
10011110 10000110 00110110 10011010 00011000 01010110 11101100 01001010.
Its Hamming weight is 69 and then there are $\ell = 68$ carry cells (the Hamming weight of $d^* = d - 2^{128}$) and $k = 128$ cells in the main register.

3 Design of F-FCSR : Filtered FCSR Automaton with a Static Filter

As for the LFSRs, a binary sequence generated by a single FCSR can not be used directly to produce a pseudorandom sequence (even if the output bits have good statistical properties and high linear complexity), since the initial state and the 2-adic structure can be recovered using a variant of the Berlekamp-Massey algorithm [8, 2]. So, we propose in this section to filter the output of an FCSR with two appropriate static functions and we prove the efficiency and the resistance against known attacks of those two constructions.

3.1 The F-FCSR-SF1 : One Output Bit

How to Filter An FCSR Automaton.
For the LFSR case many tools have been developed to mask the structure of the generator, by using boolean functions with suitable properties (see for example [12, 4]) to combine several LFSRs, by using combiners with memory or by shrinking the sequence produced by an LFSR.

It is possible to use similar methods with an FCSR generator, but with a very important difference: since an FCSR generator looks like a random generator for the non linear properties, it is not necessary to use a filter function with high non linearity.

Then the best functions for filtering an FCSR generator are linear functions:

$$f : GF(2)^n \to GF(2), \qquad f(x_1, \dots, x_n) = \bigoplus_{i=1}^{n} f_i x_i,\ f_i \in GF(2).$$

As studied previously, the sequence M_i observed on the i-th dividend register is a 2-adic fraction, with known period, good statistical properties and looks like a random sequence except from the point of view of 2-adic complexity.

The sequences C_i (with $i \in I_d^*$) produced by the carry register are not so good from a statistical point of view: these sequences are probably balanced, however, if a carry register is in the state 1 (resp. 0), it remains in the same state 1 (resp. 0) with a probability 3/4 since each of the two other entries of the corresponding addition box corresponds to 2-adic fractions and produces a 1 with a probability approximatively 1/2. It is sufficient to have only one more 1 to produce a 1 in the carry register.

These remarks lead to filter only on the k cells $m_i(t)$ of the main register, not on the cells of the carry register.

To modelize our linear filter, we consider a binary vector $F = (f_0, \ldots, f_{k-1})$ of length k.

The output sequence of our filtered FCSR is then

$$S = (s(t))_{t \in \mathbb{N}}, \quad \text{where } s(t) = \bigoplus_{i=1}^{k} f_i \cdot m_i(t).$$

The extraction of the output from the content of the main register M and the filter F can be done using the following algorithm:

$S := M \otimes F$
for $i := 6$ to 0 do
    $S := S \oplus shift_{+2^i}(S)$

Output: $par(S)$

It needs 7 shifts, 7 Xor and 1 And on 128-bit integers. So, the proposed F-FCSR is very efficient in hardware.

Design of the Static Filter for the F-FCSR-SF1 Stream Cipher. Let k_F be the integer such that $2^{k_F} \leq F < 2^{k_F+1}$. We will see in Paragraph 3.2 that it is possible to develop an attack on the initial key which needs 4^{k_F} trials.

If F is a power of 2, the output is a 2-adic sequence and is not resistant to 2-adic attacks. Moreover, if F is known, and its binary expansion contains few 1, the first equations of the algebraic attack are simpler, even if it is not possible to develop such an attack (cf. Paragraph 3.2).

A first natural solution would be to choose $F = 2^{128} - 1$, that is to xor all the cells of the main register. In this case, suppose that the output is $S = (s(t))_{t \in \mathbb{N}}$. It is easy to check that the sequence $S' = (s(t) + s(t+1))_{t \in \mathbb{N}}$ is the same that the one that would be obtained by xoring all the carry cells. Even if we do not know how to use this fact to develop a cryptanalysis, we prefer to use another filter for this reason.

In our application, we propose to choose $F = d = (|q|+1)/2$. With this filter, the output is the XOR of all cells of the main register which are just at the right of a carry cell. For the prime q proposed above in (1) the value of k_F is 128 and the Hamming weight of the filter is 69.

We propose a very simple initialization of the F-FCSR-SF1 generator: we choose a key K with 128 bits. The key K is used directly to initialize the main register M. The carry register is initialized at 0.

Statistical Properties of the Filtered Output. When two or more sequences are xored, the resulting sequence has good statistical properties as soon as one of the sequences is good, under the restriction that this sequence is not correlated with the other.

In our generator, each sequence is a 2-adic fraction with denominator q and has good statistical properties. The only problem is the fact that these sequences are not independent, since they are obtained by distinct shifts of the same periodic sequence. Note that the period of the sequence is very large ($T \geq 2^{127}$), and that a priori the 69 distinct shifts looks like random shifts. So the output sequence will have good statistical properties.

This hypothesis is comforted by the fact that our generator passes the NIST statistical test suite, as we checked.

3.2 Cryptanalysis of F-FCSR-SF1

3.2.1 2-Adic Cryptanalysis of F-FCSR-SF1

2-adic complexity of the XOR of two or more 2-adic integers
A priori, the XOR is not related with 2-adic operations (i.e. operations with carries), and then the sequence obtained by XORing two 2-adic fractions looks like a random sequence from the point of view of 2-adic complexity. Experiments support this assumption.

Moreover, due to the choice of q, in particular to the fact that $(|q|-1)/2$ is prime, the probability to have a high 2-adic complexity is greater than in the general case.

Let q be a negative prime such that 2 is of order $|q|-1$ modulo q. Consider the XOR $(p_1/q) \oplus (p_2/q)$ of the 2-adic expansions of two fractions with q as denominator and $0 < p_1, p_2 < |q|$. By Theorem 2, both summands are a sequence of period $|q|-1$ so the XOR is also a sequence of period $|q|-1$ (or dividing it). Can this latter sequence written also as a fraction p/q? (with $0 \leq p \leq q$ and possibly non reduced). Surely, the answer is yes in some cases (e.g. if $p_1 = p_2$). But in very most cases, the answer is no. Here is an heuristic argument to show this under the assumption that such an XOR gives a random sequence of period dividing $|q|-1$. The number of such sequences is $2^{|q|-1}$ and the number of sequences of the form $(p_1/q) \oplus (p_2/q)$ is at most $(|q|-1)^2/2$. So we can expect that the probability that the XOR can be written p/q is about $|q|^2/2^{|q|}$ which is very small. This remark extends to the XOR of $O(\ln|q|)$ summands.
A 2-adic attack

Theorem 2. *Assume that the filter F is known by the attacker and let k_F be an integer such that $F < 2^{k_F+1}$ (that is all cells selected by the filter belong to the rightmost k_F+1 cells of the main register). Then the attacker can discover the key of the generator at a cost $O(k^2 2^{k_F})$.*

We first state a lemma.

Lemma 1. *Assume that the attacker knows the initial values $m_i(0)$ for $0 \leq i < k_F$ (he also knows the initial values $c_i(0)$ for $0 \leq i < k_F$ which were assumed* *to be 0). Then he can compute the T first bits $m_{k_F}(t)$ (for $0 \leq t < T$) of the sequence M_{k_F} by observing the sequence outputted by the generator, in time $O(Tk_F)$.*

Proof : The attacker observes first $S(0) = \bigoplus_{i=0}^{k_F} f_i m_i(0)$. In this equality, the only unknown value is $m_{k_F}(0)$ so the attacker can compute it in time $O(k_F)$. For subsequent bits the method generalizes as follows.

Assume that the attacker has computed bits $m_{k_F}(t)$ for $0 \leq t < \tau$ and knows $m_i(t)$ and $c_i(t)$ for $0 \leq t < \tau$ and $0 \leq i < k_F$. Observing the bit $S(\tau)$ he gets

$$S(\tau) = \bigoplus_{i=0}^{k_F} f_i m_i(\tau)$$

and the only unknown value here is $m_{k_F}(\tau)$. So the attacker obtains it, also in time $O(k_F)$. He can also compute $m_i(\tau+1)$ and $c_i(\tau+1)$ for $0 \leq i < k_F$, using the transition function. The time needed to compute these $2k_F$ bits is also $O(k_F)$. We obtain the result by induction. □

The attack whose existence is asserted in Theorem 2 works following six steps.

- Choose an arbitrary new set of values for the bits $m_i(0)$ for $0 \leq i < k_F$ and put $c_i(0) = 0$ for $0 \leq i < k_F$.
- Assuming that these bits correspond to the chosen values, compute the first k bits of the sequence M_{k_F}.
- Using the transition function, compute the first $k+k_F$ bits of the sequence M_0 from the assumed values for the bits $m_i(0)$ with $0 \leq i < k_F$ and the k bits obtained in the previous step.
- Multiply the integer $\sum_{t=0}^{k-1} m_0(t) 2^t$ by q modulo 2^k to obtain a candidate p_0 for the key.
- Run a simulation of the generator with the key p_0. Stop it after generating $k + k_F$ bits. Compare the last k_F bits obtained to the ones computed in Step 3. If they don't agree, the candidate found is not the true key. Return to first step until all possibilities are exhausted.
- After all possibilities in Step 1 are exhausted, use some more bits of the generator to determine which key is the true key, if more than one good candidate remains.

Now the proof of the theorem:

Proof : From Lemma 1, the cost of Step 2 is in $O(kk_F) \leq O(k^2)$. Step 3 has also a cost of $O(kk_F)$. The cost of Step 4 is $O(k^2)$ and those of Step 5 is $O(k(k + k_F)) \leq O(k^2)$. The loop defined by Step 1 has to be iterated $O(2^{k_F})$ times. Multiplying the number of iterations by the inner cost gives the cost of the whole attack. □

With our parameters $k = 128$ and $k_F = 127$, this attack is more expensive than the exhaustive attack on the key.

Moreover, if the carries are not initialized to 0, there are 196 unknowns in the system instead of 128.

3.2.2 Linear Complexity of F-FCSR-SF1 Generator: XOR of Two or More 2-Adic Integers

Arguments for the linear complexity are similar to those yet presented for the 2-adic complexity: since each 2-adic fraction looks like a random sequence from the point of view of linear complexity, the XOR of these sequences have a high linear complexity (cf. [17]). Experiments also support this assumption.

As for the 2-adic case, the particular value chosen for the period T helps for the 2-adic complexity to be high. Let q be a negative prime such that 2 is of order $|q|-1$ modulo q. Consider the XOR $(p_1/q)\oplus(p_2/q)$ of the 2-adic expansion of two fractions with q as denominator, and numerators such that $0 < p_1, p_2 < |q|$. Similar arguments as those above about the 2-adic behavior of this XOR applies to its linear behavior.

If this XOR corresponds to the expansion of a series $P(X)/Q(X)$ (written as a fraction in reduced form), then the order of the polynomial Q must be a divisor of $T = |q| - 1$. With the value of q proposed in (1), the order of Q must be 1, 2, T, or $T/2$. The only polynomials of order 1 or 2 are the powers of $(X + 1)$. Polynomials of order T or $T/2$ must have an irreducible factor Q_1 of order T or $T/2$. But this order must be a divisor of $2^{\deg(Q_1)} - 1$, so $\deg(Q_1)$ is a multiple of the order of 2 modulo q. In the case of the above value of q, this order is $T/2$, a number of bitsize 127. Hence polynomials Q with a divisor of such a degree are not so frequent.

3.2.3 Algebraic Cryptanalysis of F-FCSR-SF1

The algebraic cryptanalysis of a pseudorandom generator is a tool developed recently (cf.[5]).

The principle is simple: we consider the bits of the initial state $m = (m_0, \ldots, m_{k-1}) = (m_0(0) \ldots, m_{k-1}(0))$ as the set of unknowns (suppose first that the initial value of the carry register is 0) and, using the transition function, we compute the successive outputs of the generator as functions of these unknowns $f_i(m_0, \ldots, m_{k-1})$. If the attacker knows the first output bits of the generator, he gets a system of (non linear) equations in k variables. We can add to this system the equations $m_i^2 = m_i$ as the unknowns are Booleans. If the system obtained is not too complicated, it can be solved using for example the Gröbner basis methods [6].

The transition function of an FCSR automaton is quadratic: the first equation is linear on 128 variables (or 196 variables if the carries are not initialized to 0), the second one is quadratic, the third is of degree 3, and so on. For example, the eleventh equation is of degree 11 in 128 variables, its size is about 2^{50} monomials and is not computable. To solve the algebraic system, we need at least 128 equations.

Note that the fact we use a known filter does not increase the difficulty of this attack. The filter is just a firewall against a 2-adic cryptanalysis.

3.2.4 The Time/Memory/Data Tradeoff Attack

There exists a recent attack on stream ciphers with inner states: the time/memory/data tradeoff attack [3]. The cost of this attack is $O(2^{n/2})$, where n is the number of inner states of the stream cipher. This cost reflects not only the time needed for the attack, but also the use of memory and the amount of data required. For the F-FCSR-SF1, the number of inner states is $n = k + \ell = 128 + 68 = 196$. Even if this attack remains impracticable, it is faster than the exhaustive one. This is why we recommend to use the dynamic filter method.

3.3 Design of F-FCSR-SF8: A Static Filter and an 8-bit Output

In order to increase the speed of the generator, we propose to use several filters to get several bits at each transition of the FCSR automaton. For example, using 8 distinct filters, it is possible to obtain an 8-bit output at each transition. However, the design of several filters may be difficult.

A first Cryptanalysis on Multiple Filters. Suppose that we use 8 filters F_1, ..., F_8 on the same state of main register M. Obviously, each of these filters must be resistant to the 2-adic attack. These 8 filters must be linearly independent to avoid a linear dependency on the 8 outputs. Moreover, by linear combinations of the 8 filters, it is possible to obtain 2^8 filters, each of them must also be resistant to the 2-adic attack.

Let $\mathcal{C}$ be the binary linear code generated by F_1, ..., F_8.

- The condition on the independence of the 8 filters is the fact that the dimension of $\mathcal{C}$ is exactly 8.
- For $F \in \mathcal{C}$, let k_F be the least integer such that $2^{k_F} > F$ (here F is viewed as an integer). The minimum over $\mathcal{C}$ of the values of k_F must be as larger as possible. Note that $min_{F \in \mathcal{C}, F \neq 0}\{k_F\} \leq k - 8 = 120$. If we choose $\mathcal{C}$ such that $min_{F \in \mathcal{C}, F \neq 0}\{k_F\} = 120$, the cost of the 2-adic attack is $O(120 \times 2^{120})$ which is approximatively the cost of the exhaustive attack.
  Note that it is easy to construct a code $\mathcal{C}$ satisfying this condition.
- We recommend to avoid the use of a code $\mathcal{C}$ with a small minimum distance d. Indeed, from a codeword of weight d, it is possible to construct a filter on d cells of the main register M. Even if we do not know how to design such an attack for $d \geq 2$, we suggest to choose $\mathcal{C}$ satisfying $d \geq 6$.

A Simple Way to Construct 8 Simultaneous Filters. In order to construct good filters with a very efficient method to extract the 8-bit output, we recommend the following method:

The filters are chosen with supports included in distinct sets. More precisely, for $i = 0$ to 7, $Supp(F_i) \subset \{j \mid j \equiv i \pmod 8\}$.

This construction ensures $dim(\mathcal{C}) = 8$, $min_{F \in \mathcal{C}, F \neq 0}\{k_F\} = min_i\{k_{F_i}\}$ and $d = min_i(w(F_i))$, where $w(F)$ is the Hamming weight of F. Moreover the extraction procedure becomes very simple:
First, set $F = \bigoplus_{i=0}^{7} F_i$. The extraction of the 8-bit output from the content of the main register M and the filter F can be done using the following algorithm:
$S := M \otimes F$
for $i := 6$ to 3 do
  $S := S \oplus shift_{+2^i}(S)$
Output: $S \otimes 255$ (the 8 lower weight bits of S)

This needs 4 shifts, 4 Xor and 2 And on 128-bit integers. This extraction is faster than the extraction of a single bit.

Note that conversely, from a 128-bit filter F, we obtain a family of 8-bit filters. As an example, for the value $F = d$ proposed for the F-FCSR-SF1 generator, we obtain a code $\mathcal{C}$ with $dim(\mathcal{C}) = 8$, $mink_F = 113$ and $d = 4$. For this choice of filter, it will be possible to design a 2-adic attack slightly more efficient than the exhaustive one.

A Possible Attack. Let $S(t) = (S_0(t), \ldots, S_7(t))$ be the 8-bit output at time t. Some entries selected by the filter on which depend $S_0(t+7)$, $S_1(t+6)$,..., $S_7(t)$ may be related. And the relations involved might be partially explicited when the state of the automaton is partially known.

So, even if we do not know how to design such an attack, we do not advice to use the 8-bit output generator with a static filter. The dynamic filter method presented in the next section will resist to such attack and will be preferred. We also propose to use an IV mode with the F-FCSR designs in order to have a high confidence on the security against be sure to resist to the different attacks.

4 Design of F-FCSR-DF1 and F-FCSR-DF8: Dynamic Filtered FCSR Stream Ciphers

Due to the fact that the filter is very simple and its quality is easy to check, it is possible to use a dynamic filter: the filter can be constructed as a function of the key K, and then, is not known by the attacker. As soon as the filter is not trivial ($F \neq 0$ and $F \neq 2^i$), it is not possible to use the algebraic attack, nor the attack exploiting the small binary size of F.

The construction of this dynamic filtered FCSR generator (DF-FCSR generator) is very simple: let g be a bijective map of $GF(2)^{128}$ onto itself. For a 128-bit key K, we construct the filter $F = g(K)$ and also we use the key to initialize the main register. The carry register is initialized at 0, since the attacker cannot find the equations for the algebraic attack.

The main interest of the use of a dynamic filter is the fact that the number n of inner state is increased of the size of the filter, i.e. $n = 2k\ell = 324$. The cost of the time/memory/data tradeoffs attack becomes higher than those of the exhaustive one.

4.1 Design of F-FCSR-DF1

This stream cipher is identical to F-FCSR-SF1 except the fact that the filter is dynamic.

We propose to use for g the Rijndael S-box (cf. [14, 15]). This S-box operates on bytes, and using it for each 16 bytes of a 128-bit key, we get a suitable function g.

It is suitable to add a quality test for the filter, for example by testing the binary size k_F of F and its Hamming weight $w(F)$. For example, if $k_F < 100$ or $w(F) < 40$, then we iterate g to obtain another filter with good quality.

The computation of this dynamic filter is very simple. The main advantages are to thwart completely the 2-adic attack (§3.2), the algebraic attack (§3.2) and to avoid the time/memory/data tradeoff attack.

However, until now, we do not find any attack faster than the exhaustive search against the static filter generator.

4.2 Design of F-FCSR-DF8

For the 8-bit output version, the use of a dynamic filter has also other justification: it avoids all possible attacks on the filter described in Paragraph 3.3.

For a practical use we recommend the following key loading procedure:

- Construction of the filter F from the 128-bit secret key K by applying the Rijndael S-box.
- Test the quality of the 8 subfilters extracted from F. Each of them must have an Hamming weight at least 6, and a binary size at least 100.
- Go to the first step until the test succeed.
- Use the key K to initialize the main register M. The carry register is initialized to 0.

The filter procedure is those of F-FCSR-SF8 (§3.3).

4.3 An Initial Vector Mode for F-FCSR Stream Ciphers

The IV Mode. There are several possibilities to add some initial vector IV to our generators. A first one will be to use it as filter F, where the main register is initialized with the key K and the carry register is initialized to 0.

In that case, we are in the situation of multiple known filters on the same initialization of the automaton. This method will be dangerous.

In fact, the good solution is to use always the same filter from a fixed key K with a static filter for 1 bit output and dynamic filter for 8-bit output. The IV is used to initialize the carry registers.

With our automaton, there are 68 bits in the carry register. It is easy to use them for IV of size 64. In order to avoid some problems related to the use of the same key K for the main register, we recommend to wait 6 cycles of the automaton before using an input after a change of IV. After these 6 cycles, every cell of the main register contains a value depending not only of K but also of IV.

We recommend to use the following protocol either with the F-FCSR-DF1 stream cipher, or with the F-FCSR-DF8 stream cipher:

Pseudocode:

1. $F := g(K)$ (dynamic construction of the filter).
2. $M := K$; $M := K$.
3. Clock 6 times the FCSR and discard the output.
4. Clock and filter the FCSR until the next change of IV.
5. If change of IV, return to step 2.

A Variant of Our Generator with a Key of Size 96 and Initial Vector of Size 64. For some purposes where the security is important only during a limited amount of time, it can be useful to define a variant with a smaller key-size (but with same IV-size). For that we propose to use the retroaction prime

$$q = -145992282562012510535118773123 = -0\text{x}1\text{D}7\text{B}9\text{FC}57\text{FE}19\text{AFEFEF}7\text{C}5\text{B}83$$

This prime has been selected according the following criteria. Its bit size is 97, so that d has bitsize 96. Also $(|q|-1)/2$ is prime. The order of 2 modulo $|q|-1$ is exactly $|q|-1$. And $d = $ 0xEBDCFE2BFF0CD7F7F7BE2DC2 has weight 65 so that there are 64 useful cells in the carries register.

Conclusion

We proposed a very fast pseudorandom generator, easy to implement especially in hardware (but also in software). It has good statistical properties and it is resistant to all known attacks. Its design can be compared to older generators (such as the summation generator [16]) for whose the heart has a linear structure, and is broken by a 2-adic device. Instead, our generator has a heart with a 2-adic structure which is destroyed by a linear filter. It might be of similar interest of these older generators (the summation generator is one of the best generator known) while being even easier to implement due to the simplicity of the filter.

Acknowledgments. Both authors would like to thank Anne Canteaut and Marine Minier for helpful comments and suggestions.

References

1. F. Arnault, T. Berger, and A. Necer. A new class of stream ciphers combining LFSR and FCSR architectures. In *Advances in Cryptology - INDOCRYPT 2002*, number 2551 in Lecture Notes in Computer Science, pp 22–33. Springer-Verlag, 2002.
2. F. Arnault, T.P. Berger, A. Necer. Feedback with Carry Shift Registers synthesis with the Euclidean Algorithm. *IEEE Trans. Inform. Theory*, Vol 50, n. 5, may 04, pp. 910–917
3. A. Biryukov and A. Shamir *Cryptanalytic time/memory/data tradeoffs for stream ciphers* LNCS 1976 (Asiacrypt 2000), pp 1–13, Springer, 2000.
4. D. Coppersmith, H Krawczyk, Y. Mansour. *The Shrinking Generator*, Lecture notes in computer science (**773**), Advances Cryptology, CRYPTO'93. Springer Verlag 1994, 22-39
5. N. Courtois, W. Meier *Algebraic attack on stream ciphers with linear feedback* LNCS 2656 (Eurocrypt'03), Springer, pp 345–359
6. J.C. Faugère *A new efficient algorithm for computing Gröbner bases without reduction to zero (F_5)* Proceedings of International Symposium on Symbolic and Algebraic Computation, ISSAC'02, Villeneuve d'Ascq, pp. 75–83

7. A. Klapper and M. Goresky. 2-adic shift registers, fast software encryption. In *Proc. of 1993 Cambridge Security Workshop*, volume 809 of *Lecture Notes in Computer Science*, pages 174–178, Cambridge, UK, 1994. Springer-Verlag.
8. A. Klapper and M. Goresky. Cryptanalysis based on 2-adic rational approximation. In *Advances in Cryptology, CRYPTO'95*, volume 963 of *Lecture Notes in Computer Science*, pages 262–274. Springer-Verlag, 1995.
9. A. Klapper and M. Goresky. Feedback shift registers, 2-adic span, and combiners with memory. *Journal of Cryptology*, 10:11–147, 1997.
10. A. Klapper and M. Goresky. Fibonacci and Galois representation of feedback with carry shift registers. *IEEE Trans. Inform. Theory*, 48:2826–2836, 2002.
11. N. Koblitz. *p-adic Numbers, p-adic analysis and Zeta-Functions.* Springer-Verlag, 1997.
12. Alfred J. Menezes, Paul C. van Oorschot and Scott A. Vanstone *Handbook of Applied Cryptography*, CRC Press, 1996.
13. *"A Statistical Test Suite for the Validation of Random Number Generators and Pseudo Random Number Generators for Cryptographic Applications"*,`http://csrc.nist.gov/rng/`
14. J. Daemen, V. Rijmen *The Block Cipher Rijndael*, Smart Card Research and Applications, LNCS 1820, J.-J. Quisquater and B. Schneier, Eds., Springer-Verlag, 2000, pp. 288-296.
15. `http://csrc.nist.gov/CryptoToolkit/aes/`
16. R.A. Rueppel, *Correlation immunity and the summation generator*, Lecture Notes in Computer Science (**218**), Advances in Cryptology, CRYPTO'85, Springer-Verlag 1985, 260-272.
17. R.A. Rueppel, *Linear complexity of random sequences*, Lecture Notes in Computer Science (**219**, Proc. of Eurocrypt'85, 167–188)

Cryptographically Significant Boolean Functions: Construction and Analysis in Terms of Algebraic Immunity

Deepak Kumar Dalai, Kishan Chand Gupta, and Subhamoy Maitra

Applied Statistics Unit, Indian Statistical Institute,
203, B T Road, Calcutta 700 108, India
{deepak_r, kishan_t, subho}@isical.ac.in

Abstract. Algebraic attack has recently become an important tool in cryptanalysing different stream and block cipher systems. A Boolean function, when used in some cryptosystem, should be designed properly to resist this kind of attack. The cryptographic property of a Boolean function, that resists algebraic attack, is known as Algebraic Immunity ($\mathcal{AI}$). So far, the attempt in designing Boolean functions with required algebraic immunity was only ad-hoc, i.e., the functions were designed keeping in mind the other cryptographic criteria, and then it has been checked whether it can provide good algebraic immunity too. For the first time, in this paper, we present a construction method to generate Boolean functions on n variables with highest possible algebraic immunity $\lceil \frac{n}{2} \rceil$. Such a function can be used in conjunction with (using direct sum) functions having other cryptographic properties.

In a different direction we identify that functions, having low degree subfunctions, are weak in terms of algebraic immunity and analyse some existing constructions from this viewpoint.

Keywords: Algebraic Attacks, Algebraic Immunity, Annihilators, Boolean Functions, Correlation Immunity, Nonlinearity.

1 Introduction

Recent literature shows that algebraic attack has gained a lot of attention in cryptanalysing stream and block cipher systems. The attack uses overdefined systems of multivariate equations to recover the secret key [1, 2, 10, 11, 12, 13, 14, 18, 17]. Given a Boolean function f on n-variables, different kinds of scenarios related to low degree multiples of f have been studied in [13, 18]. The core of the analysis is to find out minimum (or low) degree annihilators of f and $1 + f$, i.e., to find out minimum (or low) degree functions g_1, g_2 such that $f * g_1 = 0$ and $(1 + f) * g_2 = 0$. To mount the algebraic attack, one needs only the low degree linearly independent annihilators [13, 18] of $f, 1 + f$.

So far very little attempt has been made to provide construction of Boolean functions that can resist algebraic attacks. In [15], some existing construction

H. Gilbert and H. Handschuh (Eds.): FSE 2005, LNCS 3557, pp. 98–111, 2005.

methods have been analysed that can provide Boolean functions with some other cryptographic properties to see how good they are in terms of algebraic immunity.

Algebraic immunity of certain constructions have also been studied in [3, 4, 5]. In [5], the authors have proved that the algebraic immunity of the n-variable functions constructed by Tarannikov's method [21, 19] attain $\Omega(\sqrt{n})$ algebraic immunity. This presents a sharper result than what presented in [15] in terms of analysing Tarannikov's construction [21, 19]. Construction of cryptographically significant Boolean functions with improved algebraic immunity has also been presented in [7].

However, so far there is no existing construction method that can achieve maximum possible algebraic immunity. In this paper, for the first time, we provide a construction method where the algebraic immunity is the main concern. We show that given a Boolean function on $n - 2d$ variables having algebraic immunity 1, we can always construct a Boolean function on n variables with algebraic immunity at least $d + 1$. The construction is iterative in nature (a function with two more variables is constructed in each step) and we need to apply it d times to get an n-variable function from an $(n - 2d)$-variable initial function. We also show that the construction preserves the order of resiliency and increases the nonlinearity by more than 2^{2d} times in d-steps (as it can be seen as a direct sum of a function with good nonlinearity and resiliency with another function with good algebraic immunity). Also using our construction one can generate n-variable functions with highest possible algebraic immunity $\lceil \frac{n}{2} \rceil$ and good nonlinearity. For this one needs to start with 1 or 2-variable nonconstant functions.

Further, in a different direction, we show that if a Boolean function has low degree subfunctions then it is not good in terms of algebraic immunity. This result generalizes the analysis on Maiorana-McFarland type functions presented in [18]. Further our analysis answers some of the questions presented in [15] regarding the algebraic immunity of the functions presented in [20].

2 Preliminaries

A Boolean function on n variables may be viewed as a mapping from $\{0, 1\}^n$ into $\{0, 1\}$ and define B_n as the set of all n-variable Boolean functions. One of the standard representation of a Boolean function $f(x_1, \ldots, x_n)$ is by the output column of its *truth table*, i.e., a binary string of length 2^n,

$$f = [f(0, 0, \ldots, 0), f(1, 0, \ldots, 0), f(0, 1, \ldots, 0), \ldots, f(1, 1, \ldots, 1)].$$

The set of $x \in \{0, 1\}^n$ for which $f(x) = 1$ (respectively $f(x) = 0$) is called the onset (respectively offset), denoted by 1_f (respectively 0_f). We say that a Boolean function f is balanced if the truth table contains an equal number of 1's and 0's.

The Hamming weight of a binary string S is the number of ones in the string. This number is denoted by $wt(S)$. The Hamming distance between two strings,

S_1 and S_2 is denoted by $d(S_1, S_2)$ and is the number of places where S_1 and S_2 differ. Note that $d(S_1, S_2) = wt(S_1 + S_2)$ (by abuse of notation, we also use $+$ to denote the $GF(2)$ addition, i.e., the XOR). By $S_1||S_2$ we mean the concatenation of two strings. By $\overline{S}$ we mean the complement of the string S.

Any Boolean function has a unique representation as a multivariate polynomial over $GF(2)$, called the algebraic normal form (ANF),

$$f(x_1, \ldots, x_n) = a_0 + \sum_{1 \leq i \leq n} a_i x_i + \sum_{1 \leq i < j \leq n} a_{i,j} x_i x_j + \ldots + a_{1,2,\ldots,n} x_1 x_2 \ldots x_n,$$

where the coefficients $a_0, a_{i,j}, \ldots, a_{1,2,\ldots,n} \in \{0, 1\}$. The algebraic degree, $\deg(f)$, is the number of variables in the highest order term with non zero coefficient. A Boolean function is affine if there exists no term of degree > 1 in the ANF and the set of all affine functions is denoted A_n. An affine function with constant term equal to zero is called a linear function.

It is known that a Boolean function should be of high algebraic degree to be cryptographically secure [16]. Further, it has been identified recently, that it should not have a low degree multiple [13]. The algebraic attack (see [13, 18] and the references in these papers) is getting a lot of attention recently. To resist algebraic attacks, the Boolean functions used in the cryptosystems should be chosen properly. It is shown [13] that given any n-variable Boolean function f, it is always possible to get a Boolean function g with degree at most $\lceil \frac{n}{2} \rceil$ such that $f * g$ is of degree at most $\lceil \frac{n}{2} \rceil$. Here the functions are considered to be multivariate polynomials over $GF(2)$ and $f * g$ is the polynomial multiplication over $GF(2)$. Thus while choosing an f, the cryptosystem designer should be careful that it should not happen that degree of $f * g$ falls much below $\lceil \frac{n}{2} \rceil$.

Towards defining algebraic immunity [13, 18, 15], one needs to consider the multiples of both f and $1 + f$.

1. Take $f, g, h \in B_n$. Assume that there exists a nonzero function g of low degree such that $f * g = h$ or $(1 + f) * g = h$, where h is a nonzero function of low degree and without loss of generality, $\deg(g) \leq \deg(h)$. Among all such h's we denote the lowest degree h (may be more than one and then we take any one of them) by $ldgm_n(f)$.
2. Assume there exists a nonzero function g of low degree such that $f * g = 0$ or $(1 + f) * g = 0$. Among all such g's we denote the lowest degree g (may be more than one and then we take any one of them) by $ldga_n(f)$.

It can be checked that [18, 15] for $f \in B_n$, $\deg(ldgm_n(f)) = \deg(ldga_n(f))$ and in this line the following definition of algebraic immunity has been presented in [15].

Definition 1. *The algebraic immunity of an n-variable Boolean function f is denoted by* $\mathcal{AI}_n(f)$ *which is basically* $\deg(ldgm_n(f))$ *or* $\deg(ldga_n(f))$.

Later we also need the following definition related to the annihilator set of a function.

Definition 2. *Given $f \in B_n$, define $AN(f) = \{g \in B_n | g \text{ nonzero}, \ f * g = 0\}$.*

The nonlinearity of an n-variable function f is the minimum distance from the set of all n-variable affine functions, i.e.,

$$nl(f) = \min_{g \in A(n)} (d(f, g)).$$

Boolean functions used in crypto systems must have high nonlinearity to prevent linear attacks [16].

Many properties of Boolean functions can be described by the Walsh transform. Let $x = (x_1, \ldots, x_n)$ and $\omega = (\omega_1, \ldots, \omega_n)$ both belonging to $\{0,1\}^n$ and $x \cdot \omega = x_1\omega_1 + \ldots + x_n\omega_n$. Let $f(x)$ be a Boolean function on n variables. Then the *Walsh transform* of $f(x)$ is an integer valued function over $\{0,1\}^n$ which is defined as

$$W_f(\omega) = \sum_{x \in \{0,1\}^n} (-1)^{f(x) + x \cdot \omega}.$$

A Boolean function f is balanced iff $W_f(0) = 0$. The nonlinearity of f is given by $nl(f) = 2^{n-1} - \frac{1}{2} \max_{\omega \in \{0,1\}^n} |W_f(\omega)|$. Correlation immune functions and resilient functions are two important classes of Boolean functions. A function is m-resilient (respectively mth order correlation immune) iff its Walsh transform satisfies

$$W_f(\omega) = 0, \text{ for } 0 \le wt(\omega) \le m \text{ (respectively } 1 \le wt(\omega) \le m).$$

The paper is organized as follows. In the next section we present the construction and the following section discusses the analysis of algebraic immunity of a function in terms of the degree of its subfunctions.

3 Construction to Get $\mathcal{AI}$ as Required

In this section we present a construction to get Boolean function of $n+2$ variables with algebraic immunity $d + 2 \le \lceil \frac{n+2}{2} \rceil$. The construction is iterative in nature and it starts from an initial function of $n+2-2(d+1) = n-2d$ variables having algebraic immunity 1 (the minimum possible value). In each step, 2 variables are added and algebraic immunity gets increased by 1. Let us now formalize the construction.

Construction 1 . *Let $f \in B_n$ such that $f = E||F||G||H$ where $E, F, G, H \in B_{n-2}$. Let $n - 2d > 0$ and $d \ge 0$. Take an initial function $f_{n-2d} \in B_{n-2d}$ with $\mathcal{AI}_{n-2d}(f_{n-2d}) = 1$. Suppose after d-th step $f_n \in B_n$ has been constructed. The next function $f_{n+2} \in B_{n+2}$ is constructed in following manner:*

$$f_{n+2} = f_n||f_n||f_n||f_n^1, \text{ where } f^k = E^{k-1}||F^k||G^k||H^{k+1},$$

for any function f_j,

$$f_j^0 = f_j,$$

and for the initial function f_{n-2d},

$$f^s_{n-2d} = \overline{f}_{n-2d} \ (and\ \overline{f}^s_{n-2d} = f_{n-2d}),$$

for $s > 0$.

To understand the recursion in the Construction 1, we present an example up to some depths.

- $f^1_n = f_{n-2}||f^1_{n-2}||f^1_{n-2}||(f^1_{n-2})^2$ as
  $f_n = f_{n-2}||f_{n-2}||f_{n-2}||f^1_{n-2}$,
- $(f^1_{n-2})^2 = f^1_{n-4}||(f^1_{n-4})^2||(f^1_{n-4})^2||((f^1_{n-4})^2)^3$ as
  $f^1_{n-2} = f_{n-4}||f^1_{n-4}||f^1_{n-4}||(f^1_{n-4})^2$,
- $((f^1_{n-4})^2)^3 = (f^1_{n-6})^2||((f^1_{n-6})^2)^3||((f^1_{n-6})^2)^3||(((f^1_{n-6})^2)^3)^4$ as
  $(f^1_{n-4})^2 = f^1_{n-6}||(f^1_{n-6})^2||(f^1_{n-6})^2||((f^1_{n-6})^2)^3$.

This goes on unless we reach at the level of the $(n-2d)$-variable initial function. For $m \geq 2$, denote $((f^1)^{2\cdots})^m$ as $f^{1,m}$. As example, $((f^1_{n-6})^2)^3 = f^{1,3}_{n-6}$. Also, for notational consistency, we take $(f^1)^1 = f^1$ and $f^{1,0} = f^0 = f$.

Take an initial function f_l (the 2^l length binary string which is the truth table of the function) on l variables. Below we present the construction idea as truth table concatenation.

Step 1: $f_{l+2} = f_l f_l f_l \overline{f}_l$
Step 2: $f_{l+4} = f_{l+2} f_{l+2} f_{l+2} f_l \overline{f}_l \overline{f}_l f_l = f_l f_l f_l \overline{f}_l f_l f_l f_l \overline{f}_l f_l f_l f_l \overline{f}_l f_l \overline{f}_l \overline{f}_l f_l$
Step 3: $f_{l+6} = f_{l+4} f_{l+4} f_{l+4} f_{l+2} f_l \overline{f}_l \overline{f}_l f_l f_l \overline{f}_l \overline{f}_l f_l \overline{f}_l \overline{f}_l f_l f_l \overline{f}_l$
Step 4: $f_{l+8} = f_{l+6} f_{l+6} f_{l+6} f_{l+4} f_{l+2} f_l \overline{f}_l \overline{f}_l f_l f_l \overline{f}_l \overline{f}_l f_l \overline{f}_l f_l f_l \overline{f}_l f_{l+2} f_l \overline{f}_l \overline{f}_l f_l$
$f_l \overline{f}_l \overline{f}_l f_l \overline{f}_l f_l f_l \overline{f}_l f_l \overline{f}_l \overline{f}_l f_l \overline{f}_l \overline{f}_l f_l f_l \overline{f}_l \overline{f}_l f_l f_l \overline{f}_l f_l \overline{f}_l \overline{f}_l f_l$

Thus after the k-th step, the function f_{l+2k} is the concatenation of 2^{2k} numbers of f_l and $\overline{f_l} = 1 + f_l$. That is, the subfunctions of f_{l+2k} at $2k$-depth are only f_l and $\overline{f_l}$. That is, f_{l+2k} can be seen as direct sum of f_l and a $2k$-variable function.

To prove the main theorem we first present the following results. In the proofs we will use the fact that for any $f \in B_n$ and any subset $V \in \{0,1\}^n$, the restriction of any annihilator g of f to V is an annihilator of the restriction of f to V. For technical reasons (see also Remark 1 after the proof of Lemma 2), during our proofs we will encounter certain situations when degree of a function is < 0. As such functions cannot exist, we will replace those functions by 0 (function).

Lemma 1. *Consider that the function $f_{n+2} \in B_{n+2}$ has been generated by Construction 1 after $(d+1)$ many steps, $d \geq 1$, taking f_{n-2d} as the initial function. Take $g, h \in B_{n-2}$. We assume that if $g' \in AN(f_{n-2d+2j})$, $h' \in AN(f^1_{n-2d+2j})$ for $0 \leq j \leq d-1$ and $\deg(g'+h') \leq j-1$, then $g' = h' = 0$. If*

1. $g \in AN(f^{1,i}_{n-2})$ *and* $h \in AN(f^{1,i+1}_{n-2})$ *for any* i, $i \geq 1$ *and*
2. $\deg(g+h) \leq d-2-i$,

then $g = h = 0$.

Proof. We prove it by induction. For the base step $d = 1$. Here $\deg(g+h) \le 1 - 2 - i \le -2$ implies such a function cannot exist (see also Remark 1), i.e., $g + h$ is identically 0, which gives $g = h$.

Now $g \in AN(f_{n-2d}^{1,i})$ and $h \in AN(f_{n-2d}^{1,i+1})$. Since f_{n-2d} is the initial function, by Construction 1, $f_{n-2d}^{1,i+1} = (f_{n-2d}^{1,i})^{i+1} = \overline{f_{n-2d}^{1,i}}$. Hence $g \in AN(f_{n-2d}^{1,i})$ and $h \in AN(\overline{f_{n-2d}^{1,i}})$. Thus g, h, being nonzero, cannot be same. So $g = h = 0$. This proves the base step.

Now we prove the inductive step. Consider that the function $f_n \in B_n$ has been generated by Construction 1 after d many steps, $d \ge 1$, taking f_{n-2d} as the initial function. For any $g', h' \in B_{n-4}$ with $g' \in AN(f_{n-4}^{1,i})$ and $h' \in AN(f_{n-4}^{1,i+1})$ and for any i, $i \ge 1$, if $\deg(g' + h') \le (d-1) - 2 - i$, then $g' = h' = 0$.

Suppose that f_{n+2} is constructed by Construction 1 and there exists $g \in AN(f_{n-2}^{1,i})$ and $h \in AN(f_{n-2}^{1,i+1})$ with $\deg(g+h) \le d - 2 - i$. By construction, we have

$$f_{n-2}^{1,i} = f_{n-4}^{1,i-1} || f_{n-4}^{1,i} || f_{n-4}^{1,i} || f_{n-4}^{1,i+1},$$
$$f_{n-2}^{1,i+1} = f_{n-4}^{1,i} || f_{n-4}^{1,i+1} || f_{n-4}^{1,i+1} || f_{n-4}^{1,i+2}.$$

Take,

$$g = v_1 || v_2 || v_3 || v_4,$$
$$h = v_5 || v_6 || v_7 || v_8,$$

This gives, $v_1 \in AN(f_{n-4}^{1,i-1})$, $v_2, v_3, v_5 \in AN(f_{n-4}^{1,i})$, $v_4, v_6, v_7 \in AN(f_{n-4}^{1,i+1})$ and $v_8 \in AN(f_{n-4}^{1,i+2})$. Since $\deg(g+h) \le d - 2 - i$, from ANF of $g + h = (v_1 + v_5) + x_{n-3}(v_1 + v_5 + v_2 + v_6) + x_{n-2}(v_1 + v_5 + v_3 + v_7) + x_{n-3}x_{n-2}(v_1 + \cdots + v_8)$ we deduce the following.

- $\deg(v_1 + v_5) \le d - 2 - i = (d-1) - 2 - (i-1)$, implying that $v_1 = v_5 = 0$, for $i \ge 2$. For $i = 1$, we have $v_1 \in AN(f_{n-4})$, $v_5 \in AN(f_{n-4}^1)$ with $\deg(v_1 + v_5) \le d - 3$. Following the assumption in the statement of the lemma, we get $v_1 = v_5 = 0$.
- $\deg(v_2 + v_6) \le d - 3 - i = (d-1) - 2 - i$, implying that $v_2 = v_6 = 0$.
- $\deg(v_3 + v_7) \le d - 3 - i = (d-1) - 2 - i$, implying that $v_3 = v_7 = 0$.
- $\deg(v_4 + v_8) \le d - 4 - i = (d-1) - 2 - (i+1)$, implying that $v_4 = v_8 = 0$.

Hence we get $g = h = 0$ for $i \ge 1$. □

Lemma 2. *Consider that $f_{n+2} \in B_{n+2}$ has been generated using the Construction 1 after $(d+1)$-th step with initial function f_{n-2d}. Let $\mathcal{AI}(f_{n-2d+2i}) = i+1$ for $0 \le i \le d$. Consider that*

1. *$g_{n+2} \in AN(f_{n+2})$,*
2. *$\deg(g_{n+2}) \le d+1$, and*
3. *g_{n+2} is of the form $g_{n+2} = g_n + x_{n+2}x_{n+1}(g_n + h_n)$, where $g_n \in AN(f_n)$, $h_n \in AN(f_n^1)$.*

If $\deg(g_n + h_n) \leq d - 1$, *then* $g_n = h_n = 0$.

Proof. We will use induction on d. For the base step (i.e., $d = 0$) we have $f_n^1 = \overline{f}_n$ as f_n the initial function when $d = 0$. Here g_n and h_n are annihilators of f_n and $f_n^1 = \overline{f}_n$ respectively. Since $\deg(g_{n+2}) \leq 1$, and $g_{n+2} = g_n + x_{n+2}x_{n+1}(g_n + h_n)$, $g_n + h_n = 0$, which gives $g_n = h_n$. Since $g_n \in AN(f_n)$ and $h_n \in AN(\overline{f}_n)$, being non zero functions, they cannot be same, i.e., $g_n = h_n = 0$. Then $g_{n+2} = 0$.

Now we prove the inductive step. Assume the induction assumption holds till d steps, $d \geq 0$. Now we will prove the lemma statement at $(d+1)$-th step. That is $f_{n+2} \in B_{n+2}$ has been generated by Construction 1 after $(d+1)$-th step with $\mathcal{AI}(f_{n+2}) \leq d+1$ and $\mathcal{AI}(f_{n-2d+2i}) = i+1$ for $0 \leq i \leq d$. Here $g_{n+2} = g_n + x_{n+2}x_{n+1}(g_n + h_n) \in AN(f_{n+2})$ of degree $\leq d+1$, where $g_n \in AN(f_n)$ and $h_n \in AN(f_n^1)$. Suppose $\deg(g_n + h_n) \leq d - 1$. Then here, we will prove that $g_n = h_n = 0$. Here

$$f_n = f_{n-2}||f_{n-2}||f_{n-2}||f_{n-2}^1,$$
$$f_n^1 = f_{n-2}||f_{n-2}^1||f_{n-2}^1||(f_{n-2}^1)^2.$$

Let

$$g_n = A||B||C||D,$$
$$h_n = E||F||G||H, \text{ where}$$

$A, B, C, E \in AN(f_{n-2})$, $D, F, G \in AN(f_{n-2}^1)$ and $H \in AN((f_{n-2}^1)^2)$. Since $A, E \in AN(f_{n-2})$, we have $A + E \in AN(f_{n-2})$ and hence $\deg(A + E) \geq d$ or $A + E = 0$. Since $\deg(g_n + h_n) \leq d - 1$, $A + E = 0$. Then $\deg(B + F) \leq d - 2$ and $\deg(C + G) \leq d - 2$. Thus, using the induction hypothesis we have $B = C = F = G = 0$. So,

$$g_n = A||0||0||D,$$
$$h_n = E||0||0||H.$$

So, $\deg(D + H) \leq d - 3 = d - 2 - 1$.

We have assumed the inductive steps upto d-th step. That gives that if $g_{n-2d+2i} \in AN(f_{n-2d+2i})$, $h_{n-2d+2i} \in AN(f_{n-2d+2i}^1)$ for $0 \leq i \leq d - 1$ and $\deg(g_{n-2d+2i} + h_{n-2d+2i}) \leq i$, then $g_{n-2d+2i} = h_{n-2d+2i} = 0$. Note that, this satisfies the assumption considered in the statement of Lemma 1 and now we can apply it.

Since $D \in AN(f_{n-2}^1)$ and $H \in AN((f_{n-2}^1)^2)$ with $\deg(D + H) \leq d - 3$, following Lemma 1 we get $D = H = 0$. So we have $g_n = A||0||0||0$, $h_n = E||0||0||0$, and hence

$$\begin{aligned} g_{n+2} &= g_n + x_{n+2}x_{n+1}(g_n + h_n) \\ &= (1 + x_{n-1} + x_n + x_{n-1}x_n)A \\ &\quad + x_{n+1}x_{n+2}((1 + x_{n-1} + x_n + x_{n-1}x_n)(A + E)), \end{aligned}$$

i.e., $g_{n+2} = (1 + x_{n-1} + x_n + x_{n-1}x_n)A$, since $A + E = 0$. Then $\deg(g_{n+2}) \geq d+2$, since $\deg(A) \geq d$ as $A \in AN(f_{n-2})$. As $\deg(g_{n+2}) \leq d+1$, we have $A = 0$. This gives the proof. □

Remark 1. In the proof of Lemma 2 above, if $\deg(D + H) \le d - 3 < 0$, we have $(D + H) = 0$, because here $g_{n+2} = g_n + x_{n+2}x_{n+1}x_n x_{n-1}(D + H)$ and $\deg(g_{n+2}) \le d + 1$. Since there is no negative degree function, we have to take the term $x_{n+2}x_{n+1}x_n x_{n-1}(D + H)$ as 0.

Now we present the main result.

Theorem 1. *Refer to Construction 1. Let the algebraic immunity of the initial function f_{n-2d} be 1. Then after $(d + 1)$-th step the algebraic immunity of the constructed function f_{n+2} is $d + 2$.*

Proof. We have to prove that any nonzero function g_{n+2} such that $g_{n+2}f_{n+2} = 0$ has degree at least $d+2$. Suppose that such a function g_{n+2} with degree at most $d + 1$ exists. Then, g_{n+2} can be decomposed as

$$g_{n+2} = g_n || g'_n || g''_n || h_n,$$

where $g_n, g'_n, g''_n \in AN(f_n)$, and $h_n \in AN(f_n^1)$. The algebraic normal form of g_{n+2} is then

$$\begin{aligned} g_{n+2}(x_1, \ldots, x_{n+2}) = g_n + x_{n+1}(g_n + g'_n) + x_{n+2}(g_n + g''_n) \\ +x_{n+1}x_{n+2}(g_n + g'_n + g''_n + h_n) \ . \end{aligned}$$

If g_{n+2} has degree at most $d + 1$, then $(g_n + g'_n)$ and $(g_n + g''_n)$ have degree at most d. Because both functions lie in $AN(f_n)$ and $\mathcal{AI}(f_n) = d + 1$, we deduce that $g_n + g'_n = 0$ and $g_n + g''_n = 0$, which give, $g_n = g'_n = g''_n$. Therefore, $g_{n+2} = g_n + x_{n+1}x_{n+2}(g_n + h_n)$. So, $\deg(g_n + h_n) \le d - 1$. Now following the Lemma 2 we have $g_n = h_n = 0$, that gives, $g_{n+2} = 0$.

Similarly one can check that there cannot be any nonzero annihilator of $1 + f_{n+2}$ having degree $\le d + 1$. This completes the proof. □

Using this Construction 1, one can generate a function on n variables whose algebraic immunity is the highest possible, i.e., $\lceil \frac{n}{2} \rceil$. In this case one has to start from 1 or 2-variable nonconstant function. Then after each step we will get a function on two more variables and the algebraic immunity will increase by 1.

Example 1. First we present the case for odd n. One can start from $f_1 = x_1$.

Step 1: $f_1 = 01$
Step 2: $f_3 = f_1 f_1 f_1 \overline{f}_1 = 01010110$
Step 3: $f_5 = f_3 f_3 f_3 01101001 = 01010110010101100101011001101001$
Step 4: $f_7 = f_5 f_5 f_5 01010110011010010110100110010110$
Step 5: $f_9 = f_7 f_7 f_7 f_5 01010110011010010110100110010110010110$
$01101001011010011001011001101001100101101001011001101001$

Then we present the case for even n. One can start from nonlinear function $f_2 = x_1 x_2$ as the initial function.

Step 1: $f_2 = 0001$
Step 2: $f_4 = f_2 f_2 f_2 \overline{f}_2 = 0001000100011110$
Step 3: $f_6 = f_4 f_4 f_4 f_2 111011100001$
Step 4: $f_8 = f_6 f_6 f_6 f_4 f_2 111011100001 f_2 11101110000111100001000111110$
Step 5: $f_{10} = f_8 f_8 f_8 f_6 f_4 f_2 111011100001 f_2 11101110000111100001000111110$
$f_4 f_2 111011100001 f_2 11101110000111100001000111110 f_2 11101110$
$0001111000010001111011100001000111100001111011100001$

Note that the algebraic immunity stays the same if a function is subjected to linear transformation on input variables. Thus, taking any function presented in the above example, one can apply linear transformation to get number of functions. Further the nonlinearity and algebraic degree also stays same after linear transformation.

Now we will discuss some other cryptographic properties of the functions generated using Construction 1 after k-th step.

Corollary 1. *Let $f_{d+2k} \in B_{d+2k}$ is constructed by Construction 1 taking $f_d \in B_d$ as the initial function, i.e., $f_{d+2k} = f_d + \phi_{2k}$, the direct sum.*

1. $nl(f_{d+2k}) = 2^d nl(\phi_{2k}) + 2^{2k} nl(f_d) - 2nl(\phi_{2k})nl(f_d) > 4^k nl(f_d)$.
2. *Let f_d be an r-resilient function. Then f_{d+2k} is also r-resilient.*
3. $\deg(f_{d+2k}) = \max\{\deg(f_d), \deg(\phi_{2k})\}$.

Proof. The proof of item 1 follows from [20–Proposition 1(d)] and the proof of item 2 follows from [20–Proposition 1(c)]. The result related to algebraic degree is also easy to see. □

In Item 1 of Corollary 1 we are using $nl(\phi_{2k})$. We have observed that $nl(\phi_{2k})$ is equal to the number of 1's in its truth table. We have checked that the values of $nl(\phi_{2k})$ are 1, 5, 22, 93, 386, 1586, 6476, 26333 for $k = 1, \ldots, 8$. Using this, here we present the nonlinearity of the functions given in Example 1. The initial function is the $f_1 = x_1$ which is a linear function. So, $nl(f_1) = 0$. Therefore, $nl(f_3) = 2, nl(f_5) = 10, nl(f_7) = 44, nl(f_9) = 186, nl(f_{11}) = 772, nl(f_{13}) = 3172, nl(f_{15}) = 12952, nl(f_{17}) = 52666$. Similarly if one starts with a 5-variable 1-resilient function with nonlinearity 12, one gets a 7-variable 1-resilient function with nonlinearity 56 (as $nl(\phi_2) = 1$), then a 9-variable 1-resilient function with nonlinearity 232 (as $nl(\phi_4) = 5$) and so on. We like to point out once again that the nonlinearity remains very good in this construction and the order of resiliency is also not disturbed as it is a direct sum construction of a function f_d with good properties in terms of nonlinearity and resiliency and a function ϕ_{2k} which is good in terms of algebraic immunity. When the weight (also nonlinearity) of the function ϕ_{2k} is odd, then clearly its algebraic degree is $2k$. We have also checked upto $k = 6$, that when the weight (also nonlinearity) is even then the algebraic degree is $2k - 1$. The exact nonlinearity and algebraic degree of ϕ_{2k} is still open at this stage and we are working on it. Certain ideas in this area have also been provided by Carlet [9].

Note that if one starts with an initial function $f_{n-2d} \in B_{n-2d}$ having algebraic immunity D, it is not guaranteed that after d steps f_n will have algebraic immunity $d + D$; the only guarantee is that it will be $\geq d + 1$ (following similar arguement as in the proof of Theorem 1). It will be interesting to see what is the exact algebraic immunity of f_n.

4 Functions with Low Degree Subfunctions

In this section we discuss why a Boolean function with low degree subfunction is not good in terms of algebraic immunity. This result is a generalization of the result presented in [18], where the authors have shown that certain kind of Maiorana-McFarland constructions are not good in terms of algebraic immunity.

Proposition 1. *Let $f \in B_n$. Let $g \in B_{n-r}$ be a subfunction of $f(x_1, \ldots, x_n)$ after fixing r many distinct inputs $x_{i_1}, \ldots, x_{i_r} \in \{x_1, \ldots, x_n\}$. If the algebraic degree of g is d, then $\mathcal{AI}_n(f) \leq d + r$.*

Proof. Let $x_{i_1}, \ldots, x_{i_r}$ are fixed at the values $a_{i_1}, \ldots, a_{i_r} \in \{0, 1\}$. Thus g is a function on the variables $\{x_1, \ldots, x_n\} \setminus \{x_{i_1}, \ldots, x_{i_r}\}$. It can be checked that $(1 + a_{i_1} + x_{i_1}) \ldots (1 + a_{i_r} + x_{i_r})(1 + g)$ is an annihilator of f. The algebraic degree of $(1 + a_{i_1} + x_{i_1}) \ldots (1 + a_{i_r} + x_{i_r})(1 + g)$ is $d + r$. Thus the result. □

The Maiorana-McFarland construction can be seen as concatenation of affine functions on $n - r$ variables to construct an n-variable functions. Clearly we have affine subfunctions of the constructed function in this case and hence $\deg(g) = 1$ following the notation of Proposition 1. Thus there will be annihilators of degree $1 + r$. Note that if r is small, then one can get annihilators at low degree [18–Theorem 2, Example 1]. This situation for Maiorana-McFarland construction is only a subcase of our proposition. Our result works on any function, it need not be of Maiorana-McFarland type only. We present an example below.

Example 2. Let us consider a 20-variable function, with a subfunction of degree 2 on 17-variables, i.e., we fix 3 inputs. In that case the 20-variable function will have an annihilator at degree $2 + 3 = 5$.

Maiorana-McFarland type of constructions are used in design of resilient functions. One idea in this direction is to concatenate k-variable affine functions (repetition may be allowed) non degenerate on at least $m + 1$ variables to generate an m-resilient function f on n-variables. For such a function f, it is easy to find an annihilator g of degree $n - k + 1$ as described in [18]. However, it should be noted that in construction of resilient functions, there are lot of techniques [20] that use concatenation of k-variable affine functions where $k < \frac{n}{2}$. In such a case, the annihilators described in [18–Theorem 2] will be of degree greater than $\frac{n}{2}$ and will not be of practical use as there are other annihilators of degree $\leq \frac{n}{2}$ which are not of the form given in [18–Theorem 2]. We will show that even in such a case, Proposition 1 can provide further insight. We will show that a well known construction of resilient function [20–Theorem 10(b)] on n-variables (n odd) can never achieve the algebraic immunity $\lceil \frac{n}{2} \rceil$. At the best, it can only achieve the value $\lfloor \frac{n}{2} \rfloor$. To explain this construction we briefly present some notations from [20].

Take a bit b and a bit string $s = s_0 \ldots s_{n-1}$. Then the string b AND $s = s'_0 \ldots s'_{n-1}$, where $s'_i = b$ AND s_i. Take two bit strings $x = x_0 \ldots x_{n-1}$ and $y = y_0 \ldots y_{m-1}$. The Kronecker product $x \otimes y = (x_0 \text{ AND } y) \ldots (x_{n-1} \text{ AND } y)$, which is a string of length nm. The direct sum of two bit strings x, y is $x\$y =$

$(x \otimes y^c) \oplus (x^c \otimes y)$, where x^c, y^c are bitwise complement of x, y respectively. As an example presented in [20], if $f = 01$, and $g = 0110$, then $f\$g = 01101001$. Now we present the construction for $(2p+1, 1, 2p-1, 2^{2p} - 2^p)$ function as presented in [20] for $p \geq 4$.

Construction 2 . *[20–Theorem 10(b)] Let $\lambda_1, \lambda_2, \lambda_3, \lambda_4$ be the 3-variable linear functions non degenerate on two variables (i.e., the functions $x_1 + x_2, x_2 + x_3, x_1 + x_3, x_1 + x_2 + x_3$) and μ_1, μ_2, μ_3 be the 3-variable linear functions non degenerate on 1 variable (i.e., the functions x_1, x_2, x_3). Let g_i be the concatenation of the 3-variable function μ_i and its complement μ_i^c, for $1 \leq i \leq 3$. That is g_i's are basically 4-variable functions. Let h_1, h_2 be bent functions on $2p-4$ variables, and h_3, h_4, h_5 be bent functions of $2p-6$ variables and h_6, h_7 be two strings of lengths $2^{2p-6}+1$ and $2^{2p-6}-1$ which are prepared by properly adding and removing 1 bit from the truth table of $(2p-6)$-variable bent functions respectively. Let f be a concatenation of the following sequence of functions. $h_1\$\lambda_1, h_2\$\lambda_2, h_3\$g_1, h_4\$g_2, h_5\$g_3, h_6\$\lambda_3, h_7\$\lambda_4$. This is a $(2p+1, 1, 2p-1, 2^{2p} - 2^p)$ function.*

Proposition 2. *The $(2p+1)$-variable function presented in Construction 2 has a subfunction of degree at most $p-1$ when $x_{2p+1} = 0$.*

Proof. Consider the subfunction when $x_{2p+1} = 0$. The subfunction (call it g) in concatenation form is $h_1\$\lambda_1, h_2\$\lambda_2$. Since h_1, h_2 are bent functions on $2p-4$ variables, they can have algebraic degree at most $p-2$. Further λ_1, λ_2 are 3-variable linear functions. The algebraic normal form of g is $(1 + x_{2p})(h_1 + \lambda_1) + x_{2p}(h_2 + \lambda_2)$. So the degree of g is $\leq 1 + (p-2) = p-1$. □

Theorem 2. *For a function $f \in B_n$ (n odd) generated out of Construction 2, $\mathcal{AI}_n(f) \leq \lfloor \frac{n}{2} \rfloor$.*

Proof. Here $n = 2p+1$. We take $g \in B_{n-1}$, i.e., $r = 1$ according to Proposition 1. Further from Proposition 2, $\deg(g) \leq p-1 = \frac{n-1}{2} - 1$. Thus, $\mathcal{AI}_n(f) \leq \frac{n-1}{2} - 1 + 1 = \lfloor \frac{n}{2} \rfloor$. □

Thus using our technique we can show that the construction proposed in [20–Theorem 10(b)] can not achieve the maximum possible algebraic immunity $\lceil \frac{n}{2} \rceil$. The maximum value it can achieve is $\leq \lfloor \frac{n}{2} \rfloor$. This can be seen only by Proposition 1 which generalizes the result of [18–Theorem 2, Example 1]. This also answers a question presented in [15–Example 2] for $n = 9$. There Construction 2 has been exploited for $p = 4$ and the functions constructed are as follows.

1. $h_1 = 0000010100110110, h_2 = 0000010100110110, h_3 = 0001, h_4 = 0001, h_5 = 0001, h_6 = 00010, h_7 = 001$. In this case, one gets a $(9, 1, 7, 240)$ function f_1 with $\mathcal{AI}_9(f_1) = 3$.
2. If one changes $h_2 = 0000010100110110$ by $h_2 = 0000010100111001$, then we get a $(9, 1, 7, 240)$ function f_2 with $\mathcal{AI}_9(f_2) = 4$.

The question raised in [15] was why the algebraic immunity of these two function are different? The reason is in the first case the functions h_1, h_2 are same with the ANF $x_1x_3+x_2x_4$. Thus the subfunction g (i.e., $h_1\$\lambda_1, h_2\$\lambda_2$) is a degree 2 function. So the maximum algebraic immunity, according to Proposition 1 can be $2+1=3$. That is the value achieved in [15]. In the second case, h_1 is different from h_2 and the algebraic degree of g (i.e., $h_1\$\lambda_1, h_2\$\lambda_2$) becomes 3 and it achieves the value $3+1=4$. Thus Proposition 1 helps in answering this question. It is important to note that this technique can be employed to study the upper bound of algebraic immunity for various constructions by analysing their subfunctions and in particular, directly for the constructions proposed in [20, 6].

It should be noted that the converse of Proposition 1 is not always true. That is, a function having low degree annihilator does not imply it always has some low degree subfunction by fixing a few variables. As example, one may refer to the 5-variable function $f = x_1+x_2+x_2x_4+x_3x_4+(x_2+x_3+x_1x_4+x_2x_4+x_3x_4)x_5$. This function has algebraic immunity 2 and the only annihilator of degree 2 is $1+x_1+x_2+x_1x_4+x_3x_4+(x_2+x_3+x_4)x_5$. If one verifies all possible subfunctions of of f after fixing 1 and 2 variables, it is not possible to get subfunctions of degree 1 and 0 respectively.

It will be interesting to extend our idea on the Boolean functions that can be seen as concatenation of indicators of flats [8].

5 Conclusion

In this paper we study the algebraic immunity of Boolean functions since the property becomes a necessary requirement in Boolean functions to be used as cryptographic primitives. For the first time we present a construction where one can get Boolean functions with maximum possible algebraic immunity. Also the construction can be used in conjunction with Boolean functions with other cryptographic properties to have functions which are suitable for different cryptographic applications. Further we also point out that functions having low degree subfunctions are not good in terms of algebraic immunity and study some well known existing constructions from this approach.

Acknowledgment: The authors like to thank the anonymous reviewers for their excellent comments that improved both the technical and editorial quality of this paper. In particular, the current proof of Theorem 1 has been outlined by one of the reviewers and it looks more compact than our long initial proof using systems of homogeneous linear equations. We also like to thank Prof. Claude Carlet for carefully reading our construction and providing a correction in the expression of f_{l+8}.

References

1. F. Armknecht. Improving Fast Algebraic Attacks In *FSE 2004*, number 3017 in Lecture Notes in Computer Science, pages 65–82. Springer Verlag, 2004.
2. L. M. Batten. Algebraic Attacks over GF(q). In *Progress in Cryptology - INDOCRYPT 2004*, pages 84–91, number 3348, Lecture Notes in Computer Science, Springer-Verlag.

3. A. Botev. On algebraic immunity of some recursively given sequence of correlation immune functions. In Proceedings of *XV international workshop on Synthesis and complexity of control systems*, Novosibirsk, October 18-23, 2004, pages 8-12 (in Russian).
4. A. Botev. On algebraic immunity of new constructions of filters with high nonlinearity. In Proceedings of *VI international conference on Discrete models in the theory of control systems*, Moscow, December 7-11, 2004, pages 227-230 (in Russian).
5. A. Botev and Y. Tarannikov. Lower bounds on algebraic immunity for recursive constructions of nonlinear filters. Preprint 2004.
6. C. Carlet. A larger class of cryptographic Boolean functions via a study of the Maiorana-McFarland construction. In *CRYPTO 2002*, number 2442 in Lecture Notes in Computer Science, pages 549–564. Springer Verlag, 2002.
7. C. Carlet. Improving the algebraic immunity of resilient and nonlinear functions and constructing bent functions. IACR ePrint server, http://eprint.iacr.org, 2004/276.
8. C. Carlet. Concatenating indicators of flats for designing cryptographic functions. To appear in *Design, Codes and Cryptography.*
9. C. Carlet. Personal communications, 2005.
10. J. H. Cheon and D. H. Lee. Resistance of S-boxes against Algebraic Attacks. In *FSE 2004*, number 3017 in Lecture Notes in Computer Science, pages 83–94. Springer Verlag, 2004.
11. J. Y. Cho and J. Pieprzyk. Algebraic Attacks on SOBER-t32 and SOBER-128. In *FSE 2004*, number 3017 in Lecture Notes in Computer Science, pages 49–64. Springer Verlag, 2004.
12. N. Courtois and J. Pieprzyk. Cryptanalysis of block ciphers with overdefined systems of equations. In *Advances in Cryptology - ASIACRYPT 2002*, number 2501 in Lecture Notes in Computer Science, pages 267–287. Springer Verlag, 2002.
13. N. Courtois and W. Meier. Algebraic attacks on stream ciphers with linear feedback. In *Advances in Cryptology - EUROCRYPT 2003*, number 2656 in Lecture Notes in Computer Science, pages 345–359. Springer Verlag, 2003.
14. N. Courtois. Fast algebraic attacks on stream ciphers with linear feedback. In *Advances in Cryptology - CRYPTO 2003*, number 2729 in Lecture Notes in Computer Science, pages 176–194. Springer Verlag, 2003.
15. D. K. Dalai, K. C. Gupta and S. Maitra. Results on Algebraic Immunity for Cryptographically Significant Boolean Functions. In *INDOCRYPT 2004*, pages 92–106, number 3348, Lecture Notes in Computer Science, Springer-Verlag.
16. C. Ding, G. Xiao, and W. Shan. *The Stability Theory of Stream Ciphers.* Number 561 in Lecture Notes in Computer Science. Springer-Verlag, 1991.
17. D. H. Lee, J. Kim, J. Hong, J. W. Han and D. Moon. Algebraic Attacks on Summation Generators. In *FSE 2004*, number 3017 in Lecture Notes in Computer Science, pages 34–48. Springer Verlag, 2004.
18. W. Meier, E. Pasalic and C. Carlet. Algebraic attacks and decomposition of Boolean functions. In *Advances in Cryptology - EUROCRYPT 2004*, number 3027 in Lecture Notes in Computer Science, pages 474–491. Springer Verlag, 2004.
19. E. Pasalic, S. Maitra, T. Johansson and P. Sarkar. New constructions of resilient and correlation immune Boolean functions achieving upper bounds on nonlinearity. In *Workshop on Coding and Cryptography - WCC 2001*, Paris, January 8–12, 2001. Electronic Notes in Discrete Mathematics, Volume 6, Elsevier Science, 2001.

20. P. Sarkar and S. Maitra. Construction of nonlinear Boolean functions with important cryptographic properties. In *EUROCRYPT 2000*, number 1807 in Lecture Notes in Computer Science, pages 485–506. Springer Verlag, May 2000.
21. Y. V. Tarannikov. On resilient Boolean functions with maximum possible nonlinearity. In *Progress in Cryptology - INDOCRYPT 2000*, number 1977 in Lecture Notes in Computer Science, pages 19–30. Springer Verlag, 2000.

The ANF of the Composition of Addition and Multiplication mod 2^n with a Boolean Function

An Braeken[1] and Igor Semaev[2]

[1] Department Electrical Engineering, ESAT/COSIC,
Katholieke Universiteit Leuven, Kasteelpark Arenberg 10,
B-3001 Heverlee-Leuven, Belgium
an.braeken@esat.kuleuven.ac.be
[2] Selmer Center, Inst. for Informatikk,
University of Bergen, Bergen 5020 Norway
Igor.Semaev@ii.uib.no

Abstract. Compact formulas are derived to represent the Algebraic Normal Form (ANF) of $f(\overline{x} + \overline{a} \mod 2^n)$ and $f(\overline{x} \times \overline{a} \mod 2^n)$ from the ANF of f, where f is a Boolean function on $\mathbb{F}_2^n$ and $\overline{a}$ is a constant of $\mathbb{F}_2^n$. We compare the algebraic degree of the composed functions with the algebraic degree of the original function f. As an application, the formula for addition modulo 2^n is applied in an algebraic attack on the summation generator and the E_0 encryption scheme in the Bluetooth keystream generator.

1 Introduction

Addition and multiplication modulo 2^n are operations which are very often used in cryptosystems like e.g. in the block ciphers Safer [20] and Idea [22], in the key stream generators natural sequence generator [12], summation generator [25] and E_0 encryption scheme of the Bluetooth keystream generator, and in the stream ciphers Turing [24] and Helix [14].

Recently, algebraic attacks [5, 6] have been applied successfully to stream ciphers and to some block ciphers. The central idea in the algebraic attacks is to find low degree equations or approximations of the cipher and then to solve an over-determined system of nonlinear multivariate equations of low degree by efficient methods such as XL [5], simple linearization [7] or by Gröbner Bases techniques [11].

By having compact formulas for representing the algebraic normal form of the composition of a Boolean function f with addition and multiplication modulo 2^n, we can better understand the structure of the polynomial equations of the cipher and also the consequences of mixing operations from different rings. Moreover, we give a precise criteria to avoid that the degree of the composed functions will not decrease with respect to the degree of f. As an example, we apply our formulas in order to derive the algebraic relations used in an algebraic attack

H. Gilbert and H. Handschuh (Eds.): FSE 2005, LNCS 3557, pp. 112–125, 2005.

on the summation generator and the E_0 encryption scheme of the Bluetooth keystream generator.

The paper is organised as follows. Some definitions and preliminaries that will be used later in the paper are described in Sect. 2. In Sect. 3, we derive the compact formulas for the algebraic normal forms of $f(\overline{x}+\overline{a} \mod 2^n)$ and $f(\overline{x}\times\overline{a} \mod 2^n)$ from the algebraic normal form of f, where f is a Boolean function on $\mathbb{F}_2^n$ and $\overline{a}, \overline{b}$ are constants of $\mathbb{F}_2^n$. In Sect. 4, we compare the algebraic degree of the composed functions with the algebraic degree of the original function f. In Sect. 5, the formula for addition modulo 2^n is applied in order to find the algebraic equations for the summation generator and the E_0 encryption scheme of the Bluetooth keystream generator. Finally, we present some conclusions and open problems in Sect. 6.

2 Definitions and Preliminaries

For the sake of clarity, we use "$\oplus$" for the addition in characteristic 2 and "$+$" for the addition modulo 2^n or in $\mathbb{R}$. The multiplication modulo 2^n is represented by "$\times$".

Let $\mathbb{F}_2^n$ be the set of all n-tuples of elements in the field $\mathbb{F}_2$ (Galois field with two elements), endowed with the natural vector space structure over $\mathbb{F}_2$. The correspondence between $\mathbb{F}_2^n$ and $\mathbb{Z}_{2^n}$ is defined by

$$\psi : \mathbb{F}_2^n \rightarrow \mathbb{Z}_{2^n} : \overline{u} = (u_0, \ldots, u_{n-1}) \mapsto u = \sum_{i=0}^{n-1} u_i 2^{i-1}.$$

The partial ordering $\overline{x} \preceq \overline{a}$ means that x precedes a or also $x_i \leq a_i$ for all $i \in \{0, \ldots, n-1\}$.

Let $f(\overline{x})$ be a Boolean function on $\mathbb{F}_2^n$. Any Boolean function f can be uniquely expressed in the algebraic normal form (ANF). Namely,

$$f(\overline{x}) = \bigoplus_{u \in \mathbb{Z}_{2^n}} h_u \overline{x}^u, \qquad h_u \in \mathbb{F}_2,$$

where $\overline{x}^u$ denotes $x_0^{u_0} \cdots x_{n-1}^{u_{n-1}}$. The coefficients h_u are defined by the Möbius inversion principle, $h_u = h(\overline{u}) = \sum_{\overline{x} \preceq \overline{u}} f(\overline{x})$ for any $u \in \mathbb{Z}_{2^n}$. The *algebraic degree* of f, denoted by $\deg(f)$, is equal to to the number of variables in the longest term $x_0^{u_0} \cdots x_{n-1}^{u_{n-1}}$ in the ANF of f, or simply as the maximum Hamming weight of $\overline{u}$ (denoted as $\mathrm{wt}(\overline{u})$) for which $h_u \neq 0$. The Hamming weight of a binary vector is equal to the number of nonzero components.

A vectorial Boolean function $F : \mathbb{F}_2^n \rightarrow \mathbb{F}_2^m$ (also called (n, m) S-box or shortly S-box) can be represented by the vector $(f_1, f_2, \ldots, f_m)$, where f_i are Boolean functions from $\mathbb{F}_2^n$ into $\mathbb{F}_2$ for $1 \leq i \leq m$. The functions $(f_i)_{1 \leq i \leq m}$ are called the component functions of the S-box.

The composition $f \circ F$ of a Boolean function f on $\mathbb{F}_2^m$ with an (n, m) S-box F leads to a Boolean function on $\mathbb{F}_2^n$. Here, we will study the composition of an arbitrary Boolean function on $\mathbb{F}_2^n$ and addition respectively multiplication modulo 2^n with a fixed constant $\overline{a} \in \mathbb{F}_2^n$. The addition modulo 2^n, i.e., $\overline{r} = \overline{x} + \overline{a} \mod 2^n$ is defined by

$$\begin{aligned} & r_0 + r_1 \cdot 2 + \cdots + r_{n-1} \cdot 2^{n-1} = \\ & (x_0 + x_1 \cdot 2 + \cdots + x_{n-1} \cdot 2^{n-1}) + (a_0 + a_1 \cdot 2 + \cdots + a_{n-1} \cdot 2^{n-1}) \mod 2^n, \end{aligned} \tag{1}$$

with components $(r_0, \ldots, r_{n-1})$ recursively defined by

$$\begin{aligned} r_0 &= x_0 \oplus a_0 \oplus c_0, \qquad c_0 = 0, \\ r_i &= x_i \oplus a_i \oplus c_i, \qquad c_i = x_{i-1}a_{i-1} \oplus x_{i-1}c_{i-1} \oplus a_{i-1}c_{i-1}, \\ & \forall i \in \{1, \ldots, n-1\}. \end{aligned}$$

The multiplication $\overline{s} = \overline{x} \times \overline{a} \mod 2^n$ is defined by

$$\begin{aligned} & s_0 + s_1 \cdot 2 + \cdots + s_{n-1} \cdot 2^{n-1} = \\ & (x_0 + x_1 \cdot 2 + \cdots + x_{n-1}2^{n-1}) \times (a_0 + a_1 \cdot 2 + \cdots + a_{n-1}2^{n-1}) \mod 2^n, \end{aligned} \tag{2}$$

with components $(s_0, \ldots, s_{n-1})$ equal to

$$\begin{aligned} s_0 &= x_0 a_0, \\ s_1 &= x_1 a_0 \oplus x_0 a_1 \oplus c_1(x_0, a_0), \\ & \vdots \\ s_{n-1} &= x_{n-1}a_0 \oplus x_{n-2}a_1 \oplus \cdots \oplus x_0 a_{n-1} \oplus c_{n-1}(x_0, \ldots, x_{n-1}, a_0, \ldots, a_{n-1}), \end{aligned}$$

where $c_i()$ is a function of its arguments which defines the carry bit. The number of terms of c_i grows exponentially for increasing i. We write for instance c_i for $i = 1, 2, 3$ explicitely:

$$\begin{aligned} c_1 &= c_1(x_0, a_0) = 0, \\ c_2 &= c_2(x_0, x_1, a_0, a_1) = a_0 a_1 x_0 x_1, \\ c_3 &= c_3(x_0, \ldots, x_2, a_0, \ldots, a_2) = a_0 a_1 x_0 x_1 \oplus a_0 a_1 x_1 x_2 \oplus a_0 a_1 x_0 x_1 x_2 \\ & \qquad \oplus a_0 a_2 x_0 x_2 \oplus a_1 a_2 x_0 x_1 \oplus a_0 a_1 a_2 x_0 x_1. \end{aligned}$$

In this paper we will study the equations formed by the composition of addition or multiplication with a Boolean function. This corresponds with studying the ANF of $f(\overline{x} + \overline{a})$ and $f(\overline{x} \times \overline{a})$. For instance, if the Boolean function is defined by the ANF $\overline{x}^u = x_0^{u_0} \cdots x_{n-1}^{u_{n-1}}$, then the corresponding equation of $f(\overline{x} + \overline{a})$ is equal to to $r_0^{u_0} \cdots r_{n-1}^{u_{n-1}}$, where $r_0, \ldots, r_{n-1}$ are functions in variables $(x_0, \ldots, x_{n-1})$ defined by (1).

3 Algebraic Normal Form of $\cdot\,(\overline{\cdot} + \overline{\cdot})$ and $\cdot\,(\overline{\cdot} \times \overline{\cdot})$

In this section, we deduce compact formulas for representing the ANF of the functions $f(\overline{x} + \overline{a})$ and $f(\overline{x} \times \overline{a})$ using the ANF of f.

3.1 The ANF of $\bullet\,(\bar{\bullet} + \bar{\bullet})$

Theorem 1. *If the ANF of $f : \mathbb{F}_2^n \to \mathbb{F}_2 : \overline{x} \mapsto f(\overline{x})$ is given by the monomial $\overline{x}^u$ ($u \in \mathbb{Z}_{2^n}$), then the ANF of $f(\overline{x}+\overline{a})$ with $\overline{a} \in \mathbb{F}_2^n$ a fixed constant is given by*

$$f(\overline{x}+\overline{a}) = \bigoplus_{c=0}^{u} \overline{x}^{u-c}\overline{a}^{c}, \tag{3}$$

where $u - c$ represents subtraction in $\mathbb{R}$.

Proof. To prove the theorem, we need two lemmas which can be proven by induction.

Lemma 1. *For $\overline{x}, \overline{a}_0 \in \mathbb{F}_2^n$, with $\overline{a}_0 = (a_0, 0, \ldots, 0)$, we have that*

$$(\overline{x}+\overline{a}_0)^u = \overline{x}^u \oplus \overline{x}^{u-1}\overline{a}_0.$$

Proof. (*Lemma 1*)If $n = 1$, the lemma is trivial. Suppose the lemma is true for dimension less or equal than $n-1$. We will show that the lemma holds for dimension n. If $u < 2^{n-1}$, the lemma is true by induction, otherwise write u as $2^{n-1} + u_1$, where $0 < u_1 < 2^{n-1}$, and thus

$$(\overline{x}+\overline{a}_0)^{2^{n-1}+u_1} = (\overline{x}+\overline{a}_0)^{2^{n-1}}(\overline{x}+\overline{a}_0)^{u_1}.$$

On the second term of the product, we apply induction. For the first term, we use the definition of addition (1) to compute $(\overline{x}+\overline{a}_0) = (x_0 \oplus a_0, x_1 \oplus a_0x_0, x_2 \oplus a_0x_0x_1, \ldots, x_{n-1} \oplus a_0x_0 \cdots x_{n-2})$. Taking the 2^{n-1}-th power is equal to selecting the $(n-1)$-th component in the binary representation. As a result we have

$$\begin{aligned}(\overline{x}+\overline{a}_0)^u &= (\overline{x}^{2^{n-1}} \oplus \overline{a}_0\overline{x}^{2^{n-1}-1})(\overline{x}^{u_1} \oplus \overline{x}^{u_1-1}\overline{a}_0)\\ &= \overline{x}^{2^{n-1}+u_1} \oplus \overline{x}^{2^{n-1}+u_1-1}\overline{a}_0,\end{aligned}$$

where we used the fact that $\overline{a}_0\overline{x}^{2^{n-1}-1}\overline{x}^{u_1} = \overline{a}_0\overline{x}^{2^{n-1}-1}\overline{x}^{u_1-1} = \overline{a}_0\overline{x}^{2^{n-1}-1}$ in the last reduction step. This equality is due to the fact that $u_1 \preceq 2^{n-1}-1$ and $u_1 - 1 \preceq 2^{n-1}-1$. □

Lemma 2. *Denote $\overline{x} = \overline{x}_0 + 2 \times \overline{x}'$ with $\overline{x}_0 = (x_0, 0, \ldots, 0)$ and $\overline{x}' = (x_1, \ldots, x_{n-1}, 0)$. Similarly, denote $\overline{a} = \overline{a}_0 + 2 \times \overline{a}'$ with $\overline{a}_0 = (a_0, 0, \ldots, 0)$ and $\overline{a}' = (a_1, \ldots, a_{n-1}, 0)$, then*

$$(2 \times (\overline{x}' + \overline{a}'))^u = \begin{cases} 0 & \textit{if } u \textit{ is odd,}\\ \bigoplus_{v=0}^{\frac{u}{2}}(2\times\overline{x}')^{u-2v}(2\times\overline{a}')^{2v} & \textit{if } u \textit{ is even.}\end{cases}$$

Proof. (*Lemma 2*) We prove the lemma by induction on the number n of variables. Because multiplication by 2 only shifts the vector over one position, it follows that

$$(2\times\overline{x}')^u = \begin{cases} 0 & \text{if } u \text{ is odd,}\\ (\overline{x}')^{\frac{u}{2}} & \text{if } u \text{ is even.}\end{cases} \tag{4}$$

By induction on n, we have for even u that

$$(\overline{x}' + \overline{a}')^{\frac{u}{2}} = \bigoplus_{v=0}^{\frac{u}{2}} \overline{x}'^{\frac{u}{2}-v}\overline{a}'^{v}.$$

If we rescale the previous formula using (4), we get the formula of the lemma. □

By using the previous lemmas, we are now able to prove the theorem. We start with applying Lemma 1 repeatedly.

$$\begin{aligned}(\overline{x} + \overline{a})^u &= (\overline{x}_0 + 2 \times \overline{x}' + \overline{a}_0 + 2 \times \overline{a}')^u \\ &= (2 \times \overline{x}' + 2 \times \overline{a}' + \overline{x}_0)^u \oplus (2 \times \overline{x}' + 2 \times \overline{a}' + \overline{x}_0)^{u-1}\overline{a}_0 \\ &= (2 \times \overline{x}' + 2 \times \overline{a}')^u \oplus (2 \times \overline{x}' + 2 \times \overline{a}')^{u-1}\overline{x}_0 \\ &\quad \oplus \overline{a}_0((2 \times \overline{x}' + 2 \times \overline{a}')^{u-1} \oplus (2 \times \overline{x}' + 2 \times \overline{a}')^{u-2}\overline{x}_0). \end{aligned} \quad (5)$$

Note that multiplication modulo 2^n is distributive with respect to addition modulo 2^n, i.e. $(2 \times \overline{x}' + 2 \times \overline{a}') = 2 \times (\overline{x}' + \overline{a}')$. As a consequence, we can apply Lemma 2 on Equation (5). This implies that we need to distinguish the case u is odd and the case u is even. We give here the proof for u odd. The proof for u even is similar.

$$\begin{aligned}(\overline{x} + \overline{a})^u &= \overline{x}_0(2 \times \overline{x}' + 2 \times \overline{a}')^{u-1} \oplus \overline{a}_0(2 \times \overline{x}' + 2 \times \overline{a}')^{u-1} \\ &= \overline{x}_0 \bigoplus_{v=0}^{\frac{u-1}{2}} (2 \times \overline{x}')^{u-2v-1}(2 \times \overline{a}')^{2v} \oplus \overline{a}_0 \bigoplus_{v=0}^{\frac{u-1}{2}} (2 \times \overline{x}')^{u-2v-1}(2 \times \overline{a}')^{2v}. \end{aligned} \quad (6)$$

The following equalities hold

$$\begin{aligned}(2 \times \overline{x}')^{u-2v-1} &= \overline{x}^{u-2v-1}, \\ \overline{x}_0(2 \times \overline{x}')^{u-2v-1} &= \overline{x}^{u-2v}, \end{aligned} \quad (7)$$

because $2 \times \overline{x}' = (0, x_1, \ldots, x_{n-1})$ and $u - 2v - 1$ is even. The same argument holds for $2 \times \overline{a}'$ and thus

$$\begin{aligned}(2 \times \overline{a}')^{2v} &= \overline{a}^{2v}, \\ \overline{a}_0(2 \times \overline{a}')^{2v} &= \overline{a}^{2v+1}. \end{aligned} \quad (8)$$

After substituting the equalities (7) and (8) in Equation (6) and collecting the terms, we find Formula (3). □

Remark 1. If $u = 2^i$, Formula (3) expresses the i-th component of the sum $\overline{x}+\overline{a}$. Similarly, if $u = 2^i + 2^j$, Formula (3) expresses the product of the i-th and j-th component of the sum $\overline{x} + \overline{a}$. Note that Formula (3) consists only of all terms for which the integer sum of the exponents of $\overline{x}$ and $\overline{a}$ is exactly equal to u. The formula can be easily generalized for the addition of n elements $\overline{y}_1, \ldots, \overline{y}_n$ of $\mathbb{F}_2^n$

by applying Formula (3) recursively. Again, the result is equal to the sum of all terms with sum of exponents equal to u.

$$f(\overline{y}_1 + \cdots + \overline{y}_n) = \bigoplus_{\substack{k_0,\ldots,k_{n-1} \geq 0 \\ k_0 + \cdots + k_{n-1} = u}} \overline{y}_1^{k_0} \overline{y}_2^{k_1} \cdots \overline{y}_n^{k_{n-1}} \tag{9}$$

We now generalize Theorem 1 for Boolean functions where the ANF consists of an arbitrary number of terms. By collecting the terms in a right way, we obtain the following formula.

Corollary 1. *If the ANF of $f : \mathbb{F}_2^n \to \mathbb{F}_2$ is given by $\bigoplus_{u \in \mathbb{Z}_{2^n}} h_u \overline{x}^u, h_u \in \mathbb{F}_2$, the ANF of $f(\overline{x} + \overline{a})$ is given by*

$$f(\overline{x} + \overline{a}) = \bigoplus_v (\bigoplus_{u \geq v} h_u \overline{a}^{u-v}) \overline{x}^v, \tag{10}$$

where $u - v$ represents the subtraction modulo 2^n.

Example 1. Consider the ANF of the function $f(x_0, x_1, x_2) = \overline{x}^5 \oplus \overline{x}^1$. The ANF of $f(\overline{x} + \overline{a})$ is then determined by the previous corollary:

$$f(\overline{x} + \overline{a}) = (\overline{a}^1 \oplus \overline{a}^5) \oplus \overline{x}^1(\overline{a}^0 \oplus \overline{a}^4) \oplus \overline{x}^2\overline{a}^3 \oplus \overline{x}^3\overline{a}^2 \oplus \overline{x}^4\overline{a}^1 \oplus \overline{x}^5,$$

which can also be written as

$$f(\overline{x} + \overline{a}) = (a_0 \oplus a_0a_2) \oplus x_0(1 \oplus a_2) \oplus x_1a_0a_1 \oplus x_0x_1a_1 \oplus x_2a_0 \oplus x_0x_2.$$

3.2 The ANF of $\bullet(\overline{\bullet} \times \overline{\bullet})$

Theorem 2. *If the ANF of $f : \mathbb{F}_2^n \to \mathbb{F}_2 : \overline{x} \mapsto f(\overline{x})$ is given by the monomial $\overline{x}^u$ $(u \in \mathbb{Z}_{2^n})$, then the ANF of $f(\overline{x} \times \overline{a})$ with $\overline{a} \in \mathbb{F}_2^n$ a fixed constant is given by*

$$f(\overline{x} \times \overline{a}) = \bigoplus_{k^u = [k_0,\ldots,k_{n-1}]} \overline{a}^{r^k} \overline{x}^{s^k}, \tag{11}$$

where $k^u = [k_0, \ldots, k_{n-1}]$ satisfies

$$\begin{aligned} &k_0 \geq 0, \ldots, k_{n-1} \geq 0; \\ &k_0 + 2k_1 + \cdots + 2^{n-1}k_{n-1} = u. \end{aligned} \tag{12}$$

The integers $r^k = r_0^k + 2r_1^k + \cdots + r_{n-1}^k$ and $s^k = s_0^k + 2s_1^k + \cdots + s_{n-1}^k$ are defined by the following $(n+1) \times (n+1)$-table:

s_{n-1}^k	$k_{0,n-1}$	0	$\cdots$	0
s_{n-2}^k	$k_{0,n-2}$	$k_{1,n-2}$	$\cdots$	0
		$\ddots$		
s_0^k	$k_{0,0}$	$k_{1,0}$	$\cdots$	$k_{n-1,0}$
	r_0^k	r_1^k	$\cdots$	r_{n-1}^k

For each $k^u = [k_0, \ldots, k_{n-1}]$ that satisfies the properties as described by (12), we fill the table with the binary representation of $k_0 = (k_{0,0}, \ldots, k_{0,n-1}), \ldots, k_{n-1} = (k_{n-1,0}, \ldots, k_{n-1,n-1})$. The digit r_i^k (resp. s_i^k) for all $i \in \{0, \ldots, n-1\}$ is equal to to 1 if the corresponding column (resp. row) is different from the all-zero vector and is equal to to 0 otherwise. The integer r^k is then defined by the binary representation $(r_0^k, \ldots, r_{n-1}^k)$ and the integer s_k by $(s_0^k, \ldots, s_{n-1}^k)$.

Proof. Note that the multiplication $\overline{a} \times \overline{x}$ can be written as a sum:

$$\overline{a} \times \overline{x} = a_0 \cdot \overline{x} + a_1 \cdot (2 \times \overline{x}) + \cdots + a_{n-1} \cdot (2^{n-1} \times \overline{x}).$$

By formula (9) for the addition of n points, we obtain

$$f(\overline{x} \times \overline{a}) = \bigoplus_{\substack{k_0,\ldots,k_{n-1} \geq 0 \\ k_0+\cdots+k_{n-1}=u}} (a_0 x)^{k_0} (a_1 \cdot (2 \times \overline{x}))^{k_1} \cdots (a_{n-1} \cdot (2^{n-1} \times \overline{x}))^{k_{n-1}}.$$

As explained in the proof of Lemma 2, we have for the general case i, $i \in \{0, \ldots, n-1\}$ that $(2^i \times \overline{x})^{k_i}$ shifts the components of x over i positions, which means that

$$(2^i \times \overline{x})^{k_i} = \begin{cases} \overline{x}^{\frac{k_i}{2^i}} & \text{if } k_i \equiv o(2^i) \\ 0 & \text{otherwise.} \end{cases}$$

Consequently, we can write the above equation for $f(\overline{x} \times \overline{a})$ as:

$$f(\overline{x} \times \overline{a}) = \bigoplus_{\substack{k_0,\ldots,k_{n-1} \geq 0 \\ k_1 \equiv o(2),\ldots,k_{n-1} \equiv o(2^{n-1}) \\ k_0+\cdots+k_{n-1}=u}} (a_0 \overline{x})^{k_0} (a_1 \overline{x})^{\frac{k_1}{2}} \cdots (a_{n-1} \overline{x})^{\frac{k_{n-1}}{2^{n-1}}},$$

where $k_i \equiv o(2^i)$ means that 2^i is a divisor of k_i. This representation contains mixed terms, i.e. terms which consists of powers of vector x and powers of components of a. Moreover because $x_i^2 = x_i$, we can very often reduce the powers of x. However by translating this form in the representation given by (11), we avoid these disadvantages. This can be seen by the definition of the vectors r^k and s^k. The value r_i^k (resp s_i^k) for all $i \in \{0, \ldots, n-1\}$ is equal to to 1 if the corresponding column (resp. row) is different from the all-zero vector and is equal to to 0 otherwise. □

Remark 2. We note that the formula of multiplication, unlike the formula of addition, does not immediately give the full reduced form of the ANF because some terms can cancel out. For instance (see also Example 2), if the pattern $\begin{smallmatrix} 1\,0 \\ 1\,1 \end{smallmatrix}$ appears in the represenation table of the exponents of a term, then also the pattern $\begin{smallmatrix} 0\,1 \\ 1\,0 \end{smallmatrix}$ satisfies the same conditions of (12) and will give the same exponents. Consequently both terms will cancel out. Another clear example is the pattern $\begin{smallmatrix} 1\,1 \\ 1\,0 \end{smallmatrix}$ which is equivalent with the pattern $\begin{smallmatrix} 0\,1 \\ 1\,1 \end{smallmatrix}$. However, the formula is still much more practical than using explicitly the definition of multiplication.

We now generalize Theorem 2 for Boolean functions where the ANF consist of an arbitrary number of terms.

Corollary 2. *If the ANF of* $f : \mathbb{F}_2^n \rightarrow \mathbb{F}_2$ *is given by* $\bigoplus_{u \in \mathbb{Z}_{2^n}} h_u \overline{x}^u, h_u \in \mathbb{F}_2$*, the ANF of* $f(\overline{x} \times \overline{a})$ *is given by*

$$f(\overline{x} \times \overline{a}) = \bigoplus_{u \in \mathbb{Z}_{2^n}} h_u \left(\bigoplus_{k^u = [k_0, \ldots, k_{n-1}]} \overline{a}^{r_{k,u}} \overline{x}^{s_{k,u}} \right),$$

where k^u*,* $r_{k,u}$ *and* $s_{k,u}$ *for each* u *(corresponding with a non-zero* h_u *in the ANF of* f*) are defined as in Theorem 2.*

Example 2. Consider the ANF of the function $f(x_0, x_1, x_2) = \overline{x}^5$. To compute the ANF of $f(\overline{x} \times \overline{a})$, we first determine all k^5 that satisfy the properties of (12). There are 4 different possibilities for $k = (k_0, k_1, k_2)$, i.e. $k = (5,0,0), k = (3,1,0), k = (1,2,0), k = (1,0,1)$. For each k, we compute the corresponding exponent of $\overline{a}$ and $\overline{x}$ by computing its corresponding table:

1	1 0 0
0	0 0 0
1	1 0 0
	1 0 0

0	0 0 0
1	1 0 0
1	1 1 0
	1 1 0

0	0 0 0
1	0 1 0
1	1 0 0
	1 1 0

0	0 0 0
0	0 0 0
1	1 0 1
	1 0 1

$k = (5,0,0), k = (3,1,0), k = (1,2,0), k = (1,0,1)$

As a consequence, we get the following ANF of f:

$$\begin{aligned} f(\overline{x} \times \overline{a}) &= \overline{a}_1 \overline{x}^5 \oplus \overline{a}_3 \overline{x}^3 \oplus \overline{a}_3 \overline{x}^3 \oplus \overline{a}_5 \overline{x}^1. \\ &= \overline{a}_1 \overline{x}^5 \oplus \overline{a}_5 \overline{x}^1. \end{aligned}$$

4 Comparison of the Degrees

In this section, we compare the degrees of $f(\overline{x})$ at the one side with the degrees of $f(\overline{x} + \overline{a})$ and $f(\overline{x} \times \overline{a})$ at the other side.

4.1 Degrees of $\bullet(\overline{\bullet})$ and $\bullet(\overline{\bullet} + \overline{\bullet})$

Theorem 3. *If* $f(\overline{x}) = \bigoplus_{u \in \mathbb{Z}_{2^n}} h_u \overline{x}^u$*, define* $u_m = \max_{\substack{h_u \neq 0 \\ u \in \mathbb{Z}_{2^n}}} u$*. The degree of the function* $f_{\overline{a}} : \mathbb{F}_2^n \rightarrow \mathbb{F}_2 : \overline{x} \rightarrow f(\overline{x} + \overline{a})$ *will be for all values of* $\overline{a} \in \mathbb{F}_2^n$ *in the interval*

$$\begin{aligned} &[\mathrm{wt}(u_m), \lfloor \log_2 u_m \rfloor] \;\; \textit{if}\; \mathrm{wt}(u_m) \leq \lfloor \log_2 u_m \rfloor \;\textit{or}\; u_m \neq 2^{\lceil \log_2 u_m \rceil} - 1, \\ &[\mathrm{wt}(u_m), \mathrm{wt}(u_m)] \;\; \textit{otherwise.} \end{aligned}$$

Proof. From Corollary 1, we have that

$$f(\overline{x}+\overline{a}) = \bigoplus_{v}\left(\bigoplus_{u\geq v} h_u \overline{a}^{u-v}\right)\overline{x}^{v}$$

$$= \overline{x}^{u_m} \oplus \bigoplus_{v<u_m}\left(\bigoplus_{u\geq v} h_u \overline{a}^{u-v}\right)\overline{x}^{v}. \tag{13}$$

From (13), the lowerbound for the degree of $f_{\overline{a}}$ is equal to to $\mathrm{wt}(u_m)$, because the term $\overline{x}^{u_m}$ always appears and does not cancel out in the ANF of $f_{\overline{a}}$. The degree of the function $f_{\overline{a}}$ is exactly equal to $\mathrm{wt}(u_m)$ if the term $\bigoplus_{u\geq v} h_u \overline{a}^{u-v}$ from (13) is equal to to zero for all v with weight greater than $\mathrm{wt}(u_m)$.

The upperbound is equal to to the maximum weigth of v for which $v < u_m$. This is equal to $\lfloor \log_2 u_m \rfloor$ for $u \neq 2^{\lceil \log_2 u_m \rceil} - 1$. The degree of $f_{\overline{a}}$ is exactly equal to $\lfloor \log_2 u_m \rfloor$ if the term $\bigoplus_{u\geq v} h_u \overline{a}^{u-v}$ from (13) is equal to to one for at least one v with weight equal to $\lfloor \log_2 u_m \rfloor$. □

Example 3. Consider the function $f : \mathbb{F}_2^7 \to \mathbb{F}_2 : \overline{x} \mapsto \overline{x}^{64} \oplus \overline{x}^{62}$. The degree of this function is 5, while u_m is equal to to 64 with weight one. We now show that the degree of the function $f_{\overline{a}}$ is between one and six according to Theorem 3 and depending on the value $\overline{a}$.

For odd a, i.e. $a_0 = 1$, the term x^{63} appears in the ANF of $f_{\overline{a}}$, and thus the corresponding functions have degree 6. If $a_0 = 0, a_1 = 0$, the function $f_{\overline{a}}$ has degree 5 because of the term x^{62} in the ANF of the function. For functions $f_{\overline{a}}$ with $a_0 = 0, a_1 = 1, a_2 = 0$, the resulting degree is equal to to 4. If $a_0 = 0, a_1 = 1, a_2 = 1, a_3 = 0$ the degree of $f_{\overline{a}}$ is 3, and if $a_0 = 0, a_1 = 1, a_2 = 1, a_3 = 1, a_4 = 0$ the degree of $f_{\overline{a}}$ becomes 2. Finally for $a = (0,1,1,1,1,0,1)$ and $a = (0,1,1,1,1,0,0)$ the function $f_{\overline{a}}$ has degree 1.

In order to diminish the degeneration of the degree of the function $f(\overline{x}+\overline{a})$ for $\overline{a} \in \mathbb{F}_2^n$ with respect to the degree of the function $f(\overline{x})$, we need to take care that $|\mathrm{wt}(u_m) - \deg(f)|$ is small. The condition that a function satisfies $\mathrm{wt}(u_m) = \deg(f)$ will appear for instance if f is of degree d and contains the monomial $x_{n-d-1}\cdots x_{n-1}$.

4.2 Degrees of $\bullet(\overline{\bullet})$ and $\bullet(\overline{\bullet}\times\overline{\bullet})$

Theorem 4. *If $a_0 \neq 0$ then the degree of $f(\overline{x}\times\overline{a})$ will be greater or equal than the weight of u_m^0, where $u_m^0 = \max_{\substack{h_u\neq 0\\ u\in\mathbb{Z}_{2^n}}} u$. If $a_0 = 0$ and $a_1 \neq 0$, then the degree of $f(\overline{x}\times\overline{a})$ will be greater or equal than the weight of u_m^1, where $u_m^1 = \max_{\substack{h_u\neq 0\\ u\in\mathbb{Z}_{2^n}\\ u\equiv o(2)}} u$. In general, if $a_0 = \cdots a_{i-1} = 0$ and $a_i \neq 0$, then the degree of $f(\overline{x}\times\overline{a})$ will be greater or equal than the weight of u_m^i, where $u_m^i = \max_{\substack{h_u\neq 0\\ u\in\mathbb{Z}_{2^n}\\ u\equiv o(2^i)}} u$.*

Proof. We give the proof for the degree of $f(\overline{x} \times \overline{a})$ and for the case $a_0 \neq 0$. All other cases can be proven in the same way. In the following, we denote by $g(\overline{x})$ the function $g(\overline{x}) = (a_0\overline{x})^{k_0}(a_1\overline{x})^{\frac{k_1}{2}} \cdots (a_{n-1}\overline{x})^{\frac{k_{n-1}}{2^{n-1}}}$. From Theorem 2, we can write

$$\begin{aligned} f(\overline{x} \times \overline{a}) &= \bigoplus_{u \in \mathbb{Z}_{2^n}} h_u \bigoplus_{\substack{k_0,\ldots,k_{n-1} \geq 0 \\ k_1 \equiv o(2),\ldots k_{n-1} \equiv o(2^{n-1}) \\ k_0 + \cdots k_{n-1} = u}} g(\overline{x}) \\ &= \bigoplus_{u \in \mathbb{Z}_{2^n}} h_u(a_0\overline{x}^u) \oplus \bigoplus_{u \in \mathbb{Z}_{2^n}} h_u \bigoplus_{\substack{k_0,\ldots,k_{n-1} \geq 0 \\ k_0 \neq u, k_1 \equiv o(2),\ldots k_{n-1} \equiv o(2^{n-1}) \\ k_0 + \cdots k_{n-1} = u}} g(\overline{x}) \\ &= a_0\overline{x}^{u_{m_0}} \oplus \bigoplus_{u \in \mathbb{Z}_{2^n}} h_u \bigoplus_{\substack{k_0,\ldots,k_{n-1} = 0 \\ k_0 \neq u_{m_0}, k_1 \equiv o(2),\ldots k_{n-1} \equiv o(2^{n-1}) \\ k_0 + \cdots k_{n-1} = u}} g(\overline{x}) \end{aligned}$$

□

5 Algebraic Attacks

Algebraic attacks exploit the existence of low degree equations. Once a system of nonlinear multivariate equations of low degree is obtained, it is solved by efficient methods such as XL [5], simple linearization [7] or by Gröbner Bases techniques [11]. We will derive in this section low degree equations for the summation generator and the E_0 encryption scheme in the Bluetooth key stream generator.

5.1 Algebraic Attack on the Summation Generator

Consider a summation generator, proposed by Rueppel [25], that consists of n binary Linear Feedback Shift Registers (LFSR). The output bit of the j-th LFSR at time t will be denoted by x_j^t. The binary output bit z^t is defined by

$$z^t = x_1^t \oplus \cdots \oplus x_n^t \oplus c_0^t, \tag{14}$$

where c_0^t is the 0-th bit of the carry $\overline{c}^t = (c_0^t, \ldots, c_{k-1}^t)$ with $k = \lceil \log_2 n \rceil$. The carry for the next stage $t+1$ is computed by

$$\overline{c}^{t+1} = \left\lfloor (x_1^t + \cdots + x_n^t + \overline{c}^t)/2 \right\rfloor. \tag{15}$$

The summation generator is an (n, k)-combiner, which is a stream cipher that combines n LFSRs and has k bits of memory. The summation generator produces a key stream with linear complexity close to its period, which is equal to the product of the periods of the n LFSRs. Moreover, the generator has maximum algebraic degree and maximum order of correlation-immunity (cf Siegenthaler's inequality $t \leq n-d-1$ for combiners without memory). For this reason, summation generators are very interesting building blocks in stream ciphers. We here describe the algebraic attack as presented in [17], but by using the formulas for addition modulo 2^n as given is Subsection 3.1, which makes the analysis and the proofs from [17] much shorter.

To simplify notations, we denote by σ_i^t for $1 \leq i \leq n$, the symmetric polynomial that contains all terms of degree i in the variables $x_1^t, \ldots, x_n^t$, i.e.

$$\begin{aligned}
\sigma_1^t &= \oplus_{i=1}^n x_i^t, \\
\sigma_2^t &= \oplus_{1 \leq i_1 < i_2 \leq n}^n x_{i_1}^t x_{i_2}^t, \\
&\vdots \\
\sigma_n^t &= x_1^t x_2^t \cdots x_n^t.
\end{aligned}$$

We now show how we can use Formula (3) in order to simplify the proof of the main theorem in [17].

Theorem 5. *For a summation generator of $n = 2^k$ LFSRs we can write an algebraic equation connecting LFSR output bits and $k+1$ consecutive key stream bits of degree upperbounded by 2^k in the LFSR output bits.*

Proof. By using formula (9), we immediately determine $c_0^{t+1}, \ldots, c_{k-1}^{t+1}$, which are by defintion (15) the first until the $(k-1)$-th components of the sum $x_1^t + \cdots + x_n^t + \overline{c}^t$.

$$\begin{aligned}
c_0^{t+1} &= \sigma_2^t \oplus c_0^t \sigma_1^t \oplus c_1^t, & (16)\\
c_1^{t+1} &= \sigma_4^t \oplus c_0^t \sigma_3^t \oplus c_1^t \sigma_2^t \oplus c_0^t c_1^t \sigma_1^t \oplus c_2^t, & (17)\\
c_2^{t+1} &= \sigma_8^t \oplus c_0^t \sigma_7^t \oplus c_1^t \sigma_6^t \oplus c_0^t c_1^t \sigma_5^t \oplus c_2^t \sigma_4^t \\
&\quad \oplus c_0^t c_2^t \sigma_3^t \oplus c_1^t c_2^t \sigma_2^t \oplus c_0^t c_1^t c_2^t \sigma_1^t \oplus c_3^t, & (18)\\
&\vdots \\
c_{k-1}^{t+1} &= \sigma_{2^k}^t \oplus c_0^t \sigma_{2^k-1}^t \oplus \cdots \oplus c_0^t \cdots c_{k-1}^t \sigma_1^t. & (19)
\end{aligned}$$

As a consequence, c_i^{t+1} for $0 \leq i \leq k-1$ can be expressed by an equation of degree 2^{i+1} in the LFSR output bits $x_1^t, \ldots, x_n^t$ because it contains the term $\sigma_{2^{i+1}}^t$.

From (14), we derive an equation for c_0^t of degree one in the variables $x_1^t, \ldots, x_n^t$,

$$c_0^t = \sigma_1^t \oplus z^t. \tag{20}$$

Substitution of the equations for c_0^t (20) and c_0^{t+1} ((20) shifted over one position) in Equation (16), results in an equation for c_1^t of degree 2 in the variables $x_1^t, \ldots, x_n^t, x_1^{t+1}, \ldots, x_n^{t+1}$, i.e.,

$$c_1^t = \sigma_2^t \oplus (z^t \oplus 1)\sigma_1^t \oplus \sigma_1^{t+1} \oplus z^{t+1}. \tag{21}$$

Substitution of the equations for c_0^t (20), c_1^t and c_1^{t+1} (21) in Equation (17), results in an equation for c_2^t of degree 4 in the variables $x_1^t, \ldots, x_n^t, x_1^{t+1}, \ldots, x_n^{t+1}$,

$x_1^{t+2}, \ldots, x_n^{t+2}$. This process is repeated and in the last step we substitute the equations for $c_0^t, \ldots, c_{k-1}^t, c_{k-1}^{t+1}$ in Equation (19) which results in an equation in the LFSR output bits $x_1^t, \ldots, x_n^t, \ldots, x_1^{t+k}, \ldots, x_n^{t+k}$. □

We want to note that a similar approach for deriving the equations can also be used on two versions of stream ciphers which are derived from the summation generator: the improved summation generator with 2-bit memory [18] and the parallel stream cipher for secure high-speed communications [19].

5.2 Algebraic Attack on Bluetooth Key Stream Generator

The E_0 encryption system used in the Bluetooth specification [3] for wireless communication is derived from the summation generator and consists of 4 LFSRs. The variables z^t, x_i^t, σ_i^t have the same meaning as explained by the summation generator. Now the initial state consists of 4 memory bits, denoted by $(c_0^{t+1}, c_1^{t+1}, S_0^{t+1}, S_1^{t+1})$. In order to obtain the output and the initial state, the following equations are derived:

$$\begin{aligned} z^t &= \sigma_1^t \oplus c_0^t \\ c_0^{t+1} &= S_0^{t+1} \oplus c_0^t \oplus c_0^{t-1} \oplus c_1^{t-1} \\ c_1^{t+1} &= S_1^{t+1} \oplus c_1^t \oplus c_0^{t-1} \\ (S_0^{t+1}, S_1^{t+1}) &= \left\lfloor \frac{x_1^t + x_2^t + x_3^t + x_4^t + c_0^t + 2c_1^t}{2} \right\rfloor \end{aligned}$$

Using our formula for addition, we immediately find the algebraic equations for S_0^{t+1}, S_1^{t+1}:

$$\begin{aligned} S_0^{t+1} &= \sigma_4^t \oplus \sigma_3^t c_0^t \oplus \sigma_2^t c_1^t \oplus \sigma_1^t c_0^t c_1^t \\ S_1^{t+1} &= \sigma_2^t \oplus \sigma_1^t c_0^t \oplus c_1^t \end{aligned}$$

In [1], these equations are justified by comparing the truth tables of both sides, but no formal proof was given. The next step is to manipulate the equations in such a way that an equation is obtained where all memory bits are eliminated. These equations have degree 4 and are used in the algebraic attack.

6 Conclusions

We have computed compact formulas for representing the ANF of the composition of a Boolean function with addition modululo 2^n, multiplication modulo 2^n and a combination of both, from the ANF of the original function. We have shown that comparing the degrees of the compositions and the original function is not possible in general. If the function satisfies the property that its degree is equal to the weight of the highest coefficient modulo 2^n in its ANF representation, then the degree of the composition with addition modulo 2^n and multiplication with odd constants modulo 2^n will always be higher or equal than the degree of the original function. Finally, we have used our formula of addition modulo 2^n for finding low degree equations of the summation generator and the E_0 encryption scheme in the Bluetooth key stream generator.

An open problem is to further simplify the formula for multiplication modulo 2^n. Further research is required to investigate if those formulas could be used for finding efficient low degree equations in other cryptosystems. For instance, an application of our formulas on the T-functions of Shamir and Klimov [15] seems to be possible for getting better insight in the algebraic equations.

Acknowledgments

The authors would like to thank Bart Preneel and the anonymous referees for the helpful remarks and comments. An Braeken is research assistant of the Fund for Scientific research - Flanders (Belgium) and Igor Semaev was staying at the University of Leuven under the project Flanders FWO G.0186.02.

References

1. F. Armknecht, A Linearization Attack on the Bluetooth Key Stream Generator, Cryptology ePrint Archive, Report 2002/191, http://eprint.iacr. org/2002/191, 2002.
2. F. Armknecht, M. Krause, Algebraic Attacks on Combiners with Memory, Crypto 2003, LNCS 2729, Springer-Verlag, pp. 162–175, 2003.
3. Bluetooth SIG, Specification of the Bluetooth System, Version 1.1, 1 Febrary 22, 2001, available at http://www.bluetooth.com.
4. D.H. Bailey, K. Lee, H.D. Simon, Using Strassen's Algorithm to Accelerate the Solution of Linear Systems, J. of Supercomputing, Vol. 4, pp. 357–371, 1990.
5. N. Courtois, Higher Order Correlation Attacks, *XL* Algorithm and Cryptanalysis of *Toyocrypt*, Asiacrypt 2002, LNCS 2587, Springer-Verlag, pp. 182–199, 2002.
6. N. Courtois, W. Meier, Algebraic Attacks on Stream Ciphers with Linear Feedback, *Eurocrypt'03*, LNCS 2656, Springer-Verlag, pp. 345-359, 2003.
7. N. Courtois, A. Klimov, J. Patarin, A. Shamir, Efficient Algorithms for Solving Overdefined Systems of Multivariate Polynomial Equations, Eurocrypt'00, LNCS 1807, Springer-Verlag, pp. 392-407, 2000.
8. N. Courtois, Fast Algebraic Attacks on Stream Cipher with Linear Feedback, Crypto 2003, LNCS 2729, Springer-Verlag, pp. 176-194, 2003.
9. N. Courtois, Algebraic Attacks on Combiners with memory and Several Outputs, eprint archive, 2003/125
10. N. Courtois, J. Pieprzyk, Cryptanalysis of Block Ciphers with Overdefined Systems of Equations, Asiacrypt 2002, LNCS 2501, Springer-Verlag, pp. 267–287, 2002.
11. A. Joux, J.-C. Faugére, Algebraic Cryptanalysis of Hidden Field Equation (HFE) Cryptosystems Using Grobner Bases, Crypt 2003, LNCS 2729, Springer-Verlag 2003, pp. 44–60, 2003.
12. T.W. Cusick, C. Ding, A. Renvall, Stream Ciphers and Number Theory, Elsevier, Amsterdam, 1998.

13. C. Ding, The Differential Cryptanalysis and Design of the Natural Stream Ciphers, Fast Software Encryption 1994, LNCS 809, Springer-Verlag, pp. 101–115, 1994.
14. N. Ferguson, D. Whiting, B. Schneier, J. Kelsey, S. Lucks, T. Kohno, Helix: Fast Encryption and Authentication in a Single Cryptographic Primitive, Fast Software Encryption 2003, LNCS 2887, Springer-Verlag, pp. 345–362, 2003.
15. A. Klimov, A. Shamir, New Cryptographic Primitives Based on Multiword T-Functions, Fast Software Encryption 2004, LNCS 3017, Springer-Verlag, pp. 1–15, 2004.
16. A. Klapper, M. Goresky, Cryptanalysis Based on 2-adic Rational Approximation, Crypto 1995, LNCS 963, Springer-Verlag, pp. 262–273, 1995.
17. D.H. Lee, J. Kim, J. Hong, J.W. Han, D. Moon, Algebraic Attacks on Summation Generators, Fast Software Encryption 2004, LNCS 3017, Srpinger-Verlag, pp. 3448, 2004.
18. H. Lee, S. Moon, On an Improved Summation Generator with 2-bit Memory, Signal Processing 80, pp. 211-217, 2000.
19. H. Lee, S. Moon, Parallel Stream Cipher for Secure High-Speed Communications, Signal Processing 82, pp. 259–265, 2002.
20. J.L. Massey, SAFER K-64: A Byte-Oriented Block-Ciphering Algorithm, Fast Software Encryption, Cambridge Security Workshop Proceedings, Springer-Verlag, 1994, pp. 1–17.
21. W. Meier, O. Staffelbach, Correlation Properties of Combiners with Memory in Stream Cipher, Journal of Cryptology, Vol. 5, pp. 67–86, 1992.
22. X. Lai, J.L. Massey, A Proposal for a New Block Encryption Standard, Eurocrypt 1990, LNCS 473, Springer-Verlag 1990, pp. 389–404, 1991.
23. D.J. Newman, Analytic Number Theory, Springer-Verlag New York, 1998.
24. G. Rose, P. Hawkes, Turing: a Fast Stream Cipher, Fast Software Encryption 2003, Fast Software Encryption 2003, LNCS 2887, Springer-Verlag, pp. 307-324, 2003.
25. R.A. Rueppel, Correlation Immunity and the Summation Generator, Crypto 1985, LNCS 218, Springer-Verlag 1986, pp. 260–272, 1985.
26. B. Schneier, Applied Cryptography, Wiley, New York, 1996.

New Combined Attacks on Block Ciphers

Eli Biham[1], Orr Dunkelman[1,*], and Nathan Keller[2]

[1] Computer Science Department, Technion,
Haifa 32000, Israel
{biham, orrd}@cs.technion.ac.il
[2] Einstein Institute of Mathematics, Hebrew University,
Jerusalem 91904, Israel
nkeller@math.huji.ac.il

Abstract. Differential cryptanalysis and linear cryptanalysis are the most widely used techniques for block ciphers cryptanalysis. Several attacks combine these cryptanalytic techniques to obtain new attacks, e.g., differential-linear attacks, miss-in-the-middle attacks, and boomerang attacks.

In this paper we present several new combinations: we combine differentials with bilinear approximations, higher-order differentials with linear approximations, and the boomerang attack with linear, with differential-linear, with bilinear, and with differential-bilinear attacks. We analyze these combinations and present examples of their usefulness. For example, we present a 6-round differential-bilinear approximation of s^5DES with a bias of 1/8, and use it to attack 8-round s^5DES using only 384 chosen plaintexts. We also enlarge a weak key class of IDEA by a factor of 512 using the higher-order differential-linear technique. We expect that these attacks will be useful against larger classes of ciphers.

1 Introduction

In a differential attack [5], the attacker seeks a fixed input difference that propagates through the nonlinear parts of the cipher to some fixed output difference with usually high (or zero) probability. Such pair of differences with the corresponding probability is called a *differential*. In the attack, the attacker asks for the encryption of pairs of plaintexts with the input difference given by the differential, and checks whether the output difference predicted by the differential occurs (with the predicted probability).

In a linear attack [30], the attacker seeks a linear approximation between the parity of a subset of the plaintext bits and the parity of a subset of the ciphertext bits with a biased probability. The attacker asks for the encryption of many plaintexts, and checks whether the linear relation predicted by the linear approximation is satisfied or not.

* The research presented in this paper was supported by the Clore scholarship programme.

H. Gilbert and H. Handschuh (Eds.): FSE 2005, LNCS 3557, pp. 126–144, 2005.

In 1994, Langford and Hellman [28] showed that both kinds of cryptanalysis can be combined together by a technique called *differential-linear cryptanalysis*, in which the differential is used to obtain a linear approximation (between two encryptions) with bias 1/2. The technique was improved in [8, 27], allowing the usage of differentials with probability lower than 1, thus making the technique applicable to a larger set of block ciphers.

The differential-linear technique was applied to analyze several (reduced versions of) block ciphers, such as: DES [32] (attacked in [28, 8]), IDEA [26] (attacked in [13, 20]), Serpent [1] (attacked in [9]), and COCONUT98 [35] (attacked in [8]). Some of the attacks are the best known attacks against the respective versions of the ciphers. It was also shown that the ciphertext-only extensions of differential and linear cryptanalysis work with differential-linear cryptanalysis as well [10].

Langford and Hellman's technique is an example for devising the distinguisher (to be used in the attack) as a combination of two much simpler parts. In this case, a combination of a differential and a linear approximation. Such combinations were later used in other cryptanalytic techniques, e.g., cryptanalysis using impossible differentials [6, 7] (miss in the middle), and boomerang attacks [36], both using combinations of differentials.

In this paper we present several new combinations of the differential, the higher-order differential, the boomerang, the linear, and the bilinear techniques. All of these combinations treat the distinguished part of the cipher as a cascade of two (or even three) sub-ciphers.

First, we show how to combine the differential cryptanalysis with the bilinear cryptanalysis [14]. Bilinear cryptanalysis is a generalization of linear cryptanalysis specially designed for Feistel block ciphers. In bilinear cryptanalysis the attacker studies relations between bilinear functions of the bits of the plaintext and bilinear functions of the bits of the ciphertext. Usually, the results of bilinear cryptanalysis are comparable with those of ordinary linear cryptanalysis. However, there are ciphers that are relatively strong against linear cryptanalysis but are vulnerable to bilinear cryptanalysis. For example, s^5DES [21] is stronger than DES against linear cryptanalysis while the best 3-round bilinear approximation of s^5DES has a bias of 1/4, which is much larger than the corresponding linear approximation for DES.

We show that bilinear approximations can be combined with differentials essentially in the same way as ordinary linear approximations are combined. However, there are some differences between a regular differential-linear attack and a differential-bilinear attack. We explore the similarities and the differences between the two attacks, and apply the differential-bilinear technique to attack 8-round s^5DES.

The next combination we discuss is the higher order differential-linear attack. Higher-order differential cryptanalysis [2, 22, 25] is a generalization of differential cryptanalysis that uses differentials of more than two plaintexts. In the higher-order differential attack the attacker analyses the development of the XOR of the intermediate data during the encryption of a set of plaintexts satisfying some conditions. Attacks which resemble higher-order differential attack, such as SQUARE-like attacks [12, 18, 24, 29], can also be combined with linear cryptanalysis.

We show that higher-order differentials (and SQUARE-like properties) can also be used as a building block in a two-phase attack. In higher-order differential-linear cryptanalysis, the attacker examines sets of plaintexts that have the input difference of the higher-order differential. The higher-order differential predicts the XOR value of all the intermediate encryption value after the higher-order differential. Then, the linear approximation can be applied to the entire set to predict the parity of a subset of the ciphertext bits (of all the ciphertexts).

The data complexity of the higher-order differential-linear attack is proportional to $2^{2m}/p^2q^{2m}$, where p is the probability of the higher-order differential, q is the bias of the linear approximation, and m is the number of plaintexts in each set. Therefore, the attack can be used only if either the structure is small enough or the linear approximation is very good (e.g., with bias 1/2). Such instances can occur in block ciphers, especially in weak key classes for which very strong and unexpected properties hold. For example, in the linear weak key class of IDEA [17], a specially built approximation has a bias of 1/2. We show that in the case of IDEA the size of the linear weak key class is increased from 2^{23} keys in the class of a regular linear attack to 2^{32} keys using a higher-order differential-linear attack.

The last combination we discuss in this paper is the differential-linear boomerang technique. The boomerang attack [36] treats the cipher as a cascade of two sub-ciphers, and exploits two differentials, one for each sub-cipher, in order to obtain some information on the differences using an adaptive chosen plaintext and ciphertext process. In a differential-linear boomerang attack, the attacker constructs a pair of encryptions whose difference in the intermediate encryption value is known by means of the boomerang technique. This pair can then be analyzed by means similar to those of the differential-linear cryptanalysis. Moreover, it appears that the linear boomerang is a special case of a more general attack. By decomposing the first sub-cipher into two sub-sub-ciphers (and the cipher into three sub-ciphers in total), we can apply the differential-linear (or the differential-bilinear) attack to the cipher.

One interesting feature of the (differential-)(bi)linear boomerang attack is that this is the first attack that treats the cipher as a cascade of three sub-ciphers successfully, while all previous works treat the cipher as a cascade of at most two sub-ciphers.

The paper is organized as follows: In Section 2 we shortly sketch the basic differential-linear attack. In Section 3 we present differential-bilinear cryptanalysis and apply it to DES and s^{5}DES. In Section 4 we discuss higher-order differential-linear cryptanalysis and present several applications of the attack, including increasing the linear weak class of IDEA. In Section 5 we introduce (differential-)(bi)linear boomerang attacks. This set of attacks are combinations of the boomerang technique with the (differential-)(bi)linear attack. We concentrate on the differential-bilinear boomerang attack, as this attack is the most general one (while the other variants can be treated as special cases of this attack). Finally, Section 6 concludes this paper.

2 Preliminaries

2.1 Notations

We use notations based on [3, 5] for differential and linear cryptanalysis, respectively. In our notations Ω_P, Ω_T are the input and the output differences of the differential, and λ_P, λ_C are the input and the output subsets (denoted by bit masks) of the linear approximation. We also use λ_T to denote the input subset in some cases.

Let $E = E_1 \circ E_0$ be a block cipher, i.e., $C = E_k(P) = E_{1_k}(E_{0_k}(P))$. For example, if E is DES, then E_0 can be the first eight rounds of DES, while E_1 are the last eight rounds. For sake of simplicity, we omit the key, as it is clear that encryption is done using a secret key. We denote the partial encryption of P (and the partial decryption of C) by T, i.e., $T = E_0(P) = E_1^{-1}(C)$.

The last notation is the scalar product of two strings x and y and is denoted by $x \cdot y$.

2.2 Differential-Linear Cryptanalysis

Langford and Hellman [28] show that a concatenation of a differential and a linear approximation is feasible. The main idea in the combination is to encrypt pairs of plaintexts, and check whether the corresponding ciphertext pairs have the same parity of the output mask or not.

Let $\Omega_P \to \Omega_T$ be a differential of E_0 with probability 1. Let $\lambda_T \to \lambda_C$ be a linear approximation of E_1 with bias $\pm q$. We start with a pair of plaintexts P_1 and $P_2 = P_1 \oplus \Omega_P$. After the partial encryption through E_0, the intermediate encryption values are T_1 and $T_2 = T_1 \oplus \Omega_T$, respectively. For any intermediate encryption value T and its corresponding ciphertext C, $\lambda_T \cdot T = \lambda_C \cdot C$ with probability $1/2 + q$. Therefore, each of the relations $\lambda_C \cdot C_1 = \lambda_T \cdot T_1$ and $\lambda_C \cdot C_2 = \lambda_T \cdot T_2 = \lambda_T \cdot T_1 \oplus \lambda_T \cdot \Omega_T$ is satisfied with probability $1/2 \pm q$. Hence, with probability $1/2 + 2q^2$ the relation $\lambda_C \cdot C_1 = \lambda_C \cdot C_2 \oplus \lambda_T \cdot \Omega_T$ holds.

We note that λ_T and Ω_T are known, and thus, we have constructed a condition on C_1 and C_2 which has probability $1/2 + 2q^2$, while for a random pair of ciphertexts, this condition is satisfied with probability $1/2$. This fact can be used in distinguishers and in key recovery attacks. Hellman and Langford also noted that it is possible to use truncated differentials [22] as long as $\lambda_T \cdot \Omega_T$ is predictable.

As both difference and parity are linear operations, the two linear approximations in E_1 in both encryptions can be combined into an approximation of E of the form

$$E_{1_1}\text{–differential–}E_{1_2},$$

where the lower subscript denotes whether the sub-cipher is in the first encryption or in the second, and "differential" refers to the differential combiner that ensures that the parities of the data before transition from E_0 to E_1 in both encryptions are always equal (or always differ).

This led to the introduction of a differential-linear approximation for 6-round DES which was composed of a 3-round differential and a 3-round linear approximation. The differential-linear approximation was then used to attack 8-round DES. The attack requires 768 chosen plaintexts, and has the lowest data requirements between all attacks on 8-round DES.

Later research [8, 27] showed that it is possible to have $\lambda_T \cdot \Omega_T$ unknown but fixed. Also, it was shown that when the differential-linear technique is applicable when the differential has probability $p \neq 1$. In that case the probability that $\lambda_T \cdot T_1 = \lambda_T \cdot T_2 \oplus \lambda_T \cdot \Omega_T$ is $1/2 + p'$, where $p' = p/2$, and thus the event $\lambda_C \cdot C_1 = \lambda_C \cdot C_2 \oplus \lambda_T \cdot \Omega_T$ holds with probability $1/2 + 4p'q^2 = 1/2 + 2pq^2$.

As we demonstrate later, in some of the attacks that we present this property does not hold. That is, the attacker has to know the exact value of the difference Ω_T, and in some cases, only certain values of the difference Ω_T can be used in the combined attack.

Moreover, even if $\Omega_T \cdot \lambda_P$ is unknown to the attacker but constant for a given key, the attack still succeeds. In that case we know that the value $\lambda_C \cdot C_1 \oplus \lambda_C \cdot C_2$ is either 0 or 1, with a bias of $2q^2$. This case is similar to the case in linear cryptanalysis, when $\lambda_K \cdot K$ is unknown, and can be either 0 or 1.

3 Differential-Bilinear Attack

3.1 Bilinear Cryptanalysis

The bilinear attack [14] is a generalization of linear cryptanalysis aimed at Feistel ciphers. The attack considers approximations involving bilinear terms of the input, the output, and the key. The reason this attack aims at Feistel ciphers is that it is easier to find such bilinear approximations for Feistel ciphers.

For the description of the bilinear approximations we adopt the notations used in [14]. We also put aside the probabilistic nature of some of the steps for sake of clarity (of course, when we use the approximations we take the probabilities back into account). Let the input value of the r-th round in a Feistel cipher be $(L_r[0, 1, ..., n-1], R_r[0, 1, ..., n-1])$, where L stands for the left half of the data and R stands for the right half (note that R_0 and L_0 compose the plaintext). Furthermore, we denote the input and the output values of the F-function in the r-th round by $I_r[0, 1, ..., n-1]$, and $O_r[0, 1, ..., n-1]$, respectively. Due to the structure of a Feistel cipher $I_r = R_r$, $R_{r+1} = L_r \oplus O_r$, and $L_{r+1} = R_r$.

Let α be a subset of $\{0, 1, ..., n-1\}$, then $L_r[\alpha] = \oplus\{L_r[s] | s \in \alpha\} = \oplus_{s\in\alpha} L_r[s]$, i.e., $L_r[\alpha]$ is the parity of all bits in the left half masked by α. Similarly $R_r[\beta]$ is the parity all bits in the right half masked by β.

According to the Feistel round, for any mask α, β and any round r:

$$L_{r+1}[\beta] \cdot R_{r+1}[\alpha] \oplus R_r[\beta] \cdot L_r[\alpha] = I_r[\beta] \cdot O_r[\alpha].$$

Such 1-round bilinear approximations can be concatenated to obtain bilinear approximations of several rounds. Concatenation requires some additional conditions, and also introduces some probability to the whole approximation. We note that in some cases the relations involve key bits in bilinear terms as well, e.g., $L_r[12] \cdot K_r[15]$. One-round approximations can also be extended such that they include linear terms in addition to the bilinear ones. In this case, the concatenation is more complex and can be achieved only if the linear terms fulfill some additional requirements. The full description of bilinear approximations is given in [14]. The general form of the obtained bilinear approximation is

$$\begin{aligned} L_0[\alpha_0] \cdot R_0[\beta_0] \oplus R_0[\gamma_0] \oplus L_0[\delta_0] \oplus L_n[\alpha_n] \cdot R_n[\beta_n] \oplus R_n[\gamma_n] \oplus L_n[\delta_n] = \\ L_0[\epsilon_0] \cdot K[\epsilon_1] \oplus R_0[\zeta_0] \cdot K[\zeta_1] \oplus L_n[\eta_0] \cdot K[\eta_1] \oplus R_n[\theta_0] \cdot K[\theta_1] \oplus K[\iota_1] \end{aligned} \quad (1)$$

where K is the key (or more precisely, the list of subkeys), and all Greek letters represent some mask.

Given the above approximation, the bilinear attack resembles the linear attack. Many plaintext/ciphertext pairs are gathered, and for any guess of $K[\epsilon_1]$, $K[\zeta_1]$, $K[\eta_1]$, $K[\theta_1]$, and $K[\iota_1]$, the attacker counts how many pairs satisfy the approximation. The guess for which the above approximation holds with the expected probability of $1/2+q$ is assumed to be the right guess.

We note that in a bilinear approximation there might be bilinear expressions involving the subkey. This fact has implications on the differential-bilinear attack which we explore later.

3.2 Differential-Bilinear Cryptanalysis

Roughly speaking, the differential-bilinear attack encrypts many pairs of plaintexts, and examines Whether the obtained pair of ciphertexts satisfy some bilinear approximation or not. This is very similar to the way that differential and linear cryptanalysis are combined.

We shall assume, without loss of generality, that the bilinear approximation has the form presented in Equation (1), and that the probability of the approximation is $1/2+q$. We note that it is possible to have several bilinear terms in the approximation, but this fact does not change our analysis. We denote the differential to be concatenated by $\Omega_P \rightarrow \Omega_T$, and assume that the differential has probability p.

The attacker chooses pairs of plaintexts P_1 and $P_2 = P_1 \oplus \Omega_P$. With probability p the the intermediate encryption values T_1 and T_2, respectively, have a difference that satisfies the equality

$$T_{1L}[\alpha_0] \cdot T_{1R}[\beta_0] \oplus T_{1L}[\gamma_0] \oplus T_{1R}[\delta_0] = T_{2L}[\alpha_0] \cdot T_{2R}[\beta_0] \oplus T_{1L}[\gamma_0] \oplus T_{2R}[\delta_0], \quad (2)$$

where T_{iL} is the left half of T_i, and similarly T_{iR} is the right half of T_i. We note that under the random distribution[1] assumption, in the $(1-p)$ of the cases where

[1] We note that whether this assumption holds for a given cipher needs to be throughly investigated, and if possible verified as done in [9].

the differential does not hold, Equation (2) holds in half of the times. Thus, the probability that Equation (2) holds is $p+(1-p)/2 = 1/2+p/2$, and the bias is $p' = p/2$.

Then, similarly to the differential-linear case, the pair of ciphertexts C_1 and C_2 satisfies the following equation

$$C_{1L}[\alpha_n]\cdot C_{1L}[\beta_n]\oplus C_{1L}[\gamma_n]\oplus C_{1R}[\delta_n] = C_{2L}[\alpha_n]\cdot C_{2R}[\beta_n]\oplus C_{2L}[\gamma_n]\oplus C_{2R}[\delta_n] \quad (3)$$

with probability $1/2+4p'q^2 = 1/2+2pq^2$.

However, unlike differential-linear cryptanalysis where any differential can be used for the combined attack, in the bilinear case the situation is more complicated. This is due to the fact that bilinear approximations require more knowledge about the data than linear approximations. In some cases, the required information is not given by the differential.

It appears that the knowledge of the difference $L_{T_1}[\alpha_0]\oplus L_{T_2}[\alpha_0]$ and $R_{T_1}[\beta_0]\oplus R_{T_2}[\beta_0]$ in the two encryptions does not imply the knowledge of the difference between the $L_T[\alpha_0]\cdot R_T[\beta_0]$ values. Thus, the attacker is restricted to the cases where the knowledge suggested by the difference Ω_T suffices to know the difference of the $L_T[\alpha_0]\cdot R_T[\beta_0]$ values. This is clearly the case when $\alpha\cdot\Omega_{TL} = \beta\cdot\Omega_{TR} = 0$, i.e., if the parity of the differences in the bits masked by α and β is zero. Another example is when there are six active bits in the output of the differential a, b, c, d, e and f, and the bilinear approximation is $a\cdot b+c\cdot d+e\cdot f+a\cdot f+c\cdot b+e\cdot d$. For an arbitrary bilinear relation $\sum_{\alpha,\beta} L_{T_i}[\alpha]\cdot R_{T_i}[\beta]$, where α and β are masks, the difference between the two sums can be predicted (to be zero) whenever the following two conditions hold simultaneously: (1) Each $L_{T_i}[\alpha]$ appears an even number of times in products with $R_{T_i}[\beta]$'s whose difference is 1, and (2) Each $R_{T_i}[\beta]$ appears an even number of times in products with $L_{T_i}[\alpha]$'s whose difference is 1.

We note that the linear terms of the approximation behave in the same way as in differential-linear cryptanalysis. This is due to the way the attack works — the attacker examines the difference in the output mask of two encryptions, and as long as the linear terms do not affect the bias of the difference in the output mask, the linear terms do not change the attack.

A more formal way to describe a differential-bilinear approximation is: Assume that the cipher E can be decomposed to two sub-ciphers $E = E_1\circ E_0$, where the differential $\Omega_P\to\Omega_T$ (and probability p) is used in E_0, and a bilinear approximation is used for E_1. Also assume that the bits predicted in Ω_T are sufficient to know the difference in the $L_T[\alpha_0]\cdot R_T[\beta_0]$ values with bias $p/2$. Let b_1 and b_2 denote the outputs of the bilinear approximation in the first and the second encryptions, respectively. The combination between the differential and the bilinear approximation can be represented by the following extended bilinear approximation:

$$b_1\text{–differential–}b_2,$$

where "differential" refers to the differential combiner. A distinguishing attack or a key recovery attack based on the differential-bilinear property is similar to an ordinary differential-linear attack — the attacker encrypts many plaintext pairs, and checks in how many of the pairs satisfy Equation (3).

The probability that a pair of ciphertexts (C_1, C_2), originating from a pair of plaintexts $(P_1, P_2 = P_1 \oplus \Omega_T)$, to satisfy Equation (3) is $1/2 + 4p'q^2 = 1/2 + 2pq^2$.

An interesting fact that will be demonstrated in the bilinear approximation of DES is that the subkey may be a part of the bilinear approximation. While in a linear approximation the linear factors of the key are independent of the plaintext (or the ciphertext), and can be treated like such, in a bilinear approximation the key may have a bilinear term involving the plaintext (or the ciphertext). Thus, Equation (3) might involve unknown key terms. When the equation involves unknown key terms, the attacker has to try all possible combinations for these key terms in the attack.

3.3 Applying Differential-Bilinear Cryptanalysis to DES and to s^{5}DES

In [14] a 3-round bilinear approximation of DES is presented. The approximation has a bias of $q = 1.66 \cdot 2^{-3}$ which is slightly better than the best 3-round linear approximation (that has a bias of $1.56 \cdot 2^{-3}$). The bilinear approximation is as follows:

$$\begin{aligned} &L_0[3,8,14,25] \oplus R_0[17] \oplus L_0[3] \cdot R_0[16,17,20] \oplus \\ &L_3[3,8,14,25] \oplus R_3[17] \oplus L_3[3] \cdot R_3[16,17,20] \\ &= K[sth] \oplus L_0[3] \cdot K[sth'] \oplus L_3[3] \cdot K[sth''], \end{aligned}$$

where (L_0, R_0) is the plaintext (or in our case the intermediate encryption value), (L_3, R_3) is the ciphertext, and $K[sth], K[sth']$, and $K[sth'']$ are subsets of the key bits.

We can concatenate the above bilinear approximation to a differential that predicts a zero difference in $L_0[3] \cdot R_0[16,17,20]$. The best 3-round differential that satisfies the requirements for concatenating the differential and the bilinear parts is presented in Figure 1. It has probability 46/64, and has the following structure: The first round has a zero input difference. The second round has an input difference with one active S-box — $S3$. The input difference of 4_x to $S3$ may cause an output difference whose bit 2 (of $S3$) is inactive with probability 28/64. If this is the case, then the masked bits of the input of the bilinear approximation are guaranteed to have a zero difference after the third round. Otherwise (with probability 36/64), bit 2 of the output of $S3$ is active. This bit enters $S4$ in the third round, and with probability 1/2 the output difference of $S4$ does not affect the bits masked by the input mask of the bilinear approximation, and thus, with probability $28/64 + 1/2 \cdot 36/64 = 46/64$ a pair with input difference $\Omega_P = (0_x, 00\ 20\ 00\ 00_x)$ has a zero difference in Ω_T in the bits masked by the bilinear approximation.

According to the previous analysis, the bias of the 6-round differential-bilinear approximation that starts with the above input difference is

$$2pq^2 = 2\frac{46}{64}(1.662^{-3})^2 = 1.98 \cdot 2^{-5}.$$

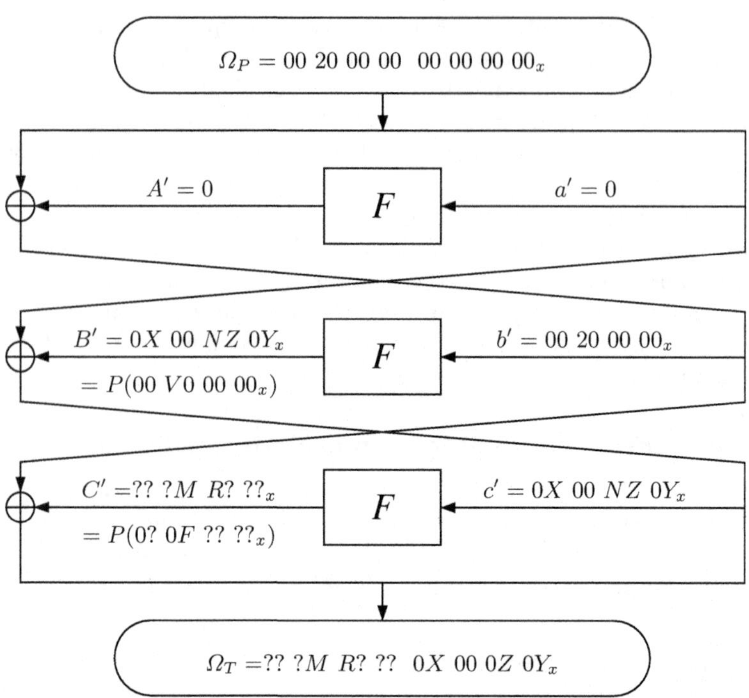

(where $X, Y \in \{0, 4\}$, $Z \in \{0, 1\}$, $M \in \{0, 2, 4, \ldots, E_x\}$, $R \in \{2, 4, 6\}$, $F \in \{0, 1, 2, 3, 8, 9, A_x, B_x\}$, $N \in \{0, 8\}$, $V \in \{3, 5, 6, 7, 9, A_x, B_x, C_x, D_x, E_x, F_x\}$ and where ? is any arbitrary value.)

Fig. 1. A 3-Round Differential of DES with Probability 46/64

This bias is slightly lower than the bias of the best 6-round differential-linear approximation (that equals to $2.43 \cdot 2^{-5}$), and thus, the differential-bilinear attack on 8-round DES requires more data than the corresponding differential-linear attack.

An example that illustrates the advantages of the differential-bilinear cryptanalysis over a regular differential-linear attack is s^{5}DES [21]. In [15] the following bilinear approximation with bias $q = 1/4$ is presented:

$$L_0[17, 23, 31] \oplus R_0[1, 5] \oplus L_0[9] \cdot R_0[5] \oplus$$
$$L_3[17, 23, 31] \oplus R_3[1, 5] \oplus L_3[9] \cdot R_3[5] = K[sth],$$

where $K[sth]$ is a subset of the key bits. This bilinear approximation can be concatenated to the 3-round differential with probability 1 presented in Figure 2. The differential assures that the difference in the input bits of the bilinear term of the bilinear approximation is zero with probability 1. Thus, the bias of the differential-bilinear approximation is:

$$2pq^2 = 2(1/4)^2 = 1/8$$

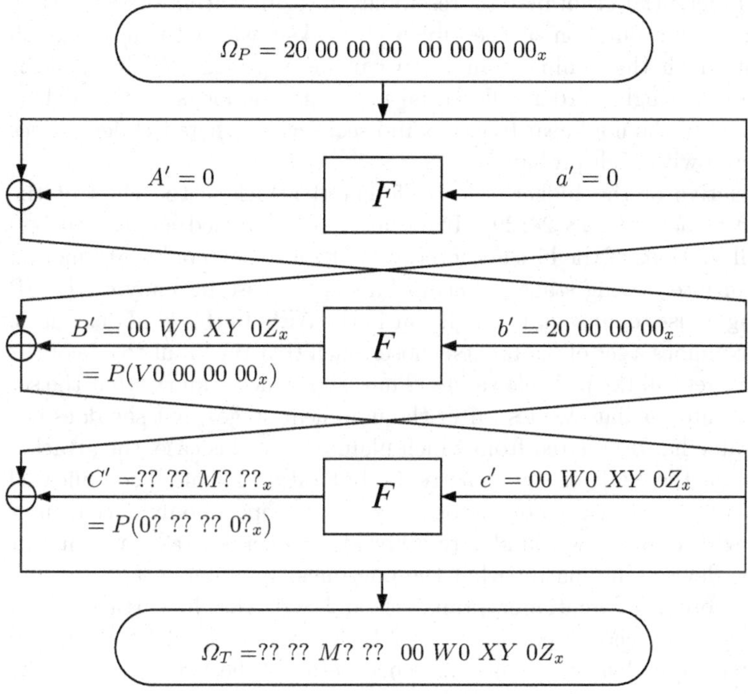

(where $V \in \{1, \dots, F_x\}$, $W \in \{0, 8\}$, $X \in \{0, 8\}$, $Y \in \{0, 2\}$, $Z \in \{0, 2\}$, $M \in \{0, \dots, 7\}$, and ? is any arbitrary value)

Fig. 2. A 3-Round Differential of s^{5}DES with Probability 1

This differential-bilinear approximation can be used to attack 8-round s^{5}DES using 384 chosen plaintexts and time complexity of $2^{20.2}$ encryptions. The attack finds about 90 suggestions for 16 bits of the key, where the right value is among the suggested values with probability of 65.5%.

4 Combining Higher-Order Differential and Linear Attacks

4.1 Higher-Order Differential Cryptanalysis and SQUARE-like Attacks

Higher-order differential cryptanalysis [2, 22, 25] is a generalization of differential cryptanalysis that exploits the algebraic structure of the cipher. In a higher-order differential attack the attacker asks for the encryption of a structured set of chosen plaintexts and analyses the XOR value (or some other function) of the ciphertexts. The motivation of the attack is the fact that while it is well known that linear relations between sets of bits during encryption should be avoided, in some instances higher-order relations between sets of bits can be found.

Ordinary differential cryptanalysis resembles an examination of the derivative of the nonlinear function of the cipher. It seeks cases with high enough probability in which the nonlinear function can be approximated by a linear function. Similarly, higher-order differential cryptanalysis looks at the higher-order derivatives of the nonlinear function and seeks cases where the derivatives can be predicted with high probability.

A close relative of the higher-order differential attack is the class of the SQUARE-like attacks [12, 18, 24, 29]. These attacks are aimed against ciphers in which small portions of the bits are interleaved by a strong nonlinear function while the main interleaving stage is linear. This is the case in many of the SP networks being in use today, and in particular in the AES. In this kind of attacks, the attacker examines a set of plaintexts, chosen such that the input to one of the non-linear part gets all the possible values. Thus, the attacker knows that the set contain all the intermediate values (after the nonlinear stage), but she does not know which value has originated from which plaintext. In this case, the attacker does not look for the XOR of the ciphertexts, but rather for more complicated functions, such as whether each of the possible values appears only once or not. SP networks with only a few rounds are especially vulnerable, as very efficient attacks can be devised, no matter what the non-linear function is [12].

Both higher-order differential cryptanalysis and SQUARE-like attacks, start with a set of specially chosen plaintexts, and look for some special structure in the obtained set of ciphertexts. The difference between the two attacks is the form of the special structure we expect/look for in the ciphertexts set.

4.2 The Higher-Order Differential-Linear Attack

The combination of higher-order differentials with linear approximations is similar to ordinary differential-linear cryptanalysis. The attacker uses the higher-order differential (or the SQUARE property) to predict the XOR value of the sets of masked bits in all of the elements of the structure, and then uses the linear approximation to compare this value with the XOR of the masked ciphertext bits in all of the encryptions.

Let *Set* be a set of plaintexts $\{P_1, P_2, \ldots, P_m\}$ such that the higher-order differential predicts (with some probability p) the value $\oplus_{i=1}^{m} T_i$ where the T_i's are the intermediate encryption values. Under standard independence assumptions, this means that the parity of any subset of bits taken over all intermediate encryption values is biased with a bias of $p' = p/2$. We also assume that there is a linear approximation that predicts the value of $\lambda_T \cdot T \oplus \lambda_C \cdot C$ with probability $1/2 + q$.

Lemma 1. *Let the event I be*

$$I = \{\lambda_P \cdot (T_1 \oplus \ldots \oplus T_m) = \lambda_C \cdot (C_1 \oplus \ldots \oplus C_m)\}.$$

Then (under standard independence assumptions) $\Pr[I] = 1/2 + 2^{m-1} q^m$.

Before the proof we note that I is actually the event that the XOR of the input mask, taken over all intermediate encryption values, is equal to the XOR of the output mask, taken over all ciphertexts.

Proof. The proof of the lemma is by induction on m, and is very similar to the proof of Matsui's Piling-up Lemma [30]. If $m = 1$, there is only one approximation and thus the probability equals to $1/2 + q$. Assume that the claim holds for structures of size k and consider a structure of size $k + 1$. We divide the structure into two structures, one consisting of k ciphertexts, and the other consisting of one ciphertext. The division into two structures can be done at random. Consider the probabilities of the events I in the two structures, i.e., consider each structure as an independent structure and consider the probability of the events I corresponding to these new structures. Clearly, the event I occurs for the whole structure if and only if the corresponding events I_1, I_k occur either for both structures or for none of them. By the induction hypothesis, the probability of such an event equals to:

$$(1/2 + 2^{k-1}q^k)(1/2 + q) + (1/2 - 2^{k-1}q^k)(1/2 - q) =$$
$$1/4 + 2^{k-2}q^k + 2^{k-1}q^{k+1} + q/2 + 1/4 - 2^{k-2}q^k + 2^{k-1}q^{k+1} - q/2 = 1/2 + 2^k q^{k+1}$$

Thus, by induction, the lemma is proven. Q.E.D.

Lemma 2. *Given a set of plaintexts with the input requirements of the higher-order differential, the bias of the event that the XOR of the output mask in all the ciphertexts equal to the value predicted by the linear approximation is*

$$\hat{b} = 2^{m-1}pq^m. \tag{4}$$

Proof. The proof is a combination of the result of the previous lemma with the probability of the higher-order differential. Let Z_1, Z_2 be the boolean variables defined as $Z_1 = \lambda_P \cdot (T_1 \oplus ... \oplus T_m)$, and $Z_2 = \lambda_C \cdot (C_1 \oplus ... \oplus C_m)$. We are interested in the probability $P(Z_2 = 0)$. If this probability differs from $1/2$, then we can use this property for the attack. Combining the higher-order differential with the results on the linear approximation obtained above, we get that $P(Z_1 = 0) = 1/2 + p/2$ and $P(Z_1 = Z_2) = 1/2 + 2^{m-1}q^m$. Therefore,

$$P(Z_2 = 0) = P(Z_1 = 0) \cdot P(Z_2 = Z_1) + P(Z_1 = 1) \cdot P(Z_2 \neq Z_1) =$$
$$(1/2 + p/2)(1/2 + 2^{m-1}q^m) + (1/2 - p/2)(1/2 - 2^{m-1}q^m) = 1/2 + 2^{m-1}pq^m.$$

Q.E.D.

Note that differential-linear cryptanalysis can be considered as a special case of higher-order differential-linear cryptanalysis, where the size of the structure is 2. Using Formula (4), the bias of the approximation is $\hat{b} = 2pq^2$.

4.3 Applications of Higher-Order Differential-Linear Cryptanalysis

Our first application of the higher-order differential-linear cryptanalysis is a generic attack. Let E be a Feistel block cipher with a bijective round function F. Denote the block size of E by $2n$. Assume that E has an r-round linear approximation with bias $1/2$. We combine this r-round linear approximation with a 3-round higher-order differential that exists with probability 1 for all such ciphers.

Let a word that is constant for all plaintexts in the structure be denoted by C. Let a word that assumes all possible values (a permutation) for a given structure be denoted by P, and let a word in which the XOR value of all the plaintexts in the structure is zero be denoted by B. For example (P, P) is a structure of 2^n plaintexts, where every possible value of the left half appears once, as well as every possible value of the right half (and we assume no relation between these instances). Another example is (B, C) — a structure of 2^n plaintexts where the right half is fixed in all the plaintexts, and the XOR of all the values in the left half is zero.

For the Feistel cipher described above, the following 3-round higher-order differential holds with probability 1:

$$(P, C) \xrightarrow{F} (C, P) \xrightarrow{F} (P, P) \xrightarrow{F} (P, B).$$

(This kind of property was first used in [4] with different attack methods). As can be seen from the higher-order differential, the attacker knows for certain that the XOR of the texts in the structure at the end of round 3 is 0, and the same is true for the XOR value in any specific bit as well. The 3-round higher-order differential can be combined with the linear approximation to devise a $(k + 3)$-round higher-order differential-linear approximation of the cipher. The overall bias of the approximation is 1/2, and thus the approximation requires several structures of 2^n chosen plaintexts to distinguish between the cipher and a random permutation.

This generic attack can be applied to FEAL [33]. FEAL is a 64-bit Feistel block cipher, with a bijective round function. There exists a linear approximation for three rounds of the cipher with bias 1/2 (see [31] for details). We can combine this linear approximation with the 3-round higher-order differential to devise a 6-round higher-order differential-linear approximation with bias 1/2 (and a set size of 2^{32} plaintexts), and use it to distinguish between FEAL-6 and a random permutation. This distinguisher can be used in a key recovery attacks on FEAL-7 and FEAL-8. Even though these attacks are far from being the best known attacks, they demonstrate the feasibility of higher-order differential-linear cryptanalysis.

Another application of this technique is a weak key class of the block cipher IDEA [26]. IDEA has a 64-bit block size and it consists of 8.5 rounds. It is based on operations on four words of 16-bit each.

There is a weak key class of 2^{32} keys, each having zero in 96 positions, that can be detected using a higher-order differential-linear attack. The underlying linear approximation is the one used in the linear weak key class of IDEA of 2^{23} keys in [17]. The approximation has bias 1/2, and it propagates through IDEA by exploiting the fact that for the weak key class the multiplication operation can be approximated with bias 1/2.

Our weak key class uses a 3-round higher-order differential that starts with sets of the form (P, C, P, C), for which after three rounds the XOR of the least significant bits of the first and the second words are zero. The linear approximation is used in the remaining 5.5 rounds, and it has a bias of 1/2. Thus, for this weak key class, the output mask of all ciphertexts in a given set is the same. We can use this fact and about 100 sets to identify whether the key used in the encryption is in the weak key class.

We conclude that our new weak key class contains 2^{32} keys, 512 times more keys than the original linear weak key class. The membership tests requires about 2^{23} chosen plaintexts with a negligible amount of computation time.

We conclude that the higher-order differential-linear attack is feasible, and that in some cases it can be used to improve existing attacks and to devise new attacks. At this stage we have not found a published cipher for which our new technique yields the best attack, even though it is clear that one can easily "engineer" a dedicated cipher with this property.

4.4 Related Work

We first note that the higher-order differential-linear attack was developed independently in [34] under the name square-nonlinear attack. The attack combines a SQUARE property with a nonlinear approximation whose input is linear. Thus, the analysis can be reproduced, and despite the non-linear nature of the attack, the biases behave in the same way. The square-nonlinear attack was used to attack reduced round version of SHACAL-2.

Another related work is the chosen plaintext linear attack [23]. In the chosen plaintext linear attack, the attacker encrypts structures of plaintexts, chosen such that the input mask is the same for all values in the structure. An alternative description would say that the set is chosen such that the difference of the intermediate encryption values is 0 in the bits considered by the approximation. In such a case the attacker can examine only the output parities. This method can be used to either eliminate rounds from the approximation, or to reduce the number of candidate subkeys (as rounds before the approximation no longer play an active role in determining whether the approximation holds or not).

While there are similarities between the chosen plaintext linear attack and our higher-order differential-linear attack, there are also major differences. Our proposed technique looks for the XOR of all ciphertexts in the set, while the chosen plaintext linear attack examines the approximation in each ciphertext separately.

Actually, chosen plaintext linear attack will usually lead to a better attack, as it takes into consideration each plaintext/ciphertext pair, rather than performs an operation that "cancels" the information conveyed in 2^{16} (or even more) plaintext/ciphertext pairs. On the other hand, the chosen plaintext linear attack fixes bits of the plaintext, leading to a smaller number of possible plaintext/ciphertext values. Another advantage of our attack is its ability to "correct" wrong structures, i.e., assume that the input mask is biased with some probability (rather than fixed).

5 Combining the Boomerang Attack with Linear and Bilinear Techniques

5.1 The Boomerang Attack

The main idea behind the boomerang attack [36] is to use two short differentials with relatively high probabilities instead of one long differential with very low probability. The attack treats the block cipher $E\colon \{0,1\}^n \times \{0,1\}^k \rightarrow \{0,1\}^n$ as a cascade $E = E_1 \circ E_0$, such that for E_0 there exists a differential $\alpha \rightarrow \beta$ with probability p_0, and for E_1 there exists a differential $\gamma \rightarrow \delta$ with probability p_1. The distinguisher performs the following boomerang process:

- Ask for the encryption of a pair of plaintexts (P_1, P_2), such that $P_1 \oplus P_2 = \alpha$, and denote the corresponding ciphertexts by (C_1, C_2).
- Calculate $C_3 = C_1 \oplus \delta$ and $C_4 = C_2 \oplus \delta$, and ask for the decryption of the pair (C_3, C_4). Denote the corresponding plaintexts by (P_3, P_4).
- Check whether $P_3 \oplus P_4 = \alpha$.

We denote the intermediate encryption value of P_i (or the intermediate decryption value of C_i) between E_0 and E_1 by X_i, i.e., $X_i = E_0(P_i) = E_1^{-1}(C_i)$. If (P_1, P_2) is a right pair with respect to the first differential, then $X_1 \oplus X_2 = \beta$. If both pairs (C_1, C_3) and (C_2, C_4) are right pairs with respect to the second differential, then $X_1 \oplus X_3 = \gamma = X_2 \oplus X_4$. If all these conditions are satisfied then $X_3 \oplus X_4 = \beta$. The boomerang attack uses the obtained β value by decrypting the pair (X_3, X_4), which with probability p_0 leads to $P_3 \oplus P_4 = \alpha$. The overall probability of such a quartet is $p_0^2 p_1^2$.

The attack can be mounted for all possible β's and γ's simultaneously (as long as $\beta \neq \gamma$). Thus, a right quartet for E is encountered with probability no less than $(\hat{p}_0 \hat{p}_1)^2$, where:

$$\hat{p}_0 = \sqrt{\sum_{\beta} \Pr{}^2[\alpha \rightarrow \beta]}, \qquad \text{and} \qquad \hat{p}_1 = \sqrt{\sum_{\gamma} \Pr{}^2[\gamma \rightarrow \delta]}.$$

The complete analysis is given in [36]. In particular it is possible to show that for a specific value of β, and the corresponding probability p_0 and all γ's simultaneously, the probability for $X_3 \oplus X_4 = \beta$ is $p_0 \hat{p_1}^2$. We shall use this fact later.

5.2 Differential-Bilinear-Boomerang Attack (and Relatives)

We first note that linear, differential-linear, and bilinear approximations, are special cases of differential-bilinear approximations (up to whether we consider pairs of plaintexts or plaintext/ciphertext pairs). Hence, if we can combine the differential-bilinear attack with some other attack, we can actually combine any of the linear, the differential-linear, or the bilinear attacks as well.

Our newly proposed attacks exploit the β difference between the intermediate decryption values X_3 and X_4 of the encryptions whose ciphertexts are C_3 and C_4. If there is a differential-bilinear approximation for E_0^{-1} (the decryption through

E_0), then the pair (X_3, X_4) has the required input difference, and thus, there is some bilinear relation between X_3 and X_4 whose probability (or bias) is non-trivial.

More formally, let (X_3, X_4) (generated by the partial decryption of C_3 and C_4 during the boomerang process) be with difference β. Assume that there exists a differential-bilinear approximation with bias $2pq^2$ for E_0^{-1} with input difference β. Thus, it is possible to analyze the corresponding plaintexts as in the differential-bilinear attack, just like as suggested in Section 3.

However, the pair (X_3, X_4) does not always have the required difference β, which occurs with probability $p_0\hat{p}_1^2$. By performing the analysis of the differential-bilinear attack again, and taking into consideration the probability that the β difference occurs, we conclude that the differential-bilinear relation has a bias of $2\hat{p}_1^2 p_0 p q^2$.

Actually, we treat the first sub-cipher E_0 as a cascade of two sub-sub-ciphers, i.e, $E_0 = E_{0_1} \circ E_{0_0}$. The differential is used in the the first part of the backward direction, i.e., in $E_{0_1}^{-1}$, while the bilinear approximation is used in the second par of $E_{0_0}^{-1}$ (also in the backward direction).

The differential-bilinear boomerang attack tries to obtain a difference between two intermediate encryption values in the transition between the first sub-sub-cipher and the second sub-sub-cipher (both are parts of the first sub-cipher). This is a somewhat "asymmetric" boomerang, where for the first pair (P_1, P_2) we have a different number of rounds in the first sub-cipher than for the pair (P_3, P_4).

As the bias of the differential-bilinear boomerang is very low, it might seem that using other techniques based on decomposing the cipher into sub-cipher is always better than this attack. Even though currently we have no example where this attack is better than other combinations, we believe such cases exist.

We start with showing that there are cases where the proposed attack can be better than the boomerang attack. At a first glance, even if we assume that the bias of the differential-bilinear approximation of E_0 is $1/2$, then the bias of the whole differential-bilinear boomerang approximation is $\hat{p}_1^2 p_0$. Thus, the data complexity of the differential-bilinear boomerang attack is expected to be at least $O(\hat{p}_1^{-4} p_0^{-2})$, while a regular boomerang attack requires a usually smaller data complexity of $O(\hat{p}_0^{-2}\hat{p}_1^{-2})$. However, this is true only for a boomerang attack that uses regular differentials. In such case, the probability of the differential in the decryption direction is equal to the probability in the encryption direction. But in some boomerang attacks, truncated differential are used, and for these kind of differentials the probability depends on the direction. Thus, it might lead to an attack which is better than the boomerang attack, if for example, there is a truncated differential that is used in the forward direction of E_0, but cannot be used in the backward direction due to low probability.

Another attack that can be used instead of the differential-bilinear boomerang is the differential-(bi)linear attack. As mentioned before, there is a good differential in the backward direction, and a good bilinear approximation. The reason why this process might yield a better attack is that the difference predicted by the differential after the partial decryption may not be suitable for concatenation with a bilinear approximation. In this case, the boomerang process is used to change the difference to a more "friendly" one.

For linear (or differential-linear) cryptanalysis, where the exact difference has a much smaller effect, the answer is different. Usually, it is assumed that the approximation has an independent random behavior for any two plaintexts, even if there is some constant difference between them. The chosen ciphertext linear cryptanalysis [23] has shown that this is not the case, and that the actual values encrypted can alter the probabilities related to the approximation. Hence, the bias of the linear approximation may increase if there is a specific difference, instead of some random difference. Such an increase would lead to an higher biases, which in turn would mean better attacks.

6 Summary

In this paper we presented several new combined attacks. Each of these combinations has scenarios where it yields an attack that may be better than differential-linear attacks, differential attacks, or linear attacks for some ciphers.

The differential-bilinear attack, the higher-order differential-linear attack, and the (differential-)(bi)linear boomerang attack, are examples of attacks based on treating the cipher as a cascade of sub-ciphers. This kind of treatment allows us to present a a differential-bilinear approximation for 6-round s^5DES with a bias of 1/8. The decomposition into sub-ciphers can be used to enlarge the linear weak-key class of IDEA by a factor of 512.

We conclude that new designs have to take into consideration combined attacks, including the well-known ones such as differential-linear and boomerang attacks, as well as the new ones presented in this paper.

Acknowledgments

We would like to thank Nicolas Courtois for the information he provided to us, that helped us a lot during the work on this paper, and the anonymous referees for suggesting valuable ideas and improvements.

References

1. Ross Anderson, Eli Biham, Lars R. Knudsen, *Serpent: A Proposal for the Advanced Encryption Standard*, NIST AES Proposal, 1998.
2. Eli Biham, *Higher Order Differential Cryptanalysis*, unpublished paper, 1994.
3. Eli Biham, *On Matsui's Linear Cryptanalysis*, Advances in Cryptology, proceedings of EUROCRYPT '94, Lecture Notes in Computer Science 950, pp. 341–355, Springer-Verlag, 1994.
4. Eli Biham, *Cryptanalysis of Ladder-DES*, proceedings of Fast Software Encryption 4, Lecture Notes in Computer Science 1267, pp. 134–138, Springer-Verlag, 1997.

5. Eli Biham, Adi Shamir, *Differential Cryptanalysis of the Data Encryption Standard*, Springer-Verlag, 1993.
6. Eli Biham, Alex Biryukov, Adi Shamir, *Miss in the Middle Attacks on IDEA and Khufu*, proceedings of Fast Software Encryption 6, Lecture Notes in Computer Science 1636, pp. 124–138, Springer-Verlag, 1999.
7. Eli Biham, Alex Biryukov, Adi Shamir, *Cryptanalysis of Skipjack Reduced to 31 Rounds Using Impossible Differentials*, Advances in Cryptology, proceedings of EUROCRYPT '99, Lecture Notes in Computer Science 1592, pp. 12–23, Springer-Verlag, 1999.
8. Eli Biham, Orr Dunkelman, Nathan Keller, *Enhanced Differential-Linear Cryptanalysis*, Advances in Cryptology, proceedings of ASIACRYPT '02, Lecture Notes in Computer Science 2501, pp. 254–266, Springer-Verlag, 2002.
9. Eli Biham, Orr Dunkelman, Nathan Keller, *Differential-Linear Cryptanalysis of Serpent*, proceedings of Fast Software Encryption 10, Lecture Notes in Computer Science 2887, pp. 9–21, Springer-Verlag, 2003.
10. Alex Biryukov, Eyal Kushilevitz, *From Differential Cryptoanalysis to Ciphertext-Only Attacks*, Advances in Cryptology, proceedings of CRYPTO '98, Lecture Notes in Computer Science 1462, pp. 72–88, Springer-Verlag, 1998.
11. Alex Biryukov, Jorge Nakahara, Bart Preneel, Joos Vandewalle, *New Weak-Key Classes of IDEA*, proceedings of ICICS '02, Lecture Notes in Computer Science 2513, pp. 315–326, Springer-Verlag, 2002.
12. Alex Biryukov, Adi Shamir, *Structural Cryptanalysis of SASAS*, Advances in Cryptology, proceedings of EUROCRYPT '01, Lecture Notes in Computer Science 2045, pp. 394–405, Springer-Verlag, 2001.
13. Johan Borst, Lars R. Knudsen, Vincent Rijmen, *Two Attacks on Reduced Round IDEA*, Advances in Cryptology, proceedings of EUROCRYPT '97, Lecture Notes in Computer Science 1233, pp. 1–13, Springer-Verlag, 1997.
14. Nicolas T. Courtois, *Feistel Schemes and Bi-Linear Cryptanalysis*, Advances in Cryptology, proceedings of CRYPTO '04, Lecture Notes in Computer Science 3152, pp. 23–40, Springer-Verlag, 2004.
15. Nicolas T. Courtois, *Feistel Schemes and Bi-Linear Cryptanalysis (extended version)*, private communications, 2004.
16. Nicolas T. Courtois, *The Inverse S-box, Non-linear Polynomial Relations and Cryptanalysis of Block Ciphers*, private communications, 2004.
17. Joan Daemen, René Govaerts, Joos Vandewalle, *Weak Keys for IDEA*, Advances in Cryptology, proceedings of CRYPTO '93, Lecture Notes in Computer Science 773, pp. 224–231, Springer-Verlag, 1994.
18. Joan Daemen, Lars R. Knudsen, Vincent Rijmen, *The Block Cipher Square*, proceedings of Fast Software Encryption 4, Lecture Notes in Computer Science 1267, pp. 149–165, Springer-Verlag, 1997.
19. Niels Ferguson, John Kelsey, Stefan Lucks, Bruce Schneier, Mike Stay, David Wagner, Doug Whiting, *Improved Cryptanalysis of Rijndael*, proceedings of Fast Software Encryption 7, Lecture Notes in Computer Science 1978, pp. 213–230, Springer-Verlag, 2001.
20. Philip Hawkes, *Differential-Linear Weak Keys Classes of IDEA*, Advances in Cryptology, proceedings if EUROCRYPT '98, Lecture Notes in Computer Science 1403, pp. 112-126, Springer-Verlag, 1998.
21. Kwangjo Kim, Sangjun Lee, Sangjun Park, Daiki Lee, *How to Strengthen DES against Two Robust Attacks*, proceedings of Joint Workshop on Information Security and Cryptology, 1995.

22. Lars Knudsen, *Truncated and Higher Order Differentials*, proceedings of Fast Software Encryption 2, Lecture Notes in Computer Science 1008, pp. 196–211, Springer-Verlag, 1995.
23. Lars R. Knudsen, John E. Mathiassen, *A Chosen-Plaintext Linear Attack on DES*, proceedings of Fast Software Encryption 7, Lecture Notes in Computer Science 1978, pp. 262–272, Springer-Verlag, 2001.
24. Lars R. Knudsen, David Wagner, *Integral Cryptanalysis*, proceedings of Fast Software Encryption 9, Lecture Notes in Computer Science 2365, pp. 112–127, Springer-Verlag, 2002.
25. Xuejia Lai, *Higher Order Derivations and Differential Cryptanalysis*, in *Communications and Cryptography: Two Sides of One Tapestry*, Kluwer Academic Publishers, pp. 227-233, 1994.
26. Xuejia Lai, James L. Massey, *A Proposal for a New Block Cipher Encryption Standard*, Advances in Cryptology, proceedings of EUROCRYPT '90, Lecture Notes in Computer Science 473, pp. 389–404, Springer-Verlag, 1991.
27. Susan K. Langford, *Differential-Linear Cryptanalysis and Threshold Signatures*, Ph.D. thesis, 1995.
28. Susan K. Langford, Martin E. Hellman, *Differential-Linear Cryptanalysis*, Advances in Cryptology, proceedings of CRYPTO '94, Lecture Notes in Computer Science 839, pp. 17–25, Springer-Verlag, 1994.
29. Stefan Lucks, *The Saturation Attack — A Bait for Twofish*, proceedings of Fast Software Encryption 8, Lecture Notes in Computer Science 2355, pp. 1–15, Springer-Verlag, 2002.
30. Mitsuru Matsui, *Linear Cryptanalysis Method for DES Cipher*, Advances in Cryptology, proceedings of EUROCRYPT '93, Lecture Notes in Computer Science 765, pp. 386–397, Springer-Verlag, 1994.
31. Mitsuru Matsui, Atsuhiro Yamagishi, *A new method for known plaintext attack of FEAL cipher*, Advances in Cryptology, proceedings of EUROCRYPT '92, Lecture Notes in Computer Science 658, pp. 81–91, Springer-Verlag, 1993.
32. US National Bureau of Standards, *Data Encryption Standard*, Federal Information Processing Standards Publications No. 46, 1977.
33. Akihiro Shimizu, Shoji Miyaguchi, *Fast Data Encipherment Algorithm FEAL*, Advances in Cryptology, proceedings of EUROCRYPT '87, Lecture Notes in Computer Science 304, pp. 267–278, Springer-Verlag, 1988.
34. Yongsup Shin, Jongsung Kim, Guil Kim, Seokhie Hong, Sangjin Lee, *Differential-Linear Type Attacks on Reduced Rounds of SHACAL-2*, proceedings of ACISP 2004, Lecture Notes in Computer Science 3108, pp. 110–122, Springer-Verlag, 2004.
35. Serge Vaudenay, *Provable Security for Block Ciphers by Decorrelation*, Journal of Cryptology, Vol 16, Number 4, pp. 249–286, Springer-Verlag, 2003.
36. David Wagner, *The Boomerang Attack*, proceedings of Fast Software Encryption 6, Lecture Notes in Computer Science 1636, pp. 156–170, Springer-Verlag, 1999.

Small Scale Variants of the AES

C. Cid*, S. Murphy, and M.J.B. Robshaw

Information Security Group,
Royal Holloway, University of London,
Egham, Surrey, TW20 0EX, U.K.
{Carlos.Cid, S.Murphy, M.Robshaw}@rhul.ac.uk

Abstract. In this paper we define small scale variants of the AES. These variants inherit the design features of the AES and provide a suitable framework for comparing different cryptanalytic methods. In particular, we provide some preliminary results and insights when using off-the-shelf computational algebra techniques to solve the systems of equations arising from these small scale variants.

1 Introduction

The potential for algebraic attacks [1, 2, 8] on the AES [4, 11] has been the source of recent speculation. Two important (and complementary) approaches to the algebraic analysis of the AES were provided in [2] and [8]. In [2] it was shown how recovering an AES encryption key could be viewed as solving a set of sparse overdefined multivariate quadratic equations over $GF(2)$ and a method—the XSL method—for solving this set of AES-specific equations was proposed. In [8] the AES was embedded in a related cipher (called the BES) and it was shown how recovering an AES encryption key could be viewed as solving a similar set of equations over $GF(2^8)$. The highly structured equation systems that result from this approach may well be more tractable than those arising from a $GF(2)$ perspective [8, 9].

Currently, however, it is unknown whether the XSL—or any other proposed method of solution—works on the AES. For most types of cryptanalysis it is straightforward to perform experiments on reduced versions of the cipher to understand how an attack might perform. However this is not so easy for the AES, and while some experiments have been conducted [2], the equation systems used were very different from those that might actually arise from the AES.

With this goal in mind, we specify a family of small scale variants of the AES. Previous variants have been used before as an educational tool [10, 12], but our aim is to provide a fully parameterised framework for the analysis of AES equation systems. We describe how to construct the equation systems corresponding to these small scale variants of the AES, and give an example of such a system

* This author was supported by EPSRC Grant GR/S42637.

H. Gilbert and H. Handschuh (Eds.): FSE 2005, LNCS 3557, pp. 145–162, 2005.

derived using the BES-style embedding. We report on some preliminary analysis of a number of small scale variants and provide the first experimental insight into the behaviour of algebraic cryptanalysis on AES-like ciphers.

2 Small Variants of the AES

We define two sets of small scale variants of the AES; these differ in the form of the final round. These two sets of variants will be denoted by $\mathrm{SR}(n, r, c, e)$ and $\mathrm{SR}^*(n, r, c, e)$.

2.1 Small Scale AES Parameters

Both $\mathrm{SR}(n, r, c, e)$ and $\mathrm{SR}^*(n, r, c, e)$ are parameterised in the following way:

- n is the number of (encryption) rounds;
- r is the number of "rows" in the rectangular arrangement of the input;
- c is the number of "columns" in the rectangular arrangement of the input;
- e is the size (in bits) of a word.

$\mathrm{SR}(n, r, c, e)$ and $\mathrm{SR}^*(n, r, c, e)$ both have n rounds and a block size of rce bits, where a data block is viewed as an array of $(r \times c)$ "words" of e bits. We will see that the full AES is equivalent to $\mathrm{SR}^*(10, 4, 4, 8)$.

Number of Rounds • . The AES is an iterated block cipher consisting of 10 rounds. The typical round uses four different operations. The small scale variants $\mathrm{SR}(n, r, c, e)$ and $\mathrm{SR}^*(n, r, c, e)$ consist of n rounds, with $1 \leq n \leq 10$, using small scale variants of these operations. These operations are specified in Section 2.2.

Data Block Array Size (• × •). Each element of the data array is a word of size e bits. The array itself has r rows and c columns. We consider small scale variants of the AES with both r and c restricted to $1, 2$, or 4. Some examples are given below. Note that we adopt the AES-style of numbering "words" within an array and work by column first.

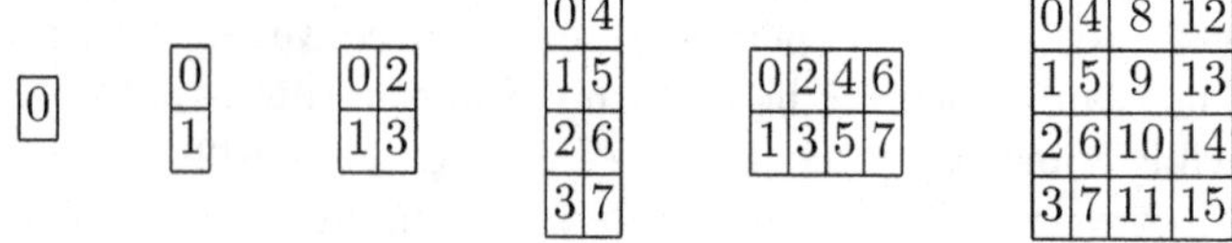

Word Size •. We define small scale variants of the AES for word sizes $e = 4$ and $e = 8$. It is natural within the context of the AES to regard a word of size e as an element of the field $GF(2^e)$. Thus we define small scale variants of the AES with respect to the two fields $GF(2^4)$ and $GF(2^8)$.

The small scale variants $\mathrm{SR}(n, r, c, 4)$ and $\mathrm{SR}^*(n, r, c, 4)$ use the field $GF(2^4)$. We use the primitive polynomial $X^4 + X + 1$ over $GF(2)$ to define this field. We let ρ be a root of this polynomial, so

$$GF(2^4) = \frac{GF(2)[X]}{(X^4 + X + 1)} = GF(2)(\rho).$$

When referring to elements of $GF(2^4)$, we sometimes use hexadecimal notation, so that $\texttt{D} = \rho^3 + \rho^2 + 1$ and so on.

The small scale variants of the AES with word size 8, $\text{SR}(n, r, c, 8)$ and $\text{SR}^*(n, r, c, 8)$, use the field $GF(2^8)$. The Rijndael polynomial $X^8 + X^4 + X^3 + X + 1$ over $GF(2)$ is used to define this field. We let θ be a root of this polynomial, so

$$GF(2^8) = \frac{GF(2)[X]}{(X^8 + X^4 + X^3 + X + 1)} = GF(2)(\theta).$$

When referring to elements of $GF(2^8)$, we again sometimes use hexadecimal notation, so that $\texttt{D1} = \theta^7 + \theta^6 + \theta^4 + 1$ and so on.

2.2 Small Scale Round Operations

Each round of the AES consists of some combination of the following operations:

1. `SubBytes`
2. `ShiftRows`
3. `MixColumns`
4. `AddRoundKey`

A round of the small scale variants of the AES consists of small scale variants of these operations. For the last round of the AES, the operation `MixColumns` is omitted. Similarly, for $\text{SR}^*(n, r, c, e)$ the final round does not use `MixColumns`, whereas `MixColumns` is retained for the final round of $\text{SR}(n, r, c, e)$. The AES is thus identical to $\text{SR}^*(10, 4, 4, 8)$.

Note that the two ciphertexts produced by $\text{SR}(n, r, c, e)$ and $\text{SR}^*(n, r, c, e)$ when encrypting the same plaintext under the same key are related by an affine mapping. A solution of the system of equations for one cipher would immediately give a solution for the other and so, without loss of generality, we only consider $\text{SR}(n, r, c, e)$ for the remainder of this paper.

`SubBytes`. The operation `SubBytes` uses an S-Box and is defined to be the simultaneous application of the S-Box to each element of the data array. For the small scale variants $\text{SR}(n, r, c, 4)$ based on the field $GF(2^4)$, we define the S-Box by analogy with the AES. Thus this S-box consists of the following three (sequential) operations.

1. *Inversion.* The first operation of the S-Box is an inversion in the field $GF(2^4)$ (with $0 \mapsto 0$), using the representation defined in Section 2.1. The look-up table for this inversion map is given below.

Inversion in $GF(2^4)$																
Input	0	1	2	3	4	5	6	7	8	9	A	B	C	D	E	F
Output	0	1	9	E	D	B	7	6	F	2	C	5	A	4	3	8

2. *GF*(2)*-linear map.* The output of the inversion is the input to a *GF*(2)-linear map. This *GF*(2)-linear map is given by (the pre-multiplication by) the circulant *GF*(2)-matrix

$$\begin{pmatrix} 1&1&1&0 \\ 0&1&1&1 \\ 1&0&1&1 \\ 1&1&0&1 \end{pmatrix}$$

with respect to the "FIPS component ordering" [11]. The look-up table for the *GF*(2)-linear map is given below.

$GF(2)$-linear map in $GF(2^4)$																
Input	0	1	2	3	4	5	6	7	8	9	A	B	C	D	E	F
Output	0	D	B	6	7	A	C	1	E	3	5	8	9	4	2	F

Note that this *GF*(2)-linear map can also be expressed as the linearised polynomial $f(X) = \lambda_0 X^{2^0} + \lambda_1 X^{2^1} + \lambda_2 X^{2^2} + \lambda_3 X^{2^3}$, where $(\lambda_0, \lambda_1, \lambda_2, \lambda_3) = (5, 1, \mathtt{C}, 5)$. Thus we have

$$\begin{aligned} f(X) &= (\rho^2+1)X + X^2 + (\rho^3+\rho^2)X^4 + (\rho^2+1)X^8 \\ &= 5X + 1X^2 + \mathtt{C}X^4 + 5X^8 \end{aligned}$$

3. *S-Box constant.* The S-Box constant 6 (or equivalently $\rho^2 + \rho$) is added (as an element of $GF(2^4)$) to the output of the *GF*(2)-linear map. This result is the output of the S-Box.

The look-up table for the entire S-Box is given below.

S-Box over $GF(2^4)$																
Input	0	1	2	3	4	5	6	7	8	9	A	B	C	D	E	F
Output	6	B	5	4	2	E	7	A	9	D	F	C	3	1	0	8

For the small scale variants SR$(n, r, c, 8)$ we use the AES S-Box. The values of the S-box operation over $GF(2^8)$ are available in the AES specification [11].

S-Box Summary	$GF(2^4)$	$GF(2^8)$
Irreducible polynomial	$X^4 + X + 1$	$X^8 + X^4 + X^3 + X + 1$
GF(2)*-linear map*	$\begin{pmatrix} 1&1&1&0 \\ 0&1&1&1 \\ 1&0&1&1 \\ 1&1&0&1 \end{pmatrix}$	$\begin{pmatrix} 1&0&0&0&1&1&1&1 \\ 1&1&0&0&0&1&1&1 \\ 1&1&1&0&0&0&1&1 \\ 1&1&1&1&0&0&0&1 \\ 1&1&1&1&1&0&0&0 \\ 0&1&1&1&1&1&0&0 \\ 0&0&1&1&1&1&1&0 \\ 0&0&0&1&1&1&1&1 \end{pmatrix}$
Constant	6	63

`ShiftRows`. The `ShiftRows` operation is defined to be the simultaneous (left) rotation of the row i of the data array, $0 \le i \le r-1$, by i positions. This is independent of the number of columns and the top row is fixed by this operation.

`MixColumns`. The `MixColumns` operation pre-multiplies each column of the data array by an invertible circulant $GF(2^e)$-matrix with row (and column) sum 1. These matrices are all MDS matrices (see [4]) and the choice of matrix in a small scale variant depends on the number of rows in the data array.

Number of Rows	$GF(2^4)$	$GF(2^8)$
$r=1$	$\begin{pmatrix}1\end{pmatrix}$	$\begin{pmatrix}1\end{pmatrix}$
$r=2$	$\begin{pmatrix}\rho+1 & \rho \\ \rho & \rho+1\end{pmatrix}$	$\begin{pmatrix}\theta+1 & \theta \\ \theta & \theta+1\end{pmatrix}$
$r=4$	$\begin{pmatrix}\rho & \rho+1 & 1 & 1 \\ 1 & \rho & \rho+1 & 1 \\ 1 & 1 & \rho & \rho+1 \\ \rho+1 & 1 & 1 & \rho\end{pmatrix}$	$\begin{pmatrix}\theta & \theta+1 & 1 & 1 \\ 1 & \theta & \theta+1 & 1 \\ 1 & 1 & \theta & \theta+1 \\ \theta+1 & 1 & 1 & \theta\end{pmatrix}$

`AddRoundKey`. The key schedule (described in Section 2.3) for an n-round small scale variant of the AES produces $n+1$ subkey blocks. `AddRoundKey` simultaneously adds (as elements of $GF(2^e)$) each element of the subkey block to some intermediate data block. Since the AES begins with an initial `AddRoundKey`, the small scale variants $\mathrm{SR}(n,r,c,e)$ also begin with this operation.

2.3 Small Scale Key Schedule

The structure of one round of the AES key schedule is illustrated below left (we only consider equal block and key sizes). Each vertical line represents one column of bytes, and F_i is the non-linear key schedule function for round i applied to one column of the array holding the previous round key. The function F_i consists of the application of the AES S-Box to all r components of the column, along with a word-based rotation and addition of a constant. When considering small scale analogues, we use this standard key schedule structure for four columns ($c=4$). For two columns or one column ($c=2$ or $c=1$) we use the key schedule structures given by the diagrams below center and below right respectively.

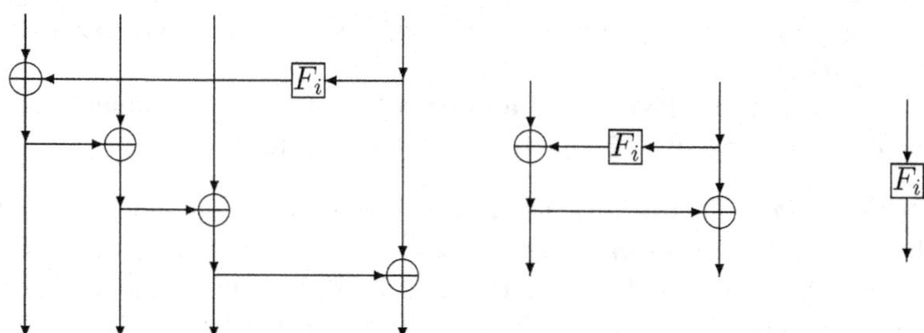

Using this framework we can define key schedules for the small scale variants. SR(n, r, c, e) has a user-provided key of size rce bits, which is considered to be an array of $(r \times c)$ e-bits words. This key forms the initial subkey. Each subkey is then used to define the succeeding subkeys. We provide a full description of the small scale key schedules in Appendix A.

3 Multivariate Quadratic Equation Systems

The existence of a sparse multivariate quadratic equation system over $GF(2^8)$ for an AES encryption was shown by defining a new block cipher, the *Big Encryption System (BES)*, as well as a "BES"-embedding of the AES [8]. The main idea of the BES-embedding is to use the vector conjugate mapping ϕ to embed the AES into the larger cipher BES [8].

This technique can be used to derive sparse multivariate quadratic equation systems over $GF(2^e)$ for the small scale variants of the AES. This is based on the vector conjugate mapping ϕ for $GF(2^4)$ defined by

$$\phi(a) = \left(a^{2^0}, a^{2^1}, a^{2^2}, a^{2^3}\right)^T .$$

Any element $a \in GF(2^4)$ can be embedded as an element $(a, a^2, a^4, a^8)^T \in GF(2^4)^4$ under ϕ. We now describe how operations on a data block $\mathbf{a}$ in small scale variants SR$(n, r, c, 4)$ can be replicated by operations on the vector conjugate $\phi(\mathbf{a})$. Further details and the justification for these operations are given in [8].

`SubBytes`. Consider the three component operations of the S-box separately.

1. *Inversion.* The operation of inversion can be replicated by the componentwise inversion of the vector conjugate.
2. *GF(2)-linear map.* The effect of the $GF(2)$-linear map can by replicated by pre-multiplying the (column) vector conjugate by the matrix
$$\begin{pmatrix} \lambda_0 & \lambda_1 & \lambda_2 & \lambda_3 \\ \lambda_3^2 & \lambda_0^2 & \lambda_1^2 & \lambda_2^2 \\ \lambda_2^4 & \lambda_3^4 & \lambda_0^4 & \lambda_1^4 \\ \lambda_1^8 & \lambda_2^8 & \lambda_3^8 & \lambda_0^8 \end{pmatrix} = \begin{pmatrix} \mathtt{5} & \mathtt{1} & \mathtt{C} & \mathtt{5} \\ \mathtt{2} & \mathtt{2} & \mathtt{1} & \mathtt{F} \\ \mathtt{A} & \mathtt{4} & \mathtt{4} & \mathtt{1} \\ \mathtt{1} & \mathtt{8} & \mathtt{3} & \mathtt{3} \end{pmatrix},$$
where $(\lambda_0, \lambda_1, \lambda_2, \lambda_3)$ are the coefficients of the linearised polynomial given previously.
3. *S-Box constant.* The effect of adding $\mathtt{6}$ to a data array element can be replicated by adding $(\mathtt{6}, \mathtt{7}, \mathtt{6}, \mathtt{7})^T$ to its vector conjugate.

`ShiftRows`. The effect of `ShiftRows` on the conjugate embedding can be easily replicated for the small variants SR$(n, r, c, 4)$. For example, the operation on SR$(n, 1, 1, 4)$, SR$(n, 2, 1, 4)$ and SR$(n, 2, 2, 4)$ is given by the following matrices, where I_4 denotes the 4×4 identity matrix over $GF(2^4)$:

$$(I_4), \begin{pmatrix} I_4 & 0 \\ 0 & I_4 \end{pmatrix}, \text{ and } \begin{pmatrix} I_4 & 0 & 0 & 0 \\ 0 & 0 & 0 & I_4 \\ 0 & 0 & I_4 & 0 \\ 0 & I_4 & 0 & 0 \end{pmatrix}.$$

`MixColumns`. The effect of multiplying a field element by some other field element z can be replicated by pre-multiplying its (column) vector conjugate by the diagonal matrix

$$D_z = \begin{pmatrix} z & 0 & 0 & 0 \\ 0 & z^2 & 0 & 0 \\ 0 & 0 & z^4 & 0 \\ 0 & 0 & 0 & z^8 \end{pmatrix}.$$

Clearly `MixColumns` is a trivial operation when $r = 1$. For $r = 2$ and $r = 4$ the effect of `MixColumns` can be replicated by pre-multiplying the vector conjugates of the column of the corresponding data array by the matrices

$$\left(\begin{array}{c|c} D_{\rho+1} & D_\rho \\ \hline D_\rho & D_{\rho+1} \end{array}\right) \text{ and } \left(\begin{array}{c|c|c|c} D_\rho & D_{\rho+1} & 1 & 1 \\ \hline 1 & D_\rho & D_{\rho+1} & 1 \\ \hline 1 & 1 & D_\rho & D_{\rho+1} \\ \hline D_{\rho+1} & 1 & 1 & D_\rho \end{array}\right) \text{ respectively.}$$

`AddRoundKey`. This operation can be replicated by adding the appropriate vector conjugates. As the key schedule essentially uses the same operations as the encryption process, it can also be easily replicated using vector conjugates.

Since inversion is the only non-linear part of the round function, we can move the S-Box constant into a slightly modified key schedule and construct an augmented linear diffusion layer consisting of the $GF(2)$-linear map, `ShiftRows` and `MixColumns` [7]. This augmented linear diffusion layer is given by an $(rc \times rc)$ matrix; examples of this useful representation are given in Appendix B.

In Appendix C, we give an example of a multivariate quadratic equation system for SR$(2, 2, 2, 4)$, as well as a link to a website where systems for other small scale variants can be downloaded from. The given systems are constructed using the "BES-style" embedding. If the plaintext and ciphertext are known, which we assume, then the given equations are sufficient. If they are unknown, then the plaintext and ciphertext can be treated as variables. We note that whilst the system of equations is systematic, it does not form a minimal system. Furthermore, the systems of equations presented are correct only if no 0-inversion is performed either in the key schedule or in the encryption rounds. The probability of any particular inversion being a 0-inversion is 2^{-e}, so the probability that the entire equation system is free from 0-inversions can be easily estimated. In general, equations for the small scale variants can be easily established and the number of equations and variables for different variants are given here:

	Equations	State Variables	Key Variables
Encryption	$(4n+1)rce$	$2nrce$	$(n+1)rce$
Key Schedule	$2nrce + 2nre + rce$	-	nre (additional)
Total	$(6n+1)rce + 2nre + rce$	$2nrce$	$(n+1)rce + nre$

An alternative approach would be to work with equation systems over $GF(2)$, as originally proposed in [2]. If $w, x \in GF(2^e)$ are the input and output respectively of the "AES inversion", the relation $wx = 1$ gives rise to e bilinear expressions over $GF(2)$. Furthermore, we also have the "associated inversion" relations $w^2x = w$ and $wx^2 = x$, each of which give e further independent relations at the bit level. While there are also further relations of the form $w^4x = w^2 \cdot w$ and $wx^4 = x \cdot x^2$, we do not consider these. Using these expressions and those derived from the linear layer, we can construct a system of multivariate quadratic equations over $GF(2)$ in a similar manner to that given by [2]. In Section 4 we present the results of some experiments using both the $GF(2)$ and the BES-style representations.

4 Experimental Results

In this section we describe some experimental results concerning the solution of the equation systems that arise for these small scale variants of AES. These are basic timing experiments for the solution of the relevant system of equations by computing the Gröbner basis of the related polynomial ideal. The computations were made using the MAGMA 2.11-1 computer algebra package [6], which includes a highly efficient (particularly for $GF(2)$) implementation of Faugère's F4 algorithm [5]. In general, the Gröbner bases were computed with respect to the graded reverse lexicographic monomial ordering. All experiments were performed on a HP workstation, with Pentium 4 - 3GHz processor, 1 GB RAM, running Windows XP.

We are aware of the limitations of performing simple timings experiments using off-the-shelf software with limited computing resources. However we believe that such experiments can still be helpful in a preliminary assessment of algebraic attacks as cryptanalytic techniques. And while particularly degenerate small scale variants might not exhibit all the features of the AES, a comparison of attacks on such variants will help to provide an understanding of how various components and representations of the cipher contribute to the complexity of algebraic attacks.

4.1 SR(• • 1• 1• •)

We ran experiments with the simple variants SR$(n, 1, 1, 4)$ and SR$(n, 1, 1, 8)$ for different number of rounds. We performed computations using the BES-style equation system over $GF(2^e)$, as well as the equation system over $GF(2)$. The $GF(2)$ equation systems are similar to that given in [2] with the addition of all field polynomials of the form $z^2 + z$. Table 1 shows the results for the

Table 1. Time (in seconds) for Gröbner Basis computation of the $GF(2^e)$ and $GF(2)$ equation systems that arise from SR$(n, 1, 1, e)$ (using graded reverse lexicographic monomial ordering)

Cipher	Variables	Equations $GF(2^e)$	Monomials $GF(2^e)$	Time	Equations $GF(2)$	Monomials $GF(2)$	Time
SR(2,1,1,4)	36	72	89	0.11	104	137	0.03
SR(3,1,1,4)	52	104	129	0.75	152	201	0.11
SR(4,1,1,4)	68	136	169	2.02	200	265	0.28
SR(5,1,1,4)	84	168	209	7.47	248	339	0.97
SR(6,1,1,4)	100	200	249	23.71	296	393	4.30
SR(7,1,1,4)	116	232	289	56.74	344	457	11.26
SR(8,1,1,4)	132	264	329	43.70	392	521	16.56
SR(9,1,1,4)	148	296	369	219.38	440	585	46.05
SR(10,1,1,4)	164	328	409	340.31	488	649	74.06
SR(2,1,1,8)	72	144	177	43.55	172	365	118.45
SR(3,1,1,8)	104	208	257	N/A	252	541	N/A

computations; timings are given in seconds and N/A denotes insufficient memory to complete the computation.

In view of their simple form, we would expect to solve such equation systems for many rounds. This happened for SR$(n, 1, 1, 4)$, where we ran tests for up to 10 rounds. However the time required varied greatly when we changed the ordering of variables. When working with the cipher SR$(n, 1, 1, 8)$, we had problems with insufficient memory as early as three rounds. In particular, we note that the system for SR$(3, 1, 1, 8)$ has a similar number of variables, monomials and equations as the system for SR$(6, 1, 1, 4)$. Thus we might expect a similar performance for these two systems. However, our results show that this is not the case. This suggests that the underlying field equations, which are implicitly included in the BES-style equations, may play an important role in the computations for solving the system. However, this is yet to be established.

By comparing the results in Table 1, it is clear that the timings of computations for SR$(n, 1, 1, 4)$ over $GF(2)$ are much better than those over $GF(2^4)$. However it is not clear whether this means that bit-level equations offer a better representation than BES-style equations in general, since MAGMA's implementation of the F4 algorithm is heavily optimised for operations over $GF(2)$ [13]. (In fact we see the opposite behaviour occurring for the cipher SR$(n, 1, 1, 8)$.) Given the highly structured form of the BES-style systems we would expect computations using equation sets over $GF(2^e)$ to be generally more efficient than those over $GF(2)$.

4.2 SR(• •2• 1• 4) and SR(• •2• 2• 4)

Some basic timing experiments with the systems derived from the variants SR$(n, 2, 1, 4)$ and SR$(n, 2, 2, 4)$ are given below. For these two variants, we also used MAGMA's implementation of Buchberger's algorithm in addition to com-

Table 2. Time (in seconds) for Gröbner Basis computation of $GF(2^4)$ equation systems with F4 and Buchberger's algorithm (using graded reverse lexicographic monomial ordering)

Cipher	Variables	Equations	Monomials	Time F4	Time Buchberger
SR(1,2,1,4)	40	80	97	0.22	1.11
SR(2,2,1,4)	72	144	177	24.55	40.58
SR(3,2,1,4)	104	208	257	519.92	2649.90
SR(4,2,1,4)	136	272	337	N/A	28999.41
SR(1,2,2,4)	72	144	169	27.73	444.07
SR(2,2,2,4)	128	256	305	N/A	N/A

putations of the Gröbner bases using the F4 algorithm. While we would expect Buchberger's algorithm to be slower, it should require less memory than the F4 algorithm. As before, timings are given in seconds, with N/A meaning insufficient memory to complete the computation (see Table 2).

By comparing the results in Table 2, we note that the equation system derived from SR(4, 2, 1, 4) has a similar number of variables, monomials and equations as the equation system arising from SR(2, 2, 2, 4). Therefore we might expect a similar performance in the computation for these two systems. However, our results confirm the important role played by the inter-word diffusion in the complexity of the computations. The diffusion of SR(n, 2, 1, 4) is limited, whereas SR(n, 2, 2, 4) has a similar diffusion pattern to that seen in the AES.

4.3 *Meet-in-the-Middle* Approach

Our experiments used the exact equation systems discussed in this paper; no pre-computation was performed and we did not explore any special structure. However it is well-known that the equation systems derived from the AES are highly structured, especially when represented as the set of BES-style equations over $GF(2^e)$. In particular, these systems might be viewed as "iterated" systems of equations, with similar blocks of multivariate quadratic equations repeated for every round. These blocks are connected to each other via the input and output variables, as well as the key schedule. When working with systems with such structure, a promising technique to find the overall solution is, in effect, a *meet-in-the-middle* approach: rather than attempting to solve the full system of equations for n rounds (we assume that n is even), we can try to solve two subsystems with $\frac{n}{2}$ rounds, by considering the output of round $\frac{n}{2}$ (which is also the input of round $\frac{n}{2}+1$) as variables. By choosing an appropriate monomial ordering we obtain two sets of equations (each covering half of the encryption operation) that relate these variables with the round subkeys. These two systems can then be combined along with some other equations relating the round subkeys. This gives a third smaller system which can be solved to obtain the encryption key.

Table 3. Time (in seconds) for the *meet-in-the-middle* approach using F4 Gröbner Basis computation of equation systems arising from SR(10, 1, 1, 4), SR(4, 1, 1, 8) and SR(4, 2, 1, 4) using lexicographic ordering

Cipher	Variables	Equations	Monomials	Time
SR(10,1,1,4) - 5 rounds ↓	88	172	217	19.22
SR(10,1,1,4) - 5 rounds ↑	76	148	189	22.41
Solve	16	40	52	0.02
			Total:	41.65
SR(4,1,1,8) - 2 rounds ↓	80	152	193	15466.37
SR(4,1,1,8) - 2 rounds ↑	56	104	137	4603.89
Solve	32	80	576	215.92
			Total:	20286.18
SR(4,2,1,4) - 2 rounds ↓	80	152	193	667.17
SR(4,2,1,4) - 2 rounds ↑	56	104	137	2722.43
Solve	80	176	524	14.87
			Total:	3404.47

We have tried this approach with some of the AES variants and compared the results with the timings obtained earlier. Our experiments suggest that this approach may be more efficient. For example, we were able to solve the system for SR(10, 1, 1, 4) using this approach in 42 seconds compared with 340 seconds using the naive approach. We also obtained better results for SR(4, 1, 1, 8) and SR(4, 2, 1, 4) using this approach (see Table 3).

This technique is cryptographically intuitive and is in fact a simple application of Elimination Theory [3], in which the Gröbner bases are computed with respect to the appropriate monomial ordering to eliminate the variables that do not appear in rounds $\frac{n}{2}$ and $\frac{n}{2}+1$. One problem with this approach is that computations using elimination orderings (such as lexicographic) are usually less efficient than those with degree orderings (such as graded reverse lexicographic). Thus, for more complex systems, we might expect that using lexicographic ordering in the two main subsystems would yield only limited benefit when compared with graded reverse lexicographic ordering for the full system. As an alternative,

Table 4. Time (in seconds) for the *meet-in-the-middle* approach using F4 Gröbner Basis computation of equation systems arising from SR(4, 2, 1, 4) using graded reverse lexicographic monomial ordering

Cipher	Variables	Equations	Monomials	Time
SR(4,2,1,4) - 2 rounds ↓	112	216	273	553.63
SR(4,2,1,4) - 2 rounds ↑	104	200	257	1501.41
Solve	136	1197	918	12.68
			Total:	2067.72

we could simply compute the Gröbner bases for the two subsystems (using the most efficient ordering) and combine both results to compute the solution of the full set equations. While this approach was more expensive for the variant $SR(10, 1, 1, 4)$, it was more efficient for the cipher $SR(4, 2, 1, 4)$ (see Table 4).

These results suggest the applicability of a more general *divide-and-conquer* approach to this problem, in which some form of (perhaps largely symbolic) pre-computation could be performed and then combined to produce the solution of the full system. This might be a promising direction and more research will assess whether this approach might increase the efficiency of algebraic attacks against the AES and related ciphers.

5 Conclusions

We have defined a family of small scale variants of the AES. This provides a common framework for the analysis of AES-like equation systems. We also present some basic experimental results when using off-the-shelf computational algebra techniques to solve these systems. These provide some preliminary insight into the behavior of algebraic attacks and future work can now take place within a framework for the systematic analysis of small scale AES variants.

References

1. C. Cid, S. Murphy, and M.J.B. Robshaw. Computational and Algebraic Aspects of the Advanced Encryption Standard. In V. Ganzha *et al.*, editors, Proceedings of the *Seventh International Workshop on Computer Algebra in Scientific Computing - CASC 2004*, St. Petersburg, Russia, pages 93–103, Technische Universität München. 2004.
2. N. Courtois and J. Pieprzyk. Cryptanalysis of block ciphers with overdefined systems of equations. In Y. Zheng, editor, Proceedings of *Asiacrypt 2002*, LNCS 2501, pages 267–287, Springer-Verlag, 2002.
3. D. Cox, J. Little and D. O'Shea. Ideals, Varieties, and Algorithms. Undergraduate Texts in Mathematics, Second Edition, Springer-Verlag, 1997.
4. J. Daemen and V. Rijmen. *The Design of Rijndael: AES – The Advanced Encryption Standard*, Springer-Verlag, 2002.
5. J.-C. Faugère. A new efficient algorithm for computing Gröbner bases (F4). Journal of Pure and Applied Algebra, 139, pages 61–88, 1999.
6. Magma V2.11-1, Computational Algebra Group, School of Mathematics and Statistics, University of Sydney. Website: `http://magma.maths.usyd.edu.au`. 2004.
7. S. Murphy and M.J.B. Robshaw. New Observations on Rijndael. Submitted to NIST. Available via `csrc.nist.gov`. 7 August 2000.
8. S. Murphy and M.J.B. Robshaw. Essential Algebraic Structure within the AES. In M. Yung, editor, Proceedings of *CRYPTO 2002*, LNCS 2442, pages 11–16, Springer-Verlag, 2002.
9. S. Murphy and M.J.B. Robshaw. Comments on the Security of the AES and the XSL Technique. *Electronics Letters* Vol. 39, pp 36-38, 2002.

10. M.A. Musa, E.F. Schaefer, and S. Wedig. A simplified AES algorithm and its linear and differential cryptanalysis. Cryptologia, Vol. XXVII (2), pages 148-177, 2003.
11. National Institute of Standards and Technology. Advanced Encryption Standard. FIPS 197. November 26, 2001.
12. R.C.-W. Phan. Mini Advanced Encryption Standard (Mini-AES): A Testbed for Cryptanalysis Students, Cryptologia, Vol. XXVI (4), pages 283-306, 2002.
13. A. Steel. (Magma Development Team). Personal communication, October 2004.

Appendix A: Key Schedule Equations

SR(n, r, c, e) has a user-provided key of rce bits, which is considered to be an array of $(r \times c)$ e-bits words. This key forms the initial subkey. Each subkey is then used to define the next subkey as described below. This description uses constants and functions which depend on the field $GF(2^e)$. All constants and functions (apart from the round constant κ_i) have been discussed elsewhere and are summarised in the following table.

		$GF(2^4)$	$GF(2^8)$
Round constant	κ_i	$\rho^{(i-1)}$	$\theta^{(i-1)}$
S-Box constant	d	6	63
Inversion	$z \mapsto z^{(-1)}$	Inversion in $GF(2^4)$	Inversion in $GF(2^8)$
$GF(2)$-linear map	$z \mapsto L(z)$	L for $GF(2^4)$	L for $GF(2^8)$

We regard each round subkey as a column $GF(2^e)$-vector of length rc. In order to define the key schedule, we effectively divide the round subkey vector into c subvectors of length r. Thus the subkey vectors are given below.

$$\begin{array}{rl} \text{Initial Subkey} & (k_{0,0}, \ldots, k_{0,r-1}, \ldots, k_{0,r(c-1)}, \ldots, k_{0,rc-1})^T \\ \text{Round 1 Subkey} & (k_{1,0}, \ldots, k_{1,r-1}, \ldots, k_{1,r(c-1)}, \ldots, k_{1,rc-1})^T \\ \vdots & \qquad\qquad\qquad \vdots \\ \text{Round } n \text{ Subkey} & (k_{n,0}, \ldots, k_{n,r-1}, \ldots, k_{n,r(c-1)}, \ldots, k_{n,rc-1})^T \end{array}$$

The definition of the round subkeys now depends on the number of rows (r) and columns (c) in the array. The round subkeys are defined recursively for each round $1 \leq i \leq n$.

Key Schedule for One Row (• = 1).

$$s_0 = k_{i-1,c-1}^{(-1)}.$$

– One column ($r = 1, c = 1$).

$$(k_{i,0}) = (L(s_0)) + (d) + (\kappa_i).$$

– More than one column ($r = 1, c > 1$). For $0 \leq q \leq c-1$

$$(k_{i,q}) = (L(s_0)) + (d) + (\kappa_i) + \sum_{t=0}^{q} (k_{i-1,t}).$$

Key Schedule for Two Rows (• = 2).

$$s_0 = k^{(-1)}_{i-1,2c-1}, \qquad s_1 = k^{(-1)}_{i-1,2c-2}.$$

– One column ($r = 2, c = 1$).

$$\begin{pmatrix} k_{i,0} \\ k_{i,1} \end{pmatrix} = \begin{pmatrix} L(s_0) \\ L(s_1) \end{pmatrix} + \begin{pmatrix} d \\ d \end{pmatrix} + \begin{pmatrix} \kappa_i \\ 0 \end{pmatrix}.$$

– More than one column ($r = 2, c > 1$). For $0 \leq q \leq c-1$

$$\begin{pmatrix} k_{i,rq} \\ k_{i,rq+1} \end{pmatrix} = \begin{pmatrix} L(s_0) \\ L(s_1) \end{pmatrix} + \begin{pmatrix} d \\ d \end{pmatrix} + \begin{pmatrix} \kappa_i \\ 0 \end{pmatrix} + \sum_{t=0}^{q} \begin{pmatrix} k_{i-1,rt} \\ k_{i-1,rt+1} \end{pmatrix}.$$

Key Schedule for Four Rows (• = 4).

$$s_0 = k^{(-1)}_{i-1,4c-1}, \quad s_1 = k^{(-1)}_{i-1,4c-2}, \quad s_2 = k^{(-1)}_{i-1,4c-3}, \quad s_3 = k^{(-1)}_{i-1,4c-4}.$$

– One column ($r = 4, c = 1$).

$$\begin{pmatrix} k_{i,0} \\ k_{i,1} \\ k_{i,2} \\ k_{i,3} \end{pmatrix} = \begin{pmatrix} L(s_0) \\ L(s_1) \\ L(s_2) \\ L(s_3) \end{pmatrix} + \begin{pmatrix} d \\ d \\ d \\ d \end{pmatrix} + \begin{pmatrix} \kappa_i \\ 0 \\ 0 \\ 0 \end{pmatrix}.$$

– More than one column ($r = 4, c > 1$). For $0 \leq q \leq c-1$

$$\begin{pmatrix} k_{i,rq} \\ k_{i,rq+1} \\ k_{i,rq+2} \\ k_{i,rq+3} \end{pmatrix} = \begin{pmatrix} L(s_0) \\ L(s_1) \\ L(s_2) \\ L(s_3) \end{pmatrix} + \begin{pmatrix} d \\ d \\ d \\ d \end{pmatrix} + \begin{pmatrix} \kappa_i \\ 0 \\ 0 \\ 0 \end{pmatrix} + \sum_{t=0}^{q} \begin{pmatrix} k_{i-1,rt} \\ k_{i-1,rt+1} \\ k_{i-1,rt+2} \\ k_{i-1,rt+3} \end{pmatrix}.$$

Appendix B: Augmented Linear Diffusion Layer

For small scale variants of the AES, we can construct an augmented linear diffusion layer that consists of the $GF(2)$-linear map, `ShiftRows` and `MixColumns` [7]. This helps to provide a natural set of equations. The augmented linear diffusion layer can be represented by an $(rc \times rc)$ matrix. If we replace every entry z of this matrix by D_z given earlier, we obtain an $(rce \times rce)$ matrix which replicates the augmented linear diffusion layer for vector conjugates. We provide these matrices for different array sizes over $GF(2^4)$. For array sizes (1×1) and (2×1), the augmented linear diffusion layers for vector conjugates are given by the matrices

$$\begin{pmatrix} 5\,1\,C\,5 \\ 2\,2\,1\,F \\ A\,4\,4\,1 \\ 1\,8\,3\,3 \end{pmatrix} \text{ and } \left(\begin{array}{c|c} F\,3\,7\,F & A\,2\,B\,A \\ A\,A\,5\,6 & 8\,8\,4\,9 \\ 7\,8\,8\,2 & D\,C\,C\,3 \\ 4\,6\,C\,C & 5\,E\,F\,F \\ \hline A\,2\,B\,A & F\,3\,7\,F \\ 8\,8\,4\,9 & A\,A\,5\,6 \\ D\,C\,C\,3 & 7\,8\,8\,2 \\ 5\,E\,F\,F & 4\,6\,C\,C \end{array}\right), \text{ respectively.}$$

For the (2×2)-array, the augmented diffusion layer is given by

$$\left(\begin{array}{c|c|c|c} F\,3\,7\,F & 0\,0\,0\,0 & 0\,0\,0\,0 & A\,2\,B\,A \\ A\,A\,5\,6 & 0\,0\,0\,0 & 0\,0\,0\,0 & 8\,8\,4\,9 \\ 7\,8\,8\,2 & 0\,0\,0\,0 & 0\,0\,0\,0 & D\,C\,C\,3 \\ 4\,6\,C\,C & 0\,0\,0\,0 & 0\,0\,0\,0 & 5\,E\,F\,F \\ \hline A\,2\,B\,A & 0\,0\,0\,0 & 0\,0\,0\,0 & F\,3\,7\,F \\ 8\,8\,4\,9 & 0\,0\,0\,0 & 0\,0\,0\,0 & A\,A\,5\,6 \\ D\,C\,C\,3 & 0\,0\,0\,0 & 0\,0\,0\,0 & 7\,8\,8\,2 \\ 5\,E\,F\,F & 0\,0\,0\,0 & 0\,0\,0\,0 & 4\,6\,C\,C \\ \hline 0\,0\,0\,0 & A\,2\,B\,A & F\,3\,7\,F & 0\,0\,0\,0 \\ 0\,0\,0\,0 & 8\,8\,4\,9 & A\,A\,5\,6 & 0\,0\,0\,0 \\ 0\,0\,0\,0 & D\,C\,C\,3 & 7\,8\,8\,2 & 0\,0\,0\,0 \\ 0\,0\,0\,0 & 5\,E\,F\,F & 4\,6\,C\,C & 0\,0\,0\,0 \\ \hline 0\,0\,0\,0 & F\,3\,7\,F & A\,2\,B\,A & 0\,0\,0\,0 \\ 0\,0\,0\,0 & A\,A\,5\,6 & 8\,8\,4\,9 & 0\,0\,0\,0 \\ 0\,0\,0\,0 & 7\,8\,8\,2 & D\,C\,C\,3 & 0\,0\,0\,0 \\ 0\,0\,0\,0 & 4\,6\,C\,C & 5\,E\,F\,F & 0\,0\,0\,0 \end{array}\right).$$

Appendix C: Equation System for SR(2•2•2•4)

We illustrate the kind of equation systems that arise by listing the BES-style relations between variables in an SR(2, 2, 2, 4) encryption and key schedule. If neither the encryption nor the key schedule require a 0-inversion, then each of these relations is identically 0. Under this assumption, the following relations give a multivariate quadratic equation system for SR(2, 2, 2, 4) over $GF(2^4)$. The probability that the encryption rounds do not require any 0-inversions is about $\left(\frac{15}{16}\right)^8 \approx 0.60$. The probability that the key schedule does not require any 0-inversions is about $\left(\frac{15}{16}\right)^4 \approx 0.77$. Systems for other small scale variants can be downloaded from the site `http://www.isg.rhul.ac.uk/aes/index.html`.

Component j and conjugate l for the plaintext, ciphertext and the key (also used as the initial subkey) are denoted by p_{jl}, c_{jl} and k_{0jl} respectively. We regard the two rounds as round one and round two. We denote the input and output of the inversion and the subkey used in round i for component j and conjugate l by w_{ijl}, x_{ijl} and k_{ijl} respectively.

	$GF(2^4)$ Variable	Round i	Component j	Conjugate l
Plaintext	p_{jl}		$0,1,2,3$	$0,1,2,3$
Ciphertext	c_{jl}		$0,1,2,3$	$0,1,2,3$
State				
Inversion Input	w_{ijl}	$1,2$	$0,1,2,3$	$0,1,2,3$
Inversion Output	x_{ijl}	$1,2$	$0,1,2,3$	$0,1,2,3$
Key				
Subkey	k_{ijl}	$0,1,2$	$0,1,2,3$	$0,1,2,3$
Dummy	s_{ijl}	$0,1$	$0,1$	$0,1,2,3$

Initial Subkey Relations

$w_{100}+p_{00}+k_{000}$	$w_{110}+p_{10}+k_{010}$	$w_{120}+p_{20}+k_{020}$	$w_{130}+p_{30}+k_{030}$
$w_{101}+p_{01}+k_{001}$	$w_{111}+p_{11}+k_{011}$	$w_{121}+p_{21}+k_{021}$	$w_{131}+p_{31}+k_{031}$
$w_{102}+p_{02}+k_{002}$	$w_{112}+p_{12}+k_{012}$	$w_{122}+p_{22}+k_{022}$	$w_{132}+p_{32}+k_{032}$
$w_{103}+p_{03}+k_{003}$	$w_{113}+p_{13}+k_{013}$	$w_{123}+p_{23}+k_{023}$	$w_{133}+p_{33}+k_{033}$

Inversion and Conjugacy Relations: Rounds 1 and 2

$w_{100}^2+w_{101}$	$w_{100}x_{100}+1$	$x_{100}^2+x_{101}$	$w_{200}^2+w_{201}$	$w_{200}x_{200}+1$	$x_{200}^2+x_{201}$
$w_{101}^2+w_{102}$	$w_{101}x_{101}+1$	$x_{101}^2+x_{102}$	$w_{201}^2+w_{202}$	$w_{201}x_{201}+1$	$x_{201}^2+x_{202}$
$w_{102}^2+w_{103}$	$w_{102}x_{102}+1$	$x_{102}^2+x_{103}$	$w_{202}^2+w_{203}$	$w_{202}x_{202}+1$	$x_{202}^2+x_{203}$
$w_{103}^2+w_{100}$	$w_{103}x_{103}+1$	$x_{103}^2+x_{100}$	$w_{203}^2+w_{200}$	$w_{203}x_{203}+1$	$x_{203}^2+x_{200}$
$w_{110}^2+w_{111}$	$w_{110}x_{110}+1$	$x_{110}^2+x_{111}$	$w_{210}^2+w_{211}$	$w_{210}x_{210}+1$	$x_{210}^2+x_{211}$
$w_{111}^2+w_{112}$	$w_{111}x_{111}+1$	$x_{111}^2+x_{112}$	$w_{211}^2+w_{212}$	$w_{211}x_{211}+1$	$x_{211}^2+x_{212}$
$w_{112}^2+w_{113}$	$w_{112}x_{112}+1$	$x_{112}^2+x_{113}$	$w_{212}^2+w_{213}$	$w_{212}x_{212}+1$	$x_{212}^2+x_{213}$
$w_{113}^2+w_{110}$	$w_{113}x_{113}+1$	$x_{113}^2+x_{110}$	$w_{213}^2+w_{210}$	$w_{213}x_{213}+1$	$x_{213}^2+x_{210}$
$w_{120}^2+w_{121}$	$w_{120}x_{120}+1$	$x_{120}^2+x_{121}$	$w_{220}^2+w_{221}$	$w_{220}x_{220}+1$	$x_{220}^2+x_{221}$
$w_{121}^2+w_{122}$	$w_{121}x_{121}+1$	$x_{121}^2+x_{122}$	$w_{221}^2+w_{222}$	$w_{221}x_{221}+1$	$x_{221}^2+x_{222}$
$w_{122}^2+w_{123}$	$w_{122}x_{122}+1$	$x_{122}^2+x_{123}$	$w_{222}^2+w_{223}$	$w_{222}x_{222}+1$	$x_{222}^2+x_{223}$
$w_{123}^2+w_{120}$	$w_{123}x_{123}+1$	$x_{123}^2+x_{120}$	$w_{223}^2+w_{220}$	$w_{223}x_{223}+1$	$x_{223}^2+x_{220}$
$w_{130}^2+w_{131}$	$w_{130}x_{130}+1$	$x_{130}^2+x_{131}$	$w_{230}^2+w_{231}$	$w_{230}x_{230}+1$	$x_{230}^2+x_{231}$
$w_{131}^2+w_{132}$	$w_{131}x_{131}+1$	$x_{131}^2+x_{132}$	$w_{231}^2+w_{232}$	$w_{231}x_{231}+1$	$x_{231}^2+x_{232}$
$w_{132}^2+w_{133}$	$w_{132}x_{132}+1$	$x_{132}^2+x_{133}$	$w_{232}^2+w_{233}$	$w_{232}x_{232}+1$	$x_{232}^2+x_{233}$
$w_{133}^2+w_{130}$	$w_{133}x_{133}+1$	$x_{133}^2+x_{130}$	$w_{233}^2+w_{230}$	$w_{233}x_{233}+1$	$x_{233}^2+x_{230}$

Diffusion Relations: Rounds 1 and 2

$$w_{200}+\mathtt{F}x_{100}+\mathtt{3}x_{101}+\mathtt{7}x_{102}+\mathtt{F}x_{103}+\mathtt{A}x_{130}+\mathtt{2}x_{131}+\mathtt{B}x_{132}+\mathtt{A}x_{133}+k_{100}+\mathtt{6}$$
$$w_{201}+\mathtt{A}x_{100}+\mathtt{A}x_{101}+\mathtt{5}x_{102}+\mathtt{6}x_{103}+\mathtt{8}x_{130}+\mathtt{8}x_{131}+\mathtt{4}x_{132}+\mathtt{9}x_{133}+k_{101}+\mathtt{7}$$
$$w_{202}+\mathtt{7}x_{100}+\mathtt{8}x_{101}+\mathtt{8}x_{102}+\mathtt{2}x_{103}+\mathtt{D}x_{130}+\mathtt{C}x_{131}+\mathtt{C}x_{132}+\mathtt{3}x_{133}+k_{102}+\mathtt{6}$$
$$w_{203}+\mathtt{4}x_{100}+\mathtt{6}x_{101}+\mathtt{C}x_{102}+\mathtt{C}x_{103}+\mathtt{5}x_{130}+\mathtt{E}x_{131}+\mathtt{F}x_{132}+\mathtt{F}x_{133}+k_{103}+\mathtt{7}$$

$$w_{210} + \mathtt{A}x_{100} + \mathtt{2}x_{101} + \mathtt{B}x_{102} + \mathtt{A}x_{103} + \mathtt{F}x_{130} + \mathtt{3}x_{131} + \mathtt{7}x_{132} + \mathtt{F}x_{133} + k_{110} + \mathtt{6}$$
$$w_{211} + \mathtt{8}x_{100} + \mathtt{8}x_{101} + \mathtt{4}x_{102} + \mathtt{9}x_{103} + \mathtt{A}x_{130} + \mathtt{A}x_{131} + \mathtt{5}x_{132} + \mathtt{6}x_{133} + k_{111} + \mathtt{7}$$
$$w_{212} + \mathtt{D}x_{100} + \mathtt{C}x_{101} + \mathtt{C}x_{102} + \mathtt{3}x_{103} + \mathtt{7}x_{130} + \mathtt{8}x_{131} + \mathtt{8}x_{132} + \mathtt{2}x_{133} + k_{112} + \mathtt{6}$$
$$w_{213} + \mathtt{5}x_{100} + \mathtt{E}x_{101} + \mathtt{F}x_{102} + \mathtt{F}x_{103} + \mathtt{4}x_{130} + \mathtt{6}x_{131} + \mathtt{C}x_{132} + \mathtt{C}x_{133} + k_{113} + \mathtt{7}$$
$$w_{220} + \mathtt{A}x_{110} + \mathtt{2}x_{111} + \mathtt{B}x_{112} + \mathtt{A}x_{113} + \mathtt{F}x_{120} + \mathtt{3}x_{121} + \mathtt{7}x_{122} + \mathtt{F}x_{123} + k_{120} + \mathtt{6}$$
$$w_{221} + \mathtt{8}x_{110} + \mathtt{8}x_{111} + \mathtt{4}x_{112} + \mathtt{9}x_{113} + \mathtt{A}x_{120} + \mathtt{A}x_{121} + \mathtt{5}x_{122} + \mathtt{6}x_{123} + k_{121} + \mathtt{7}$$
$$w_{222} + \mathtt{D}x_{110} + \mathtt{C}x_{111} + \mathtt{C}x_{112} + \mathtt{3}x_{113} + \mathtt{7}x_{120} + \mathtt{8}x_{121} + \mathtt{8}x_{122} + \mathtt{2}x_{123} + k_{122} + \mathtt{6}$$
$$w_{223} + \mathtt{5}x_{110} + \mathtt{E}x_{111} + \mathtt{F}x_{112} + \mathtt{F}x_{113} + \mathtt{4}x_{120} + \mathtt{6}x_{121} + \mathtt{C}x_{122} + \mathtt{C}x_{123} + k_{123} + \mathtt{7}$$
$$w_{230} + \mathtt{F}x_{110} + \mathtt{3}x_{111} + \mathtt{7}x_{112} + \mathtt{F}x_{113} + \mathtt{A}x_{120} + \mathtt{2}x_{121} + \mathtt{B}x_{122} + \mathtt{A}x_{123} + k_{130} + \mathtt{6}$$
$$w_{231} + \mathtt{A}x_{110} + \mathtt{A}x_{111} + \mathtt{5}x_{112} + \mathtt{6}x_{113} + \mathtt{8}x_{120} + \mathtt{8}x_{121} + \mathtt{4}x_{122} + \mathtt{9}x_{123} + k_{131} + \mathtt{7}$$
$$w_{232} + \mathtt{7}x_{110} + \mathtt{8}x_{111} + \mathtt{8}x_{112} + \mathtt{2}x_{113} + \mathtt{D}x_{120} + \mathtt{C}x_{121} + \mathtt{C}x_{122} + \mathtt{3}x_{123} + k_{132} + \mathtt{6}$$
$$w_{233} + \mathtt{4}x_{110} + \mathtt{6}x_{111} + \mathtt{C}x_{112} + \mathtt{C}x_{113} + \mathtt{5}x_{120} + \mathtt{E}x_{121} + \mathtt{F}x_{122} + \mathtt{F}x_{123} + k_{133} + \mathtt{7}$$
$$c_{00} + \mathtt{F}x_{200} + \mathtt{3}x_{201} + \mathtt{7}x_{202} + \mathtt{F}x_{203} + \mathtt{A}x_{230} + \mathtt{2}x_{231} + \mathtt{B}x_{232} + \mathtt{A}x_{233} + k_{200} + \mathtt{6}$$
$$c_{01} + \mathtt{A}x_{200} + \mathtt{A}x_{201} + \mathtt{5}x_{202} + \mathtt{6}x_{203} + \mathtt{8}x_{230} + \mathtt{8}x_{231} + \mathtt{4}x_{232} + \mathtt{9}x_{233} + k_{201} + \mathtt{7}$$
$$c_{02} + \mathtt{7}x_{200} + \mathtt{8}x_{201} + \mathtt{8}x_{202} + \mathtt{2}x_{203} + \mathtt{D}x_{230} + \mathtt{C}x_{231} + \mathtt{C}x_{232} + \mathtt{3}x_{233} + k_{202} + \mathtt{6}$$
$$c_{03} + \mathtt{4}x_{200} + \mathtt{6}x_{201} + \mathtt{C}x_{202} + \mathtt{C}x_{203} + \mathtt{5}x_{230} + \mathtt{E}x_{231} + \mathtt{F}x_{232} + \mathtt{F}x_{233} + k_{203} + \mathtt{7}$$
$$c_{10} + \mathtt{A}x_{200} + \mathtt{2}x_{201} + \mathtt{B}x_{202} + \mathtt{A}x_{203} + \mathtt{F}x_{230} + \mathtt{3}x_{231} + \mathtt{7}x_{232} + \mathtt{F}x_{233} + k_{210} + \mathtt{6}$$
$$c_{11} + \mathtt{8}x_{200} + \mathtt{8}x_{201} + \mathtt{4}x_{202} + \mathtt{9}x_{203} + \mathtt{A}x_{230} + \mathtt{A}x_{231} + \mathtt{5}x_{232} + \mathtt{6}x_{233} + k_{211} + \mathtt{7}$$
$$c_{12} + \mathtt{D}x_{200} + \mathtt{C}x_{201} + \mathtt{C}x_{202} + \mathtt{3}x_{203} + \mathtt{7}x_{230} + \mathtt{8}x_{231} + \mathtt{8}x_{232} + \mathtt{2}x_{233} + k_{212} + \mathtt{6}$$
$$c_{13} + \mathtt{5}x_{200} + \mathtt{E}x_{201} + \mathtt{F}x_{202} + \mathtt{F}x_{203} + \mathtt{4}x_{230} + \mathtt{6}x_{231} + \mathtt{C}x_{232} + \mathtt{C}x_{233} + k_{213} + \mathtt{7}$$
$$c_{20} + \mathtt{A}x_{210} + \mathtt{2}x_{211} + \mathtt{B}x_{212} + \mathtt{A}x_{213} + \mathtt{F}x_{220} + \mathtt{3}x_{221} + \mathtt{7}x_{222} + \mathtt{F}x_{223} + k_{220} + \mathtt{6}$$
$$c_{21} + \mathtt{8}x_{210} + \mathtt{8}x_{211} + \mathtt{4}x_{212} + \mathtt{9}x_{213} + \mathtt{A}x_{220} + \mathtt{A}x_{221} + \mathtt{5}x_{222} + \mathtt{6}x_{223} + k_{221} + \mathtt{7}$$
$$c_{22} + \mathtt{D}x_{210} + \mathtt{C}x_{211} + \mathtt{C}x_{212} + \mathtt{3}x_{213} + \mathtt{7}x_{220} + \mathtt{8}x_{221} + \mathtt{8}x_{222} + \mathtt{2}x_{223} + k_{222} + \mathtt{6}$$
$$c_{23} + \mathtt{5}x_{210} + \mathtt{E}x_{211} + \mathtt{F}x_{212} + \mathtt{F}x_{213} + \mathtt{4}x_{220} + \mathtt{6}x_{221} + \mathtt{C}x_{222} + \mathtt{C}x_{223} + k_{223} + \mathtt{7}$$
$$c_{30} + \mathtt{F}x_{210} + \mathtt{3}x_{211} + \mathtt{7}x_{212} + \mathtt{F}x_{213} + \mathtt{A}x_{220} + \mathtt{2}x_{221} + \mathtt{B}x_{222} + \mathtt{A}x_{223} + k_{230} + \mathtt{6}$$
$$c_{31} + \mathtt{A}x_{210} + \mathtt{A}x_{211} + \mathtt{5}x_{212} + \mathtt{6}x_{213} + \mathtt{8}x_{220} + \mathtt{8}x_{221} + \mathtt{4}x_{222} + \mathtt{9}x_{223} + k_{231} + \mathtt{7}$$
$$c_{32} + \mathtt{7}x_{210} + \mathtt{8}x_{211} + \mathtt{8}x_{212} + \mathtt{2}x_{213} + \mathtt{D}x_{220} + \mathtt{C}x_{221} + \mathtt{C}x_{222} + \mathtt{3}x_{223} + k_{232} + \mathtt{6}$$
$$c_{33} + \mathtt{4}x_{210} + \mathtt{6}x_{211} + \mathtt{C}x_{212} + \mathtt{C}x_{213} + \mathtt{5}x_{220} + \mathtt{E}x_{221} + \mathtt{F}x_{222} + \mathtt{F}x_{223} + k_{233} + \mathtt{7}$$

Key Schedule Conjugacy Relations

$k_{000}^2 + k_{001}$	$k_{100}^2 + k_{101}$	$k_{200}^2 + k_{201}$
$k_{001}^2 + k_{002}$	$k_{101}^2 + k_{102}$	$k_{201}^2 + k_{202}$
$k_{002}^2 + k_{003}$	$k_{102}^2 + k_{103}$	$k_{202}^2 + k_{203}$
$k_{003}^2 + k_{000}$	$k_{103}^2 + k_{100}$	$k_{203}^2 + k_{200}$
$k_{010}^2 + k_{011}$	$k_{110}^2 + k_{111}$	$k_{210}^2 + k_{211}$
$k_{011}^2 + k_{012}$	$k_{111}^2 + k_{112}$	$k_{211}^2 + k_{212}$
$k_{012}^2 + k_{013}$	$k_{112}^2 + k_{113}$	$k_{212}^2 + k_{213}$
$k_{013}^2 + k_{010}$	$k_{113}^2 + k_{110}$	$k_{213}^2 + k_{210}$
$k_{020}^2 + k_{021}$	$k_{120}^2 + k_{121}$	$k_{220}^2 + k_{221}$
$k_{021}^2 + k_{022}$	$k_{121}^2 + k_{122}$	$k_{221}^2 + k_{222}$
$k_{022}^2 + k_{023}$	$k_{122}^2 + k_{123}$	$k_{222}^2 + k_{223}$
$k_{023}^2 + k_{020}$	$k_{123}^2 + k_{120}$	$k_{223}^2 + k_{220}$
$k_{030}^2 + k_{031}$	$k_{130}^2 + k_{131}$	$k_{230}^2 + k_{231}$
$k_{031}^2 + k_{032}$	$k_{131}^2 + k_{132}$	$k_{231}^2 + k_{232}$
$k_{032}^2 + k_{033}$	$k_{132}^2 + k_{133}$	$k_{232}^2 + k_{233}$
$k_{033}^2 + k_{030}$	$k_{133}^2 + k_{130}$	$k_{233}^2 + k_{230}$

Key Schedule Inversion and Conjugacy Relations

$k_{030}s_{000}+1$	$s_{000}^2+s_{001}$	$k_{130}s_{100}+1$	$s_{100}^2+s_{101}$
$k_{031}s_{001}+1$	$s_{001}^2+s_{002}$	$k_{131}s_{101}+1$	$s_{101}^2+s_{102}$
$k_{032}s_{002}+1$	$s_{002}^2+s_{003}$	$k_{132}s_{102}+1$	$s_{102}^2+s_{103}$
$k_{033}s_{003}+1$	$s_{003}^2+s_{000}$	$k_{133}s_{103}+1$	$s_{103}^2+s_{100}$
$k_{020}s_{010}+1$	$s_{010}^2+s_{011}$	$k_{120}s_{110}+1$	$s_{110}^2+s_{111}$
$k_{021}s_{011}+1$	$s_{011}^2+s_{012}$	$k_{121}s_{111}+1$	$s_{111}^2+s_{112}$
$k_{022}s_{012}+1$	$s_{012}^2+s_{013}$	$k_{122}s_{112}+1$	$s_{112}^2+s_{113}$
$k_{023}s_{013}+1$	$s_{013}^2+s_{010}$	$k_{023}s_{113}+1$	$s_{113}^2+s_{110}$

Key Schedule Diffusion Relations: Round 1

$$
\begin{array}{ll}
k_{100}+k_{000} & +\mathtt{5}s_{000}+\mathtt{1}s_{001}+\mathtt{C}s_{002}+\mathtt{5}s_{003}+\mathtt{7}\\
k_{101}+k_{001} & +\mathtt{2}s_{000}+\mathtt{2}s_{001}+\mathtt{1}s_{002}+\mathtt{F}s_{003}+\mathtt{6}\\
k_{102}+k_{002} & +\mathtt{A}s_{000}+\mathtt{4}s_{001}+\mathtt{4}s_{002}+\mathtt{1}s_{003}+\mathtt{7}\\
k_{103}+k_{003} & +\mathtt{1}s_{000}+\mathtt{8}s_{001}+\mathtt{3}s_{002}+\mathtt{3}s_{003}+\mathtt{6}\\
k_{110}+k_{010} & +\mathtt{5}s_{010}+\mathtt{1}s_{011}+\mathtt{C}s_{012}+\mathtt{5}s_{013}+\mathtt{6}\\
k_{111}+k_{011} & +\mathtt{2}s_{010}+\mathtt{2}s_{011}+\mathtt{1}s_{012}+\mathtt{F}s_{013}+\mathtt{7}\\
k_{112}+k_{012} & +\mathtt{A}s_{010}+\mathtt{4}s_{011}+\mathtt{4}s_{012}+\mathtt{1}s_{013}+\mathtt{6}\\
k_{113}+k_{013} & +\mathtt{1}s_{010}+\mathtt{8}s_{011}+\mathtt{3}s_{012}+\mathtt{3}s_{013}+\mathtt{7}\\
k_{120}+k_{020}+k_{000} & +\mathtt{5}s_{000}+\mathtt{1}s_{001}+\mathtt{C}s_{002}+\mathtt{5}s_{003}+\mathtt{7}\\
k_{121}+k_{021}+k_{001} & +\mathtt{2}s_{000}+\mathtt{2}s_{001}+\mathtt{1}s_{002}+\mathtt{F}s_{003}+\mathtt{6}\\
k_{122}+k_{022}+k_{002} & +\mathtt{A}s_{000}+\mathtt{4}s_{001}+\mathtt{4}s_{002}+\mathtt{1}s_{003}+\mathtt{7}\\
k_{123}+k_{023}+k_{003} & +\mathtt{1}s_{000}+\mathtt{8}s_{001}+\mathtt{3}s_{002}+\mathtt{3}s_{003}+\mathtt{6}\\
k_{130}+k_{030}+k_{010} & +\mathtt{5}s_{010}+\mathtt{1}s_{011}+\mathtt{C}s_{012}+\mathtt{5}s_{013}+\mathtt{6}\\
k_{131}+k_{031}+k_{011} & +\mathtt{2}s_{010}+\mathtt{2}s_{011}+\mathtt{1}s_{012}+\mathtt{F}s_{013}+\mathtt{7}\\
k_{132}+k_{032}+k_{012} & +\mathtt{A}s_{010}+\mathtt{4}s_{011}+\mathtt{4}s_{012}+\mathtt{1}s_{013}+\mathtt{6}\\
k_{133}+k_{033}+k_{013} & +\mathtt{1}s_{010}+\mathtt{8}s_{011}+\mathtt{3}s_{012}+\mathtt{3}s_{013}+\mathtt{7}
\end{array}
$$

Key Schedule Diffusion Relations: Round 2

$$
\begin{array}{ll}
k_{200}+k_{100} & +\mathtt{5}s_{100}+\mathtt{1}s_{101}+\mathtt{C}s_{102}+\mathtt{5}s_{103}+\mathtt{4}\\
k_{201}+k_{101} & +\mathtt{2}s_{100}+\mathtt{2}s_{101}+\mathtt{1}s_{102}+\mathtt{F}s_{103}+\mathtt{3}\\
k_{202}+k_{102} & +\mathtt{A}s_{100}+\mathtt{4}s_{101}+\mathtt{4}s_{102}+\mathtt{1}s_{103}+\mathtt{5}\\
k_{203}+k_{103} & +\mathtt{1}s_{100}+\mathtt{8}s_{101}+\mathtt{3}s_{102}+\mathtt{3}s_{103}+\mathtt{2}\\
k_{210}+k_{110} & +\mathtt{5}s_{110}+\mathtt{1}s_{111}+\mathtt{C}s_{112}+\mathtt{5}s_{113}+\mathtt{6}\\
k_{211}+k_{111} & +\mathtt{2}s_{110}+\mathtt{2}s_{111}+\mathtt{1}s_{112}+\mathtt{F}s_{113}+\mathtt{7}\\
k_{212}+k_{112} & +\mathtt{A}s_{110}+\mathtt{4}s_{111}+\mathtt{4}s_{112}+\mathtt{1}s_{113}+\mathtt{6}\\
k_{213}+k_{113} & +\mathtt{1}s_{110}+\mathtt{8}s_{111}+\mathtt{3}s_{112}+\mathtt{3}s_{113}+\mathtt{7}\\
k_{220}+k_{120}+k_{100} & +\mathtt{5}s_{100}+\mathtt{1}s_{101}+\mathtt{C}s_{102}+\mathtt{5}s_{103}+\mathtt{4}\\
k_{221}+k_{121}+k_{101} & +\mathtt{2}s_{100}+\mathtt{2}s_{101}+\mathtt{1}s_{102}+\mathtt{F}s_{103}+\mathtt{3}\\
k_{222}+k_{122}+k_{102} & +\mathtt{A}s_{100}+\mathtt{4}s_{101}+\mathtt{4}s_{102}+\mathtt{1}s_{103}+\mathtt{5}\\
k_{223}+k_{123}+k_{103} & +\mathtt{1}s_{100}+\mathtt{8}s_{101}+\mathtt{3}s_{102}+\mathtt{3}s_{103}+\mathtt{2}\\
k_{230}+k_{130}+k_{110} & +\mathtt{5}s_{110}+\mathtt{1}s_{111}+\mathtt{C}s_{112}+\mathtt{5}s_{113}+\mathtt{6}\\
k_{231}+k_{131}+k_{111} & +\mathtt{2}s_{110}+\mathtt{2}s_{111}+\mathtt{1}s_{112}+\mathtt{F}s_{113}+\mathtt{7}\\
k_{232}+k_{132}+k_{112} & +\mathtt{A}s_{110}+\mathtt{4}s_{111}+\mathtt{4}s_{112}+\mathtt{1}s_{113}+\mathtt{6}\\
k_{233}+k_{133}+k_{113} & +\mathtt{1}s_{110}+\mathtt{8}s_{111}+\mathtt{3}s_{112}+\mathtt{3}s_{113}+\mathtt{7}
\end{array}
$$

Unbiased Random Sequences from Quasigroup String Transformations

Smile Markovski[1], Danilo Gligoroski[1], and Ljupco Kocarev[2]

[1] "Ss Cyril and Methodius" University,
Faculty of Natural Sciences and Mathematics, Institute of Informatics,
P.O.Box 162, 1000 Skopje, Republic of Macedonia
{danilo, smile}@ii.edu.mk

[2] University of California San Diego, Institute for Nonlinear Science,
9500 Gilman Drive, La Jolla, CA 92093-0402, USA
lkocarev@ucsd.edu

Abstract. The need of true random number generators for many purposes (ranging from applications in cryptography and stochastic simulation, to search heuristics and game playing) is increasing every day. Many sources of randomness possess the property of stationarity. However, while a biased die may be a good source of entropy, many applications require input in the form of unbiased bits, rather than biased ones. In this paper, we present a new technique for simulating fair coin flips using a biased, stationary source of randomness. Moreover, the same technique can also be used to improve some of the properties of pseudo random number generators. In particular, an improved pseudo random number generator has almost unmeasurable period, uniform distribution of the letters, pairs of letters, triples of letters, and so on, and passes many statistical tests of randomness. Our algorithm for simulating fair coin flips using a biased, stationary source of randomness (or for improving the properties of pseudo random number generators) is designed by using quasigroup string transformations and its properties are mathematically provable. It is very flexible, the input/output strings can be of 2-bits letters, 4-bits letters, bytes, 2-bytes letters, and so on. It is of linear complexity and it needs less than 1Kb memory space in its 2-bits and 4-bits implementations, hence it is suitable for embedded systems as well.

1 Introduction

Random number generators (RNGs) are useful in every scientific area which uses Monte Carlo methods. It is difficult to imagine a scientific area where Monte Carlo methods and RNGs are not used. Extremely important is the application of RNGs in cryptography for generation of cryptographic keys, and random initialization of certain variables in cryptographic protocols. Countless applications in cryptography, stochastic simulation, search heuristics, and game playing rely on the use of sequences of random numbers.

H. Gilbert and H. Handschuh (Eds.): FSE 2005, LNCS 3557, pp. 163–180, 2005.

The choice of the RNG for a specific application depends on the requirements specific to the given application. If the ability to regenerate the random sequence is of crucial significance such as debugging simulations, or the randomness requirements are not very stringent (flying through space on your screen saver), or the hardware generation costs are unjustified, then one should resort to pseudo-random number generators (PRNGs). PRNGs are algorithms implemented on finite-state machines and are capable of generating sequences of numbers which appear random-like from many aspects. Though they are necessarily periodic ("Anyone who considers arithmetical methods of producing random digits is, of course, in a state of sin", John von Neumann), their periods are very long, they pass many statistical tests and can be easily implemented with simple and fast software routines.

It is widely accepted that the core of any RNG must be an intrinsically random physical process. So, it is no surprise that the proposals and implementations of RNGs range from tossing a coin, throwing a dice, drawing from a urn, drawing from a deck of cards and spinning a roulette to measuring thermal noise from a resistor and shot noise from a Zener diode or a vacuum tube, measuring radioactive decay from a radioactive source, integrating dark current from a metal insulator semiconductor capacitor, detecting locations of photoevents, and sampling a stable high-frequency oscillator with an unstable low-frequency clock. Some of the sources of randomness, such as radioactive sources [1] and quantum-mechanical sources [2], may yield data from probability distributions that are stationary. Therefore, the output of these sources does not change over time and does not depend on previous outputs. However, even if a source is stationary, it generally has a bias. In other words, the source does not give unbiased bits as direct output. It is therefore quite important to be able to extract unbiased bits efficiently from a stationary source with unknown bias.

Suppose that a reading obtained from a stationary source of randomness can be equal to any one of m different values, but that the probability of obtaining any one of these values is unknown and in general not equal to $1/m$. In other words, we assume that the source may be loaded. Our aim in this paper is to simulate unbiased coin flips using a biased source.

Previous Work – There are several references to the problem of simulating unbiased physical sources of randomness. Von Neumann [3] described the following method; flip the biased coin twice: if it comes up HT, output an H, if it comes up TH, output a T, otherwise, start over. This method will simulate the output of an unbiased coin irrespective of the bias of the coin used in the simulation. Elias [4] proposed a method of extracting unbiased bits from biased Markov chains. Stout and Warren [5] and Juels et. al. [6] presented new extensions of the technique of von Neumann. Stout and Warren suggested a method for simulating a fixed number of fair coin flips using as few rolls of a biased die as possible, while the authors of [6] proposed an algorithm for extracting, given a fixed number of rolls of a biased die, as many fair coin flips as possible. The general characteristics of the methods for simulating unbiased physical sources of randomness are: (i) all of them do not use each bit of information generated by the source, (ii) some of the methods can be implemented in computationally effective way, but for some of them corresponding algorithms are of exponential nature and then approximations should be involved, and (iii) for some of them mathematical proofs are supplied for their properties.

Our Work – In this paper we propose a method for simulating unbiased physical sources of randomness which is based on the quasigroup string transformations and some of their provable properties. Our method uses each bit of information produced by a discrete source of randomness. Moreover, our method is capable of producing a random number sequence from a very biased stationary source (for example, from a source that produces 0 with probability 1/1000 and 1 with probability 999/1000). The complexity of our algorithm is linear, i.e. an output string of length n will be produced from an input string of length n with complexity $O(n)$. Our algorithm is highly parallel. This means there exist computationally very effective software and hardware implementations of the method. Our algorithm is also very flexible: the same design can be used for strings whose letters consists of 2-bits, 4-bits, bytes, 2-bytes, and generally it can be designed for an arbitrary n-bit letters alphabet ($n \geq 2$). The method proposed in this paper can also be used to improve the quality of existing PRNGs so that they pass many statistical tests and their periods can be arbitrary large numbers. Since many of the weak PRNGs are still in use because of the simplicity of their design and the speed of producing pseudo random strings (although of bad quality), our method in fact can improve the quality of these PRNGs very effectively.

The paper is organized as follows. Needed definitions and properties of quasigroups and quasigroup string transformations are given in Section 2. The algorithm for simulating unbiased physical sources of randomness (or for improving PRNGs) is presented in Section 3. In this section we also present some numerical results concerning our method, while the proofs of the main theorems are given in the appendicitis. In Section 4 we close our paper with conclusion.

2 Quasigroup String Transformations

Here we give a brief overview of quasigroups, quasigroup operations and quasigroup string transformations (more detailed explanation the reader can find in [7], [8]).

A quasigroup is a groupoid $(Q, *)$ satisfying the laws

$$(\forall u, v \in Q)(\exists x, y \in Q)(u * x = v,\ y * u = v),$$

$$x * y = x * z \implies y = z,\ y * x = z * x \implies y = z.$$

Hence, a quasigroup satisfies the cancelation laws and the equations $a * x = b,\ y * a = b$ have unique solutions $x,\ y$ for each $a, b \in Q$. If $(Q, *)$ is a quasigroup, then $*$ is called a quasigroup operation.

Here we consider only finite quasigroups, i.e Q is a finite set. Closely related combinatorial structures to finite quasigroups are the so called Latin squares: a

Latin square L on a finite set Q (with cardinality $|Q| = s$) is an $s \times s$-matrix with elements from Q such that each row and each column of the matrix is a permutation of Q. To any finite quasigroup $(Q, *)$ given by its multiplication table it is associated a Latin square L, consisting of the matrix formed by the main body of the table, and each Latin square L on a set Q define a quasigroup $(Q, *)$.

Given a quasigroup $(Q, *)$ five new operations, so called parastrophes or adjoint operations, can be derived from the operation $*$. We will need only the following two, denoted by $\backslash$ and $/$, and defined by:

$$x * y = z \iff y = x \backslash z \iff x = z/y \tag{1}$$

Then the algebra $(Q, *, \backslash, /)$ satisfies the identities

$$x \backslash (x * y) = y, \; x * (x \backslash y) = y, \; (x * y)/y = x, \; (x/y) * y = x \tag{2}$$

and $(Q, \backslash)$, $(Q, /)$ are quasigroups too.

Several quasigroup string transformations can be defined and those of interest of us will be explained bellow. Consider an alphabet (i.e. a finite set) A, and denote by A^+ the set of all nonempty words (i.e. finite strings) formed by the elements of A. The elements of A^+ will be denoted by $a_1 a_2 \dots a_n$ rather than $(a_1, a_2, \dots, a_n)$, where $a_i \in A$. Let $*$ be a quasigroup operation on the set A. For each $l \in A$ we define two functions $e_{l,*}, \; e'_{l,*} : A^+ \longrightarrow A^+$ as follows. Let $a_i \in A, \; \alpha = a_1 a_2 \dots a_n$. Then

$$e_{l,*}(\alpha) = b_1 \dots b_n \iff b_{i+1} = b_i * a_{i+1} \tag{3}$$

$$e'_{l,*}(\alpha) = b_1 \dots b_n \iff b_{i+1} = a_{i+1} * b_i \tag{4}$$

for each $i = 0, 1, \dots, n-1$, where $b_0 = l$. The functions $e_{l,*}$ and $e'_{l,*}$ are called e- and e'-transformations of A^+ based on the operation $*$ with leader l. Graphical representations of the e- and e'-transformations are shown on Figure 1 and Figure 2.

Example 1. Take $A = \{0, 1, 2, 3\}$ and let the quasigroup $(A, *)$ be given by the multiplication scheme

*	0	1	2	3
0	2	1	0	3
1	3	0	1	2
2	1	2	3	0
3	0	3	2	1

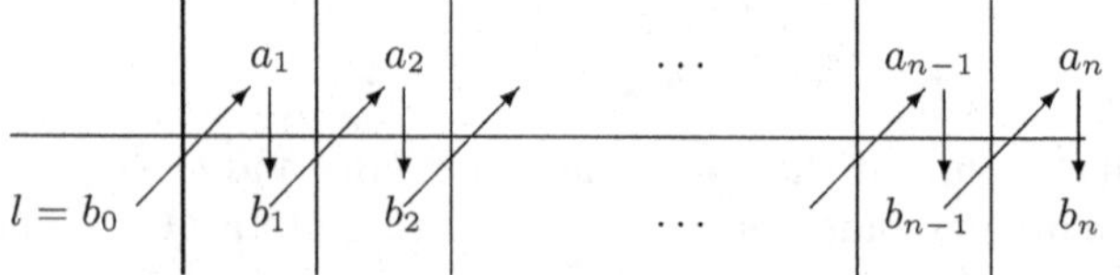

Fig. 1. Graphical representation of an e-transformation

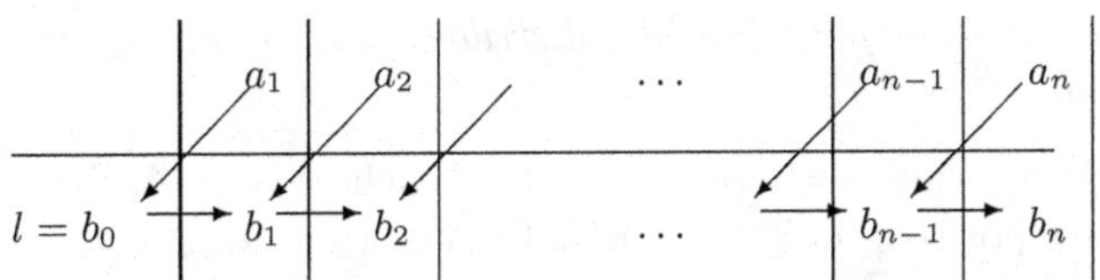

Fig. 2. Graphical representation of an e'-transformation

Consider the string $\alpha = 1\,0\,2\,1\,0\,0\,0\,0\,0\,0\,0\,0\,1\,1\,2\,1\,0\,2\,2\,0\,1\,0\,1\,0\,3\,0\,0$ and choose the leader 0. Then we have the following transformed strings

$e_{0,*}(\alpha) = 1\,3\,2\,2\,1\,3\,0\,2\,1\,3\,0\,2\,1\,0\,1\,1\,2\,1\,1\,1\,3\,3\,0\,1\,3\,1\,3\,0,$

$e'_{0,*}(\alpha) = 3\,3\,0\,3\,3\,3\,3\,3\,3\,3\,3\,3\,3\,2\,1\,2\,1\,1\,2\,3\,3\,2\,0\,3\,3\,1\,1\,1.$

Four consecutive applications of these transformations are presented below:

	1 0 2 1 0 0 0 0 0 0 0 0 1 1 2 1 0 2 2 0 1 0 1 0 3 0 0 $= \alpha$
0	1 3 2 2 1 3 0 2 1 3 0 2 1 0 1 1 2 1 1 1 3 3 0 1 3 1 3 0 $= e_{0,*}(\alpha)$
0	1 2 3 2 2 0 2 3 3 1 3 2 2 1 0 1 1 2 2 2 0 3 0 1 2 2 0 2 $= e_{0,*}{}^2(\alpha)$
0	1 1 2 3 2 1 1 2 0 1 2 3 2 2 1 0 1 1 1 1 3 1 3 3 2 3 0 0 $= e_{0,*}{}^3(\alpha)$
0	1 0 0 3 2 2 2 3 0 1 1 2 3 2 2 1 0 1 0 1 2 2 0 3 2 0 2 1 $= e_{0,*}{}^4(\alpha)$

	1 0 2 1 0 0 0 0 0 0 0 0 1 1 2 1 0 2 2 0 1 0 1 0 3 0 0 $= \alpha$
0	3 3 0 3 3 3 3 3 3 3 3 3 3 2 1 2 1 1 2 3 3 2 0 3 3 1 1 1 $= e'_{0,*}(\alpha)$
0	0 0 2 2 2 2 2 2 2 2 2 2 2 3 2 3 2 1 2 2 2 3 3 1 3 2 1 0 $= e'_{0,*}{}^2(\alpha)$
0	2 0 1 2 3 0 1 2 3 0 1 2 3 1 2 2 3 2 3 0 1 3 1 0 0 1 0 2 $= e'_{0,*}{}^3(\alpha)$
0	1 1 0 1 3 3 2 3 1 1 0 1 3 2 3 0 0 1 3 3 2 2 1 1 1 0 2 3 $= e'_{0,*}{}^4(\alpha)$

One can notice that the starting distribution of 0, 1, 2 and 3 in α : 16/28, 7/28, 4/28, 1/28 is changed to 7/28, 7/28, 10/28, 4/28 in $e_{0,*}{}^4(\alpha)$ and to 5/28, 10/28, 5/28, 8/28 in $e'_{0,*}{}^4(\alpha)$, hence the distributions became more uniform.

Several quasigroup operations can be defined on the set A and let $*_1$, $*_2$, $\dots$, $*_k$ be a sequence of (not necessarily distinct) such operations. We choose also leaders l_1, l_2, $\dots$, $l_k \in A$ (not necessarily distinct either), and then the compositions of mappings

$$E_k = E_{l_1 \dots l_k} = e_{l_1,*_1} \circ e_{l_2,*_2} \circ \dots \circ e_{l_k,*_k},$$

$$E'_k = E'_{l_1 \dots l_k} = e'_{l_1,*_1} \circ e'_{l_2,*_2} \circ \dots \circ e_{l_k,*_k},$$

are said to be E- and E'-transformations of A^+ respectively. The functions E_k and E'_k have many interesting properties, and for our purposes the most important ones are the following:

Theorem 1. ([8]) *The transformations E_k and E'_k are permutations of A^+.*

Theorem 2. ([8]) *Consider an arbitrary string $\alpha = a_1 a_2 \dots a_n \in A^+$, where $a_i \in A$, and let $\beta = E_k(\alpha)$, $\beta' = E'_k(\alpha)$. If n is large enough integer then, for each l : $1 \leq l \leq k$, the distribution of substrings of β and β' of length l is* *uniform. (We note that for $l > k$ the distribution of substrings of β and β' of length l may not be uniform.)*

We say that a string $\alpha = a_1 a_2 \dots a_n \in A^+$, where $a_i \in A$, has a period p if p is the smallest positive integer such that $a_{i+1}a_{i+2}\dots a_{i+p} = a_{i+p+1}a_{i+p+2}\dots \dots a_{i+2p}$ for each $i \geq 0$. The following property holds:

Theorem 3. ([9]) *Let $\alpha = a_1 a_2 \dots a_n \in A^+$, $a_i \in A$, and let $\beta = E_k(\alpha)$, $\beta' = E'_k(\alpha)$, where $E_k = E_{aa\dots a}$, $E'_k = E'_{aa\dots a}$, $a \in A$ and $a * a \neq a$. Then the periods of the strings β and β' are increasing at least linearly by k.*

We should note that the increasing of the periods depends of the number of quasigroup transformations k, and for some of them it is exponential, i.e. if α has a period p, then $\beta = E_k(\alpha)$ and $\beta' = E'_k(\alpha)$ may have periods greater than $p\,2^{c\,k}$, where c is some constant. We will discuss this in more details in the next section. In what follows we will usually use only E-transformations, since the results will hold for E'-transformations by symmetry.

Theorem 1 is easy to prove (and one can find the proof in [8]). The proofs of Theorem 2 and Theorem 3 are given in the Appendix I and the Appendix II, respectively.

3 Description of the Algorithms

Assume that we have a discrete biased stationary source of randomness which produces strings from A^+, i.e. the alphabet of source is A, where

$$A = \{a_0, a_1, \dots, a_{s-1}\}$$

is a finite alphabet. (However, we may also think that strings in A^+ are produced by a PRNG.)

Now we define two algorithms for simulating unbiased physical sources of randomness (or for improving PRNGs), based on E- and E'-transformations accordingly. We call them an $E-algorithm$ and an $E'-algorithm$. In these algorithms we use several internal and temporal variables b, L_1, $\dots$, L_n. The input of the algorithm is the order of the quasigroup s, a quasigroup $(A, *)$ of order s, a fixed element $l \in A$ (the leader), an integer k giving the number of applications of the transformations $e_{l,*}$ and $e'_{l,*}$ and a biased random string $b_0, b_1, b_2, b_3, \dots$. The output is an unbiased random string.

The performance of the algorithms is based on Theorems 1, 2 and 3. By Theorem 1 we have that $E-algorithm$ and $E'-algorithm$ are injective, meaning that different input string produces different output string. Theorem 2 guarantees that the algorithms generate unbiased output random strings. Theorem 3 guarantees that if the biased source has period p (such as some Pseudo Random Number Generator) the algorithm will generate unbiased output with longer period.

Both $E-algorithm$ and $E'-algorithm$ can also be used to improve the properties of PRNGs. For example, for suitable choice of the quasigroup and suitable choice of the parameter s, Theorem 3 shows that the period of the output pseudo random string can be made arbitrary large. In addition, we have checked the quality of output pseudo random strings by using available statistical tests (such as Diehard [10] and those suggested by NIST [11]) for different quasigroups, leaders, and different values of n: in all these cases the pseudo strings passed all of the tests.

E-algorithm
Phase I. Initialization
1. Choose a positive integer $s \geq 4$;
2. Choose a quasigroup $(A, *)$ of order s;
3. Set a positive integer k;
4. Set a leader l, a fixed element of A such that $l * l \neq l$;
Phase II. Transformations of the random string $b_0b_1b_2b_3\ldots,\ \ b_j \in A$
5. For $i = 1$ to k do $L_i \leftarrow l$;
6. $j \leftarrow 0$;
7. do
$b \leftarrow b_j$;
$L_1 \leftarrow L_1 * b$;
For $i = 2$ to k do $L_i \leftarrow L_i * L_{i-1}$;
Output: L_k;
$j \leftarrow j + 1$;
loop;

The $E' - algorithm$ differs of the $E - algorithm$ only in step 7:

$E' - algorithm$
7'. do
$b \leftarrow b_j$;
$L_1 \leftarrow b * L_1$;
For $i = 2$ to k do $L_i \leftarrow L_{i-1} * L_i$;
Output: L_k;
$j \leftarrow j + 1$;
loop;

Example 2. The PRNG used in GNU C v2.03 do not passed all of the statistical tests in the Diehard Battery v0.2 beta [10], but the improved PRNG passed all of them after only one application ($k = 1$) of an e-transformation performed by a quasigroup of order 256. The results are given in the next two screen dumps.

```
***** TEST SUMMARY FOR GNU C (v2.03) PRNG *****
All p-values:
0.2929,0.8731,0.9113,0.8755,0.4637,0.5503,0.9435,0.7618,0.9990,0.0106,1.0000,0.0430,0.0680,
1.0000,1.0000,1.0000,1.0000,1.0000,1.0000,1.0000,1.0000,1.0000,1.0000,1.0000,1.0000,1.0000,
1.0000,1.0000,1.0000,0.0000,1.0000,1.0000,1.0000,1.0000,1.0000,1.0000,1.0000,1.0000,1.0000,
```

1.0000,1.0000,1.0000,1.0000,1.0000,1.0000,1.0000,0.2009,0.0949,0.1939,0.0944,0.2514,0.3419,
0.5714,0.2256,0.1484,0.7394,0.0562,0.3314,0.2559,0.5677,0.3061,0.4763,0.8185,0.1571,0.2072,
0.5667,0.7800,0.6428,0.7636,0.1529,0.9541,0.8689,0.1558,0.6235,0.5275,0.6316,0.7697,0.7181,
0.7921,0.4110,0.3050,0.8859,0.4783,0.3283,0.4073,0.2646,0.0929,0.6029,0.4634,0.8462,0.3777,
0.2385,0.6137,0.1815,0.4001,0.1116,0.2328,0.0544,0.4320,0.0000,0.0000,0.0000,0.0000,0.0000,
0.0000,0.0000,0.0000,0.0000,0.0000,0.0000,0.0000,0.0000,0.0000,0.0000,0.0000,0.0000,0.0000,
0.0000,0.0000,0.0003,0.0000,0.0000,0.0000,0.0000,0.0000,0.0000,1.0000,0.0013,0.0000,0.0000,
0.0000,0.0000,0.0000,0.0000,1.0000,0.0000,0.0000,0.0000,0.0000,0.0000,0.0000,0.0000,1.0000,
0.0003,0.0000,0.0000,0.0000,0.0000,0.0000,0.0000,0.0000,0.0000,0.0000,0.0000,0.0000,0.0000,
0.0753,0.0010,0.0000,0.0000,0.0000,0.0000,0.0000,0.0000,0.0233,0.0585,0.0000,0.0000,0.0000,
0.0000,0.0000,0.0000,0.2195,0.0321,0.0000,0.0000,0.9948,0.0006,0.0000,0.0000,0.0688,0.5102,
0.6649,0.1254,0.2967,0.1218,0.8199,0.7125,0.6873,0.1663,0.7150,0.7275,0.9035,0.1946,0.7261,
0.7243,0.1083,0.4266,0.7664,0.8384,0.7317,0.8340,0.3155,0.0987,0.7286,0.6645,0.9121,0.0550,
0.6923,0.1928,0.7236,0.0159,0.4636,0.2764,0.2325,0.3406,0.3746,0.1208,0.8145,0.3693,0.7426,
0.6272,0.6139,0.4957,0.3623,0.4929,0.3628,0.5266,0.2252,0.7948,0.7327,0.2732,0.6895,0.2325,
0.2303,0.1190,0.8802,0.0377,0.6887,0.4175,0.0803,0.3687,0.7010,0.7425,0.1003,0.0400,0.5055,
0.9488,0.3209,0.5965,0.0676,0.0021,0.2337,0.5204,0.5343,0.0630,0.2008,0.6496,0.4157,0.0559,
0.9746,0.1388,0.4657,0.5793,0.6455,0.8441,0.5248,0.7962,0.8870

Overall p-value after applying KStest on 269 p-values = 0.000000

```
*** TEST SUMMARY FOR GNU C v2.03 + QUASIGROUP PRNG IMPROVER ***
All p-values:
```

0.5804,0.3010,0.1509,0.5027,0.3103,0.5479,0.3730,0.9342,0.4373,0.5079,0.0089,0.3715,0.3221,
0.0584,0.1884,0.1148,0.0662,0.8664,0.5070,0.7752,0.1939,0.9568,0.4948,0.1114,0.2042,0.4190,
0.4883,0.4537,0.0281,0.0503,0.0346,0.6085,0.1596,0.1545,0.0855,0.5665,0.0941,0.7693,0.0288,
0.1372,0.8399,0.0320,0.6930,0.3440,0.9842,0.9975,0.1354,0.8776,0.1919,0.2584,0.6437,0.1995,
0.2095,0.3298,0.5180,0.8136,0.7294,0.7560,0.0458,0.6285,0.1775,0.1546,0.0397,0.5135,0.0938,
0.6544,0.9673,0.8787,0.9520,0.8339,0.4397,0.3687,0.0044,0.7146,0.9782,0.7440,0.3042,0.3388,
0.8465,0.7123,0.8752,0.8775,0.7552,0.5711,0.3768,0.1390,0.9870,0.9444,0.6101,0.1090,0.2032,
0.8538,0.6871,0.8785,0.9159,0.4128,0.4513,0.1512,0.8808,0.7079,0.2278,0.1400,0.6461,0.4082,
0.3353,0.1064,0.6739,0.2066,0.5119,0.0558,0.5748,0.5064,0.8982,0.6422,0.7512,0.8633,0.1712,
0.4625,0.0843,0.0903,0.7641,0.6253,0.8523,0.7768,0.8041,0.5360,0.0826,0.0378,0.8710,0.4901,
0.7994,0.7748,0.8403,0.9886,0.1373,0.7082,0.8860,0.9595,0.2671,0.0038,0.7572,0.8403,0.7410,
0.5615,0.6181,0.1257,0.5960,0.2432,0.8302,0.1981,0.7764,0.2109,0.2109,0.6620,0.8938,0.0052,
0.8116,0.5196,0.0836,0.4144,0.2466,0.3298,0.8724,0.9837,0.8748,0.0930,0.5055,0.6511,0.3569,
0.2832,0.4029,0.9290,0.3470,0.6598,0.4796,0.3758,0.6077,0.4213,0.1886,0.1500,0.3341,0.0594,
0.0663,0.0946,0.8279,0.2451,0.2969,0.9297,0.0739,0.4839,0.1307,0.4527,0.0272,0.9913,0.0570,
0.0791,0.9028,0.4706,0.4020,0.7592,0.4105,0.7107,0.5505,0.7223,0.3233,0.3037,0.9924,0.5545,
0.7944,0.0854,0.5545,0.4455,0.4636,0.2613,0.2467,0.9586,0.4275,0.8175,0.5793,0.1189,0.7109,
0.2115,0.8156,0.8468,0.9429,0.8382,0.1463,0.4212,0.6948,0.4816,0.3454,0.2114,0.3493,0.1389,
0.3448,0.0413,0.2422,0.6363,0.2340,0.8404,0.0065,0.7319,0.8781,0.2751,0.5197,0.4105,0.7121,
0.0832,0.1503,0.1148,0.3008,0.0121,0.0029,0.4423,0.6239,0.0651,0.3838,0.0165,0.2770,0.0475,
0.2074,0.0004,0.7962,0.4750,0.4839,0.9152,0.1681,0.0822,0.0518

Overall p-value after applying KStest on 269 p-values = 0.018449

Example 3. In this example as a source we used a highly biased source of randomness where 0 has frequency of $\frac{1}{1000}$ and 1 has frequency of $\frac{999}{1000}$. We applied several consecutive e-transformation with a random quasigroup of order 256, monitoring the results from Diehard battery. After the fifth e-transformation we obtained the following results:

```
** TEST SUMMARY - HIGHLY BIASED SOURCE & FIVE e-Transformations **
   All p-values:
0.9854,0.8330,0.4064,0.9570,0.6597,0.5447,0.5796,0.5885,0.3482,0.1359,0.1788,0.1194,0.8588,
0.3455,0.6627,0.3610,0.5622,0.9905,0.8430,0.1259,0.0799,0.9061,0.8378,0.4313,0.7249,0.4505,
0.9192,0.1007,0.2785,0.9099,0.0422,0.7891,0.2681,0.4452,0.9389,0.5081,0.7621,0.0914,0.0066,
0.6915,0.8662,0.7176,0.5658,0.7957,0.0590,0.4287,0.5772,0.4809,0.9891,0.1439,0.0000,0.6089,
0.2351,0.2533,0.0061,0.0171,0.6894,0.5279,0.9075,0.7313,0.6401,0.8004,0.1155,0.4374,0.8159,
0.9895,0.4989,0.5433,0.6915,0.9944,0.5661,0.7771,0.5461,0.8875,0.6586,0.0340,0.4701,0.9087,
0.1412,0.4037,0.7326,0.1809,0.3157,0.0573,0.3875,0.4210,0.9403,0.9805,0.2278,0.7588,0.2840,
0.5109,0.4997,0.5554,0.1334,0.5332,0.3025,0.2139,0.4366,0.2514,0.5530,0.7288,0.7055,0.3316,
0.0870,0.0853,0.6714,0.7704,0.9582,0.8772,0.2448,0.6751,0.0658,0.1317,0.6096,0.8317,0.0234,
0.6689,0.3353,0.5257,0.9411,0.7219,0.5881,0.1103,0.5709,0.3836,0.4470,0.6104,0.3517,0.5841,
0.1097,0.0597,0.6784,0.4045,0.6929,0.5104,0.5828,0.8125,0.5481,0.0264,0.3244,0.6821,0.8731,
0.8773,0.7624,0.7748,0.7128,0.4698,0.1195,0.0842,0.3780,0.8346,0.4562,0.5745,0.9541,0.3341,
0.0480,0.0753,0.3713,0.9637,0.9479,0.2401,0.8256,0.8368,0.2636,0.8346,0.9236,0.1218,0.3859,
0.8203,0.6748,0.5384,0.6346,0.8667,0.0006,0.6346,0.3780,0.8693,0.1459,0.7995,0.0483,0.7434,
0.2872,0.2546,0.2167,0.4233,0.8091,0.0451,0.2333,0.3243,0.8374,0.0915,0.3251,0.3731,0.5076,
0.8991,0.0931,0.9258,0.2831,0.8281,0.8386,0.0906,0.0979,0.5441,0.7129,0.8298,0.8427,0.8732,
0.7236,0.9397,0.5545,0.9397,0.9544,0.8312,0.2325,0.8424,0.2325,0.0176,0.8621,0.0401,0.7033,
0.2288,0.2786,0.6751,0.3424,0.5295,0.9344,0.7879,0.9744,0.0259,0.0487,0.1014,0.8589,0.8655,
0.1008,0.8204,0.5564,0.7432,0.8604,0.2008,0.2081,0.4452,0.2352,0.5092,0.4250,0.6055,0.5262,
0.1459,0.0838,0.2735,0.9764,0.6419,0.7941,0.2412,0.6055,0.9725,0.1075,0.2903,0.5552,0.1643,
0.0813,0.8206,0.0742,0.5889,0.3077,0.4771,0.7677,0.8252,0.3248
```

Overall p-value after applying KStest on 269 p-values = 0.373599

We now discuss the choice of the quasigroup, and the parameters s and k. If $E-algorithm$ and $E'-algorithm$ are used for simulating unbiased physical sources of randomness, then the quasigroup can be chosen to be arbitrary (we recommend $4 \leq s \leq 256$) while k depends on s and how biased is the source of randomness. The number k should be chosen by the rule 'for smaller s larger k' and its choice depends on the source. For example, if a source is highly biased (it produces 0 with probability $1/1000$ and 1 with probability $999/1000$), we suggest the following rule (derived from our numerous numerical experiments): '$ks \geq 512$ and $k > 8$'. In fact, the number s is in a way predefined by the source. Let the alphabet of the source consists of all 8-bits letters. Then we have the following choices of A: $A = \{0,1,2,3\}$, $A = \{0,1,2,\ldots,7\}$, $A = \{0,1,\ldots,15\}$, $A = \{0,1,\ldots,31\}$, $A = \{0,1,\ldots,63\}$, $A = \{0,1,2,\ldots,127\}$, $A = \{0,1,2,\ldots,255\}$. Thus, the output string of the source is considered as string of bits and then the bits are grouped in two, three, and so on. We can consider in this case alphabets with two–byte letters, three–byte letters etc., but quasigroups of orders 65536 or higher need a lot of storage memory and generally the computations are slower, and we do not recommend to be used.

If $E - algorithm$ and $E' - algorithm$ are used for improving some of the properties of PRNGs, then the quasigroup should be exponential. Our theoretical results ([8], [12], [13]) and numerical experiments indicate that the class of finite quasigroups can be separated into two disjoint subclasses: the class of linear quasigroups and the class of exponential quasigroups. There are several characteristics that separate these two classes and for our purposes this one is important. Given a finite set $Q = \{q_0, q_1, \ldots, q_{s-1}\}$, let $(Q, *)$ be a quasigroup and let $\alpha = q_0 q_1 \cdots\ q_{p-1} q_0 q_1 \cdots\ q_{p-1} q_0 q_1 \cdots\ q_{p-1} \cdots$ be an enough long string of period p. Let

$$\alpha_k = \underbrace{e_{l,*} \ldots e_{l,*}}_{k-\text{times}}(\alpha).$$

If the period of the string α_k is a linear function of k, then the quasigroup $(Q, *)$ is said to be linear. On the other hand, if the period of the string α_k is an exponential function 2^{ck} (where c is some constant), then the quasigroup $(Q, *)$ is said to be exponential. The number c is called the period growth of the exponential quasigroup $(Q, *)$.

The numerical experiments presented in [14] show that the percentage of linear quasigroups decreases when the order of the quasigroup increases. Furthermore, the percentage of 'bad' quasigroups, i.e. linear quasigroups and exponential quasigroup with period growth $c < 2$, is decreasing exponentially by the order of the quasigroups. For quasigroups of order 4, 5, 6, 7, 8, 9 and 10 the results are summarized in Table 1. We stress that the above results are not quite precise (except for the quasigroups of order 4, where complete classification is obtained in [15]), since the conclusion is made when only 7 e-transformation were applied. Namely, it can happen that some of quasigroups, after more than 7 applications, will have period growth $c \geq 2$.

We made the following experiment over 10^6 randomly chosen quasigroups of order 16. We counted the period growth after 5 applications of $e_{l,*}$- transformations of each of the quasigroups on the following periodical strings with period 16: $0, 1, 2, \ldots, 14, 15, 0, 1, 2, \ldots, 14, 15, \ldots, 0, 1, 2, \ldots, 14, 15, \ldots$. The value of the leader l did not affect the results. The obtained distribution of the period growth is presented on the Table 2. It can be seen from Table 2 that 907 quasigroups have period growth $c < 2$ after 5 applications of the e-transformation. We counted the period growth after 6 applications of each of those quasigroups and we obtained that only 15 of them have period growth $c < 2$. After 7 applications, only one quasigroup has period growth $c < 2$, but after 10 applications of e-

Table 1. Percentage of 'bad' quasigroups of order 4 – 10

Order of the quasigroup	4	5	6	7	8	9	10
Percentage of 'bad' quasigroups	34.7	4.1	1.6	0.6	0.38	0.25	0.15

Table 2. Period growth of 10^6 randomly chosen quasigroups of order 16 after 5 applications of e-transformations

Value of c	Number of quasigroups with period growth $2^{c\,k}$	Value of c	Number of quasigroups with period growth $2^{c\,k}$
$0.00 \le c < 0.25$	4	$2.00 \le c < 2.25$	79834
$0.25 \le c < 0.50$	23	$2.25 \le c < 2.50$	128836
$0.50 \le c < 0.75$	194	$2.50 \le c < 2.75$	174974
$0.75 \le c < 1.00$	686	$2.75 \le c < 3.00$	199040
$1.00 \le c < 1.25$	2517	$3.00 \le c < 3.25$	175848
$1.25 \le c < 1.50$	7918	$3.25 \le c < 3.50$	119279
$1.50 \le c < 1.75$	18530	$3.50 \le c < 3.75$	45103
$1.75 \le c < 2.00$	42687	$3.75 \le c \le 4.00$	4527

transformations, this quasigroup has period growth 2. This experiment shows that it is not easy to find randomly a linear quasigroup of order 16.

4 Conclusion

We have suggested algorithms based on quasigroup string transformations for simulating unbiased coin flips using a biased source and for improving the properties of PRNGs. The performances of the algorithms are obtained from three theorems. The first theorem shows that the employed quasigroup string transformations are in fact permutations, the second theorem guarantees that the algorithms generate uniform output strings, while the third theorem proves that the period of the output pseudo random string can be arbitrary large number. We note that one have to choose an exponential quasigroup for obtaining better performances of the algorithms.

The proposed algorithms are very simple, of linear complexity and there are mathematical proofs of their properties. If quasigroups of order ≤ 16 are used the algorithms can be implemented in less than 1Kb working memory. Hence, they can be used in embedded systems as well.

The simplicity of the algorithms allows effective hardware realization. The initial results about parallel implementation of our algorithms are highly parallel and pipelined solution with delay of $O(n)$, where n is the number of e-transformations [16].

The use of the algorithms for cryptographic purposes (like designs of hash functions, synchronous, self-synchronizing and totaly asynchronous stream ciphers) is considered in several papers ([9], [17], [18], [19]), where it is emphasized that the employed quasigroups and the leaders of the transformations should be kept secret and the number n of applied e-transformations should be enough large. Note that the number of quasigroups of relatively small orders is huge one (there are 576 quasigroups of order 4, about 10^{20} of order 8, about 10^{47} of order 11 (see [20]), and much more than 10^{120} of order 16 and much more than

10^{58000} of order 256). On the other hand, by using the P. Hall's algorithm [21] for choosing a system of different representatives of a family of sets, a suitable algorithm for generating a random quasigroup of order s can be designed with complexity $O(s^3)$.

References

1. Gude, M.: Concept for a high-performance random number generator based on physical random phenomena. Frequenz, Vol. 39 (1985) 187–190
2. Agnew, G. B.: Random sources for cryptographic systems. In: D. Chaum and W. L. Price (eds.): Advances in Cryptology–Eurocrypt 87. Lecture Notes in Computer Science, Vol. 304. Springer–Verlag, Berlin Heidelberg New York (1987) 77–81
3. von Neumann, J.: Various techniques used in connection with random digits. In: Collected Works, Vol. 5. Pergamon Press (1963) 768–770
4. Elias, P.: The efficient construction of an unbiased random sequence. Ann. Math. Statist., Vol. 43, No. 3 (1972) 865–870
5. Stout, Q. F. and Warren, B. L.: Tree algorithms for unbiased coin tossing with a biased coin. Ann. Probab., Vol. 12, No. 1 (1984) 212–222
6. Juels, A., Jakobsson, M., Shriver, E. and Hillyer, B. K.: How to Turn Loaded Dice into Fair Coins. IEEE Transactions on Information Theory, Vol. 46, No. 3 (2000) 911–921
7. J. Dénes and A.D. Keedwell, Latin Squares and their Applications, English Univer. Press Ltd. (1974)
8. Markovski, S., Gligoroski, D., Bakeva, V.: Quasigroup String Processing: Part 1. Contributions, Sec. Math. Tech. Sci., MANU **XX**, 1-2 (1999) 13–28
9. Markovski, S.: Quasigroup string processing and applications in cryptography. Proc. 1-st Inter. Conf. Mathematics and Informatics for industry MII, Thessaloniki (2003) 278–290
10. http://stat.fsu.edu/~geo/diehard.html
11. http://www.nist.gov/
12. Markovski, S., Kusakatov, V.: Quasigroup String Processing: Part 2. Contributions, Sec. Math. Tech. Sci., MANU, **XXI**, 1-2 (2000) 15–32
13. Markovski, S., Kusakatov, V.: Quasigroup string processing: Part 3. Contributions, Sec. Math. Tech. Sci., MANU, **XXIII-XXIV**, 1-2 (2002-2003) 7–27
14. Dimitrova, V., Markovski, J.: On quasigroup pseudo random sequence generator. In: Manolopoulos, Y. and Spirakis, P. (eds.): Proc. of the 1-st Balkan Conference in Informatics, Thessaloniki (2004) 393–401
15. Markovski, S., Gligoroski, D., Markovski, J.: Classification of quasigroups by using random wolk on torus. Tech. Report, RISC-LINZ (2004) http://www.risc.uni-linz.ac.at/about/conferences/IJCAR-WS7
16. Gusev, M., Markovski,S., Gligoroski, G., Kocarev, Lj.: Processor Array Realization of Quasigroup String Transformations. Tech. Report, 01-2005, II-PMF Skopje (2005)
17. Markovski, S., Gligoroski, D., Bakeva, V.: Quasigroup and Hash Functions, Disc. Math. and Appl, Sl.Shtrakov and K. Denecke ed., Proceedings of the 6th ICDMA, Bansko 2001, pp . 43-50
18. Gligoroski, D., Markovski, S., Bakeva, V.: On Infinite Class of Strongly Collision Resistant Hash Functions "EDON-F" with Variable Length of Output. Proc. 1-st

Inter. Conf. Mathematics and Informatics for industry MII, Thessaloniki (2003) 302–308
19. Gligoroski, G., Markovski,S., Kocarev, Lj.: On a family of stream ciphers "EDON". Tech. Report, 10-2004, II-PMF Skopje (2004)
20. McKay, B.D., Meynert, A. and Myrvold, W.: Small Latin Squares, Quaisgroups and Loops. http://csr.uvic.ca/~wendym/ls.ps
21. Hall, M.: Combinatorial theory. Blaisdell Publishing Company, Massachusetts (1967)

Appendix 1: Proof of Theorem 2

In order to simplify the technicalities in the proof we take that the alphabet A is $\{\mathbf{0},\dots,\mathbf{s-1}\}$, where $0,1,\dots,s-1$ ($s>1$) are integers, and $*$ is a quasigroup operation on A. We define a sequence od random variables $\{Y_n|\ n\geq 1\}$ as follows. Let us have a probability distribution $(q_0,q_1,\dots,q_{s-1})$ of the letters $\mathbf{0},\mathbf{1},\dots,\mathbf{s-1}$, such that $q_i>0$ for each $i=0,1,\dots,s-1$ and $\sum\limits_{i=0}^{s-1}q_i=1$. Consider an e-transformation E and let $\gamma=E(\beta)$ where $\beta=b_1\dots b_k$, $\gamma=c_1\dots c_k\in A^+$ ($b_i,c_i\in A$). We assume that the string β is arbitrarily chosen. Then by $\{Y_m=i\}$ we denote the random event that the m-th letter in the string γ is exactly $\mathbf{i}$. The definition of the e-transformation given by(3) implies

$$P(Y_m=j|\ Y_{m-1}=j_{m-1},\dots,Y_1=j_1)=P(Y_m=j|\ Y_{m-1}=j_{m-1})$$

since the appearance of the m-th member in γ depends only of the $(m-1)$-th member in γ, and not of the $(m-2)$-th,..., 1-st ones. So, the sequence $\{Y_m|\ m\geq 1\}$ is a Markov chain, and we refer to it as a quasigroup Markov chain (qMc). Let p_{ij} denote the probability that in the string γ the letter $\mathbf{j}$ appears immediately after the given letter $\mathbf{i}$, i.e.

$$p_{ij}=P(Y_m=j|\ Y_{m-1}=i),\qquad i,j=0,1,\dots,s-1.$$

The definition of qMc implies that p_{ij} does not depend of m, so we have that qMc is a homogeneous Markov chain. The probabilities p_{ij} can be determined as follows. Let $\mathbf{i},\mathbf{j},\mathbf{t}\in A$ and let $\mathbf{i}*\mathbf{t}=\mathbf{j}$ be a true equality in the quasigroup $(A,*)$. Then

$$P(Y_m=j|\ Y_{m-1}=i)=q_t,$$

since the equation $\mathbf{i}*x=\mathbf{j}$ has a unique solution for the unknown x. So, $p_{ij}>0$ for each $i,j=0,\dots,s-1$, i.e. the transition matrix $\Pi=(p_{ij})$ of qMc is regular. Clearly, as in any Markov chain, $\sum\limits_{j=0}^{s-1}p_{ij}=1$. But for the qMc we also have

$$\sum_{i=0}^{s-1}p_{ij}=\sum_{\mathbf{t}\in A}q_t=1$$

i.e. the transition matrix Π of a qMc is doubly stochastic.

As we have shown above, the transition matrix Π is regular and doubly stochastic. The regularity of Π implies that there is a unique fixed probability vector $p = (p_0, \ldots, p_{s-1})$ such that $p\Pi = p$, and all components of p are positive. Also, since Π is a doubly stochastic matrix too, one can check that $\left(\frac{1}{s}, \frac{1}{s}, \ldots, \frac{1}{s}\right)$ is a solution of $p\Pi = p$. So, $p_i = \frac{1}{s}$ $(i = 0, \ldots, s-1)$. In such a way we have the following Lemma:

Lemma 1. *Let $\beta = b_1 b_2 \ldots b_k \in A^+$ and $\gamma = E^{(1)}(\beta)$. Then the probability of the appearance of a letter $\mathbf{i}$ at the m-th place of the string $\gamma = c_1 \ldots c_k$ is approximately $\frac{1}{s}$, for each $\mathbf{i} \in A$ and each $m = 1, 2, \ldots, k$.*

Lemma 1 tells us that the distribution of the letters in the string $\gamma = E(\beta)$ obtained from a sufficiently large string β by a quasigroup string permutation is uniform. We proceed the discussion by considering the distributions of the substrings $c_{i+1} \ldots c_{i+l}$ of the string $\gamma = E^n(\beta)$ $(\beta = b_1 b_2 \ldots b_k \in A^+)$, where $l \geq 1$ is fixed and $i \in \{0, 1, \ldots, k-l\}$. As usual, we say that $c_{i+1} \ldots c_{i+l}$ is a substring of γ of length l. Define a sequence $\{Z_m^{(n)} \mid m \geq 1\}$ of random variables by

$$Z_m^{(n)} = t \iff \begin{cases} Y_m^{(n)} = i_m^{(n)},\ Y_{m+1}^{(n)} = i_{m+1}^{(n)},\ \ldots, Y_{m+l-1}^{(n)} = i_{m+l-1}^{(n)}, \\ \\ t = i_m^{(n)} s^{l-1} + i_{m+1}^{(n)} s^{l-2} + \cdots + i_{m+l-2}^{(n)} s + i_{m+l-1}^{(n)} \end{cases}$$

where here and further on the superscripts (n) denote the fact that we are considering substrings of a string $\gamma = \mathbf{i}_1^{(n)} \mathbf{i}_2^{(n)} \ldots \mathbf{i}_k^{(n)}$ obtained from a string β by transformations of kind e^n. Thus, $Y_m^{(n)}$ is just the random variable Y_m defined as before. The mapping

$$(\mathbf{i}_m^{(n)}, \mathbf{i}_{m+1}^{(n)}, \ldots, \mathbf{i}_{m+l-1}^{(n)}) \mapsto i_m^{(n)} s^{l-1} + i_{m+1}^{(n)} s^{l-2} + \cdots + i_{m+l-2}^{(n)} s + i_{m+l-1}^{(n)}$$

is a bijection from A^l onto $\{0, 1, \ldots, s^l - 1\}$, so the sequence $\{Z_m^{(n)} \mid m \geq 1\}$ is well defined. The sequence $\{Z_m^{(n)} \mid m \geq 1\}$ is also a Markov chain (n-qMc), since the appearance of a substring $\mathbf{i}_m^{(n)} \mathbf{i}_{m+1}^{(n)} \ldots \mathbf{i}_{m+l-1}^{(n)}$ of l consecutive symbols in γ depends only of the preceding substring $\mathbf{i}_{m-1}^{(n)} \mathbf{i}_m^{(n)} \mathbf{i}_{m+1}^{(n)} \ldots \mathbf{i}_{m+l-2}^{(n)}$. Denote by t and t' the following numbers:

$$t = i_m^{(n)} s^{l-1} + i_{m+1}^{(n)} s^{l-2} + \cdots + i_{m+l-2}^{(n)} s + i_{m+l-1}^{(n)},$$

$$t' = i_{m-1}^{(n)} s^{l-1} + i'^{(n)}_m s^{l-2} + \cdots + i'^{(n)}_{m+l-3} s + i'^{(n)}_{m+l-2}.$$

Let $p_{t't}$ be the probability that in some string $\gamma = E^{(n)}(\beta)$, the substring $\mathbf{i}_m^{(n)} \ldots \mathbf{i}_{m+l-2}^{(n)} \mathbf{i}_{m+l-1}^{(n)}$ of γ (from the m-th to the $m+l-1$-th position) appears (with overlapping) after a given substring $\mathbf{i}_{m-1}^{(n)} \mathbf{i'}^{(n)}_m \ldots \ldots \mathbf{i'}^{(n)}_{m+l-3} \mathbf{i'}^{(n)}_{m+l-2}$ of γ

(from the $m-1$-th to the $m+l-2$-th position). Clearly, $p_{t't}=0$ if $\mathbf{i}_j^{(n)} \neq \mathbf{i'}_j^{(n)}$ for some $j \in \{m, m-1, \dots, m+l-2\}$. In the opposite case (when $l-1$ letters are overlapped) we have:

$$\begin{aligned}
p_{t't} &= P(Z_m^{(n)} = t \mid Z_{m-1}^{(n)} = t') \\
&= P(Y_m^{(n)} = i_m^{(n)}, \dots, Y_{m+l-1}^{(n)} = i_{m+l-1}^{(n)} \mid Y_{m-1}^{(n)} = i_{m-1}^{(n)}, Y_m^{(n)} = i_m^{(n)}, \dots \\
&\qquad \dots, Y_{m+l-2}^{(n)} = i_{m+l-2}^{(n)}) \\
&= P(\cap_{j=0}^{l-1}(Y_{m+j}^{(n)} = i_{m+j}^{(n)}) \mid \cap_{j=0}^{l-1}(Y_{m+j-1}^{(n)} = i_{m+j-1}^{(n)})) \\
&= \frac{P(\cap_{j=0}^{l}(Y_{m+j-1}^{(n)} = i_{m+j-1}^{(n)}))}{P(\cap_{j=0}^{l-1}(Y_{m+j-1}^{(n)} = i_{m+j-1}^{(n)}))} \\
&= \frac{P(\cap_{j=0}^{l-1}(Y_{m+j}^{(n)} = i_{m+j}^{(n)})) \mid Y_{m-1}^{(n)} = i_{m-1}^{(n)})}{P(\cap_{j=0}^{l-2}(Y_{m+j}^{(n)} = i_{m+j}^{(n)})) \mid Y_{m-1}^{(n)} = i_{m-1}^{(n)})}
\end{aligned} \tag{5}$$

By using an induction of the numbers n of quasigroup transformations we will prove the Theorem 2, i.e we will prove the following version of it:

Let $1 \le l \le n$, $\beta = b_1 b_2 \dots b_k \in A^+$ and $\gamma = E^{(n)}(\beta)$. Then the distribution of substrings of γ of length l is uniform.

Recall the notation $A = \{\mathbf{0}, \dots, \mathbf{s-1}\}$. For $n=1$ we have the Lemma 1, and let $n = r+1$, $r \ge 1$. By the inductive hypothesis, the distribution of the substrings of length l for $l \le r$ in $\gamma' = E^r(\beta)$ is uniform. At first, we assume $l \le r$ and we are considering substrings of length l of $\gamma = E^{r+1}(\beta) = \mathbf{i}_1^{(r+1)} \dots \mathbf{i}_k^{(r+1)}$. We take that $*_1, \dots, *_{r+1}$ are quasigroup operations on A and recall that $E^{(r+1)} = E_{r+1} \circ E^{(r)} = E_{r+1} \circ E_r \circ E^{(r-1)} = \dots$. Since $(A, *_{r+1})$ is a quasigroup, the equation $\mathbf{i}_{j-1}^{(r+1)} *_{r+1} \mathbf{x} = \mathbf{i}_j^{(r+1)}$ has a unique solution on $\mathbf{x}$, for each j, $2 \le j \le k$, and we denote it by $\mathbf{x} = \mathbf{i}_j^{(r)}$. Denote by $\mathbf{i}_1^{(r)}$ the solution of the equation $a_{r+1} *_{r+1} \mathbf{x} = \mathbf{i}_1^{(r+1)}$, where $a_{r+1} \in A$ is the fixed element in the definition of E_{r+1}. In such a way, instead of working with substrings $\mathbf{i}_m^{(r+1)} \mathbf{i}_{m+1}^{(r+1)} \dots \mathbf{i}_{m+d}^{(r+1)}$ of γ, we can consider substrings $\mathbf{i}_m^{(r)} \mathbf{i}_{m+1}^{(r)} \dots \mathbf{i}_{m+d}^{(r)}$ of $\gamma' = E^{(r)}(\beta)$, for any d, $0 \le d \le k-m$. The uniqueness of the solutions in the quasigroup equations implies that we have

$$P(\cap_{j=0}^{d}(Y_{m+j}^{(r+1)} = i_{m+j}^{(r+1)})) \mid Y_{m-1}^{(r+1)} = i_{m-1}^{(r+1)}) = P(\cap_{j=0}^{d}(Y_{m+j}^{(r)} = i_{m+j}^{(r)})) \tag{6}$$

as well. Here, $i_0^{(r+1)} = a_{r+1}$. Then, by (5) and (6) (for $d = l-1$, $d = l-2$ and $n = r+1$) we have

$$p_{t't} = \frac{P(\cap_{j=0}^{l-1}(Y_{m+j}^{(r)} = i_{m+j}^{(r)}))}{P(\cap_{j=0}^{l-2}(Y_{m+j}^{(r)} = i_{m+j}^{(r)}))} \tag{7}$$

where $l \leq r$. By the inductive hypothesis we have $P(\cap_{j=0}^{l-1}(Y_{m+j}^{(r)} = i_{m+j}^{(r)})) = \frac{1}{s^l}$, $P(\cap_{j=0}^{l-2}(Y_{m+j}^{(r)} = i_{m+j}^{(r)})) = \frac{1}{s^{l-1}}$, i.e. $p_{t't} = \frac{1}{s}$. Thus, for the probabilities $p_{t't}$ we have

$$p_{t't} = \begin{cases} 0 & \text{if } \mathbf{i}'^{(r+1)}_j \neq \mathbf{i}^{(r+1)}_j \text{ for some } j = m, \dots, m+l-2 \\ \frac{1}{s} & \text{if } \mathbf{i}'^{(r+1)}_j = \mathbf{i}^{(r+1)}_j \text{ for each } j = m, \dots, m+l-2. \end{cases}$$

This means that in each column of the $s^l \times s^l$-matrix of transitions Π of n-qMc there will be exactly s members equal to $\frac{1}{s}$ (those for which $\mathbf{i}'^{(r+1)}_j = \mathbf{i}^{(r+1)}_j$, $j = m, \dots, m+l-2$), the other members will be equal to 0 and then the sum of the members of each column of Π is equal to 1. Hence, the transition matrix Π is doubly stochastic. It is a regular matrix too, since each element of the matrix Π^l is positive. This implies that the system $p\Pi = p$ has a unique fixed probability vector $p = \left(\frac{1}{s^l}, \frac{1}{s^l}, \dots, \frac{1}{s^l}\right)$ as a solution. In other words, the distribution of substrings of γ of length $l \leq r$ is uniform. Assume now that $l = r+1$, and let the numbers t, t' and the probabilities $p_{t't}$ be defined as before. Then for $p_{t't}$ we have that (7) holds too, i.e.

$$p_{t't} = \frac{P(\cap_{j=0}^{r}(Y_{m+j}^{(r)} = i_{m+j}^{(r)}))}{P(\cap_{j=0}^{r-1}(Y_{m+j}^{(r)} = i_{m+j}^{(r)}))} = \frac{P(\cap_{j=0}^{r-1}(Y_{m+j+1}^{(r)} = i_{m+j+1}^{(r)}) \mid Y_m^{(r)} = i_m^{(r)})}{P(\cap_{j=0}^{r-2}(Y_{m+j+1}^{(r)} = i_{m+j+1}^{(r)}) \mid Y_m^{(r)} = i_m^{(r)})} \tag{8}$$

In the same way as it was done before, by using the fact that the equations $\mathbf{i}^{(u)}_{j-1} *_u \mathbf{x} = \mathbf{i}^{(u)}_j$ have unique solutions $\mathbf{x} = \mathbf{i}^{(u-1)}_j$ in the quasigroup $(A, *_u)$, where $u = r, r-1, \dots, 2, 1$, we could consider substrings of $\gamma' = E^{(r)}(\beta)$, $\gamma'' = E^{(r-1)}(\beta)$, $\dots$, $\gamma^{(r)} = E^{(1)}(\beta)$, $\gamma^{(r+1)} = E^{(0)}(\beta) = \beta$. Then, for the probabilities $p_{t't}$, by repeatedly using the equations (6) and (8), we will reduce the superscripts (r) to $(r-1)$, to $(r-2)$, $\dots$, to (1), i.e. we will have

$$\begin{aligned} p_{t't} &= \frac{P(Y_{m+r-1}^{(1)} = i_{m+r-1}^{(1)}, \; Y_{m+r}^{(1)} = i_{m+r}^{(1)})}{P(Y_{m+r-1}^{(1)} = i_{m+r-1}^{(1)})} \\ &= P(Y_{m+r}^{(1)} = i_{m+r}^{(1)} \mid Y_{m+r-1}^{(1)} = i_{m+r-1}^{(1)}) \\ &= P(Y_{m+r}^{(0)} = i_{m+r}^{(0)}) \end{aligned}$$

where $\mathbf{i}^{(0)}_{m+r} \in \beta$. Since $P(Y_{m+r}^{(0)} = i_{m+r}^{(0)}) = q^{(0)}_{i_{m+r}}$ we have

$$p_{t't} = \begin{cases} 0 & \text{if } \mathbf{i'}_j^{(r+1)} \neq \mathbf{i}_j^{(r+1)} \text{ for some } j = m, \ldots, m+r-1 \\ q_{i_{m+r}}^{(0)} & \text{if } \mathbf{i'}_j^{(r+1)} = \mathbf{i}_j^{(r+1)} \text{ for each } j = m, \ldots, m+r-1 \end{cases}$$

which implies

$$\begin{aligned} \sum_{t'=0}^{s^{r+1}-1} p_{t't} &= \sum_{i_{m-1}^{(r+1)}=0}^{s-1} \sum_{i'^{(r+1)}_{m}=0}^{s-1} \cdots \sum_{i'^{(r+1)}_{m+r-2}=0}^{s-1} p_{t't} = \sum_{i_{m-1}^{(r+1)}=0}^{s-1} q_{i_{m+r}}^{(0)} \\ &= \sum_{i_m^{(r)}=0}^{s-1} q_{i_{m+r}}^{(0)} = \sum_{i_{m+1}^{(r-1)}=0}^{s-1} q_{i_{m+r}}^{(0)} = \cdots = \sum_{i_{m+r}^{(0)}=0}^{s-1} q_{i_{m+r}}^{(0)} = 1 \end{aligned} \tag{9}$$

We should note that the equations

$$\sum_{i_{m-1}^{(r+1)}=0}^{s-1} q_{i_{m+r}}^{(0)} = \sum_{i_m^{(r)}=0}^{s-1} q_{i_{m+r}}^{(0)} = \ldots$$

hold true since the equations $\mathbf{i}_{j-1}^{(u)} *_u \mathbf{x} = \mathbf{i}_j^{(u)}$ have unique solutions in the quasigroup $(A, *_u)$, for each $u = r+1, r, \ldots, 2, 1$.
Hence, the transition matrix Π is doubly stochastic, it is regular (Π^{r+1} has positive entries) which means that the system $p\Pi = p$ has a unique fixed probability vector $p = \left(\frac{1}{s^{r+1}}, \frac{1}{s^{r+1}}, \ldots, \frac{1}{s^{r+1}}\right)$ as a solution.

Remark 1. Generally, the distribution of the substrings of lengths l for $l > n$ in a string $\gamma = E^{(n)}(\beta)$ is not uniform. Namely, for $l = n+1$, in the same manner as in the last part of the preceding proof, one can show that $p_{t't} = P(Y_{m+n+1}^{(0)} = i_{m+n+1}^{(0)} \mid Y_{m+n}^{(0)} = i_{m+n}^{(0)})$ and then (as in (9)) we have

$$\sum_{t'=0}^{s^{n+1}-1} p_{t't} = \sum_{i_{m+n}^{(0)}=0}^{s-1} P(Y_{m+n+1}^{(0)} = i_{m+n+1}^{(0)} \mid Y_{m+n}^{(0)} = i_{m+n}^{(0)}).$$

Of course, the last sum must not be equal to 1, i.e. the transition matrix Π must not be doubly stochastic. The same consideration could be made for $l = n+2, n+3, \ldots$ as well.

Appendix 2: Proof of Theorem 3

Consider a finite quasigroup $(A, *)$ of order s and take a fixed element $a \in A$ such that $a * a \neq a$. We will prove the Theorem 3 in the more extreme case and so we take a string $\alpha = a_1 \dots a_k$ of period 1 where $a_i = a$ for each $i \geq 1$. Then we apply the transformation $E = e_{a,*}$ on α several times. E^n means that E is applied n times and we denote $E^n(\alpha) = a_1^{(n)} \dots a_k^{(n)}$. The results are presented on Figure 4.

$$
\begin{array}{c|cccccc}
 & a & a & \dots & a & a & \dots \\
\hline
a & a_1' & a_2' & \dots & a_{p-1}' & a_p' & \dots \\
 & & & & & & \\
a & a_1'' & a_2'' & \dots & a_{p-1}'' & a_p'' & \dots \\
 & & & & & & \\
a & a_1''' & a_2''' & \dots & a_{p-1}''' & a_p''' & \dots \\
 & & & & & & \\
a & a_1^{(4)} & a_2^{(4)} & \dots & a_{p-1}^{(4)} & a_p^{(4)} & \dots \\
\vdots & \vdots & \vdots & & \vdots & \vdots &
\end{array}
$$

We have that $a'_p = a$ for some $p > 1$ since $a * a \neq a$ and $a'_i \in A$ (so we have that p is at least s), and let p be the smallest integer with this property. It follows that the string $E(\alpha)$ is periodical with period p. For similar reasons we have that each of the strings $E_a^n(\alpha)$ is periodical. We will show that it is not possible all of the strings $E_a^n(\alpha)$ to be of same period p . If we suppose that it is true, we will have $a_p^{(n)} = a$ for each $n \geq 1$. Then we will also have that there are $b_i \in A$ such that the following equalities hold:

$$
\begin{array}{ll}
a_{p-1}^{(n)} = b_{p-1} & \text{for } n \geq 2 \\
a_{p-2}^{(n)} = b_{p-2} & \text{for } n \geq 3 \\
\vdots & \\
a_1^{(n)} = b_1 & \text{for } n \geq p
\end{array}
$$

Then we have that $a * b_1 = b_1$, and that implies $a_1^{(n)} = b_1$ for each $n \geq 1$. We obtained $a * a = a * b_1 = b_1$, implying $a = b_1$, a contradiction with $a * a \neq a$. As a consequence we have that $a_1^{(p+1)} = a * a_1^{(p)} = a * b_1 \neq b_1$, $a_2^{(p+1)} = a_1^{(p+1)} * b_2 \neq b_2$, $\dots$, $a_{p-1}^{(p+1)} = a_{p-2}^{(p+1)} * b_{p-1} \neq b_{p-1}$, $a_p^{(p+1)} = a_{p-1}^{(p+1)} * a \neq a$. We conclude that the period of the string $E_a^{p+1}(\alpha)$ is not p.

Next we show that if a string $\beta \in A^+$ has a period p and $\gamma = E(\beta)$ has a period q, then p is a factor of q. Recall that the transformation E by Theorem 1 is a permutation and so there is the inverse transformation E^{-1}. Now, if $\gamma = b_1 \dots b_q b_1 \dots b_q \dots b_1 \dots b_q$, then $\beta = E^{-1}(\gamma) = c_1 c_2 \dots c_q c_1 c_2 \dots c_q \dots c_1 c_2 \dots c_q$ is a periodical string with period $\leq q$. So, $p \leq q$ and this implies that p is a factor of q.

Combining the preceding results, we have proved the following version of Theorem 3:

Let α be a string with period p_0. Then the strings $\beta = E_a^n(\alpha)$ are periodical with periods p_n that are multiples of p_0. The periods p_n of β satisfy the inequality

$$p p_{n-1} > p_{n-1}$$

for each $n \geq 1$.

A New Distinguisher for Clock Controlled Stream Ciphers*

Håkan Englund and Thomas Johansson

Dept. of Information Technology, Lund University,
P.O. Box 118, 221 00 Lund, Sweden

Abstract. In this paper we present a distinguisher targeting towards irregularly clocked filter generators. The attack is applied on the irregularly clocked stream cipher called LILI-II. LILI-II is the successor of the cipher LILI-128 and its design was published in [1]. There have been no known attacks better than exhaustive key search on LILI-II. Our attack is the first of this kind that distinguishes the cipher output from a random source using 2^{103} bits of keystream using computational complexity of approximately 2^{103} operations.

1 Introduction

Stream ciphers are a part of the symmetric family of encryption schemes. Stream ciphers are divided into two classes, synchronous and self-synchronous. In this paper we will consider a special class of the synchronous stream ciphers, namely irregularly clocked binary stream ciphers. The considered class of irregularly clocked stream ciphers include a filter generator from which the output is decimated in some way. A filter generator consists of a linear part and a boolean function (typically a nonlinear boolean function). To create the keystream some positions are taken from the internal state of the linear part and fed into the boolean function. The output of the boolean function is then combined with the message by an output function, typically the XOR operation.

Although there exist standardized block ciphers like AES [2], many people believe that the use of stream ciphers can offer advantages in some cases, e.g., in situations when low power consumption is required, low hardware complexity or when we need extreme software efficiency. To reinforce the trust in stream ciphers it is imperative that the security of stream ciphers are carefully studied.

Several different kinds of attacks can be considered on stream ciphers. We usually consider the plaintext to be known, i.e. the keystream is known and we try to recover the key. In 1984 Siegenthaler [3] introduced the idea of exploiting the correlations in the keystream. As a consequence of this attack, nonlinear functions must have high nonlinearity.

* The work described in this paper has been supported in part by the European Commission through the IST Programme under Contract IST-2002-507932 ECRYPT. The information in this document reflects only the author's views, is provided as is and no guarantee or warranty is given that the information is fit for any particular purpose. The user thereof uses the information at its sole risk and liability.

H. Gilbert and H. Handschuh (Eds.): FSE 2005, LNCS 3557, pp. 181–195, 2005.

This attack was later followed by the fast correlation attack by Meier and Staffelbach [4]. In a fast correlation attack one first tries to find a low weight parity check polynomial of the LFSR and then applies some iterative decoding procedure. Many improvements have been introduced on this topic, see [5, 6, 7, 8, 9, 10].

Algebraic attacks have received much interest lately. These attacks try to reduce the key recovery problem to the problem of solving a large system of algebraic equations [11, 12].

In this paper we will consider a distinguishing attack. A distinguishing attack is a known keystream attack, i.e., we have access to some amount of the keystream and from this data we try to decide whether this data origins from the cipher we consider, or if the data appears to be random data, see e.g., [13, 14, 15, 16, 17].

One of the submissions to the NESSIE project [18] was the irregularly clocked stream cipher LILI-128 [19]. Several attacks such as [20, 7, 11, 12, 17] on LILI-128 motivated a larger internal state. The improved design that became the successor of LILI-128 is called LILI-II [1], and it was first published in ACISP 2002. LILI-II was designed by Clark, Dawson, Fuller, Golić, Lee, Millan, Moon and Simpson, and uses a 128 bit key which is expanded and used with a much larger internal state, namely 255 bits instead of 128 in LILI-128.

So far no attacks on LILI-II have been published. In this paper we present a distinguishing attack on LILI-II. The attack uses a low weight multiple of one of the linear feedback shift registers (LFSR), i.e., it belongs to the class of linear distinguishers, see [21, 22]. It collects statistics from sliding windows around the positions of the keystream, where the members of this recursion are likely to appear. The strength of the attack is the updating procedure used when moving the windows, this procedure allows us to receive many new samples with very few operations. To distinguish the cipher from a random source we need 2^{103} bits of keystream and the complexity is around 2^{103} operations. This is the first attack on LILI-II faster than exhaustive key search.

The paper is organized as follows, in Section 2 we explain some theory needed for the attack. Section 3 describes the idea and the different steps in the attack. In Section 4 we describe the stream cipher LILI-II, and how the attack can be applied to this cipher step by step. To verify the correctness of the attack some simulations are presented in Section 5. The results of the attack is assembled in Section 6, and finally we conclude the paper in Section 7.

2 Preliminaries

2.1 Irregularly Clocked Filter Generators

Our attack considers irregularly clocked stream ciphers where the output is taken from some LFSR sequence in some arbitrary way. Note that many well known designs are of this form, e.g., the shrinking generator, self-shrinking generator, alternating step, LILI-128, LILI-II etc. The case we consider in this paper is illustrated in Figure 1.

The keystream generator is divided into two parts, a clock control part and a data generation part. The clock control part produces symbols, denoted by c_t

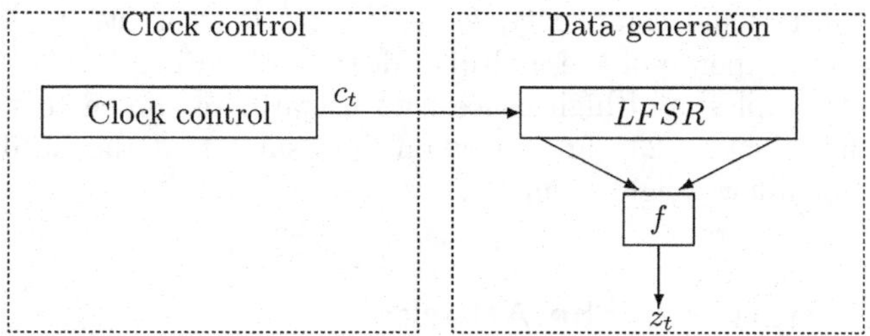

Fig. 1. A general model of an irregularly clocked filter generator

at time instant t in Figure 1, in some arbitrary way. This sequence determines how many times we should clock the LFSR in the data generation part, before we produce a new output symbol.

The data generation part is a filter generator, i.e., an LFSR producing a linear sequence, denoted by $s_t, s_{t+1}, \ldots$, from this LFSR some symbols are taken from the internal state and are used as input into a boolean function, denoted f in Figure 1.

2.2 Finding a Low Weight Multiple

In our attack we need a low weight recursion of weight w for the LFSR sequence s, i.e, a relation that sums to zero for all time instances t.

$$s_t + s_{t+\tau_1} + \ldots + s_{t+\tau_{w-1}} = 0 \mod 2. \qquad (1)$$

One technique to find such relations is to find multiples of the original feedback polynomial. Several methods to find such multiples of low weight has been proposed and they focus on optimizing different aspects, e.g., finding multiple with as low degree as possible, or accepting a higher degree but reducing the complexity to find the multiple. In our attack the degree of the multiple is of high concern.

Assume that we have a feedback polynomial $g(x)$ of degree r and search for a multiple of weight w, according to [23] the critical degree when these multiples start to appear is $(w-1)!^{1/(w-1)}2^{r/(w-1)}$. Golić [23] also describes an algorithm that focuses on finding multiples of the critical degree. The first step is to calculate the residues $x^i \mod g(x)$, then one computes the residues $x^{i_1} + \ldots x^{i_k} \mod g(x)$ for all $\binom{n}{k}$ combinations $1 \leq i_1 \leq \ldots \leq i_k \leq n$, with n being the maximum degree of the multiples. The last step is to use fast sorting to find all of the zero and one matches of the residues from the second step. The complexity of this algorithm is approximately $O(S \log S)$ with $S = \frac{(2k)!^{1/2}}{k!}2^{r/2}$ for odd multiples of weight $w = 2k+1$, and $S = \frac{(2k-1)!^{k/(2k-1)}}{k!}2^{rk/(2k-1)}$ for even multiples of weight $w = 2k$.

Wagner [24] presented a generalization of the birthday problem, i.e., given k lists of r-bit values, find a way to choose one element from each list, so that these k values XOR to zero. This algorithm finds a multiple of weight $w = k+1$ using lower computational complexity, $k \cdot 2^{r/(1+\lfloor \log k \rfloor)}$, than the method described above, on the expense of the multiples degree, which is $2^{r/(1+\lfloor \log k \rfloor)}$. Since the number of samples is of high concern to us we have chosen to work with the method described in [23]. From now on we assume that the LFSR sequence is described by a low weight recursion.

3 Description of the Attack

We consider a stream cipher as given in Figure 1, where $s_0, s_1, s_2, \ldots$ denotes the sequence from the LFSR in the data generation part, and $z_0, z_1, z_2, \ldots$ denotes the keystream sequence. The clock control mechanism c_t determines for each t how many times the LFSR is clocked before z_t is produced. After observing $z_0, z_1, \ldots, z_T$ the LFSR has been clocked $\sum_{t=0}^{T} c_t$ times. Since we are attacking irregularly clocked ciphers we will not know exactly where symbols from the LFSR sequence will be located in the output keystream, not even if they appear at all.

We will fix one position in the recurrence relation, and around the estimated location of the other symbols we will use sliding windows. When using windows of adequate size we have a high probability that all the symbols in the relation (if not removed by the irregular decimation) are included. We will then calculate how many symbols from the different windows sum to zero. Only one of these combinations contribute with a bias, the others will appear as random samples. In the following subsections we will describe the different steps we use in our attack.

The way we build the distinguisher is influenced by previous work on distinguishers, see for example [14, 21, 22, 25, 13].

3.1 Finding a Low Weight Multiple

The success of our attack depends on the use of low weight recurrence relations, hence the first step is to find a low weight multiple of the LFSR in the data generation part. In the attack we use a multiple of weight three, it is also possible to mount the attack with multiples of higher weight. Using a multiple of higher weight lowers the degree of the multiple, but it also lowers the probability that all symbols in a recurrence are included after the decimation, and in general also lowers the correlation property of the boolean function. So from now on we assume that we use a weight three recurrence relation. We will use the methods described in Section 2.2 to find the multiple.

3.2 Calculating the Correlation Property of • for a Weight Three Recursion

Assume that we have a weight three relation for the LFSR sequence according to

$$s_t + s_{t+\tau_1} + s_{t+\tau_2} = 0 \mod 2.$$

Let $S_t = (s_{t+i_1}, s_{t+i_2}, \ldots, s_{t+i_d})$ denote the input bits to f at time t taken from positions $i_1, i_2, \ldots, i_d$. The correlation property for weight three recurrence relation of the nonlinear boolean function f, denoted ε_f, is defined to be

$$\varepsilon_f = \left|\frac{1}{2} - \Pr\Big(f(S_t) \oplus f(S_{t+\tau_1}) \oplus f(S_{t+\tau_2}) \;= 0 \mid s_t \oplus s_{t+\tau_1} \oplus s_{t+\tau_2} = 0,\ \forall t\Big)\right|.$$

If the LFSR would be regularly clocked ($c_t = 1,\ \forall t$), the probability above is equivalent to

$$|\frac{1}{2} - \Pr(z_t + z_{t+\tau_1} + z_{t+\tau_2} = 0)|.$$

The correlation property can be calculated by simply trying all possible input combinations into the function. Since we use a weight three recursion some combinations will not be possible and the distribution will be biased (bias>0), see [17]. The correlation property of boolean functions has also been discussed in [20].

3.3 The Positions of the Windows

Consider again the weight three relation, but now with irregular clocking. We denote the expected value of the clocking sequence c_t by $E(C)$. The size of the windows depends on the distance from the fixed position, hence we will fix the center position in the recurrence and use windows around the two other positions. We rewrite the recurrence as

$$s_{t-\tau_1} + s_t + s_{t+\tau_2-\tau_1} = 0 \mod 2.$$

The expected distance between the output from f, corresponding to input $S_{t-\tau_1}$ and S_t, is $\tau_1/E(C)$ since the sequence is decimated, similarly the distance between S_t and $S_{t+\tau_2-\tau_1}$ is $\frac{\tau_2-\tau_1}{E(C)}$. Figure 2 illustrates how we position the windows of size r in the case of a weight three recursion.

3.4 Determining the Size of the Windows

The output sequence from the clock-control part, denoted by c_t in Figure 1, is assumed to have a fixed distribution independent of t. By using the central limit theorem we know that the sum of a large number of random variables approaches the normal distribution. So $Y_n = C_1 + C_2 + \ldots + C_n \in \mathrm{N}(n \cdot E(C), \sigma_c\sqrt{n})$, where n denotes the number of observed symbols, $E(C)$ the expected value of the clocking sequence and σ_c the standard deviation of the clocking sequence.

If the windows are sufficiently large, the correct position of the symbol will be located inside the window with a high probability,

$$\begin{aligned} P(nE(C) - \sigma_c\sqrt{n} < Y_n < nE(C) + \sigma_c\sqrt{n}) &= 0.682, \\ P(nE(C) - 2\sigma_c\sqrt{n} < Y_n < nE(C) + 2\sigma_c\sqrt{n}) &= 0.954. \end{aligned}$$

Thus we choose a window size of four standard deviations.

$$\boxed{z_{t-\frac{\tau_1}{E(C)}-\frac{r_1}{2}} \cdots z_{t-\frac{\tau_1}{E(C)}+\frac{r_1}{2}}} \quad \cdots \quad \boxed{z_t} \quad \cdots \quad \boxed{z_{t+\frac{\tau_2-\tau_1}{E(C)}-\frac{r_2}{2}} \cdots z_{t+\frac{\tau_2-\tau_1}{E(C)}+\frac{r_2}{2}}}$$

Fig. 2. Illustration of the window positions in the case with a weight three recurrence relation

3.5 Estimate the Number of Bits We Need to Observe

The main idea of the distinguishing attack is to create samples of the form

$$z_{w_1} + z_t + z_{w_2},$$

where w_1 is any position in the first window and w_2 is any position in the second window. We will run through all such possible combinations. As will be demonstrated, each sample is drawn according to a biased distribution.

To determine how many bits we need to observe to reliably distinguish the cipher from a random source, we need to make an estimate of the bias.

First we consider the case of a regularly clocked cipher. We denote the window sizes by r_1, r_2. In the following estimations we have to remember that we are calculating samples, and that for every time instant we get $r_1 \cdot r_2$ new samples. For each time instant one relation contributes with the bias ε_f, the other $r_1 \cdot r_2 - 1$ relations are random. The bias can roughly be calculated as

$$\varepsilon_f \cdot \frac{1}{r_1} \cdot \frac{1}{r_2},$$

assuming that $s_{t-\tau_1}$ and $s_{t+\tau_2-\tau_1}$ always appear inside the windows. When we have irregular clocking the output from the LFSR is decimated, i.e., some terms will not contribute to the output sequence. The probability that the end two terms of a weight three recurrence relation is included in the keystream and in the windows is denoted by p_{dec}. In the approximation we neglect the probability that the result in some cases deviates more than two standard deviations from the expected position, i.e., the component lies outside the window. This gives an estimate of the full bias showed in (2), the approximation has been compared with simulation results and works well, see Section 5.

$$\varepsilon_{full} = \varepsilon_f \cdot \frac{1}{r_1} \cdot \frac{1}{r_2} \cdot p_{dec}. \tag{2}$$

In the approximation we have estimated that the probability for the position of the taps inside the windows is uniform, the purpose is to make the updating procedure when moving the windows as efficient as possible, this is in fact the strength of the attack. A better approximation would be to weight the positions inside the window according to the normal distribution. This might decrease the number of needed symbols but would make an efficient updating procedure much more difficult.

We can now estimate how many keystream bits we need to observe, in order to make a correct decision. In [25] the *statistical distance* is used.

Definition 1. *The statistical distance, denoted* ε, *between two distributions* P_0, P_1 *defined over a finite alphabet* $\mathcal{X}$, *is defined as*

$$\varepsilon = |P_0 - P_1| = \frac{1}{2} \sum_{x \in \mathcal{X}} |P_0(x) - P_1(x)|, \tag{3}$$

where x *is an element of* $\mathcal{X}$.

If the distributions are smooth, the number of variables N we need to observe is $N \approx 1/\varepsilon^2$, see [25]. Note that the error probabilities are decreasing exponentially with N. Thus the number of samples we need to observe can be estimated as

$$\frac{r_1^2 \cdot r_2^2}{\varepsilon_f^2 \cdot p_{dec}^2}. \tag{4}$$

At each time instant we receive $r_1 \cdot r_2$ new samples, and hence the total number of bits we need for the distinguisher can be estimated by (5).

$$N \approx \frac{r_1 \cdot r_2}{\varepsilon_f^2 \cdot p_{dec}^2}. \tag{5}$$

The above Equations (2-5) assume independent samples, this is not true in our case, but the equation is still good approximation on the number of samples needed in the attack. A similar expression can be derived in a another independent way, in Appendix A it is stated that the standard deviation for the total sum of samples is $\sqrt{N\frac{r_1 r_2}{4}}$. The bias of the samples is denoted by ε_{tot}, for a successful attack $N\varepsilon_{tot} > 2\sqrt{N\frac{r_1 r_2}{4}}$ should hold, solving this equation for N gives $N > \frac{r_1 r_2}{\varepsilon_{tot}^2}$.

3.6 Complexity of Calculating the Samples

The strength of the distinguisher is that the calculation of the number of ones and zeros in the windows can be performed very efficiently, when we move the first position from z_t to z_{t+1} we also move the windows one step to the right.

We denote the number of zeros in window one at time instant t by X_t, similarly we denote the number of zeros in windows two by Y_t, the number of samples that fulfill $z_{w_1} + z_t + z_{w_2} = 0$ is denoted W_t, where w_1, w_2 are some positions in window one respectively window two . Hence when moving the windows we get the new number of zeros X_{t+1} and Y_{t+1} by subtracting the first bit in the old window and adding the new bit included in the window, e.g., for window one,

$$X_{t+1} = X_t - z_{t-\frac{r_1}{E(C)}-\frac{r_1}{2}} + z_{t-\frac{r_1}{E(C)}+\frac{r_1}{2}+1},$$

and similarly for window two. From the X_{t+1} and Y_{t+1} we can, with few basic computations calculate W_{t+1}.

We define one operation as the computations required to calculate W_{t+1} from X_t and Y_t.

Theorem 1. *The proposed distinguisher requires* $N = \frac{r_1 \cdot r_2}{\varepsilon_f^2 \cdot p_{dec}^2}$ *bits of keystream and uses a computational complexity of approximately* N *operations.*

Although the number of zeros in the windows X_{t+1}, Y_{t+1} are dependent of the previous number of zeros in the window X_t, Y_t, the covariance between the number of samples received at time instant t and $t+1$ is zero, $\mathrm{Cov}(W_{t+1}, W_t) = 0$, see Appendix A.

```
1. Find a weight three multiple of theLFSR.
2. Calculate the bias ε_f.
3. Determine the positions of the windows.
4. Calculate the sizes r_1, r_2 of the windows.
5. Estimate the number of bits N we need to observe.
6. for t from 0 to N
      if z_t = 0
          W += X_t · Y_t + (r_1 − X_t)(r_2 − Y_t)
      else if z_t = 1
          W += X_t(r_2 − Y_t) + (r_1 − X_t)Y_t
      end if
      Move window and update X_t, Y_t
   end for
7. if |W − N · (r_1·r_2)/2| > sqrt(N · r_1 · r_2)
      output "cipher" otherwise "random".
```

Fig. 3. Summary of the proposed distinguishing attack

3.7 Hypothesis Testing

The last step in the attack is to determine whether the collected data really is biased. A rough method for the hypothesis test is to check whether the result deviates more than two standard deviations from the expected result in the case when the bits are truly random. The standard deviation for a sum of these samples, can be estimated by $\sigma = \sqrt{N\frac{r_1 r_2}{4}}$, see Appendix A, where r_1 and r_2 are the sizes of the windows and N is the number of bits of keystream we observe.

3.8 Summary of the Attack

In Figure 3 we summarize the attack, where X_t denotes the number of zeros in window one, Y_t the number of zeros in window two, W denotes the total sum of the samples and r_1, r_2 the sizes of window one respectively window two.

4 LILI-II

4.1 Description of LILI-II

LILI-II [1] is the successor of the NESSIE candidate stream cipher LILI-128 [19]. Attacks such as [20, 7, 11, 12, 17] on LILI-128 motivated a larger internal state, which is the biggest difference between the two ciphers, LILI-II also use a nonlinear boolean function f_d with 12 input bits instead of 10 as in LILI-128.

Both the members of the LILI family are binary stream cipher that use irregular clocking. They consists of an $LFSR_c$, that via a nonlinear function clocks a second LFSR, called $LFSR_d$, irregularly. The structured can be viewed in Figure 4. LILI-II use a key length of 128 bits, the key is expanded and used to initialize the two LFSRs. The first shift register, $LFSR_c$ is a primitive polynomial of length 128, and hence has a period of $2^{128} - 1$. The feedback

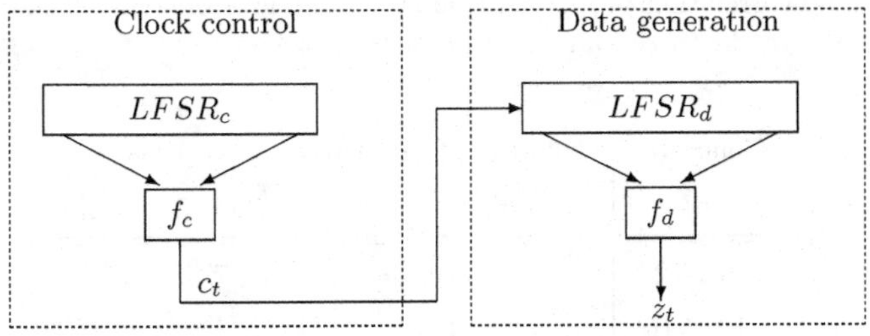

Fig. 4. Overview of LILI keystream generator

polynomial for $LFSR_c$ is given by

$$x^{128} + x^{126} + x^{125} + x^{124} + x^{123} + x^{122} + x^{119} + x^{117} + x^{115} + x^{111} + x^{108} + x^{106} + x^{105} + x^{104} + x^{103} + x^{102} + x^{96} + x^{94} + x^{90} + x^{87} + x^{82} + x^{81} + x^{80} + x^{79} + x^{77} + x^{74} + x^{73} + x^{72} + x^{71} + x^{70} + x^{67} + x^{66} + x^{65} + x^{61} + x^{60} + x^{58} + x^{57} + x^{56} + x^{55} + x^{53} + x^{52} + x^{51} + x^{50} + x^{49} + x^{47} + x^{44} + x^{43} + x^{40} + x^{39} + x^{36} + x^{35} + x^{30} + x^{29} + x^{25} + x^{23} + x^{18} + x^{17} + x^{16} + x^{15} + x^{14} + x^{11} + x^{9} + x^{8} + x^{7} + x^{6} + x^{1} + 1.$$

The Boolean function f_c takes two input bits from $LFSR_c$, it is chosen as

$$f_c(x_0, x_{126}) = 2 \cdot x_0 + x_{126} + 1 . \tag{6}$$

The output of this function is used to clock $LFSR_d$ irregularly. The output sequence from f_c is denoted c_t and $c_t \in \{1, 2, 3, 4\}$, i.e., $LFSR_d$ is clocked at least once and at most four times between consecutive outputs. On average, $LFSR_d$ is clocked $\bar{c} = 2.5$ times.

$LFSR_d$ is chosen to have a primitive polynomial of length 127 which produces a maximal-length sequence with a period of $P_d = 2^{127} - 1$. The original polynomial was found not to be primitive, see [26], and has hence been changed into

$$x^{127} + x^{121} + x^{120} + x^{114} + x^{107} + x^{106} + x^{103} + x^{101} + x^{97} + x^{96} + x^{94} + x^{92} + x^{89} + x^{87} + x^{84} + x^{83} + x^{81} + x^{76} + x^{75} + x^{74} + x^{72} + x^{69} + x^{68} + x^{65} + x^{64} + x^{62} + x^{59} + x^{57} + x^{56} + x^{54} + x^{52} + x^{50} + x^{48} + x^{46} + x^{45} + x^{43} + x^{40} + x^{39} + x^{37} + x^{36} + x^{35} + x^{30} + x^{29} + x^{28} + x^{27} + x^{25} + x^{23} + x^{22} + x^{21} + x^{20} + x^{19} + x^{18} + x^{14} + x^{10} + x^{8} + x^{7} + x^{6} + x^{4} + x^{3} + x^{2} + 1$$

Twelve bits are taken from $LFSR_d$ as input to the function f_d, these bits are taken from the positions (0,1,2,7,12,20,30,44,65,80,96,122) of the LFSR. The function f_d is given as a truth table, note that also the boolean function described in the original proposal was weak and has been replaced, see [26].

4.2 Attack Applied on LILI-II

Low Weight Multiple: According to [23] weight three multiples will start to appear at the degree 2^{64}, since the original shift register has degree 127. The complexity to find the multiple is $O(2^{70})$.

Table 1. The correlation property of boolean functions of some clock controlled generators, using weight three and weight four recursions

Generator	Number of input bits	Bias	
		$w = 3$	$w = 4$
LILI-128	10	$2^{-9.00}$	$2^{-9.07}$
LILI-II	12	$2^{-13.22}$	$2^{-12.36}$

If we instead would mount the attack with a weight four multiple the expected degree of the polynomial would be $2^{43.19}$, the complexity to find a weight four multiple is $O(2^{91.81})$.

Correlation property of $\bullet_d$: In Table 1 some examples are presented from two clock controlled ciphers, these results are based on a weight three and a weight four recursion.

In the case of LILI-128 and LILI-II the correlation property of f_d are approximately the same for a weight three relation as for a weight four relation. When using multiples of higher weight than four the correlation property of the functions decreases significantly.

Position of the Windows: When trying to find a multiple of weight three for $LFSR_d$ in LILI-II, we expect the degree of the recurrence to be 2^{64}, i.e., $\tau_1 \approx 2^{63}$ and $\tau_2 \approx 2^{64}$, and hence $\tau_2 - \tau_1 \approx 2^{63}$. The output sequence from the clock-control part denoted by c_t in Figure 4 takes the values $c_t \in \{1, 2, 3, 4\}$ with equal probability, i.e., a geometric distribution. Thus in the case of LILI-II we know that $E(C) = 2.5$ and $\sigma_c = \sqrt{7.5}$. The center positions of the windows will be positioned approximately at $t - 2^{61.68}$ and $t + 2^{61.68}$, where t denotes the position of the center symbol in the recurrence.

Determine the Size of the Windows: As stated in Section 4.2 we know that $E(C) = 2.5$ and $\sigma_c = \sqrt{7.5}$ for LILI-II. We will use a window size of four standard deviations, i.e., $r = 4\sqrt{7.5 \cdot n}$.

Using the expected positions of the windows for LILI-II from previous section the expected window sizes for a weight three relation are $r = 4\sqrt{7.5 \cdot 2^{61.68}} = 2^{34.29}$.

Estimate the Number of Bits We Need to Observe: If we use the estimated numbers from the previous section and Equation (5) we get the following estimate on the number of bits we need to observe to distinguish LILI-II from a random source. For $w = 3$,

$$N \approx \frac{2^{34.29 \cdot 2}}{2^{(-13.22) \cdot 2} \cdot 2^{(-4) \cdot 2}} \approx 2^{103.02}.$$

5 Simulations on a Scaled Down Version of LILI-II

To verify the correctness of the attack we performed the attack on a scaled down versions of LILI-II. In the scaled down version we kept the original clock control part unchanged, but used a weaker data generation part. Instead of the original $LFSR_d$ we used the primitive trinomial,

$$x^{3660} + x^{1637} + 1.$$

We fix the center member of the feedback polynomial, and the center position for window one will be positioned at $t - \tau_1/E(C) = t - \frac{3660-1637}{2.5} = t - 809$, and at $t + \frac{\tau_2 - \tau_1}{E(C)} = t - \frac{1637}{2.5} = t + 655$ for window two. We use window sizes of four standard deviations, i.e., $r_1 = 4\sqrt{7.5 \cdot 809} = 312$ and $r_2 = 4\sqrt{7.5 \cdot 655} = 280$.

The boolean function f_d was replaced with the 3-resilient 7-input plateaued function also used in [27],

$$f_d(x) = 1 + x_1 + x_2 + x_3 + x_4 + x_5 + x_6 + x_1x_7 + x_2(x_3 + x_7) + x_1x_2(x_3 + x_6 + x_7).$$

The bias of this boolean function for a weight three relation is $\varepsilon_{f_d} = 2^{-4}$. For a weight three relation the probability that all bits are included in the keystream is $p_{dec} = 2^{-4}$. The number of bits we need to observe can now be estimated as

$$N \approx \frac{r_1 \cdot r_2}{\varepsilon_{f_d}^2 \cdot p_{dec}^2} = 2^{33}.$$

We used $N = 2^{36.8054}$ in our simulated attack. The number of combinations fulfilling the recurrence equation, when simulating the attack was

$$W = 2^{52.2201} - 2^{29.2829},$$

where $2^{52.2201}$ is half of the total number of collected samples. This gives a deviation of $2^{29.2829}$ from the expected value of a random sequence and hence a simulated value of $\varepsilon_{tot} = 2^{-23.9372}$. This can be compared with the theoretically derived value which is $\varepsilon_{tot} = 2^{-24.4147}$. The standard deviation can be calculated as $\sigma = \sqrt{N\frac{r_1 r_2}{4}} = 2^{25.6101}$. We reliably distinguish the cipher from a random source.

To verify the expression on the variance (Appendix A) we also performed the attack on a random sequence of bits. The results matched the theory well.

6 Results

In this section we summarize the results of the attack applied on LILI-II, we also show the results for the attack if performed on LILI-128. Observe that there exist many better attacks on LILI-128. These attacks all use the fact that one of the LFSRs only has degree 39, if this degree would be increased the attacks would become significantly less effective, the complexity of our attack would not be affected at all.

Table 2. The number of bits needed for the distinguisher for two members of the LILI family

Function	r_1	r_2	♯ bits needed
LILI-128	$2^{25.45}$	$2^{25.45}$	$2^{76.95}$
LILI-II	$2^{34.29}$	$2^{34.29}$	$2^{103.02}$

In Table 2 we list the sizes on the windows used to attack the generator and the total number of keystream bits we need to observe to reliably distinguish the ciphers from a random source. The results in the table is calculated for a weight three recurrence relation.

7 Conclusion

In this paper we have described a distinguisher applicable to irregularly clocked stream ciphers. The attack has been applied on a member of the LILI family, namely LILI-II. The attack on LILI-II needed 2^{103} bits of keystream and a computational complexity of approximately 2^{103} operations to reliably distinguish the cipher from random data. This is the best known attack of this kind so far.

Acknowledgment

We thank the anonymous reviewer who pointed out a small improvement in our approach.

References

1. A. Clark, E. Dawson, J. Fuller, J. Golic, H-J. Lee, W. Millan, S-J. Moon, and L. Simpson. The LILI-II keystream generator. In L. Batten and J. Seberry, editors, *Information Security and Privacy: 7th Australasian Conference, ACISP 2002*, volume 2384 of *Lecture Notes in Computer Science*, pages 25–39. Springer-Verlag, 2002.
2. J. Daemen and V. Rijmen. *The Design of Rijndael.* Springer-Verlag, 2002.
3. T. Siegenthaler. Correlation-immunity of non-linear combining functions for cryptographic applications. *IEEE Transactions on Information Theory*, 30:776–780, 1984.
4. W. Meier and O. Staffelbach. Fast correlation attacks on stream ciphers. In C.G. Günter, editor, *Advances in Cryptology—EUROCRYPT'88*, volume 330 of *Lecture Notes in Computer Science*, pages 301–316. Springer-Verlag, 1988.
5. A. Canteaut and M. Trabbia. Improved fast correlation attacks using parity-check equations of weight 4 and 5. In B. Preneel, editor, *Advances in Cryptology—EUROCRYPT 2000*, volume 1807 of *Lecture Notes in Computer Science*, pages 573–588. Springer-Verlag, 2000.

6. V. Chepyzhov, T. Johansson, and B. Smeets. A simple algorithm for fast correlation attacks on stream ciphers. In B. Schneier, editor, *Fast Software Encryption 2000*, volume 1978 of *Lecture Notes in Computer Science*, pages 181–195. Springer-Verlag, 2000.
7. T. Johansson and F. Jönsson. A fast correlation attack on LILI-128. In *Information Processing Letters*, volume 81, pages 127–132, 2002.
8. T. Johansson and F. Jönsson. Fast correlation attacks through reconstruction of linear polynomials. In M. Bellare, editor, *Advances in Cryptology—CRYPTO 2000*, volume 1880 of *Lecture Notes in Computer Science*, pages 300–315. Springer-Verlag, 2000.
9. T. Johansson and F. Jönsson. Fast correlation attacks based on turbo code techniques. In M.J. Wiener, editor, *Advances in Cryptology—CRYPTO'99*, volume 1666 of *Lecture Notes in Computer Science*, pages 181–197. Springer-Verlag, 1999.
10. T. Johansson and F. Jönsson. Improved fast correlation attacks on stream ciphers via convolutional codes. In J. Stern, editor, *Advances in Cryptology—EUROCRYPT'99*, volume 1592 of *Lecture Notes in Computer Science*, pages 347–362. Springer-Verlag, 1999.
11. N. Courtois and WS. Meier. Algebraic attacks on stream ciphers with linear feedback. In E. Biham, editor, *Advances in Cryptology—EUROCRYPT 2003*, volume 2656 of *Lecture Notes in Computer Science*, pages 345–359. Springer-Verlag, 2003.
12. N. Courtois. Fast algebraic attacks on stream ciphers with linear feedback. In D. Boneh, editor, *Advances in Cryptology—CRYPTO 2003*, volume 2729 of *Lecture Notes in Computer Science*, pages 176–194. Springer-Verlag, 2003.
13. P. Ekdahl and T. Johansson. Distinguishing attacks on SOBER-t16 and SOBER-t32. In J. Daemen and V. Rijmen, editors, *Fast Software Encryption 2002*, volume 2365 of *Lecture Notes in Computer Science*, pages 210–224. Springer-Verlag, 2002.
14. J.D. Golić and R. Menicocci. A new statistical distinguisher for the shrinking generator. Available at `http://eprint.iacr.org/2003/041`, Accessed September 29, 2003, 2003.
15. P. Junod. On the optimality of linear, differential and sequential distinguishers. In *Advances in Cryptology—EUROCRYPT 2003*, volume 2656 of *Lecture Notes in Computer Science*, pages 17–32. Springer-Verlag, 2003.
16. D. Watanabe, A. Biryukov, and C. De Canniere. A distinguishing attack of SNOW 2.0 with linear masking method. In *Selected Areas in Cryptography—SAC 2003*, To be published in Lecture Notes in Computer Science. Springer-Verlag, 2003.
17. H. Englund and T. Johansson. A new simple technique to attack filter generators and related ciphers. In *Selected Areas in Cryptography—SAC 2004*, Lecture Notes in Computer Science. Springer-Verlag, 2004.
18. NESSIE. New European Schemes for Signatures, Integrity, and Encryption. Available at `http://www.cryptonessie.org`, Accessed November 10 , 2004, 1999.
19. A. Clark, E. Dawson, J. Fuller, J. Golic, H-J. Lee, William Millan, S-J. Moon, and L. Simpson. The LILI-128 keystream generator. In *Selected Areas in Cryptography—SAC 2000*, volume 2012 of *Lecture Notes in Computer Science*. Springer-Verlag, 2000.
20. H. Molland and T. Helleseth. An improved correlation attack against irregular clocked and filtered keystream generators. In *Advances in Cryptology—CRYPTO 2004*, volume 3152 of *Lecture Notes in Computer Science*, pages 373–389. Springer-Verlag, 2004.
21. J.D. Golić and L. O'Connor. A unified markow approach to differential and linear cryptanalysis. In *Advances in Cryptology—ASIACRYPT'94*, Lecture Notes in Computer Science, pages 387–397. Springer-Verlag, 1994.

22. J.D. Golić. Towards fast correlation attacks on irregularly clocked shift registers. In L.C. Guillou and J-J. Quisquater, editors, *Advances in Cryptology—EUROCRYPT'95*, volume 921 of *Lecture Notes in Computer Science*, pages 248–262. Springer-Verlag, 1995.
23. J.D. Golić. Computation of low-weight parity-check polynomials. *Electronic Letters*, 32(21):1981–1982, October 1996.
24. D. Wagner. A generalized birthday problem. In M. Yung, editor, *Advances in Cryptology—CRYPTO 2002*, volume 2442 of *Lecture Notes in Computer Science*, pages 288–303. Springer-Verlag, 2002.
25. D. Coppersmith, S. Halevi, and C.S. Jutla. Cryptanalysis of stream ciphers with linear masking. In M. Yung, editor, *Advances in Cryptology—CRYPTO 2002*, volume 2442 of *Lecture Notes in Computer Science*, pages 515–532. Springer-Verlag, 2002.
26. LILI-II design. Available at `http://www.isrc.qut.edu.au/resource/lili/lili2design.php`, Accessed November 10, 2004, 2004.
27. S. Leveiller, G. Zémor, P. Guillot, and J. Boutros. A new cryptanalytic attack for pn-generators filtered by a boolean function. In K. Nyberg and H. Heys, editors, *Selected Areas in Cryptography—SAC 2002*, volume 2595 of *Lecture Notes in Computer Science*, pages 232–249. Springer-Verlag, 2003.

A Variance of the Number of Combinations

Let X_t denote the number of zeros in window one and similarly Y_t denotes the number of zeros in window two at the time t. r_1, r_2 denotes the sizes of windows.

$$\begin{array}{ccc} E(X_t) = \frac{r_1}{2} & E(X_t^2) = \frac{r_1^2+r_1}{4} & V(X_t) = \frac{r_1}{4} \\ E(Y_t) = \frac{r_2}{2} & E(Y_t^2) = \frac{r_2^2+r_2}{4} & V(Y_t) = \frac{r_2}{4} \end{array}$$

Let Z_t denote the bit in the center position at time t, and W_t' the number of samples fulfilling the recurrence relation at time t. To make the computations a bit simpler we denote $W_t = W_t' - \frac{r_1 r_2}{2}$, i.e., we subtract the expected value of W_t', hence $E(W_t) = 0$. We also introduce the symbol $A_t = X_t Y_t + (r_1 - X_t)(r_2 - Y_t) - r_1 r_2/2$.

$$W_t = \begin{cases} \underbrace{X_t Y_t + (r_1 - X_t)(r_2 - Y_t) - r_1 r_2/2}_{A_t} & \text{if } Z_t = 0, \\ -\underbrace{(X_t Y_t + (r_1 - X_t)(r_2 - Y_t) - r_1 r_2/2)}_{A_t} & \text{if } Z_t = 1. \end{cases}$$

We define W as the sum of W_t for N bits, $W = \sum_{t=0}^{N-1} W_t$. Hence

$$E(W) = E(\sum_{t=0}^{N-1} W_t) = \sum_{t=0}^{N-1} E(W_t) = 0.$$

We are trying to calculate $V(\sum_{i=0}^{N-1} W'_t)$ $=V(\sum_{i=0}^{N-1} W_t + N\frac{r_1 r_2}{2})$ $=V(\sum_{i=0}^{N-1} W_t)$ $= V(W)$.

$$E(W^2) = E((\sum_{t_1=0}^{N-1} W_{t_1})(\sum_{t_2=0}^{N-1} W_{t_2})) = \sum_{t_1=0}^{N-1}\sum_{t_2=0}^{N-1} E(W_{t_1} \cdot W_{t_2})$$

– For $t_1 \neq t_2$

$$\begin{aligned} E(W_{t_1} W_{t_2}) = \tfrac{1}{4}\Big(& E(W_{t_1} W_{t_2} | Z_{t_1} = 0, Z_{t_2} = 0) + E(W_{t_1} W_{t_2} | Z_{t_1} = 0, Z_{t_2} = 1) + \\ & + E(W_{t_1} W_{t_2} | Z_{t_1} = 1, Z_{t_2} = 0) + E(W_{t_1} W_{t_2}) | Z_{t_1} = 1, Z_{t_2} = 1)\Big) = \\ = & \tfrac{1}{4} E\big(A_{t_1} A_{t_2} - A_{t_1} A_{t_2} - A_{t_1} A_{t_2} + (-A_{t_1})(-A_{t_2})\big) = 0. \end{aligned}$$

– For $t_1 = t_2$

$$\begin{aligned} E(W_t^2) &= \tfrac{1}{2}\big(E(W_t^2 | Z_t = 0) + E(W_t^2 | Z_t = 1)\big) = E(A^2) = \\ &= 4E(X^2)E(Y^2) + 4r_1 r_2 E(X)E(Y) - 4r_1 E(X)E(Y^2) \\ &\quad -4r_2 E(X^2)E(Y) + \tfrac{r_1^2 r_2^2}{4} - \\ &\quad -r_1^2 r_2 E(Y) - r_1 r_2^2 E(X) + r_1^2 E(Y^2) + r_2^2 E(X) = \\ &= \tfrac{r_1 r_2}{4} \end{aligned}$$

So

$$E(W^2) = \sum_{t_1=0}^{N-1}\sum_{t_2=0}^{N-1} E(W_{t_1} W_{t_2}) = \sum_{t=0}^{N-1} E(W_t^2) = \sum_{t=0}^{N-1} \frac{r_1 r_2}{4} = N \cdot \frac{r_1 r_2}{4}.$$

Finally we can give and expression or the variance.

$$V(W) = E(W^2) - E(W)^2 = N \cdot \frac{r_1 r_2}{4}.$$

Analysis of the Bit-Search Generator and Sequence Compression Techniques*

Aline Gouget[1], Hervé Sibert[1], Côme Berbain[2], Nicolas Courtois[3], Blandine Debraize[3,4], and Chris Mitchell[5]

[1] France Telecom Research and Development,
42 rue des Coutures, F-14000 Caen, France
[2] France Telecom Research and Development,
38-40 rue du Général Leclerc, F-92794 Issy-les-Moulineaux, France
[3] Axalto Cryptographic Research & Advanced Security,
36-38 rue de la Princesse, BP 45, F-78430 Louveciennes Cedex, France
[4] Versailles University, 45 avenue des Etats-Unis, F-78035 Versailles, France
[5] Information Security Group, Royal Holloway, University of London,
Egham, Surrey TW20 0EX, United Kingdom

Abstract. Algebraic attacks on stream ciphers apply (at least theoretically) to all LFSR-based stream ciphers that are clocked in a simple and/or easily predictable way. One interesting approach to help resist such attacks is to add a component that de-synchronizes the output bits of the cipher from the clock of the LFSR. The Bit-search generator, recently proposed by Gouget and Sibert, is inspired by the so-called Self-Shrinking Generator which is known for its simplicity (conception and implementation-wise) linked with some interesting properties. In this paper, we introduce two modified versions of the BSG, called MBSG and ABSG, and some of their properties are studied. We apply a range of cryptanalytic techniques in order to compare the security of the BSGs.

1 Introduction

In recent years there has been renewed interest in designing stream cipher keystream generators (KGs) capable of being implemented in small software or hardware and operating at very high rates. The *Shrinking Generator (SG)* [2] and *Self-Shrinking Generator (SSG)* [8] are schemes providing a method for irregular decimation of pseudorandom sequences such as those generated by linear feedback shift registers (LFSRs).

Recently, Gouget and Sibert [6] introduced the *Bit-Search Generator* (BSG), that is, like the SG and SSG, a scheme designed to offer attractive characteristics for both software and hardware implementation when used as a part of a KG.

* Work partially supported by the French Ministry of Research RNRT Project "X-CRYPT" and by the European Commission via ECRYPT network of excellence IST-2002-507932.

H. Gilbert and H. Handschuh (Eds.): FSE 2005, LNCS 3557, pp. 196–214, 2005.

However, similarly to the SG and the SSG, the BSG can be vulnerable to timing attacks. The BSG has the advantage over the SG and SSG that it operates at a rate of 1/3 instead of 1/4 (i.e. producing n bits of the output sequence requires, on average, $3n$ bits of the input sequence).

Given that the BSG is aimed as a building block for constructing a KG, it is essential to know how simple it is to reconstruct parts of the input sequence from the output. This arises naturally in the context of stream cipher design, where matching known plaintext and ciphertext immediately gives keystream values, i.e. subsequences of the output sequence, and where knowledge of parts of the input sequence is a prerequisite to determining the secret key used to generate the sequence. Furthermore, in order to avoid algebraic attacks (see among other [1, 3]), it is important to know how many relations that relate some outputs bits to consecutive input bits can be obtained.

The outline of the paper is as follows. In Section 2, we recall the original description of the BSG and we provide an equivalent specification which operates on the *differential* of the original sequence. In Section 3, we consider two strategies in order to reconstruct the original sequence from the output sequence of the BSG. We give a basic attack which has complexity $\mathcal{O}(L^3 2^{\frac{L}{3}})$ and requires $\mathcal{O}(L 2^{\frac{L}{3}})$ keystream bits, where L is the length of the underlying LFSR. We then improve this attack to get a complexity of $\mathcal{O}(L^3 2^{\frac{L}{4}})$. In Section 4, we propose two modified versions of the BSG designed to increase its security. Analogously to the work in [6] for the BSG, we study some properties of both the MBSG and the ABSG. In Section 5, we apply, to both the MBSG and the ABSG, the strategies of Section 3 and the FBDD attack against LFSR-based generators introduced by Krause in [7]. The best attack that we give against the ABSG and the MBSG has complexity $\mathcal{O}(2^{\frac{L}{2}})$ and requires $\mathcal{O}(L 2^{\frac{L}{2}})$ bits of keystream. Finally, we conclude in Section 6.

2 The Bit-Search Generator

One can consider that both the SG and SSG are methods for *bit-search*-based decimation. Indeed, both generators use a search for ones along an input bit sequence in order to determine the output bit. Instead of using a search for ones, the BSG uses a search for some bit b, where b varies during the process; the variations depend on the input sequence. During the search process for a bit b, a cursor moves along the input sequence. The search process ends when the next occurrence of b is reached. Then, the output bit is zero if the search process ends after just reading one bit, otherwise the output is one. The next value of the bit b corresponds to the value of the following bit of the sequence.

We recall the original description of the BSG given in [6] and we provide an equivalent specification of the BSG which operates on the *differential* of the original sequence; the *differential sequence* $d = (d_0, d_1, \dots)$ of a sequence s is defined by $d_i = s_i \oplus s_{i+1}$, $i \geq 0$, where $\oplus$ denotes bit-wise exclusive-or (or modulo 2 addition). As usual, the complement of b in $\{0, 1\}$ is denoted $\overline{b}$.

Definition 1 (BSG). *Let $s = (s_0, s_1, \dots)$ be a pseudorandom bit sequence and $d = (d_0, d_1, \dots)$ be the differential sequence. The output sequence $y = (y_0, y_1, \dots)$ of the BSG is constructed as follows:*

BSG (original)	*BSG_{diff} (differential)*
Input: $(s_0, s_1, \dots)$ *Set:* $i \leftarrow 0$; $j \leftarrow 0$; *Repeat the following steps:* 1. $e \leftarrow s_i$, $y_j \leftarrow s_i \oplus s_{i+1}$; 2. $i \leftarrow i + 1$; 3. *while* $(s_i = \overline{e})$ $i \leftarrow i + 1$; 4. $i \leftarrow i + 1$; 5. *output* y_j; 6. $j \leftarrow j + 1$;	*Input:* $(d_0, d_1, \dots)$ *Set:* $i \leftarrow 0$; $j \leftarrow 0$; *Repeat the following steps:* 1. $y_j \leftarrow d_i$; 2. *if* $(y_j = 1)$ *then* (a) $i \leftarrow i + 1$; (b) *while* $(d_i = 0)$ $i \leftarrow i + 1$; 3. $i \leftarrow i + 2$; 4. *output* y_j; 5. $j \leftarrow j + 1$;

Example 1. Let $s = 0101001110100100011101$ be a bit sequence. Then, the action of the BSG on s is described by:

$$\underbrace{010}_{1}\underbrace{1001}_{1}\underbrace{11}_{0}\underbrace{010}_{1}\underbrace{010}_{1}\underbrace{00}_{0}\underbrace{11}_{0}\underbrace{101}_{1} \quad .$$

The action of the BSG on the input sequence s consists in splitting up the sequence s into subsequences of the form $(\overline{b}, b^i, \overline{b})$ where $b \in \{0,1\}$ and $i \geq 0$. For every subsequence of the form $(\overline{b}, b^i, \overline{b})$, the output bit is 0 if $i = 0$, and 1 otherwise. The action of the BSG on the input differential sequence d consists in splitting up the subsequence d into subsequences of the form either $(0, b)$ or $(1, 0^i, 1, b)$ with $i \geq 0$; for every such subsequence, the output bit is the first bit of the subsequence.

It is simple to verify that both descriptions of the BSG are equivalent given that the output bit is zero when the search along the sequence s ends immediately and it is one otherwise. We denote the output sequence of the BSG by $BSG(s)$ or $BSG_{\text{diff}}(d)$ depending on the sequence we are focusing on.

Remark 1. Recovering elements of the sequence d is likely to be of very similar signifiance to recovering elements of s. For instance, when s is generated using an LFSR, then d can also be generated using an identical LFSR [5, 9]. Furthermore, the transformation from s to d simply shifts the position of the starting point of the sequence. In this case, recovering the entire sequence d from partial information has precisely the same difficulty as for the sequence s.

Assuming that the input sequence of the BSG is evenly distributed, then the output rate of the BSG is clearly 1/3 (the number of input bits required to produce one output bit is $1 + i$ with probability $1/2^i$ $(i \geq 1)$).

Proposition 1. *Assume that the output sequence y produced by the BSG is evenly distributed. Then, for each output bit y_j, the expected number of known input bits is 2 with an average entropy of 1.*

Proof. Every zero in y corresponds to a pair of bits $(0, b)$ in the differential sequence d of the original sequence s, and no information is available about the bit b. Thus, if an output bit is a zero, then one input bit of d is known. Every one in y corresponds to a pattern $(1, 0^i, 1, b)$ with $i \geq 0$ in d and the following possibilities exist: two bits are known with probability 1/2, three bits are known with probability 1/4, ..., that is $1+i$ bits are known with probability $1/2^i$ for $i \geq 1$. Hence the expected number of known bits is $\sum_{i=1}^{\infty}(1+i)/2^i = 3$. The associated entropy is given by $\sum_{i=1}^{\infty} 2^{-i} \log_2(2^{-i}) = \sum_{i=1}^{\infty} i2^{-i} = 2$. Thus, assuming that the output sequence is evenly distributed, for each output bit the expected number of known input bits is 2, with an average entropy of 1. □

3 How to Reconstruct the Input Sequence?

In this section, we consider two approaches, called *Strategy 1* and *Strategy 2* in order to evaluate how simple it is to reconstruct parts of either the input sequence s or its differential from the output sequence y.

For the first approach, called *Strategy 1*, we assume that we have no additionnal information on the means used to generate the input sequence. This approach is based on the random generation of *candidates* for the input sequence which are *consistent* with the information derived from the output sequence. For the second approach, called *Strategy* 2, we assume that the feedback polynomial used to generate the input sequence is known. This second approach consists of building an attack on the BSG based on the choice of the most probable case for LFSR sequences as input.

3.1 Strategy 1: Use of Random Generation of Candidates

Consider a bit sequence s, its differential sequence d and the output sequence $y = BSG(s) = BSG_{\text{diff}}(d)$. In this approach, we focus on the reconstruction of the differential sequence d that we call the *correct input string* and we assume that we have no additional information on the means used to generate the input sequence.

A sequence c is called a *differential-candidate* for the output sequence y if the equality $BSG_{\text{diff}}(c) = y$ is fulfilled. One method to search for the correct input string is to randomly generate a sequence of differential candidates for the input bits. The probability of success of such a strategy depends on the Hamming weight w of the subsequence, i.e. there are w places in the input sequence where a string of zeros of uncertain length may occur. Recall that every one in the output sequence arises from a tuple of the form $(1, 0^i, 1, b)$, where $i \geq 0$ and b is an undetermined bit. The Hamming weight of a finite sequence y is denoted by $w(y)$.

Proposition 2. *Let d be a (finite) bit sequence and y be a sequence such that $y = BSG_{diff}(d)$. Let c be a randomly chosen string with the property that $BSG_{diff}(c) = y$, where the probability distribution used to choose c reflects the* *probability that $c = d$. The probability that, for every k such that $y_k = 1$, the sequences d and c agree on the length of the tuple from which y_k arises, is $3^{-w(y)}$.*

Proof. Each differential-candidate input string should have a tuple $(1, 0^i, 1)$ inserted for every one occurring in the output sequence; i is chosen independently at random for each output bit and $i = j$ with probability 2^{-j-1}. In each of the $w(y)$ locations, a string of i zeros occurs in the correct input sequence with probability 2^{-i-1}. The probability that the candidate string and the correct string agree in any one of the $w(y)$ positions is thus $\sum_{i=0}^{\infty}(2^{-i-1})^2 = 1/3$. That is, the probability that the correct input sequence and a candidate c agree on the $w(y)$ choices of length of the tuples from which the ones of y arises is $3^{-w(y)} \simeq 2^{-1.585y}$. □

Thus, finding one output sequence with small Hamming weight yields attacks that are likely to be easier than brute force attacks. This idea is used in Strategy 2.

3.2 Strategy 2: Choice of the Most Probable Case

The goal of Strategy 2 is the reconstruction of the original input sequence s. We assume that s is generated by a maximum length LFSR of size L with a public feedback polynomial and the initial state of the LFSR is the secret key. We further suppose that the feedback polynomial has been chosen carefully, i.e. it does not have a low Hamming weight and no low weight multiple exists, in order to avoid attacks on the differential sequence similar to the distinguishing attack on the SG given in [4].

Recall that each zero of the output sequence y comes from two consecutive equal bits in the input sequence s. Thus, each zero in y provides a linear equation over the unknown LFSR sequence, namely the equality between two consecutive bits. Similarly, each one of y comes from a pattern $(\bar{b}, b^i, \bar{b})$ for some integer $i \geq 1$. Thus, by guessing i, we can construct $i+1$ linear equations involving consecutive bits of the unknown LFSR sequence which are valid with probability 2^{-i}.

Basic Attack. Let us take the first window of $2L/3$ consecutive bits in the sequence y with a Hamming weight of at most $L/3$. For a random window of size $2L/3$, this condition is satisfied with probability close to $1/2$, so that the first window can be found in negligible time. If the Hamming weight of the window is strictly lower than $L/3$, we expand it in such a way that it contains exactly $L/3$ ones (or until its size is L). We now assume that each one in the sequence y comes from a pattern of length 3, that is a pattern of the form $\bar{b}b\bar{b}$, which is the most probable case, occuring with probability $2^{-\frac{L}{3}}$. Then, we can write L equations involving consecutive bits of the LFSR sequence or, equivalently, the bits of the current state. We solve this system and we instantly check if we have found the correct values by testing whether it allows to the correct prediction of a few additional bits of the sequence y. In order to find the current state with high probability (close to $1 - \frac{1}{e}$), we have to repeat this procedure $2^{\frac{L}{3}}$ times. This attack costs $\mathcal{O}(L^3 2^{\frac{L}{3}})$ and requires $\mathcal{O}(L 2^{\frac{L}{3}})$ bits of keystream.

Improvements to the Basic Attack. We tried several alternative strategies such as finding a large enough keystream window with a low Hamming weight, or connecting two smaller windows of low weight. For instance, we can determine, in a first computation phase, 2^w windows of size ℓ bits and Hamming weight w. For each of these windows, we suppose that every one comes from a pattern $(\overline{b}, b, \overline{b})$, which gives $\ell + w$ linear equations. These equations are all valid with probability 2^{-w}. This costs:

$$\mathcal{O}\left(\frac{2^{w+\ell+1}}{\binom{\ell}{w}}\right).$$

For each pair of such windows, we know the number n_1 of ones and n_0 of zeros in the sequence y between the two windows. Considering all the possible strings $\overline{b}b^i\overline{b}$ for integer $i \geq 1$, the mean value m of i and the variance v are given by:

$$m = \sum_{k=1}^{\infty} \frac{k}{2^k} = 2 \ , \qquad v = \sum_{k=1}^{\infty} \frac{(k-2)^2}{2^k} = 2.$$

Thus, the distance between those two windows in the original sequence is likely to belong to the interval $[2n_0+4n_1-\sqrt{2n_1}, 2n_0+4n_1+\sqrt{2n_1}]$. The Central Limit Theorem gives the probability that the real distance between the two windows is outside this interval:

$$Pr\left(\frac{\sum_{i=1}^{n_1} X_i - mn_1}{\sqrt{vn_1}} \geq 1\right) = \frac{2}{\sqrt{2\pi}} \int_1^{\infty} e^{-\frac{x^2}{2}} dx \simeq 0.31.$$

Therefore, for each pair of windows, the probability of failure provided that the distance used is not correct is around 1/3. We try all the values of the distance between the two windows in this interval. If we make a correct guess, the equations associated to the two windows can be combined to provide $2(\ell+w)$ equations. We choose ℓ and w such that $2(\ell + w) = L$ and we just have to solve the system so as to test whether the obtained solution correctly predicts a few additional keystream bits.

Since n_1 is $\mathcal{O}\left(\frac{2^w 2^\ell}{\binom{\ell}{w}}\right)$, testing all the pairs of windows costs $\mathcal{O}\left(2^{2w+1}\sqrt{2\frac{2^w 2^\ell}{\binom{\ell}{w}}}\right)$, and the total complexity of the attack is:

$$\mathcal{O}\left(\frac{2^{w+\ell+1}}{\binom{\ell}{w}} + 2^{2w+1}\sqrt{2\frac{2^{w+\ell}}{\binom{\ell}{w}}}L^3\right).$$

Moreover the number of keystream bits required for the attack is:

$$\mathcal{O}\left(\ell 2^w \frac{2^\ell}{\binom{\ell}{w}}\right).$$

For practical values of L ($L \in [128, 4096]$), $\ell = \frac{25L}{58}$ and $w = \frac{7L}{116}$, this provides a complexity close to or slightly smaller than $L^3 2^{\frac{L}{4}}$ and a keystream length of $2^{\frac{L}{4}}$.

4 New BSGs to Improve the Security?

The discussion in section 3 suggests that the security of the BSG relies on the uncertainty about the length of the input tuple required to output a one. By contrast, if a zero is output, then there is no uncertainty about the length of the input string. This suggests that the security might be improved by introducing ambiguity no matter whether a zero or a one is output by the scheme.

Remark 2. Instead of aiming at an improvement in security, one may want to enhance the rate with the same level of security. Indeed, a simple modification to the BSG enables its rate to be increased from 1/3 to 1/2 by changing Step 3 in the BSG_{diff} Algorithm (Definition 1) from $i \leftarrow i+2$ to $i \leftarrow i+1$. However, an adaptation of the basic attack presented in Section 3.2 to this case (for an LFSR input) leads to an attack which costs $\mathcal{O}(2^{\frac{L}{3}})$ and requires $\mathcal{O}(L2^{\frac{L}{3}})$ bits of keystream; the security is then slightly lower than for the BSG.

4.1 BSG Variants

We give two possible modifications of the BSG that are called the MBSG and the ABSG; these two modifications are not equivalent (even if we consider the differential sequence instead of the original sequence).

Definition 2 (MBSG & ABSG). *Let $s = (s_0, s_1, \dots)$ be a pseudorandom bit sequence. The output sequences of the MBSG and of the ABSG are constructed as follows.*

MBSG algorithm	*ABSG algorithm*
Input: $(s_0, s_1, \dots)$	*Input:* $(s_0, s_1, \dots)$
Set: $i \leftarrow 0$; $j \leftarrow 0$;	*Set:* $i \leftarrow 0$; $j \leftarrow 0$;
Repeat the following steps:	*Repeat the following steps:*
1. $y_j \leftarrow s_i$;	*1.* $e \leftarrow s_i$, $y_j \leftarrow s_{i+1}$;
2. $i \leftarrow i+1$;	*2.* $i \leftarrow i+1$;
3. while $(s_i = 0)$ $i \leftarrow i+1$;	*3. while* $(s_i = \overline{e})$ $i \leftarrow i+1$;
4. $i \leftarrow i+1$;	*4.* $i \leftarrow i+1$;
5. output y_j;	*5. output* y_j
6. $j \leftarrow j+1$;	*6.* $j \leftarrow j+1$

Example 2. Let $s = 0101001110100100011101$ be the input bit sequence. Then, the action of the MBSG on s is described by:

$$\underbrace{01}_{0}\underbrace{01}_{0}\underbrace{001}_{0}\underbrace{11}_{1}\underbrace{01}_{0}\underbrace{001}_{0}\underbrace{0001}_{0}\underbrace{11}_{1}\underbrace{01}_{0},$$

and the action of the ABSG on s is described by:

$$\underbrace{010}_{1}\underbrace{1001}_{0}\underbrace{11}_{1}\underbrace{010}_{1}\underbrace{010}_{1}\underbrace{00}_{0}\underbrace{11}_{1}\underbrace{101}_{0}.$$

The action of the MBSG on the input sequence s consists in splitting up s into subsequences of the form $(b, 0^i, 1)$, with $i \geq 0$ and $b \in \{0,1\}$. For every pattern of the form $(b, 0^i, 1)$, the output bit is b. The action of the ABSG on s consists in splitting up s into subsequences of the form $(\bar{b}, b^i, \bar{b})$, with $i \geq 0$ and $b \in \{0,1\}$. For every subsequence $(\bar{b}, b^i, \bar{b})$, the output bit is b for $i = 0$, and $\bar{b}$ otherwise. Both the MBSG and the ABSG clearly have a rate of 1/3, like the BSG. Indeed, for every $i \geq 1$, an output bit is produced by $1 + i$ input bits with probability $1/2^i$.

Remark 3. The action of the ABSG on an input sequence is identical to that of the BSG, but their outputs are computed differently.

Proposition 3. *Let s be a pseudorandom bit sequence. Assume that the output sequence $y = MBSG(s)$ is evenly distributed. Then for every output bit y_j, the expected number of known bits of s is 3 with an average entropy of 2.*

Proof. If an output bit is a b, then the input sequence used to generate this output bit must have the form $(b, 0^i, 1)$, where $i \geq 0$ and $i = j$ with probability 2^{-j-1}. Thus, if an output bit is a b, then $i + 1$ bits are known with probability $1/2^i$ for $i \geq 1$. As shown in the proof of Proposition 1, the expected number of known bits is 3 and the associated entropy is 2. □

Proposition 3 also holds for the ABSG.

4.2 Filtering Periodic Input Sequences

We now describe the output of the MBSG and ABSG when applied to periodic sequences (of period greater than 1) as was done in [6] for the BSG. We will show that the BSG and the ABSG on the one hand, and the MBSG on the other hand, behave differently in this regard.

Definition 3. *For two sequences $s = (s_i)_{i\geq 0}$ and $s' = (s'_i)_{i\geq 0}$, we say that s' is* (k-)shifted *from s if there exists $k \geq 0$ such that $s'_i = s_{i+k}$ for every $i \geq 0$.*

As usual, s is said to be *eventually periodic* if there exists a shifted sequence from s which is periodic. We denote by $BSG(s, i)$ (resp. $MBSG(s, i)$, $ABSG(s, i)$) the i-shifted sequence from $BSG(s)$ (resp. $MBSG(s)$, $ABSG(s)$).

The next proposition was proved in [6] for the BSG. It also holds for the ABSG thanks to the fact that the ABSG acts like the BSG on the input sequence.

Proposition 4. *Let s be a sequence of period T. Then, the sequence $ABSG(s)$ is periodic, and there exists $k \in \{1,2,3\}$ such that $ABSG(s_0, \ldots, s_{kT-1})$ is a period of $ABSG(s)$.*

The framework introduced in [6] uses the associated permutation p to a periodic sequence s: we define two transpositions $t_0 = (\emptyset\ 0)$ and $t_1 = (\emptyset\ 1)$. Then, we associate with s the permutation $t_{s_{T-1}} \circ \cdots \circ t_{s_0}$ over the set $\{\emptyset, 0, 1\}$. The integer k in the previous proposition is the order of the permutation associated with s.

For the MBSG, the picture is slightly different. The MBSG acts on an input sequence s as follows: read a bit b, go to the next occurrence of one and start again. We give to the cursor moving along the input sequence two states: $\emptyset$ when there is no current bit looked for, and 1 otherwise. The cursor changes from state $\emptyset$ to state 1 after reading a bit. When the cursor is in state 1, it remains in state 1 if the next bit read is 0, and changes to state $\emptyset$ if the next bit read is 1.

Proposition 5. *Let s be a sequence of period T. Then, the sequence $MBSG(s)$ is eventually periodic. Moreover, if $s_{i-1}s_i$ is an occurrence of $(0,1)$ in s, then the sequence $MBSG(s,i+1)$ is periodic and a period is $MBSG(s_{i+1},\dots,s_{i+T})$.*

Proof. After reading a pattern $(0,1)$, the cursor is always in state $\emptyset$, and thus the bit s_i is the last bit read during some search process. Now, as the cursor is in state $\emptyset$ after s_i, it will also be in this state after s_{i+kT} for every k. □

In the sequel, we denote by $MBSG_P(s)$ the sequence $MBSG(s,i+1)$ where (s_{i-1},s_i) is the first occurrence of $(0,1)$ in s. Thus $MBSG_P(s)$ is a periodic shift of $MBSG(s)$.

Output Sequence Sets and Shifts. Given an input sequence s of period T, one can filter the shifted sequences (s,i) for $0 \leq i \leq T-1$, so as to obtain at most T distinct output sequences. We call the set of these output sequences the *output sequence set for input s*. We will show that these output sequences are closely related to one another. The following proposition was proved in [6] in the case of the BSG using only the action of the BSG on the input sequence. Thus, it also holds for the ABSG.

Proposition 6. *Let $s=(s_i)_{i\geq 0}$ be an infinite bit sequence and k be the minimal index such that $s_k \neq s_0$. Then, for every $i \geq 0$, the sequence $ABSG(s,i)$ is shifted from one sequence among $ABSG(s,0)$, $ABSG(s,1)$ and $ABSG(s,k+1)$.*

In the case of the MBSG, we have to consider the periodic part $MBSG_P(s)$ so as to obtain a similar proposition:

Proposition 7. *Let $s=(s_i)_{i\geq 0}$ be an infinite bit sequence where both 0 and 1 appears infinitely many times. Then, for every $i \geq 0$, the sequence $MBSG_P(s,i)$ is shifted from the sequence $MBSG_P(s)$.*

Proof. Let us consider the cursor in initial state $\emptyset$ running along the sequence s. Let $s_{k-1}s_k$ be the first occurrence of 01 in the sequence (s,i). After reading a pattern $(0,1)$, the cursor is always in state $\emptyset$. Thus, the cursor is in state $\emptyset$ after reading s_k. Therefore, $MBSG(s,k+1)$ is shifted from both $MBSG(s)$ and $MBSG(s,i)$. Now, $MBSG(s,k+1)$ is periodic, which yields the result. □

Maximum Length LFSR Sequences as Input. When the input sequence s is produced by a maximum length LFSR, the periodicity properties differ between the MBSG on the one hand, and the BSG and the ABSG on the other hand.

A lower bound on the length of $BSG(s_0,\dots,s_{kT-1})$, where $k \in \{1,2,3\}$, is the order of the permutation associated with s, was proven in [6]. The proof also holds for the ABSG:

Proposition 8. *[6] Suppose s is the output of a maximum length LFSR of degree $L \geq 3$, and let p be the associated permutation. Let k be the minimal strictly positive integer such that $p^k(\emptyset) = \emptyset$. The length of the sequences $BSG(s_0, \ldots, s_{kT-1})$ and $ABSG(s_0, \ldots, s_{kT-1})$ are both greater than $k \cdot 2^{L-3}$.*

This bound does not answer the issue of possible subperiods. A strict lower bound on the period length of $BSG(s)$ was introduced in [6]. Experimentally, for both the BSG and the ABSG, no subperiod appears when the input is produced by a maximal-length LFSR with feedback polynomial of degree $3 \leq L \leq 16$. As was done in [6] for the BSG, one can show that the output sequence set of the ABSG can be easily described from 2 distinct output sequences whose period lengths, called *short period* and *long period*, are respectively, when no subperiod appears, very close to $T/3$ and $2T/3$, and their sum is then exactly T. The results for the ABSG are given in Tables 2 of Appendix C. For the MBSG, we have:

Proposition 9. *Let s be a sequence produced by a maximum length LFSR of degree L. Consider a period of the output of the form $0^{\lambda_1}1^{\mu_1}0^{\lambda_2}1^{\mu_2}\ldots0^{\lambda_p}1^{\mu_p}$. Then, the sequence $MBSG_P(s)$ has a period $MBSG(t)$ of length T such that:*

- *for $L = 0 \mod 2$, we have $T = (2^L - 1)/3$, the number of zeros in this period is $(T-1)/2$, and the number of ones is $(T+1)/2$,*
- *for $L = 1 \mod 2$, we have $T = (2^L + 1)/3$, the number of zeros in this period is $(T+1)/2$, and the number of ones is $(T-1)/2$.*

The proof of Proposition 9 is given in Appendix A.

Like for the BSG and the ABSG, subperiods may appear in a period of $MBSG_P(s)$ of length T. Experimentally, this never happens for $L \leq 16$, so that the values in Proposition 9 are exact. The periodicity results are given in Table 4 in Appendix C.

Linear Complexity of Output Sequences. We do not have theoretical bounds for the linear complexity, but the statistics for maximum length LFSRs of degree $L \leq 16$ suggest that the linear complexity is well-behaved. The results for the linear complexity are given in Appendix C, in Tables 3 and 4 respectively for the ABSG and the MBSG.

For the ABSG, we give the average linear complexity (denoted by LC), and its minimal and maximal values for short and long output sequences. These values are to be compared with those in Table 2: indeed, they show that the linear complexity is always almost equal to the period.

For the MBSG, preliminary experiments on the linear complexity of the output sequences when filtering maximum length LFSR sequences show that the linear complexity is very close to the period. Furthermore, when the period is prime, we observe that the linear complexity is always equal to the period (for degrees up to 16, for which we tested every possible maximum length LFSR output). Therefore, further study of the MBSG seems promising.

5 Security of the MBSG and the ABSG

5.1 Strategy 1: Random Generation of Candidates

By applying Strategy 1 of subsection 3 to the MBSG and the ABSG, we get:

Proposition 10. *Let s be a (finite) sequence and y be the (finite) sequence such that $y = MBSG(s)$. Let c be a randomly chosen string with the property that $MBSG(c) = y$, where the probability distribution used to choose c reflects the probability that $c = s$. The probability that, for every output bit y_k, the sequences d and c agree on the length of the tuple from which y_k arises, is $3^{-\ell}$, where ℓ denotes the length of y.*

Proof. Each candidate input string should have a tuple $(b, 0^i, 1)$ inserted for every b occurring in the output sequence, where i is chosen independently at random for each output bit, such that $i = j$ with probability 2^{-j-1}. The probability that the candidate string and the correct string agree in any one of the ℓ positions is $\sum_{i=0}^{\infty}(2^{-i-1})^2 = 1/3$. That is, the probability that the correct input sequence and a candidate c agree on the ℓ choices of length of the tuples from which the ones of y arises is $3^{-\ell} \simeq 2^{-1.585\ell}$. □

One can show that Proposition 10 also holds for the ABSG. By assuming the knowledge of no additional information on the means used to generate the input sequence, we deduce from Proposition 10 that the Hamming weight of the output sequence does not play a part in the input sequence reconstruction problem.

5.2 Strategy 2: Choice of the Most Favourable Case

In the case of the MBSG (resp. ABSG) applied to the output sequence of a maximum length LFSR of size L with a public feedback polynomial, the following attack can be mounted: it consists of finding a window of $L/2$ bits coming from a pair of bits $(b, 1)$ (resp. (b, b) for the ABSG), which occurs with probability $2^{-\frac{L}{2}}$. This window can give instantly the L bits of the current state of the LFSR. Thus, we can instantly check if we have found the correct values. In order to find the current state with high probability, we have to repeat this procedure $2^{\frac{L}{2}}$ times. This "attack" costs $\mathcal{O}(2^{\frac{L}{2}})$ and requires $\mathcal{O}(L2^{\frac{L}{2}})$ bits of keystream. This "attack" is slightly better than the generic attack thanks to the reduction in memory required.

6 FBDD-Based Cryptanalysis

Krause [7] introduced a new type of attack against keystream generators, called the FBDD-attack (FBDD for Free Binary Decision Diagram), which is a cryptanalysis method for LFSR-based generators. A generator is said to be LFSR-based if it consists of two components, a linear bitstream generator LG wich generates for each initial state $x \in \{0,1\}^n$ a linear bitstream $LG(x)$ using one or more parallel LFSRs, and a compression function C which transforms the internal bitstream into an output keystream $y = C(LG(x))$.

The cryptanalysis method relies on two assumptions called the *FBDD Assumption* and the *Pseudorandomness Assumption* (see [7] for details). The cost of the cryptanalysis depends on two parameters of the compression function C. The first parameter is the maximal number of output bits which C produces on internal bitstreams of length m; let γ be the best case compression ratio of C. Krause cryptanalysis applies when the following property is fulfilled: for all $m > 1$, the probability that $C(z)$ is a prefix of y for a randomly chosen and uniformly distributed $z \in \{0,1\}^m$ is the same for all keystreams y. Observe that both the ABSG and the MBSG have this property but the BSG does not have (nevertheless the Krause attack is expected still to work, and later we will try to estimate its complexity). Let us denote this probability $p_C(m)$. The second parameter, called α, depends on the probability $p_C(m)$. Indeed, the probability $p_C(m)$ is supposed to behave as $p_C(m) = 2^{-\alpha m}$, with α a constant such that $0 < \alpha \leq 1$. This result comes from the following partition rule: each internal bitstream z can be divided into consecutive elementary blocks $z = z^0\, z^1 \ldots z^{s-1}$ such that $C(z) = y_0 y_1 \ldots y_{s-1}$ with $y_j = C(z^j)$ and the average length of the elementary blocks is a small constant. Then, we have $\alpha \approx -\frac{1}{m}\log(p_C(m))$ for large m.

Theorem 1. *[7] Let E be an LFSR-based keystream generator of key-length L with linear bitstream generator LG and a compression function C of information rate α and best case compression ratio γ. Let C fulfill the FBDD and the pseudorandomness assumption. Then, there is an $L^{O(1)}2^{(1-\alpha)(1+\alpha)L}$-time bounded algorithm which computes the secret initial state x from the first $\lceil \gamma\alpha^{-1}L \rceil$.*

Remark 4. The parameter α used to compute the complexity of the FBDD attack is **not** the information rate of the compression function, see appendix B.1 for details.

For both the ABSG and the MBSG, one can check that the FBDD Assumption and the Pseudorandomness Assumption are fulfilled and the value of γ is clearly 1/2. Our results on the FBDD attack are summarised in Table 1. In Appendix B we explain how these results are obtained.

Remark 5. We can see in the table that $\alpha_{MBSG} = \alpha_{ABSG}$. We deduce that the (time and space) complexity of the FBDD attack applied to both the ABSG and the MBSG is $L^{O(1)}2^{0.53L}$.

Remark 6. When the Krause attack is applied to BSG, the complexity depends (in a somewhat complex way) on the number of zeros in the current output sequence. Roughly speaking, with many zeros placed at the beginning of it, the attack will work better and one should apply the attack at such well chosen places in the output sequence. In Appendix B we show that the complexity ranges from $L^{O(1)}2^{0.33L}$ to $L^{O(1)}2^{0.62L}$. The best case cannot be obtained in practice, this would require $O(2^{\frac{2}{3}L})$ of keystream, and moreover $2^{0.33L}$ would still be worse than $2^{0.25L}$ obtained with the best attack of Section 3.2.

Table 1. Application of Krause FBDD attack to *BSG and SSG

		SSG	BSG	ABSG	MBSG
output rate		0.25	0.333	0.333	0.333
Krause rate γ		0.5	0.5	0.5	0.5
information rate		0.25	?	0.333	0.333
α		0.208	$0.238 \leq \alpha \leq 0.5$	0.306	0.306
Krause	time	$L^{O(1)}2^{0.66L}$	$L^{O(1)}2^{0.33L} < ... < L^{O(1)}2^{0.62L}$	$L^{O(1)}2^{0.53L}$	$L^{O(1)}2^{0.53L}$
Attack	memory	$L^{O(1)}2^{0.66L}$	$L^{O(1)}2^{0.33L} < ... < L^{O(1)}2^{0.62L}$	$L^{O(1)}2^{0.53L}$	$L^{O(1)}2^{0.53L}$

7 Conclusion

In this paper, we studied two bit-search based techniques derived from the bit-search generator. The three related compression techniques (BSG, MBSG and ABSG) studied in this paper have rate 1/3, and have good periodicity properties. Experiments suggest that they produce sequences with high linear complexity when given maximum length LFSR sequences as input. However, according to the cryptanalysis techniques that we have considered, the BSG seems less secure than both the MBSG and ABSG. Indeed, the main attack that we propose on the BSG has a complexity close to or slightly smaller than $\mathcal{O}(L2^{\frac{L}{4}})$ and requires $\mathcal{O}(2^{\frac{L}{4}})$ bits of keystream and the main attack that we propose on both the MBSG and the ABSG costs $\mathcal{O}(2^{\frac{L}{2}})$ and requires $\mathcal{O}(L2^{\frac{L}{2}})$ bits of keystream. It seems that the MBSG and the ABSG are attractive components that can be used for the de-synchronization of LFSR outputs in keystream generation.

References

1. F. Armknecht, M. Krause, *Algebraic Attacks on Combiners with Memory*, Advances in Cryptology – CRYPTO'03 Proceedings, LNCS **2729**, Springer-Verlag, (2003), 162–176.
2. D. Coppersmith, H. Krawczyk, Y. Mansour, *The Shrinking Generator*, Advances in Cryptology – CRYPTO'93 Proceedings, LNCS **773**, Springer-Verlag, D. R. Stinson, ed., (1993), 22–39.
3. N. Courtois, W. Meier, *Algebraic Attacks on Stream Ciphers with Linear Feedback*, Advances in Cryptology – EUROCRYPTO'03 Proceedings, LNCS **2656**, Springer-Verlag, (2003), 345–359.
4. P. Ekdahl, T. Johansson, W. Meier, *Predicting the Shrinking Generator with Fixed Connections*, Advances in Cryptology – EUROCRYPT 2003 Proceedings, LNCS **2656**, Springer-Verlag, E. Biham, ed., (2003), 330–344.
5. S. Golomb, *Shift Register Sequences*, Revised Edition, Aegean Park Press, (1982).

6. A. Gouget and H. Sibert. *The bit-search generator*, In *The State of the Art of Stream Ciphers: Workshop Record, Brugge, Belgium, October 2004*, pages 60–68, 2004.
7. M. Krause. *BDD-based Cryptanalysis of Keystream Generators*, In EUROCRYPT 2002, pp. 222-237, LNCS 2332, Springer, 2002.
8. W. Meier, O. Staffelbach, *The Self-Shrinking Generator*, Advances in Cryptology – EUROCRYPT'94 Proceedings, LNCS **950**, Springer-Verlag, A. DeSantis, ed., (1994), 205–214.
9. R. A. Rueppel, *Analysis and Design of Stream Ciphers*, Springer-Verlag, (1986).

A Proof of Proposition 9

We first prove the following lemma:

Lemma 1. *Let s be a periodic sequence with a period of the form:*

$$0^{\lambda_1}1^{\mu_1}0^{\lambda_2}1^{\mu_2}\dots0^{\lambda_p}1^{\mu_p} \ ,$$

with $\lambda_i > 0$ and $\mu_i > 0$ for every i, and $\mu_p = 1$. Then, we have:

1. *the finite sequence $MBSG(t)$ is a period of $MBSG_P(s)$,*
2. *the length of $MBSG(t)$ is equal to $p + \sum_{i=1}^{p} \left\lfloor \frac{\mu_i - 1}{2} \right\rfloor$,*
3. *the number of zeros in $MBSG(t)$ is equal to $\#\{i, \mu_i = 1 \mod 2\}$,*
4. *the number of ones in $MBSG(t)$ is equal to $\sum_{i=1}^{p} \left\lfloor \frac{\mu_i}{2} \right\rfloor$.*

Proof. As the period ends with the pattern 01, we know that a cursor with initial state $\emptyset$ before reading the period is in state $\emptyset$ after reading this period. Therefore, a period of $MBSG_P(s)$ is $MBSG(0^{\lambda_1}1^{\mu_1}\dots0^{\lambda_p}1^{\mu_p})$.
Next, the output of a bit corresponds to reading a pattern of the form $(b, 0^k, 1)$, with $b \in \{0, 1\}$ and $k \geq 0$. In the periodic part of the output, these patterns necessarily contain a maximal sequence of 0's, so 0^k is some 0^{λ_i} if $b = 1$, otherwise $(b, 0^k)$ is some 0^{λ_i}. Therefore, one output bit corresponds to each maximal sequence 0^{λ_i}. There are p such sequences in the period. The other output bits come from patterns that do not contain 0, that is, from patterns $(1, 1)$. Now, in every maximal sequence 1^{μ_i}, the first 1 is the end of a pattern containing a maximal sequence of zeros. Therefore, there remains only $\lfloor \frac{\mu_i - 1}{2} \rfloor$ complete pairs of ones in the sequence 1^{μ_i} in order to output bits from pairs of ones. Thus, $\sum_{i=1}^{p} \lfloor \frac{\mu_i - 1}{2} \rfloor$ output bits correspond to pairs of ones in the period. This completes the second result.
We now turn to the number of zeros. A zero is output if and only if the corresponding pattern in the input is of the form $(0^{\lambda_i}, 1)$. Now, this pattern can correspond to an output bit if, and only if, the cursor is in state $\emptyset$ before the maximal sequence 0^{λ_i}. This is the case if, and only if, the length of the maximal sequence $1^{\mu_{i-1}}$ is odd. This gives the next result.
The number of ones comes directly from the two previous results. □

The proof of Proposition 9 is then a straightforward computation given the well-known distribution of maximal sequences in a period of the input, that appears for example in [5].

B The FBDD Attack Applied to the BSG, the ABSG and the MBSG

B.1 Comments on Krause Article

In [7], Krause denotes by $p_C(m)$ the probability that a randomly chosen and uniformly distributed $z \in \{0,1\}^m$ is compatible with a given keystream y, i.e., that $C(z)$ is a prefix of y. He considers only sequence generators such that this probability is the same for every y. Then, he defines $\alpha = -\frac{1}{m}\log(p_C(m))$ and he claims that α is the *information rate* per bit revealed by the keystream y about the first m bits of the corresponding internal bitstream z, i.e.

$$\alpha = \frac{1}{m}(H(Z^{(m)}) - H(Z^{(m)}|Y)) = \frac{1}{m}(m - \log(p_c(m)2^m)),$$

where $Z^{(m)}$ denotes a random $z \in \{0,1\}^m$ and Y a random keystream. This would hold if $H(Z^{(m)}|Y) = -\log(p_c(m)2^m)$ which is not always true, because, given an output keystream, not all compatible inputs are equally probable.

To clarify, the complexity of the Krause attack does indeed depend on α as defined, but this α is not in general equal to the information rate. We obtain a counterexample if we compare α_{ABSG} and the information rate of its compression function.

Computation of the information rate: we computed the information rate θ for the ABSG and the MBSG. Let m be the length of the internal bitstream, and let z denote a random, uniformly distributed element from $\{0,1\}^m$. The number of z such that $C(z)$ has length $i \geq 0$ is the number of patterns of the form $\overline{b_1}\ b_1^{k_1}\ \overline{b_1}\ \ \overline{b_2}\ b_2^{k_2}\ \overline{b_2}\ \ \ldots\ \ \overline{b_i}\ b_i^{k_i}\ \overline{b_i}\ \ \overline{b_{i+1}b_{i+1}^{k_{i+1}}}$ with $k_j \geq 0$ and $\sum_{j=1}^{i} k_i = m - 2i$. We have the following possible values for $w = \overline{b_{i+1}b_{i+1}^{k_{i+1}}}$:

- if w is the empty word or one bit (which can then be both 0 or 1), the pattern occurs with probability 2^{m-i},
- if w has length at least 2, then we have $w = \overline{b}b^k$ with $k > 0$, and only one case is possible for compatibility with the next output bit. The whole pattern occurs with probability 2^{m-i-1}.

Let $N(m)$ be the number of sequences $\overline{b_1}\ b_1^{k_1}\ \overline{b_1}\ \ \overline{b_2}\ b_2^{k_2}\ \overline{b_2}\ \ \ldots\ \ \overline{b_i}\ b_i^{k_i}\ \overline{b_i}$ of length m that are a prefix for a given y. We know that $N(m)$ is the number of ways of distributing $m - 2i$ bits among i places. The number of ways of distributing p bits among q places is a known combinatorial problem and can be written as $\binom{p+q-1}{p}$. Therefore $N(m) = \binom{m-i-1}{m-2i}$.

Then we have:

$$H_m(Z|Y) = \sum_{k=2}^{m-2} \sum_{i=1}^{\lfloor \frac{k}{2} \rfloor} \binom{k-i-1}{k-2i} \frac{m-i-1}{2^{m-i-1}} + 2\sum_{i=1}^{\lfloor \frac{m-1}{2} \rfloor} \binom{m-i-2}{m-2i-1} \frac{m-i}{2^{m-i}} + \sum_{i=1}^{\lfloor \frac{m}{2} \rfloor} \binom{m-i-1}{m-2i} \frac{m-i}{2^{m-i}} + \frac{m-1}{2^{m-1}}$$

Now we compute θ_{ABSG} with the formula above: we obtain $\lim_{m\to\infty}(\theta_{ABSG}) = \frac{1}{3}$, and for $m \geq 128$, we already have $\theta_{ABSG} \approx 0.33$.

Remark 7. We can also, in a very similar way, compute θ for MBSG:

$$H_m(Z|Y) = \sum_{m=2}^{M-1} \sum_{i=1}^{\lfloor \frac{m}{2} \rfloor} \binom{m-i-1}{m-2i} \frac{M-i-1}{2^{M-i-1}} + \sum_{i=1}^{\lfloor \frac{M}{2} \rfloor} \binom{M-i-1}{M-2i} \frac{M-i}{2^{M-i}} + \frac{M-1}{2^{M-1}}$$

We also obtain $\lim_{m\to\infty}(\theta_{MBSG}) = \frac{1}{3}$, and $\theta_{MBSG} \approx 0.33$ for $m \geq 128$. At last, a similar computation for the SSG yields $\lim_{m\to\infty}(\theta_{SSG}) = \frac{1}{4}$.

B.2 The FBDD Attack Applied to the ABSG and the MBSG

Recall that the cost of the FBDD cryptanalysis depends on two parameters called α and γ. For both the ABSG and the MBSG, the best compression ratio γ is achieved when each keystream bit comes from a pattern of length 2 and we have $\gamma_{ABSG} = \gamma_{MBSG} = \frac{1}{2}$. We compute in this subsection the value of α_{ABSG} (resp. α_{MBSG}), that is, the number of possible sequences of internal bitstream z of length m such that $ABSG(z)$ (resp. $MBSG(z)$) is a prefix for a given y when z is a random and uniformly distributed element from $\{0,1\}^m$; for both the ABSG and the MBSG this number does not depends on the keystream y.

Let us consider the action of the ABSG on an input sequence z. A sequence z that produces m keystream bits, where $m \geq 0$, has two possible forms:

- $\overline{b_1}\, b_1^{k_1}\, \overline{b_1}\;\; \overline{b_2}\, b_2^{k_2}\, \overline{b_2}\;\; \ldots\;\; \overline{b_i}\, b_i^{k_i}\, \overline{b_i}$, where $k_j \geq 0$
- $\overline{b_1}\, b_1^{k_1}\, \overline{b_1}\;\; \overline{b_2}\, b_2^{k_2}\, \overline{b_2}\;\; \ldots\;\; \overline{b_i}\, b_i^{k_i}\, \overline{b_i}\;\; \overline{b_{i+1}} b_{i+1}^{k_{i+1}}$, where $k_j \geq 0$ and the last part $\overline{b_{i+1}} b_{i+1}^{k_{i+1}}$, that we call the *last word*, does not produce any bit.

Let y be an arbitrary keystream. Let B_m be the number of possible bitstream sequences z of the form $\overline{b_1}\, b_1^{k_1}\, \overline{b_1}\;\; \overline{b_2}\, b_2^{k_2}\, \overline{b_2}\;\; \ldots\;\; \overline{b_i}\, b_i^{k_i}\, \overline{b_i}$ of length m which are a prefix of y. This number does not depend on y. We know that $B_0 = 1, B_1 = 0, B_2 = 1, B_3 = 1, B_4 = 2\ldots$. For every $m > 0$, we have $B_m = B_{m-2} + B_{m-3} + \cdots + B_0$. Indeed, if we fix the length of the first pattern $\overline{b_1}\, b_1^{k_1}\, \overline{b_1}$, the number of possibilities is then B_{m-k_1-2}.

Let A_m the number of all possible bitstream sequences z such that $ABSG(z)$ is a prefix of y. We have:

$$A_m = B_m + 2B_{m-1} + \sum_{j=0}^{m-2} B_j,$$

where B_m is the number of possible z of the first form, $2B_{m-1}$ is the number of possible z of the second form with $k_{i+1} = 0$ (when the last word contains only one bit, there are two possibilities for this bit), and the B_is for $i \leq m-2$ are the number of possible z with $k_{i+1} = m - i - 1$. Therefore we have:

$$\begin{aligned} A_m - A_{m-1} &= B_m + B_{m-1} - B_{m-2} \\ &= \sum_{i=0}^{m-2} B_i + \sum_{i=0}^{m-3} B_i - \sum_{i=0}^{m-4} B_i \quad = \quad B_{m-2} + 2B_{m-3} + \sum_{i=0}^{m-4} B_i \\ &= A_{m-2} \end{aligned}$$

Thus $A_0 = 0, A_1 = 2$ and for every $m > 1$, $A_m = A_{m-1} + A_{m-2}$. Solving this recursion gives:

$$A_m = \frac{2}{\sqrt{5}} \left((\frac{1+\sqrt{5}}{2})^m - (\frac{1-\sqrt{5}}{2})^m \right) \approx \frac{2}{\sqrt{5}} (\frac{1+\sqrt{5}}{2})^m$$

Finally when m is large enough, we compute $\alpha_{ABSG} = \log(\sqrt{5} - 1) \approx 0.306$.

In the same way, one can show that $\alpha_{MBSG} = \alpha_{ABSG}$. We deduce that the (time and space) complexity of the FBDD attack applied to both the ABSG and the MBSG is $L^{O(1)} 2^{0.53L}$. All our results for the FBDD attack are summarised in Table 1.

B.3 The FBDD Attack Applied to the BSG

We have seen in part 6 that, for the BSG, the probability that $C(z)$ is a prefix of y for a randomly chosen and uniformly distributed $z \in \{0,1\}^m$ is not the same for all keystreams y. Thus, it is not clear whether the FBDD attack is still relevant. In this part, we suppose it is, and we show that still the attack would not be as effective as other attacks presented in this paper. We have at least to take into account the fact that the probability we called $p_C(m)$ does depend on y.

From an attacker's point of view, the best case is when the keystream y is uniquely composed of 0s. In this case, the value of α can be easily computed and we have $\alpha = -\frac{1}{m}\log(2^{\frac{m}{2}}) = \frac{1}{2}$.

The worst case occurs when the keystream is uniquely composed of 1s. Let B_m denote the number of bitstream sequences of length m such that the keystream is 111....1. We have: $B_0 = 0$, $B_1 = 2$, $B_2 = 2$, $B_3 = 4$. Moreover, for $m \geq 3$, if the bitstream sequence starts by $b\,\overline{b}^i\,b$, then the number of possibilities is $2 \times B_{m-2-i}$. Otherwise, the bitstream sequence starts by a sequence of the form $b\,\overline{b}^{m-1}$ and there are two possible values for the bit b. Then, we have $B_m = 2 + 2(B_0 + ... + B_{m-3})$. Let $A_m = B_0 + ... + B_m$, then we get the relation: $A_m = A_{m-1} + 2A_{m-3} + 2$. By computing the limit of the series $1 - \frac{1}{m}\log(A_m - A_{m-1})$ with Magma, we obtain $\alpha \approx 0.238$.

Thus, in the general case, α belongs to the interval $[0.238, 0.5]$. If the attack can be extended, its complexity will range from $L^{O(1)} 2^{0.33L}$ to $L^{O(1)} 2^{0.62L}$. Then the attacker should start at the most interesting place in the output sequence, but in practice he has no hope to achieve the best-case complexity.

To obtain the best case, the attacker needs to find an all-zero subsequence with length $\frac{2}{3}L$, and this can hardly be achieved without disposing of $O(2^{\frac{2}{3}L})$ bits of keystream. Moreover an FBDD attack in $2^{0.33L}$ will still be worse than $2^{0.25L}$ we obtain in Section 3.2.

C Statistical Results

Period and linear complexity statistics for m-LFSRs filtered by the BSG are given in [6].

Table 2. Period statistics for m-LFSRs filtered by the ABSG

L	Average *short* period length	Minimal *short* period length	Maximal *short* period length	Average *long* period length	Minimal *long* period length	Maximal *long* period length
8	84.63	82	88	170.38	167	173
9	169	159	183	342	328	352
10	341.1	328	358	681.9	665	695
11	682.91	657	714	1364.09	1333	1390
12	1364.08	1330	1399	2730.92	2696	2765
13	2731.34	2658	2796	5459.66	5395	5533
14	5460.08	5344	5587	10922.92	10796	11039
15	10923.04	10776	11082	21843.96	21685	21991
16	21846.16	21619	22075	43688.84	43460	43916

Table 3. Linear complexity statistics for m-LFSRs filtered by the ABSG

L	Average *short* lin. compl.	Minimal *short* lin. compl.	Maximal *short* lin. compl.	Average *long* lin. compl.	Minimal *long* lin. compl.	Maximal *long* lin. compl.
8	84	81	88	169.38	166	173
9	167.71	158	182	340.79	326	352
10	340.2	327	358	680.83	661	695
11	680.95	654	710	1363.24	1332	1390
12	1363.33	1330	1399	2729.96	2696	2761
13	2729.80	2656	2793	5458.75	5391	5532
14	5459.17	5342	5587	10921.96	10796	11038
15	10921.47	10774	11076	21843.05	21684	21991
16	21845.28	21618	22075	43687.95	43460	43912

Table 4. Period and linear complexity statistics for m-LFSRs filtered by the MBSG

L	Period	Average LC	Minimal LC	Maximal LC
8	85	84, 25	77	85
9	171	170, 46	165	171
10	341	339, 92	326	341
11	683	683	683	683
12	1365	1362, 53	1347	1365
13	2731	2731	2731	2731
14	5461	5461	5461	5461
15	10923	10923	10923	10923
16	21845	21844, 35	21833	21845

Some Attacks on the Bit-Search Generator*

Martin Hell and Thomas Johansson

Dept. of Information Technology, Lund University,
P.O. Box 118, 221 00 Lund, Sweden
{martin, thomas}@it.lth.se

Abstract. The bit-search generator (BSG) was proposed in 2004 and can be seen as a variant of the shrinking and self-shrinking generators. It has the advantage that it works at rate 1/3 using only one LFSR and some selection logic. We present various attacks on the BSG based on the fact that the output sequence can be uniquely defined by the differential of the input sequence. By knowing only a small part of the output sequence we can reconstruct the key with complexity $O(L^3 2^{0.5L})$. This complexity can be significantly reduced in a data/time tradeoff manner to achieve a complexity of $O(L^3 2^{0.27L})$ if we have $O(2^{0.27L})$ of keystream. We also propose a distinguishing attack that can be very efficient if the feedback polynomial is not carefully chosen.

1 Introduction

Lately, we have seen many new proposals for stream ciphers. The aim of binary additive stream cipher is to produce a random looking sequence and then xor this sequence with the plaintext sequence to produce the ciphertext. There are several possible approaches when designing a stream cipher. A Linear Feedback Shift Register (LFSR) with primitive feedback polynomial generates sequences that possess many of the properties which we would expect from a random sequence. Because of this it is common to use an LFSR as a building block in a stream cipher. The problem with just using an LFSR is that any output bit of the LFSR can be written as a linear function in the initial state bits. This problem is solved by introducing some nonlinearity into the cipher. There are many ways to do this and some classical approaches include letting the output of several LFSRs or some state bits of one LFSR serve as input to a nonlinear Boolean function. Another common way to introduce nonlinearity is to, by some algorithm, decimate the LFSR output sequence in some irregular way. Two well known keystream generators based on this principle are the shrinking generator [1] and the self-shrinking generator [16]. Another generator related to

* The work described in this paper has been supported in part by the European Commission through the IST Programme under Contract IST-2002-507932 ECRYPT. The information in this document reflects only the author's views, is provided as is and no guarantee or warranty is given that the information is fit for any particular purpose. The user thereof uses the information at its sole risk and liability.

H. Gilbert and H. Handschuh (Eds.): FSE 2005, LNCS 3557, pp. 215–227, 2005.

these is the alternating step generator [11]. This generator uses one LFSR to decide the clocking of two other LFSRs.

The bit-search generator (BSG) is a keystream generator intended to be used as a stream cipher. It was introduced in 2004 by Gouget and Sibert [10] and the construction is similar to the generators mentioned above. The output of the BSG is produced by a simple algorithm, taking a pseudorandom sequence as input.

In this paper we investigate some possible attacks on the bit-search generator. Throughout the paper we assume that the pseudorandom sequence is generated by a maximum length LFSR and that the (primitive) feedback polynomial is known to the attacker.

We give an alternative description of the BSG based on the differential of the input sequence and then we describe a simple but efficient algorithm to reconstruct the differential sequence with knowledge of only a few keystream bits. By reconstructing the differential sequence we can reconstruct the original input sequence and also the key. This attack works regardless of the form of the feedback polynomial and has complexity $O(L^3 2^{0.5L})$. If we know more keystream bits we show that the complexity will be significantly decreased. More specifically, with $O(2^{0.27L})$ bits of keystream we can mount the attack with time complexity $O(L^3 2^{0.27L})$ according to our simulations. Moreover, we describe the basis for a distinguishing attack on the BSG. This attack can be made very efficient if the feedback polynomial is of low weight or if it is possible to find a low degree multiple of the feedback polynomial with low weight.

The outline of the paper is the following. In Section 2 we describe the BSG and we compare the construction with similar generators. Then, in Section 3 we present an attack that reconstructs the input sequence to the BSG algorithm. By doing this we can recover the initial state of the LFSR. Section 4 gives the framework for a possible distinguishing attack and in Section 5 we summarize some previous attacks on the shrinking, self-shrinking and the alternating step generators. We also compare these attacks with the attacks on the BSG shown in this paper. In Section 6 we give our conclusions.

2 Description of the Bit-Search Generator

In this section we describe the bit-search generator in two different but equivalent ways. First we give the original description that uses a sequence s as input, as presented in [10]. Then we give an alternative description that uses the differential sequence d of s as input. We also compare the construction to similar keystream generators.

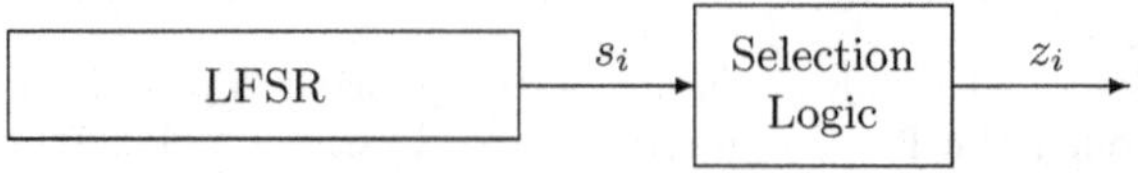

Fig. 1. Block model of the bit-search generator

Table 1. Comparison between the BSG and some well known generators

Generator	Number of LFSRs needed	Rate
Alternating Step	3	1
Shrinking	2	1/2
Self-Shrinking	1	1/4
BSG	1	1/3

The principle of the BSG is very simple. It consists only of an LFSR and some small selection logic, see Fig. 1. Consider a sequence $s = (s_0, s_1, s_2 \ldots)$ generated by the LFSR. The output sequence $z = (z_0, z_1, z_2 \ldots)$ is constructed from s by first letting $b = s_0$ be the first bit to search for. If the search ends immediately, i.e. $s_1 = b = s_0$ we output 0, otherwise we continue to search the sequence s until the bit we search for is found. When the correct bit is found we output 1 and we let the following bit be the next to search for. An output bit is produced after 2 input bits with probability 1/2, after 3 input bits with probability 1/4 etc. In general, an output bit is produced after $i+1$ input bits with probability 2^{-i} so the average number of input bits needed to produce one output bit is $\sum_{i=0}^{\infty}(i+1) \cdot 2^{-i} = 3$. This shows that the rate of the BSG is asymptotically 1/3.

To motivate why this generator is interesting we compare it to some other well known generators based on the idea of only using LFSRs and some selection logic. We base the comparison on the number of LFSRs used and the rate of the cipher. As we can see in Table 1 the BSG has lower rate than the alternating step generator and the shrinking generator but it uses only one LFSR. The self-shrinking generator has also only one LFSR but it has lower rate.

We now consider the differential sequence d of s. The differential sequence is defined as $d_i = s_i \oplus s_{i+1}$. If the sequence s is generated by an LFSR it is well known, see e.g. [14], that the differential sequence can be generated by the same LFSR. The two sequences differ only by some shift. When reconstructing s from d we need to guess the first bit in s, then the remaining bits are uniquely determined from d.

The output of the BSG can be uniquely described by knowledge of the differential sequence. Hence, if we can reconstruct the differential sequence we can predict the future outputs uniquely and we can also recover the key used to initialize the LFSR. The BSG operates on the differential sequence in the following way. If $d_i = 1$ we know that $s_i \neq s_{i+1}$ so we will output 1. Then we search the sequence d until we find the next $d_j = 1$. If instead $d_i = 0$ we know that we have two consecutive bits which are the same, hence we output 0. Now we know we have found the bit we search for in the original BSG and we skip the next bit since it does not matter which value it has. It is clear that the output of the BSG can be generated from either the original LFSR sequence or from the differential sequence. The following is an example of a sequence s and the corresponding differential sequence d. Applying the algorithms, we can see that they produce the same output.

```
Output generated from s

i = -1; j = -1;
while (1)
    i++; j++;
    b = s[i];
    i++;
    if (s[i] == b) z[j] = 0;
    else z[j] = 1;
    while (s[i] != b) i++;
```

```
Output generated from d

i = 0; j = 0;
while (1)
    z[j] = d[i];
    if (d[i] == 1)
        i++;
        while (d[i] == 0) i++;
    i += 2;
    j++;
```

Fig. 2. The original BSG algorithm and an equivalent algorithm using the differential sequence d of s as input

$$\begin{aligned} s &= 0101001001110111010110 10\ldots \Rightarrow z = 110010101\ldots \\ d &= 11110110100110011110111\ldots \quad \Rightarrow z = 110010101\ldots \end{aligned}$$

A summary of the two algorithms can be found in Fig. 2.

3 Reconstructing the Input Sequence

In this section we will describe a known plaintext attack that tries to reconstruct the differential sequence from the output sequence. In our attack we assume that we have an LFSR generating the pseudorandom sequence and that the feedback polynomial of the LFSR is known to the attacker. If we have an LFSR of length L we need to guess L bits to be able to find a candidate initial state of the LFSR. Each bit can be written as a linear function of the initial state bits and by clocking the LFSR with a candidate initial state we can see if the candidate output equals the given output.

It follows from the algorithm given in Fig. 2 that $z_i = 0$ corresponds to a 0 followed by an unknown value in the differential sequence. It is also clear that $z_i = 1$ corresponds to a 1 followed by $j \geq 0$ 0s followed by a single 1 and an unknown value. In short,

$$\begin{aligned} z_i = 0 \quad &\Rightarrow \quad (0, -) \\ z_i = 1 \quad &\Rightarrow \quad (1, 0^j, 1, -) \end{aligned}$$

The probability of having j zeros is 2^{-j-1}, i.e. $z_i = 1$ corresponds to (1,1,–) with probability 1/2, (1,0,1,–) with probability 1/4, (1,0,0,1,–) with probability 1/8 etc. The expected number of inserted zeros is $\sum_{i=0}^{\infty} i \cdot 2^{-i-1} = 1$.

In the following we will denote by a the number of ones that we observe in an output sequence, b is the number of zeros in the output sequence and k is the number of zeros that are inserted in the candidate differential sequence, stemming from a set of a ones in the output sequence.

```
Search algorithm

Pick a part z' of z s.t. 2a+b=L;
k=0;
while (k <= k_max)
    Try all ways to insert k zeros in z';
    Delete last bit in z';
    if (Deleted bit == 0) k = k + 1;
    else  k = k + 2;
```

Fig. 3. The algorithm used to find the correct differential sequence

Now, assume that we have a set of a ones. There is one way to insert a total of $k = 0$ zeros and this happens with probability 2^{-a}. The number of ways to insert a total of $k = 1$ zero is $\binom{a}{1}$ and each has a probability of $2^{-a+1} \cdot 2^{-2} = 2^{-a-1}$. The number of ways to insert k zeros into a set consisting of a ones is a well known combinatorial problem and can be written as $\binom{a-1+k}{k}$, Hence, the probability of having a total of k zeros inserted will be

$$\binom{a-1+k}{k} 2^{-a-k}.$$

We construct a simple search algorithm based on these observations. The easiest way to find the correct differential sequence is to just guess the number of inserted zeros.

When we try to insert k zeros we need to look at an output sequence that satisfies $2a + b + k = L$. This is clear since every one in the output will give us two known bits and every zero will give us one known bit in the differential sequence. If we insert k extra zeros we will have a total of L bits which is enough to find the initial state. This leads us immediately to the algorithm in Fig. 3.

We start by just picking a part z' of the output sequence such that the length of z' satisfies $2a + b = L$. Then we insert $k = 0$ zeros. If this candidate is not the correct differential sequence, we delete the last bit in z'. If a 0 is deleted we try $k = 1$ next time since $b \leftarrow b - 1$ and we still require $2a + b + k = L$ to hold. For the same reason, if a 1 is deleted we try $k = 2$ next time. Every time $k \leftarrow k + 2$ we will miss some possible combinations and, hence, not the full space will be searched.

3.1 Analysis of the Algorithm

The complexity of the algorithm and the probability of success will depend on two factors. First, the ratio between the number of zeros and the number of ones in the sequence. If we have found a z' which has many more zeros than ones, the complexity will be lower. This will also give us a higher success probability since we will delete a 0 more often than we will delete a 1. The second factor is the maximum number of zeros we will try to insert into the sequence before we give up. This is the value k_{max} in the algorithm in Fig 3. Choosing a high value for k_{max} will increase the success probability but it will also increase the complexity.

We consider the case when we choose a sequence z' at random. We expect the number of zeros in the sequence to be equal to the number of ones. We also expect that the deleted bit is 1 every second time. Moreover, when z' is of odd length, we consider the pessimistic case when $a = b + 1$. We have the following equations

$$\left.\begin{array}{r} 2a + b + k = L \\ a = b \end{array}\right\} \Rightarrow a = \left\lceil \frac{L-k}{3} \right\rceil$$

The probability of success will be

$$\sum_{k=0}^{k_{max}} \binom{\lceil \frac{L-k}{3} \rceil - 1 + k}{\lceil \frac{L-k}{3} \rceil} 2^{-\lceil \frac{L-k}{3} \rceil - k}$$

and we have a total complexity of

$$\sum_{k=0}^{k_{max}} \binom{\lceil \frac{L-k}{3} \rceil - 1 + k}{\lceil \frac{L-k}{3} \rceil}.$$

Similar equations can easily be found also if $a \neq b$. We choose k_{max} as the smallest integer such that the probability of success is > 0.5. Focusing on the expected case when $a = b$, we summarize the complexity of an attack in Table 2 with respect to the length of the LFSR (keylength). It is clear that the complexity of the attack is very close to $2^{0.5L}$ tests for all cases.

Table 2. The attack complexity when the number of zeros equals the number of ones in z'

Keylength	k_{max}	Complexity
64	19	$2^{31.74}$
96	27	$2^{47.50}$
128	36	$2^{63.96}$
160	44	$2^{79.82}$
192	52	$2^{95.71}$
224	61	$2^{112.37}$
256	69	$2^{128.29}$

We can approximate the least number of plaintext bits needed in the expected case. With $L/3$ ones and $L/3$ zeros we will have knowledge of at least $2 \cdot L/3 + L/3 = L$ bits in the input sequence. With about $2^{0.5L}$ different sequences to test we need to compare the candidate sequence with an extra $0.5L$ bits to see if the candidate is correct. Hence, by knowing about $2L/3 + L/2 = 7L/6$ bits of the keystream sequence we can reconstruct the input sequence with a complexity of $O(2^{0.5L})$ tests. In the 128 bit case we need approximately 150 bits of keystream.

3.2 A Data/Time Tradeoff

As mentioned in the previous section, it is clear that the complexity of the attack depends on the number of ones that we observe in the keystream. With a large amount of keystream we can find sequences with few ones and, hence, the attack complexity is decreased. This provides a data/time tradeoff in the attack. Assume that we want to find a part z' of z that contains at most a ones and at least b zeros, where $b > a$. Looking at a random sequence of $a + b$ bits, the probability that we find a sequence with at most a ones is given by

$$P(\#ones \leq a) = \frac{\sum_{i=0}^{a} \binom{a+b}{i}}{2^{a+b}}$$

using the approximation that sequences are independent. The number of tries needed before a desired sequence is found is geometrically distributed with an expected value of

$$\frac{2^{a+b}}{\sum_{i=0}^{a} \binom{a+b}{i}} = \frac{2^{L-a}}{\sum_{i=0}^{a} \binom{L-a}{i}}.$$

In the equality we use $2a + b = L$. Table 3 demonstrates this data/time tradeoff for the case when $L = 128$, i.e. the keylength is 128 bits. Simulations show that the time complexity and the amount of keystream needed intersect at around $2^{0.27L}$ for all L between 64 and 1024 bits.

Table 3. The data/time tradeoff based on the number of ones and zeros in z' using a 128 bit key

Number of zeros (b) and ones (a) in z'	k_{max}	Complexity	Keystream
b = 2a	29	$2^{51.09}$	$2^{10.46}$
b = 3a	24	$2^{42.21}$	$2^{21.59}$
b = 4a	21	$2^{36.31}$	$2^{31.32}$
b = 5a	19	$2^{32.14}$	$2^{39.75}$
b = 6a	17	$2^{28.46}$	$2^{48.70}$

The complexities in Table 2 and Table 3 are given as the number of tests. To test if a candidate sequence is correct, a constant time is needed. This time can be divided into two parts. First we need to find the initial state of the LFSR by solving a system of L unknowns and L equations. This system can be solved in time L^{ω}. In theory $\omega \leq 2.376$, see [2], but the constant factor in this algorithm is expected to be very big. The fastest practical algorithm is Strassen's algorithm [19], which requires about $7 \cdot L^{\log_2 7}$ operations. For simplicity we write the time complexity for this step as L^3. We also need to clock the LFSR a sufficient number of times to compare our candidate output sequence with the observed output sequence. This second constant would also be needed in an exhaustive key search. Thus, the total time complexity for our key recovery attack is $O(L^3 2^{0.5L})$ knowing only $7L/6$ bits of the keystream. With the data/time tradeoff the complexity of the attack is $O(L^3 2^{0.27L})$ if we know $O(2^{0.27L})$ bits of the keystream. Note that these complexities are not formally derived but simulations show that they are valid (at least) up to keylengths of 1024 bits. The memory complexity of the attack is limited to the memory needed to solve the system of linear equations.

4 Distinguishing Attack

In this section we describe a possible distinguishing attack on the BSG. A distinguishing attack does not try to recover the key or any part of the input sequence. Instead, the aim is to distinguish the keystream from a purely random sequence. In the attack we assume that we have found a multiple of the feedback polynomial that is of weight w and degree h. Any multiple of a feedback polynomial will produce the same output sequence as the original polynomial. The well known fast correlation attack, see [15], depends on the existence of low weight multiples of modest degree of the LFSR feedback polynomial. Due to the importance of finding low weight multiples this subject has been studied in several papers, see [6, 20]. In [6], Golić estimates that the critical degree when polynomial multiples of weight w start to appear is $(w-1)!^{1/(w-1)} 2^{L/(w-1)}$, where L is the degree of the polynomial. Hence, a feedback polynomial of degree L is expected to have a multiple of weight w that is of degree approximately $2^{\frac{L}{w-1}}$. Now, assume that we have found a multiple of weight w that is of degree h.

The linear recurrence of the LFSR can be written as

$$0 = d_i + d_{i+\tau_1} + d_{i+\tau_2} + \ldots + d_{i+\tau_{w-1}} \qquad (1)$$

where $\tau_{w-1} = h$ and $\tau_j < \tau_k,\ j < k$. A zero in the output sequence z corresponds to a zero in the differential sequence and a one in the output corresponds to a one in the differential sequence. Since the BSG has rate $1/3$ we can consider the following sums of symbols from the output sequence

$$B_i = z_i + z_{i+\frac{\tau_1}{3}} + z_{i+\frac{\tau_2}{3}} + \ldots + z_{i+\frac{\tau_{w-1}}{3}}. \qquad (2)$$

We know that $B_i = 0$ if we have the correct synchronization ($d_{i+\tau_1}$ appears as $z_{i+\frac{\tau_1}{3}}$, $d_{i+\tau_2}$ appears as $z_{i+\frac{\tau_2}{3}}$ etc.) in the positions. We give an approximate value of the probability that we have synchronization in one position. With a multiple of low weight and high degree h the distance between z_i and any $z_{i+\frac{\tau_j}{3}}$ is in the order of h. Using the central limit theorem we say that the total number of inserted zeros after h outputs is normally distributed with standard deviation $\sigma \cdot \sqrt{h}$, where σ is the standard deviation for the number of inserted zeros after one output. Now, we approximate the probability that we have the correct synchronization as $h^{-\frac{1}{2}}$.

Hence, the probability that $z_{i+\frac{\tau_j}{3}}$, $1 \le j \le w-1$ is synchronized with z_i is approximately $h^{-\frac{1}{2}}$. The probability that all $w-1$ positions are synchronized, denoted $P(sync)$, is

$$P(sync) = (h^{-\frac{1}{2}})^{w-1} = h^{-\frac{w-1}{2}}$$

and the probability that $B_i = 0$ can be calculated as

$$\begin{aligned} P(B_i = 0) &= P(B_i = 0 \mid sync) \cdot P(sync) \\ &\quad + P(B_i = 0 \mid no\ sync) \cdot P(no\ sync) \\ &= 1 \cdot h^{-\frac{w-1}{2}} + 1/2 \cdot (1 - h^{-\frac{w-1}{2}}) \\ &= 1/2 + 1/2 \cdot h^{-\frac{w-1}{2}}. \end{aligned} \tag{3}$$

With a bias of $h^{-\frac{w-1}{2}}$ we will need about h^{w-1} samples of the output sequence to distinguish it from random. The complexity of the distinguishing attack depends on the degree of the multiple and if the degree is the expected degree, $h = 2^{\frac{L}{w-1}}$, our distinguisher needs about 2^L samples. However, if the feedback polynomial is not carefully chosen and we instead can find a multiple of low weight that is of much lower degree than expected, then the attack can be very efficient. This distinguisher can be improved in several ways. One way is to consider blocks of bits instead of individual bits.

We can consider a feedback polynomial with $h << 2^{\frac{L}{w-1}}$ as being a weak polynomial and ciphers using a weak polynomial can be efficiently attacked. Another class of weak feedback polynomials and an attack on these can be found in [5]. One can do a similar attack on the bit-search generator.

The values in the previous attack are approximated but for large h they are quite accurate. In the case where the feedback polynomial itself is of low weight, the values are not very accurate. We now describe how this attack can be mounted if the LFSR uses a feedback polynomial of some low weight w. Equation (1) will always hold for the differential sequence. To find the optimum guess for $z_{i+\frac{\tau_j}{3}}$, $1 \le j \le w-1$ in (2) we use the generating function for the probability of the number of clockings after λ outputs. Recall that the BSG will produce a keystream bit after two clockings with probability 1/2, after 3 clockings with probability 1/4 etc. The generating function can be written as

$$\left(\sum_{n=0}^{\infty} \frac{1}{2^n} z^{n+1} \right)^{\lambda}. \tag{4}$$

The coefficient of z^n is the probability that the LFSR has been clocked n times when the BSG has generated λ keystream bits.

By choosing the λ_j for which the coefficient of z^{τ_j} is highest we can determine which guess will give us the best probability of synchronization and we will also get the exact probability of a correct guess. We denote the probability that we guess λ_j correctly by p_{λ_j}. If p_{λ_j}, $1 \le j \le w-1$ are independent the probability that $B_i = 0$ can be written, similarly to (3), as

$$P(B_i = 0) = 1 \cdot \prod_{j=1}^{w-1} p_{\lambda_j} + 1/2 \cdot (1 - \prod_{j=1}^{w-1} p_{\lambda_j}) \\ = 1/2 + 1/2 \cdot \prod_{j=1}^{w-1} p_{\lambda_j}.$$

With a bias of $\prod_{j=1}^{w-1} p_{\lambda_j}$ we need about

$$\frac{1}{\prod_{j=1}^{w-1} p_{\lambda_j}^2}$$

samples for a successful distinguishing attack. We end this section with a small numerical example showing the performance of this distinguisher on a low weight feedback polynomial.

Example 1. Consider the weight 5 primitive feedback polynomial $1+x^{29}+x^{66}+x^{95}+x^{128}$. Write the linear recurrence in the differential sequence as

$$0 = d_i + d_{i+29} + d_{i+66} + d_{i+95} + d_{i+128}.$$

Using (4) we find that the highest coefficient for z^{29}, z^{66}, z^{95} and z^{128} is achieved when we have $\lambda_1 = 10$, $\lambda_2 = 22$, $\lambda_3 = 32$ and $\lambda_4 = 43$ respectively. The best possible approximation of (2) is then $B_i = z_i + z_{i+10} + z_{i+22} + z_{i+32} + z_{i+43}$. The probability that each of these terms are synchronized with z_i is the coefficient for each term in (4), i.e.,

$$p_{\lambda_1} = 2^{-3.43}, \quad p_{\lambda_2} = 2^{-4.06}, \quad p_{\lambda_3} = 2^{-4.31}, \quad p_{\lambda_4} = 2^{-4.53}.$$

This gives us a total bias of $\prod_{j=1}^{w-1} p_{\lambda_j} = 2^{-16.33}$ and, hence, our distinguisher needs approximately $2^{32.66}$ bits to succeed.

This shows that low weight feedback polynomials can be easily and efficiently attacked. Note that the attack described above can be further improved, using slightly more advanced techniques.

5 Comparison with the Alternating Step, Shrinking and the Self-shrinking Generator

The shrinking generator, the self-shrinking generator and the alternating step generator are similar to the BSG in that they only contain one or more LFSRs and some selection logic. There is no Boolean function used as a nonlinear combiner or as a nonlinear filter. In section 2 we compared the number of LFSRs used in and the rate of these generators. Here we summarize a small selection of the attacks proposed for the 3 well known generators and we compare them to our attacks on the BSG.

The alternating step generator is the oldest generator and based on 3 LFSRs in such a way that L_3 controls the clocking of L_1 and L_2. In the original paper [11] a divide-and-conquer attack with complexity $O(2^{L_3+2\log_2 \min(L_1,L_2)})$ was shown. 1997, in [8], Golić and Menicocci showed a correlation attack with complexity $O(2^{L_1+L_2+2\log_2(L_1+L_2)})$ and the year after [9] they improved this attack significantly to $O(2^{\max(L_1,L_2)+2\log_2 \max(L_1,L_2)})$.

The shrinking generator uses two LFSRs, denoted A and S. The sequence generated by S is used to select bits in the A-sequence. These selected bits are the output bits. In the original paper [1], an attack with known feedback polynomials was proposed that has complexity $O(2^{L_S} \cdot L_A^3)$. In 1998, Simpson, Golić and Dawson [18] presented a correlation attack that can recover the initial state of A with complexity $O(2^{L_A} \cdot L_A^2)$ using about $20 \cdot L_A$ bits. In 1998, Johansson [12] gave another correlation attack that is based on finding weak sequences. The complexity of this attack is better than previous attacks but it is still exponential in $|A|$. Distinguishing attacks on the shrinking generator have also been presented in [3, 7].

The BSG is probably most related to the self-shrinking generator since they both consist of only one LFSR. Because of this the attacks on the self-shrinking generator are easy to compare to the attacks on the BSG. Several key recovery attacks have been proposed for the self-shrinking generator. In the original paper [16] a key recovery attack was proposed that has average complexity $O(2^{0.75L})$. In 1996, Mihaljevic [17] presented an attack that has a complexity that varies between $O(2^{0.5L})$ and $O(2^{0.75L})$ but the required length of the keystream varies between $O(2^{0.5L})$ and $O(2^{0.25L})$ respectively. In 2001, Zenner, Krause and Lucks [21] described an attack that uses a search tree. This attack needs very few keystream bits and has complexity $O(2^{0.69L})$. The attack was later improved by Krause [13] to a complexity of $O(2^{0.66L})$. The problem with these two attacks is that they require a large amount of memory. In 2003, Ekdahl, Johansson and Meier [4] presented an attack that is much more efficient than the previous attacks if the polynomial is of a certain form.

Our attack that reconstructs the input sequence of the bit-search generator is equivalent to a key recovery attack. We can reconstruct the initial state of the LFSR that produces the differential sequence d. From this sequence we can reconstruct the original sequence. We propose an attack with complexity $O(L^3 2^{0.5L})$ that uses very few keystream bits. Knowing more keystream bits will reduce the complexity significantly. Using a data/time tradeoff we show that we can mount the attack using $O(2^{0.27L})$ keystream bits with a time complexity of $O(L^3 2^{0.27L})$. Finally, we suggest a distinguishing attack that can be very efficient if the feedback polynomial is not carefully chosen. It has the complexity $O(h^{w-1})$ for a degree h multiple of weight w. The framework of this distinguishing attack can be used also to attack the other generators. Hence, for all the generators considered here it is important to choose a feedback polynomial that has no low weight multiples of degree much lower than expected.

Finally, we would like to mention that the BSG, as well as the other generators in this section, can be vulnerable to various side channel attacks. Though, we have not pursued any work in this direction.

6 Conclusion

The bit-search generator, recently proposed by Gouget and Sibert has been considered and an equivalent description based on the differential of the input sequence has been given. We propose an efficient attack that recovers the differential sequence, and hence, the key. The construction as well as the security of the generator has been compared to similar generators. The self-shrinking generator is very similar to the BSG and we find that the key recovery attacks presented here are more efficient than any known key recovery attack on the self-shrinking generator. The basis for a distinguishing attack is also described and we show that if the feedback polynomial is not carefully chosen, the BSG may be prone to efficient distinguishing attacks.

References

1. D. Coppersmith, H. Krawczyk and Y. Mansour. The Shrinking Generator. In D. Stinson, editor, *Advances in Cryptology–CRYPTO'93*, pages 22–39. Springer-Verlag, 1993. Lecture Notes in Computer Science Volume 773.
2. D. Coppersmith and S. Winograd. Matrix Multiplication via Arithmetic Progressions, *J. Symbolic Computation (1990)*, 9, pp. 251–280.
3. P. Ekdahl, W. Meier and T. Johansson. Predicting the Shrinking Generator with Fixed Connections. In E. Biham, editor, *Advances in Cryptology–EUROCRYPT 2003*, volume 2656 of Lecture Notes in Computer Science, pages 330–344. Springer-Verlag, 2003.
4. P. Ekdahl, T. Johansson and W. Meier. A Note on the Self-Shrinking Generator. In *Proceedings of International Symposium on Information Theory*, page 166. IEEE, 2003.
5. H. Englund, M. Hell and T. Johansson. Correlation Attack Using a New Class of Weak Feedback Polynomials. In, B. Roy and W. Meier, editors, *Fast Software Encryption 2004*, volume 3017 of Lecture Notes in Computer Science, pages 127–142. Springer-Verlag 2004.
6. J. D. Golić. Computation of Low-Weight Parity Check Polynomials. *Electronic Letters* 32(21):1981-1982, October 1996.
7. J. D. Golić. Correlation Analysis of the Shrinking Generator. In J. Kilian, editor, *Advances in Cryptology–CRYPTO 2001*, volume 2139 of Lecture Notes in Computer Science, pages 440–457. Springer-Verlag, 2001.
8. J. D. Golić and R. Menicocci. Edit Distance Correlation Attack on the Alternating Step Generator. In, B. S. Kaliski, editor, *Advances in Cryptology–CRYPTO'97*, Volume 1294 of Lecture Notes in Computer Science, pages 499–512. Springer-Verlag 1997.
9. J. D. Golić and R. Menicocci. Edit Probability Correlation Attack on the Alternating Step Generator. In C. Ding, T. Helleseth and H. Niederreiter, editors, *Sequences*

and their Applications–SETA'98. Discrete Mathematics and Theoretical Computer Science, pages 213–227. Springer-Verlag 1999.

10. A. Gouget, H. Sibert. The Bit-Search Generator. In *The State of the Art of Stream Ciphers: Workshop Record, Brugge, Belgium, October 2004*, pages 60–68, 2004.
11. C. G. Günther. Alternating Step Generators Controlled by de Bruijn Sequences. In D. Chaum and W. L. Price, editors, *Advances in Cryptology–EUROCRYPT'87*, pages 5–14. Springer-Verlag, 1988. Lecture Notes in Computer Science Volume 304.
12. T. Johansson. Reduced Complexity Correlation Attacks on Two Clock-Controlled Generators. In K. Otha and D. Pei, editors, *Advances in Cryptology–ASIACRYPT'98*, volume 1541 of Lecture Notes in Computer Science, pages 342–357. Springer-Verlag, 1998.
13. M. Krause. BDD-Based Cryptanalysis of Keystream Generators. In L. R. Knudsen, editor, *Advances in Cryptology–EUROCRYPT 2002*, Volume 304 of Lecture Notes in Computer Science, pages 222–237. Springer-Verlag, 2002.
14. R. J. McEliece. Finite Fields for Computer Scientists and Engineers. Kluwer Academic Publishers. 1987.
15. W. Meier and O. Staffelbach. Fast Correlation Attacks on Stream Ciphers. *Advances in Cryptology-EUROCRYPT'88*, volume 330 of *Journal of Cryptology*, pages 310-314. Springer-Verlag, 1988.
16. W. Meier and O. Staffelbach. The Self-Shrinking Generator. In A. De Santis, editor, *Advances in Cryptology–EUROCRYPT'94*, pages 205–214. Springer-Verlag, 1995. Lecture Notes in Computer Science Volume 950.
17. M. Mihaljevic. A Faster Cryptanalysis of the Self-Shrinking Generator. In J. Pieprzyk and J. Seberry, editors, *First Australasian Conference on Information Security and Privacy ACISP'96*, volume 1172 of Lecture Notes in Computer Science, pages 182–189, Springer-Verlag, 1996.
18. L. Simpson, J. D. Golić and E. Dawson. A Probabilistic Correlation Attack on the Shrinking Generator. In C. Boyd and E. Dawson, editors, *Information security and privacy '98*, volume 1438 of Lecture Notes in Computer Science, pages 147–158, Springer-Verlag, 1998.
19. V. Strassen. Gaussian Elimination is Not Optimal, *Numerische Mathematik*, vol. 13, pages 354-356, 1969.
20. D. Wagner. A Generalized Birthday Problem. In M. Yung, editor, *Advances in Cryptology–CRYPTO 2002*, volume 2442 of Lecture Notes in Computer Science, pages 288–303, Springer-Verlag, 2002.
21. E. Zenner, M. Krause and S. Lucks. Improved Cryptanalysis of the Self-Shrinking Generator. In *ACIPS'2001*, Volume 2119 of Lecture Notes in Computer Science, pages 21–35. Springer-Verlag, 2001.

SMASH - A Cryptographic Hash Function

Lars R. Knudsen

Department of Mathematics, Technical University of Denmark

Abstract. [1]This paper presents a new hash function design, which is different from the popular designs of the MD4-family. Seen in the light of recent attacks on MD4, MD5, SHA-0, SHA-1, and on RIPEMD, there is a need to consider other hash function design strategies. The paper presents also a concrete hash function design named SMASH. One version has a hash code of 256 bits and appears to be at least as fast as SHA-256.

1 Introduction

A *cryptographic hash function* takes as input a binary string of arbitrary length and returns a binary string of a fixed length. Hash functions which satisfy some security properties are widely used in cryptographic applications such as digital signatures, password protection schemes, and conventional message authentication. In the following let $H : \{0,1\}^* \rightarrow \{0,1\}^n$ denote a hash function which returns a string of length n. Most hash functions in use today are so-called *iterated* hash functions, based on iterating a compression function. Examples of iterated hash functions are MD4[19], MD5[20], SHA[13] and RIPEMD-160[7]. For a cryptographic hash function H, one is interested in the complexity of the following attacks[16]:

- **Collision:** find x and x' such that $x' \neq x$ and $H(x) = H(x')$,
- **2nd preimage:** given an x and $y = H(x)$ find $x' \neq x$ such that $H(x') = y$,
- **Preimage:** given $y = H(x)$, find x' such that $H(x') = y$.

Clearly the existence of a 2nd preimage implies the existence of a collision. In a brute-force attack preimages and 2nd preimages can be found after about 2^n applications of H, and a collision can be found after about $2^{n/2}$ applications of H. It is usually the goal in the design of a cryptographic hash function that no attacks perform better than the brute-force attacks.

Often hash functions define an initial value, iv. The hash is then denoted $H(\text{iv}, \cdot)$ to explicitly denote the dependency on the iv. Attacks like the above, but where the attacker is free to choose the value(s) of the iv are called pseudo-attacks. The following assumptions are well-known and widely used in cryptology (where $\oplus$ is the exclusive-or operation).

[1] After the presentation of SMASH at FSE 2005, the proposal was broken[15].

H. Gilbert and H. Handschuh (Eds.): FSE 2005, LNCS 3557, pp. 228–242, 2005.

Assumption 1. *Let $g : \{0,1\}^n \to \{0,1\}^n$ be a randomly chosen mapping. Then the complexities of finding a collision, a 2nd preimage and a preimage are of the order $2^{n/2}$, 2^n, respectively 2^n. Let $f : \{0,1\}^n \to \{0,1\}^n$ be a randomly chosen, bijective mapping. Define the function $h : \{0,1\}^n \to \{0,1\}^n$ by $h(x) = f(x) \oplus x$ for all x. It is assumed that the expected complexity of finding collisions, 2nd preimages and preimages for h is roughly the same as for g.*

Most popular hash functions are based on iterating a compression function, which processes a fixed number of bits. The message to be hashed is split into blocks of a certain length where the last block is possibly padded with extra bits. Let $h : \{0,1\}^n \times \{0,1\}^\ell \to \{0,1\}^n$ denote the compression function, where n and ℓ are positive integers. Let $m = m_0 \mid m_1 \mid \ldots \mid m_t$ be the message to be hashed, where $|m_i| = \ell$ for $0 \leq i \leq t$. Then the hash value is taken as h_t, where

$$h_i = h(h_{i-1}, m_i),$$

for $h_0 = \text{iv}$ an initial, fixed value. The values $\{h_i\}$ are called the chaining variables. If a message m cannot be split into blocks of equal length n, i.e., if the last block consists of less than n bits, then a collision-free padding rule is used. If x and y are two arbitrary different strings, then it must hold that the corresponding padded strings are different.

For iterated hash functions the MD-strengthening (after Merkle [11] and Damgård [6]) is as follows. One fixes the iv of the hash function and appends to a message some additional, fixed number of blocks at the end of the input string containing the length of the original message. Then it can be shown that attacks on the resulting hash function implies a similar attack on the compression function.

There has been much progress in recent years in cryptanalysis of iterated hash functions and attacks have been reported on MD4, MD5, SHA-0, reduced SHA-1 and RIPEMD[2, 18, 21]. For these hash functions and for most other popular iterated hash functions, the compression function takes a rather long message and compresses this together with a shorter chaining variable (containing the internal state) to a new value of the chaining variable. E.g., in SHA-0 and SHA-1 the message is 512 bits and the chaining variable 160 bits. One way of viewing this is, that the compression function defines 2^{160} functions from 512 bits to 160 bits (from message to output), but at the same time it defines 2^{512} functions (bijections) from 160 bits to 160 bits (from chaining variable to output). If just a few of these functions are cryptographically weak, this could give an attacker the open door for an attack.

In this paper we consider compression functions built from one, fixed bijective mapping $f : \{0,1\}^n \to \{0,1\}^n$. A related but different approach is in [17]. In our model this leads to hash functions where the compression functions themselves are not cryptographically strong, thus a result similar to the one by Merkle and Damgård, cf. above, cannot be proved. However, the constructions have other advantages and it is conjectured that the resulting hash functions are not easy to break, despite the fact that the compression functions are "weak".

2 Compression Functions from One Bijective Mapping

Our approach is to build an iterated hash function $H : \{0,1\}^* \to \{0,1\}^n$ from one fixed, bijective mapping $f : \{0,1\}^n \to \{0,1\}^n$. If this bijection is chosen carefully, the goal or hope is that such a hashing construction is hard to attack. Such constructions could potentially be built from using a block cipher with a fixed value of the key.

2.1 Motivation for Design

Consider iterated hash functions with compression functions $h : \{0,1\}^n \times \{0,1\}^n \to \{0,1\}^n$ for which the computation of chaining variables is as follows: $h_i = h(A, B) = f(A) \oplus B$. Here $f : \{0,1\}^n \to \{0,1\}^n$ is a bijective mapping and the inverse of f is assumed to be as easy to compute as f itself. A and B are variables which depend on the chaining variable h_{i-1} and on the message block m_i. Ideally we would like to have an efficient (easy-to-compute) transformation $e(h_{i-1}, m_i) = (A, B)$. We do not want e to cause collisions so we require that it is invertible. Since we want e to be an invertible function (very) easy to compute, we shall also assume that the inverse of e is easy to compute.

For such compression functions it is possible to invert also h. Given an h_i, simply choose a random value of B, compute $A = f^{-1}(B \oplus h_i)$, then by inverting e, find (h_{i-1}, m_i) which hash to h_i. We shall assume that the complexity of one application of e is small compared to one application of f and thus that inverting h takes roughly time one, where one unit is one application of f (or its inverse). It follows that it is easy to find both collisions and preimages for the compression function. Next we examine what this means for similar attacks on the hash functions (where a fixed value of h_0 is assumed) induced by these compression functions.

Inverting h as above enables a (2nd) preimage attack on H by a meet-in-the-middle approach[10] of complexity about $2^{n/2+1}$, i.e., compute the values of "h_{i-1}" for $2^{n/2}$ messages (of each $i-1$ blocks) and store them. For a fixed value of h_i choose $2^{n/2}$ random values of A, B (as above), which yield $2^{n/2}$ "random" values of "h_{i-1}". The birthday paradox gives the (2nd) preimage. If f is a truly randomly chosen bijection on n bits (which is the aim for it to be) then this (2nd) preimage attack is always possible on the constructions we are considering. So the best we can do regarding (2nd) preimages is try to make sure that the attacker does not have full control over the message blocks when inverting h, in which case such preimages may be of lesser use in practice. Thus, we want to avoid that given (any) h_{i-1}, m_i (and thereby h_i) and m'_i, it is easy to find h'_{i-1} such that $(h_{i-1}, m_i) \neq (h'_{i-1}, m'_i)$ and $h_i = h'_i$, since in this case one can find preimages for the hash function for meaningful messages also in time roughly $2^{n/2}$.

This meet-in-the-middle attack is "irrelevant" regarding collisions, since the complexity of a brute-force attack is $2^{n/2}$ regardless of the nature of the compression function. For collisions it is important that when inverting h the attacker does not have full control over the chaining variable(s) h_{i-1}. If given (any) h_{i-1}, h'_{i-1}, it is easy to find m_i, m'_i such that $(h_{i-1}, m_i) \neq (h'_{i-1}, m'_i)$ and

Table 1. Six compression functions

Scheme	Attack I	Attack II
$h_i = f(h_{i-1}) \oplus m_i$	easy	easy
$h_i = f(m_i) \oplus h_{i-1}$	easy	easy
$h_i = f(h_{i-1}) \oplus m_i \oplus h_{i-1}$	easy	?
$h_i = f(m_i) \oplus m_i \oplus h_{i-1}$	?	easy
$h_i = f(h_{i-1} \oplus m_i) \oplus h_{i-1}$	easy	?
$h_i = f(h_{i-1} \oplus m_i) \oplus m_i$	?	easy

$h_i = h'_i$ then one can find a collision easily also for the hash function. Simply choose two messages $m = m_1, \ldots, m_{i-1}$ and $m' = m'_1, \ldots, m'_{i-1}$ (e.g., with $h_{i-1} \neq h'_{i-1}$), where $i \geq 2$, then the two i-block messages $M = m \mid m_i$ and $M' = m' \mid m'_i$ yield a collision for the hash function.

The above is the motivation for examining the compression functions with respect to the following two attacks:

- I: Given h_{i-1}, h'_{i-1} find m_i, m'_i such that $(h_{i-1}, m_i) \neq (h'_{i-1}, m'_i)$ and $h_i = h'_i$.
- II: Given h_{i-1}, m_i and m'_i, find h'_{i-1} such that $(h_{i-1}, m_i) \neq (h'_{i-1}, m'_i)$ and $h_i = h'_i$.

Consider first the simple e-functions where $A, B \in \{m_i, h_{i-1}, m_i \oplus h_{i-1}\}$. With the requirements for e above, this yields six possibilities for the compression function, see the first column in Table 1. It follows that in all six cases either the first or the second attack is easy to implement, in some cases both. So one needs to consider more complex e-functions to achieve better resistance against the two attacks. There may be many possible ways to build such functions; we believe to have found a simple one.

First we note that there is a natural one-to-one correspondence between bit vectors of length s and elements in the finite field of 2^s elements. We introduce "multiplication by θ" as follows.

Definition 1. *Consider $a \in GF(2)^s$. Let θ be an element of $GF(2^s)$ such that $\theta \notin \{0,1\}$. Define the multiplication of a by θ as follows. View a as an element of $GF(2^s)$, compute $a\theta$ in $GF(2^s)$, then view the result as an s-bit vector.*

Let $f : \{0,1\}^n \to \{0,1\}^n$ be a bijective mapping and let $\oplus$ denote the exclusive-or operation. Consider the compression function $h : \{0,1\}^n \times \{0,1\}^n \to \{0,1\}^n$:

$$h(h_{i-1}, m_i) = h_i = f(h_{i-1} \oplus m_i) \oplus h_{i-1} \oplus \theta m_i, \tag{1}$$

where θ is as in Definition 1. Multiplication with certain values of θ can be done very efficiently as we shall demonstrate later. Consider Attacks I and II from before.

Attack I: Given h_{i-1} and h'_{i-1} the attacker faces the task of finding m_i and m'_i such that

$$f(h_{i-1} \oplus m_i) \oplus h_{i-1} \oplus \theta m_i = f(h'_{i-1} \oplus m'_i) \oplus h'_{i-1} \oplus \theta m'_i. \tag{2}$$

Or in other words, with $h_{i-1} \oplus h'_{i-1} = \alpha$ and $m_i \oplus m'_i = \beta$ one needs to find two inputs to f of difference $\alpha \oplus \beta$ which yield outputs of difference $\alpha \oplus \theta\beta$ for a fixed value of θ. But if f is "as good as" a randomly chosen mapping, the attacker has no control over the relation between the outputs for two different inputs to f, and he has no better approach than the birthday attack. Note that with $m_i \oplus m'_i = h_{i-1} \oplus h'_{i-1} = \alpha \neq 0$ one never has a collision for h, since in this case the difference in the outputs of f is zero and the difference in the outputs of h is $(\theta + 1)\alpha \neq 0$.

Attack II: For fixed values of h_{i-1}, m_i and m'_i, the attacker faces the task of finding h'_{i-1} such that Eq. 2 is satisfied. But in this case (1) has the form of $g(h_{i-1}) \oplus h_{i-1} \oplus c_1$, where $g(x) = f(x \oplus c_2)$ and where c_1, c_2 are constants. Thus, under Assumption 1 (with sufficiently large n) attacks using a fixed value of m_i seem to be hard to mount.

Although the two attacks above do not seem to be easy to do for the proposed compression function, it is clear that there are properties of it which are not typical for compression functions. These are already discussed above but we highlight them here again.

Inversion: (1) can be inverted. Given h_i, choose an arbitrary value of a, compute $b = f^{-1}(h_i \oplus a) = h_{i-1} \oplus m_i$, then solve for h_{i-1} and m_i. With θ as in Definition 1 this can be accomplished by solving

$$(a \quad b) = (h_{i-1} \quad m_i) \begin{pmatrix} 1 & 1 \\ \theta & 1 \end{pmatrix}$$

which always succeeds, since $\theta \neq 1$.

Forward prediction: Let h_{i-1} and h'_{i-1} be two inputs to (1) where $\alpha = h_{i-1} \oplus h'_{i-1}$. Choose a value for m_i and compute $m'_i = m_i \oplus \alpha$. Then

$$\begin{aligned} h_i \oplus h'_i &= f(h_{i-1} \oplus m_i) \oplus h_{i-1} \oplus \theta m_i \oplus f(h'_{i-1} \oplus m'_i) \oplus h'_{i-1} \oplus \theta m'_i \\ &= \theta\alpha \oplus \alpha. \end{aligned}$$

The following is a list of potential problems of hash functions based on the proposed compression function.

1. Collisions for the compression function.
2. Pseudo (2nd) preimages for the hash function.
3. (2nd) preimages for the hash function in time roughly $2^{n/2}$.
4. Non-random, predictable properties for the compression function.

Ad 1: It is easy to find collisions for the compression function, so it is not possible to prove a result similar to that of Merkle and Damgård, cf., the introduction. However the simple approach, presented above, does not give the attacker any control over the values of h_{i-1} and h'_{i-1} and it does not appear to be directly useful in attempts to find a collision for the hash function (with a fixed iv).
Ad 2: Since h can be inverted it is trivial to find a (2nd) message and an $\widetilde{\text{iv}}$ which is hashed to a given hash value. However, this approach given h_i does not give

an attacker control over the value of h_{i-1} and this approach will not directly lead to (2nd) preimages for the hash function (with a fixed iv). Moreover the attacker has no control over m_i.
Ad 3: Let there be given a hash value and an iv. Then since the compression function is easily inverted, it was shown that (2nd) preimages can be found in time roughly $2^{n/2}$ using a meet-in-the-middle attack. One can argue that this is a weakness, however since for any hash function of size n there is a collision attack of complexity $2^{n/2}$ based on the birthday paradox, one can also argue that if this level of security is too low, then a hash function with a larger hash result should be used anyway.
Ad 4: Consider the "Forward prediction" property above with some $\alpha \neq 0$. It follows that given the difference in two chaining variables one can find two message blocks such that the values of the corresponding outputs of the compression function is $\gamma = \alpha(\theta + 1)$. This approach (alone) will never lead to a collision since $\gamma \neq 0$. Note that the approach extends to longer messages. E.g., assume that for a pair of messages one has $h_{i-1} \oplus h'_{i-1} = \alpha$. Then with $m_{i+s} \oplus m'_{i+s} = h_{i-1+s} \oplus h'_{i-1+s}$ for $s = 0, \ldots, t$ one gets that $h_{i+s} \oplus h'_{i+s} = \alpha(\theta+1)^{s+1}$. Note that although $\alpha(\theta+1)^{s+1} \neq 0$ for any s, one can compute a long list of (intermediate) hash values without evaluating h. Also there are applications of hash functions where it is assumed that the output is "pseudorandom" (e.g., HMAC[4]).

2.2 The Proposed Hash Function

To avoid some of the problems of the compression function as listed above, we add some well-known elements in the design of the hash function. Let m be the message to the hashed and assume that it includes padding bits and the message length. Let $m = m_0, m_1, \ldots, m_t$, where each m_i is of n bits. Let iv be initial value to the hash function, compute

$$h_0 = f(\text{iv}) \oplus \text{iv} \tag{3}$$

$$h_i = f(h_{i-1} \oplus m_i) \oplus h_{i-1} \oplus \theta m_i \qquad \text{for } i = 1, \ldots, t \tag{4}$$

$$h_{t+1} = f(h_t) \oplus h_t \tag{5}$$

As seen, we have introduced two applications of a secure compression function based on f, namely one which from the user-selected iv computes h_0 in a secure fashion, and one which from h_t computes the final hash result in a secure fashion.

It is conjectured that this hash function protects against pseudo-attacks, since the attacker has no control over h_0. Moreover because of the final application of a secure compression function it is not possible to predict the final hash value (using the approach of item 4 above). Also, the inclusion of the message length in the padding bits complicates the utilization of long message attacks, e.g., using the approach of item 4 above, see also [16, 9]. Finally, the construction complicates preimage attacks, since the hash results are outputs of a (conjectured) one-way function.

It is claimed that if f is (as good as) a randomly chosen bijective mapping on n bits, then the complexity of the best approach for a preimage, 2nd preimage or a collision attack on the proposed hash function is at least $2^{n/2}$.

2.3 $\theta = 0$ and $\theta = 1$

Consider the compression function above with $\theta = 0$. Then

$$h_i = f(h_{i-1} \oplus m_i) \oplus h_{i-1} \qquad \text{for } i = 1, \ldots, t$$

This variant of the compression function is easy to break. Choose two different messages $m_1, \ldots, m_{i-1}$ and $m'_1, \ldots, m'_{i-1}$ such that $h_{i-1} \neq h'_{i-1}$. Choose a value of $h_i = h'_i$, and compute $m_i = f^{-1}(h_{i-1} \oplus h_i) \oplus h_{i-1}$ and $m'_i = f^{-1}(h'_{i-1} \oplus h'_i) \oplus h'_{i-1}$. Then there is a collision for the messages $m_1, \ldots, m_i$ and $m'_1, \ldots, m'_i$. Therefore, the proposed hash function should not be used with $\theta = 0$. With $\theta = 1$ it follows that the pairs (h_{i-1}, m_i) and (h'_{i-1}, m'_i) collide when $h_{i-1} \oplus m_i = h'_{i-1} \oplus m'_i$.

3 SMASH

In this section a concrete hash function proposal is presented which has been named SMASH.[2] The version presented here has a 256-bit output, hence we refer to it as SMASH-256. Another version with a 512-bit output is named SMASH-512. These are therefore candidate alternatives to SHA-256 and SHA-512 [14]. The designs of SMASH-256 and SMASH-512 are very similar but where the former works on 32-bit words and the latter on 64-bit words. We focus on SMASH-256 next, the details of SMASH-512 is in an appendix.

3.1 SMASH-256

SMASH-256 is designed particularly for implementation on machines using a 32-bit architecture. A 256-bit string y is then represented by eight 32-bit words, $y = y_7, \ldots, y_0$. We shall refer to y_7 and y_0 as the most significant respectively least significant words.

SMASH-256 takes a bit string of length less than 2^{128} and produces a 256-bit hash result. The outline of the method is as follows. Let m be a u-bit message. Apply a padding rule to m (see later), split the result into blocks of 256 bits, $m_1, m_2, \ldots, m_t$ and do the following:

$$h_0 = g_1(\text{iv}) = f(\text{iv}) \oplus \text{iv} \tag{6}$$
$$h_i = h(h_{i-1}, m_i) = f(h_{i-1} \oplus m_i) \oplus h_{i-1} \oplus \theta m_i \qquad \text{for } i = 1, \ldots, t \tag{7}$$
$$h_{t+1} = g_2(h_t) = f(h_t) \oplus h_t, \tag{8}$$

where iv is an initial value. The hash result of a message m is then defined as $\text{Hash}(\text{iv}, m) = h_{t+1}$. The subfunctions g_1, g_2, and f all take a 256-bit input and produce a 256-bit output and h takes a 512-bit input and produces a 256-bit output. g_1 is called the input transformation, g_2 the output transformation, h is

[2] **smash** /smaesh/: to break (something) into small pieces by hitting, throwing, or dropping, often noisily.

called the compression function and f the "core" function, which is a bijective mapping. g_1 and g_2 are of the same form, constructed under Assumption 1.

As a target value of the iv use the all zero 256-bit string.

Padding Rule. Let m be a t-bit message for $t > 0$. The padding rule is as follows: append a '1'-bit to m, then append u '0'-bits, where $u \geq 0$ is the minimum integer value satisfying

$$(t+1) + u \equiv 128 \bmod 256.$$

Append to this string a 128-bit string representing the binary value of t.

The Compression Function, •. The function takes two arguments of each 256 bits, h_{i-1} and m_i. The two arguments are exclusive-ored and the result evaluated through f. The output of f is then exclusive-ored to h_{i-1} and to θm_i.

"Multiplication" by •. This section outlines one method to implement the multiplication of a particular value of θ. As already mentioned there is a natural one-to-one correspondence between bit vectors of length 256 with elements in the finite field $GF(2^{256})$. Consider the representation of the finite field defined via the irreducible polynomial $q(\theta) = \theta^{256} \oplus \theta^{16} \oplus \theta^3 \oplus \theta \oplus 1$ over $GF(2)$. Then multiplication of a 256-bit vector y by θ can be implemented with a linear shift by one position plus an exclusive-or. Let $z = \theta y$, then

$$z = \begin{cases} \text{ShiftLeft}(y, 1), & \text{if } \text{msb}(y) = 0 \\ \text{ShiftLeft}(y, 1) \oplus \text{poly}_1, & \text{if } \text{msb}(y) = 1 \end{cases},$$

where poly_1 is the 256-bit representation of the element $\theta^{16} \oplus \theta^3 \oplus \theta \oplus 1$, that is, eight words (of each 32 bits) where the seven most significant ones have values zero and where the least significant word is $0001000b_x$ in hexadecimal notation. In a 32-bit architecture the multiplication can be implemented as follows. Let $y = (y_7, y_6, y_5, y_4, y_3, y_2, y_1, y_0)$, where $|y_i| = 32$, then $\theta y = z = (z_7, z_6, z_5, z_4, z_3, z_2, z_1, z_0)$, where for $i = 1, \ldots, 7$

$$z_i = \begin{cases} \text{ShiftLeft}(y_i, 1), & \text{if } \text{msb}(y_{i-1}) = 0 \\ \text{ShiftLeft}(y_i, 1) \oplus 1, & \text{if } \text{msb}(y_{i-1}) = 1, \end{cases}$$

and where

$$z_0 = \begin{cases} \text{ShiftLeft}(y_0, 1), & \text{if } \text{msb}(y_7) = 0 \\ \text{ShiftLeft}(y_0, 1) \oplus 0001000b_x, & \text{if } \text{msb}(y_7) = 1. \end{cases}$$

The Core Function, •. The core function in SMASH-256 consists of several rounds, some called H-rounds and some called L-rounds, see Figure 1. There are three different H-rounds. In each of them a 4×4 bijective S-box is used together with some linear diffusion functions. The S-box is used in "bit-slice" mode, which

$$H_1 \circ H_3 \circ H_2 \circ L \circ H_1 \circ H_2 \circ H_3 \circ L \circ H_2 \circ H_1 \circ H_3 \circ L \circ H_3 \circ H_2 \circ H_1(\cdot)$$

Fig. 1. SMASH-256: Outline of f, the core function

$$
\begin{array}{|l|}
\hline
(a_7, a_6, a_5, a_4) = \text{Sbs}(a_7, a_6, a_5, a_4) \\
a_{i+4} = a_{i+4} + a_i^{<<r_i} \quad \text{for } i = 0, \ldots, 3 \\
(a_3, a_2, a_1, a_0) = \text{Sbs}(a_3, a_2, a_1, a_0) \\
a_i = a_i + a_{i+4}^{<<r_{i+4}} \quad \text{for } i = 0, \ldots, 3 \\
(a_7, a_6, a_5, a_4) = \text{Sbs}(a_7, a_6, a_5, a_4) \\
a_{i+4} = a_{i+4} + a_i^{<<r_{i+8}} \quad \text{for } i = 0, \ldots, 3 \\
(a_3, a_2, a_1, a_0) = \text{Sbs}(a_3, a_2, a_1, a_0) \\
a_i = a_i + a_{i+4}^{<<r_{i+12}} \quad \text{for } i = 0, \ldots, 3, \\
\hline
\end{array}
$$

Fig. 2. SMASH-256: Outline of an H-round

was used also in the block cipher designs Three-way[5] and Serpent[3]. In the following let $\mathbf{a} = (a_7, a_6, a_5, a_4, a_3, a_2, a_1, a_0)$ be the 256-bit input to an H-round, where each a_i is of 32 bits. The outline of all H-rounds is the same, see Figure 2, where $a^{<<r}$ is the word a rotated r positions to the left. $(x, y, z, w) = \text{Sbs}(x, y, z, w)$ means that for all $i = 0, \ldots, 31$, the four ith bits from x, y, z, w are evaluated through a 4-bit bijective S-box (Sbs is short for S-box bit-slice) using the convention that the bit from x is the most significant bit. In one H-round the same particular S-box is used in all four bitslice applications. The differences between H_1, H_2, and H_3 are in the S-box used and in the rotations used. For H_i the S-box used is S_i, and the rotations are R_i, see Figures 3 and 4.

The L-round (there is only one) is defined as in Figure 5, where ShiftLeft$(x, 8)$ is the 32-bit quantity x shifted eight positions to the left and ShiftRight$(x, 8)$ is x shifted eight positions to the right.

3.2 Some Ideas Behind the Design

In this section some further details of the design of the core function of SMASH-256 are explained. Let $\mathbf{a} = (a_7, a_6, a_5, a_4, a_3, a_2, a_1, a_0)$ be a 256-bit variable, where the a_is are of 32 bits each. $\mathbf{a}$ represents the internal state of the compres-

	s_0	s_1	s_2	s_3	s_4	s_5	s_6	s_7	s_8	s_9	s_{10}	s_{11}	s_{12}	s_{13}	s_{14}	s_{15}
S_1 :	6	13	12	7	15	1	3	10	8	11	5	0	2	4	14	9
S_2 :	1	11	6	0	14	13	5	10	12	2	9	7	3	8	15	4
S_3 :	4	2	9	12	8	1	14	7	15	5	0	11	6	10	3	13

Fig. 3. The SMASH-256 S-boxes

	r_0	r_1	r_2	r_3	r_4	r_5	r_6	r_7	r_8	r_9	r_{10}	r_{11}	r_{12}	r_{13}	r_{14}	r_{15}
R_1 :	19	18	17	7	1	7	26	20	0	16	20	5	28	2	20	4
R_2 :	22	29	12	4	18	2	13	29	26	20	16	29	18	4	10	9
R_3 :	4	21	19	5	24	20	12	16	14	30	3	4	23	15	13	12

Fig. 4. The SMASH-256 rotations

$$\begin{aligned} a_3 &= a_3 \oplus \text{ShiftLeft}(a_7, 8) \\ a_2 &= a_2 \oplus \text{ShiftLeft}(a_6, 8) \\ a_1 &= a_1 \oplus \text{ShiftRight}(a_5, 8) \\ a_0 &= a_0 \oplus \text{ShiftRight}(a_4, 8) \end{aligned}$$

Fig. 5. SMASH-256: Outline of an L-round

sion function. We concentrate first of the design of the H-rounds and L-rounds in the core function. Arrange the 256 bits of the internal state in a matrix as follows.

a_7	a_6	a_5	a_4
a_3	a_2	a_1	a_0

Consider Figure 2. First a bitslice S-box is applied to the top row. Rotated versions of the words in the top row are then added to words in the second row. Then a bitslice S-box is applied to the second row, and rotated versions of words of this result added to words in the top row. This is repeated once, such that in total in one H-round, four bitslice S-box applications and four diffusion layers are performed. The rotations in the H-rounds have been chosen such that each of the 256 bits in the internal memory is mixed with all other bits as quickly as possible (relative to this design!). It is clear that with rotations and modular additions all bits depend on all bits after a few steps. However, the H-rounds were designed such that the dependencies between the bits are stronger than just via the carry bits of the addition.

The diffusion layer. For the purpose of studying optimum diffusion functions replace all additions in the H-rounds by exclusive-ors. Also, it shall be assumed that each of the four output bits of an S-box depend on all four input bits. Because of the bitslice method and the assumption on the S-box (that all output bits depend on all input bits), it is convenient to consider only the 32 bit positions in words of the top row and 32 bit positions in the words of the second row when discussing dependencies of bits. Consider one bit in the top row of the input to an H-round. After the first bitslice S-box, this bit affects still only one bit position (of 32 in total) in the top row. After the diffusion layer, the bit affects in the best cases four bit positions in the second row (if the four rotations r_1, r_2, r_3, r_4 are all different). After the bitslice of the second row, the bit still only affects four bit positions in the second row, however after the next diffusion layer, the bit affects up to 17 positions in the top row, where we have counted also the initial single position from the beginning. After the subsequent bitslice of the top row together with the diffusion layer, the bit affects in the best cases all 32 positions in the second row, but still only 17 positions in the top row. The fourth and last bitslice and diffusion layer of the H-round ensures that the initial bit affects in the best cases all 32 positions of both the top and the second row.

Consider next one bit in the second row of the input to an H-round. After the first bitslice S-box and diffusion layer, this bit affects still only one bit position in the second row and zero in the top row. After the next bitslice S-box and diffusion layer, the bit affects (in the best case) four bit positions in the top row (if the four rotations r_5, r_6, r_7, r_8 are all different) and one bit position in the second row. After the subsequent bitslice of the top row together with the diffusion layer, the bit affects 17 positions in the second row in the best case, but still only 4 positions in the top row. After the fourth and last bitslice and diffusion layer of the H-round the initial bit affects in the best cases all 32 positions of the top row and 17 positions of the second row. It is a simple matter to implement a search algorithm which finds values of $r_0, \ldots, r_{15}$ such that the diffusion is optimum as outlined here. All H-rounds in SMASH-256 are designed according to this strategy.

The L-round. Consider variants of SMASH-256 where the modular additions are replaced by exclusive-ors. Let a be the 256-bit input to the core function f. Then the following property holds for the H-rounds:

$$H(a)^{<<c} = H(a^{<<c}).$$

This property does not always hold when modular additions are used in the H-rounds, that is,

$$(a + b)^c = a^{<<c} + b^{<<c}, \qquad (9)$$

does not always hold, since there is no carry bit in the least significant bit of a modular addition and since the carry in the most significant bit of a modular addition is thrown away. However, empirical results show that equality holds in (9) with a probability of about 1/4. Therefore we introduce the L-round, which uses the shift operation. The shift operation is not invariant under rotations. We believe that the L-round together with modular additions prevent exploitable properties like (9) for the core functions in SMASH-256.

The S-boxes. The S-boxes are chosen as in the design of the block cipher Serpent [3]. These are 4-bit permutations with the following properties:

- each differential characteristic has a probability of at most 1/4, and a one-bit input difference will never lead to a one-bit output difference;
- each linear characteristic has a probability in the range $1/2 \pm 1/4$, and a linear relation between one single bit in the input and one single bit in the output has a probability in the range $1/2 \pm 1/8$;

The three S-boxes used in SMASH-256 are derived as linear variants of the S-boxes S_0, S_2, and S_4 from Serpent [3]. An implementation of SMASH-256 [1] uses the bitslice implementations of the Serpent S-boxes from [12], which were modified slightly to reduce the number of variables used in the program.

4 Short Analysis of SMASH

There is very little theory in the design of cryptographic hash functions today and it is very difficult to prove much about the security of these. Therefore it is not possible to give a precise analysis of cryptographic hash functions like SMASH. In this section we consider a few general attacks and (try to) argue that they are unlikely to succeed.

SMASH-256 consists of a total of 48 S-box layers. Differential characteristics with one active S-box per S-box layer are not possible due to the above design criteria. A very crude estimate is that there are at least three active S-boxes per every two S-box layers. Since the most likely differential characteristic for one layer has probability 2^{-2} this leads to a complexity of $((2^{-6})^{24})^{-1} = 2^{144}$ for a differential characteristic for the function f. A linear characteristic for one S-box layer has a bias of at most 2^{-2}. An analogue crude estimate for linear cryptanalysis gives a complexity of 2^{144}. Since the aim for SMASH-256 is a security level of 2^{128} it is believed that (traditional) approaches in differential and linear cryptanalysis are unlikely to be very efficient when applied to SMASH-256.

Dobbertin's attacks on MD4 and MD5 as well as the recent attacks[2, 18] on SHA-0 and SHA-1 exploit that the attacker has much freedom to influence many of the individual steps of the respective compression functions, namely through the message blocks. SMASH is different from the SHA-designs in that the message is input at the beginning (at step 0) only, and it seems this gives an attacker much less room to play. This is not a proof that these attacks will not work and the readers are invited to apply them (or variants of them) to SMASH.

5 Performance

An implementation of SMASH-256 [1] shows a performance of about 30 cycles per byte in a pure C-implementation. For comparison the implementation of SHA-256 by B. Gladman [8] produced a speed of 40 cycles per byte on the same platform using the same compiler. Speeds of about 21 cycles per byte for SHA-256 have been reported in an assembler implementation. It is expected that an assembler implementation of SMASH-256 would likewise increase the performance.

6 Finishing Remarks

We have presented a new approach in hash function design together with a concrete proposal for a hash function. The proposal deviates from the most popular hash function designs in use today, in that only one, fixed and bijective, (supposedly strong) cryptographic mapping is used. After the presentation at FSE 2005 SMASH was broken. In [15] it is shown that it is possible to find messages with 256 blocks which collide when compressed through SMASH-256. There appears to be a similar attack on SMASH-512 for messages of 512 blocks. The attack makes use of the "forward prediction" together with some differential techniques. It appears that there are several ways to modify SMASH to thwart the new attacks. One is to use different f functions for every iteration[15]. Another is to use a secure compression function not only in the first and last iteration (see (3)-(5)) but after the processing of every n blocks of the message for, say, $n = 8$ or $n = 16$.

One interesting avenue for further research is compression function designs using two (or more) fixed, bijective mappings.

Acknowledgments

The author would like to thank Martin Clausen for many discussions regarding this paper and for implementing SMASH-256.

References

1. Martin Clausen. An implementation of SMASH-256. Private communications.
2. E. Biham, R. Chen. Near-Collisions of SHA-0. In Matt Franklin, editor, *Advances in Cryptology: CRYPTO'2004, Lecture Notes in Computer Science 3152*. Springer Verlag, 2004.
3. R.J. Anderson, E. Biham, and L.R. Knudsen. SERPENT - a 128-bit block cipher. A candidate for the Advanced Encryption Standard. Documentation available at `http://www.ramkilde.com/serpent`.
4. M. Bellare, R. Canetti, and H. Krawczyk. Keying hash functions for message authentication. In Neal Koblitz, editor, *Advances in Cryptology: CRYPTO'96, Lecture Notes in Computer Science 1109*, pages 1–15. Springer Verlag, 1996.
5. J. Daemen. A new approach to block cipher design. In R. Anderson, editor, *Fast Software Encryption - Proc. Cambridge Security Workshop, Cambridge, U.K., Lecture Notes in Computer Science 809*, pages 18–32. Springer Verlag, 1994.
6. I.B. Damgård. A design principle for hash functions. In G. Brassard, editor, *Advances in Cryptology: CRYPTO'89, Lecture Notes in Computer Science 435*, pages 416–427. Springer Verlag, 1990.
7. H. Dobbertin, A. Bosselaers, and B. Preneel. RIPEMD-160: A strenghened version of RIPEMD. In Gollmann D., editor, *Fast Software Encryption, Third International Workshop, Cambridge, UK, February 1996, Lecture Notes in Computer Science 1039*, pages 71–82. Springer Verlag, 1996.
8. Brian Gladman. Available at `http://fp.gladman.plus.com/cryptography_technology/sha/index.htm`
9. X. Lai. On the design and security of block ciphers. In J.L. Massey, editor, *ETH Series in Information Processing*, volume 1. Hartung-Gorre Verlag, Konstanz, 1992.
10. X. Lai and J.L. Massey. Hash functions based on block ciphers. In *Advances in Cryptology - EUROCRYPT'92, Lecture Notes in Computer Science 658*, pages 55–70. Springer Verlag, 1993.
11. R. Merkle. One way hash functions and DES. In G. Brassard, editor, *Advances in Cryptology - CRYPTO'89, Lecture Notes in Computer Science 435*, pages 428–446. Springer Verlag, 1990.

12. D.A. Osvik. Speeding Up Serpent, *Third Advanced Encryption Standard Candidate Conference, April 13–14, 2000, New York, USA*, pp. 317-329, NIST, 2000.
13. NIST. Secure hash standard. FIPS 180-1, US Department of Commerce, Washington D.C., April 1995.
14. NIST. Secure hash standard. FIPS 180-2, US Department of Commerce, Washington D.C., August 2002.
15. N. Pramstaller, C. Rechberger, and V. Rijmen. *Smashing SMASH.* The IACR Eprint Archive, 2005/081.
16. B. Preneel. *Analysis and Design of Cryptographic Hash Functions.* PhD thesis, Katholieke Universiteit Leuven, January 1993.
17. B. Preneel, R. Govaerts, and J. Vandewalle. On the power of memory in the design of collision resistant hash functions. In J. Seberry and Y. Zheng, editors, *Advances in Cryptology: AusCrypt 92, Lecture Notes in Computer Science 718*, pages 105–121. Springer Verlag, 1993.
18. V. Rijmen. Update on SHA-1. Accepted for presentation at CT-RSA'2005.
19. R.L. Rivest. The MD4 message digest algorithm. In S. Vanstone, editor, *Advances in Cryptology - CRYPTO'90, Lecture Notes in Computer Science 537*, pages 303–311. Springer Verlag, 1991.
20. R.L. Rivest. The MD5 message-digest algorithm. Request for Comments (RFC) 1321, Internet Activities Board, Internet Privacy Task Force, April 1992.
21. X. Wang, D. Feng, X. Lai, H. Yu. Collisions for Hash Functions MD4, MD5, HAVAL-128 and RIPEMD. Cryptology ePrint Archive, Report 2004/199. Available at `eprint.iacr.org/2004/199`.

SMASH-512

SMASH-512 takes a bit string of length less than 2^{256} and produces a 512-bit hash result. The outline of the method is as follows. Let m be a u-bit message. Apply a padding rule to m (see later), split the result into blocks of 512 bits, $m_1, m_2, \ldots, m_t$ and do as in (6), (7), and (8). The hash result of a message m is the defined as $\text{Hash}(\text{iv}, m) = h_{t+1}$. The subfunctions g_1, g_2, and f all take a 512-bit input and produce a 512-bit output and h takes a 1024-bit input and produces a 512-bit output. As a target value of the iv use the all zero 512-bit string. The design is very similar to that of SMASH-256, the main difference is that the latter is designed for 32-bit architectures whereas SMASH-512 is for best suited for 64-bit architectures.

Consider the representation of the finite field $GF(2^{512})$ defined via the irreducible polynomial $q(\theta) = \theta^{512} \oplus \theta^8 \oplus \theta^5 \oplus \theta^2 \oplus 1$ over $GF(2)$. Then multiplication by θ can be defined by a linear shift by one position and an exclusive-or. In a 64-bit architecture the multiplication can be implemented as follows. Let $y = (y_7, y_6, y_5, y_4, y_3, y_2, y_1, y_0)$, where $|y_i| = 64$, then $\theta y = z = (z_7, z_6, z_5, z_4, z_3, z_2, z_1, z_0)$, where for $i = 1, \ldots, 7$

$$z_i = \begin{cases} \text{ShiftLeft}(y_i, 1), & \text{if } \text{msb}(y_{i-1}) = 0 \\ \text{ShiftLeft}(y_i, 1) \oplus 1, & \text{if } \text{msb}(y_{i-1}) = 1, \end{cases}$$

and where

$$z_0 = \begin{cases} \text{ShiftLeft}(y_0, 1), & \text{if } \text{msb}(y_7) = 0 \\ \text{ShiftLeft}(y_0, 1) \oplus 0000000000000125_x, & \text{if } \text{msb}(y_7) = 1. \end{cases}$$

$$H_{1,3,2} \circ L \circ H_{2,3,1} \circ L \circ H_{1,2,3} \circ L \circ H_{2,1,3} \circ L \circ H_{3,2,1} \circ L \circ H_{3,1,2}(\cdot)$$

Fig. 6. SMASH-512: Outline of f, the core function, where $H_{a,b,c}$ denotes $H_a \circ H_b \circ H_c$

	r_0	r_1	r_2	r_3	r_4	r_5	r_6	r_7	r_8	r_9	r_{10}	r_{11}	r_{12}	r_{13}	r_{14}	r_{15}
R_1 :	56	40	24	8	55	48	61	14	37	13	25	17	61	29	13	45
R_2 :	24	8	48	32	12	62	57	35	1	45	33	13	4	60	12	20
R_3 :	8	56	48	0	22	21	7	44	34	30	62	2	58	50	34	10

Fig. 7. The SMASH-512 rotations

The core function in SMASH-512 consists of a mix of 18 H-rounds and five L-rounds, see Figure 6. The differences between the H-rounds of SMASH-256 and of SMASH-512 are in the rotations used. The outline is the same as for SMASH-256, see Figure 2, as are the S-boxes. The rotations for SMASH-512 are in Figure 7. The definition of the L-round is the same as the one for SMASH-256, see Figure 5.

Padding rule. Let m be a t-bit message for $t > 0$. The padding rule is as follows: append a '1'-bit to m, then append u '0'-bits, where $u \geq 0$ is the minimum integer value satisfying

$$(t + 1) + u \equiv 256 \bmod 512.$$

Append to this string a 256-bit string representing the binary value of t.

Security Analysis of a 2• 3-Rate Double Length Compression Function in the Black-Box Model

Mridul Nandi[1], Wonil Lee[2], Kouichi Sakurai[2], and Sangjin Lee[3]

[1] Applied Statistics Unit,
Indian Statistical Institute, Kolkata, India
mridul_r@isical.ac.in
[2] Faculty of Information Science and Electrical Engineering,
Kyushu University, Fukuoka, Japan
wonil@itslab.csce.kyushu-u.ac.jp
sakurai@csce.kyushu-u.ac.jp
[3] Center for Information Security Technologies (CIST),
Korea University, Seoul, Korea
sangjin@cist.korea.ac.kr

Abstract. In this paper, we propose a 2/3-rate double length compression function and study its security in the black-box model. We prove that to get a collision attack for the compression function requires $\Omega(2^{2n/3})$ queries, where n is the single length output size. Thus, it has better security than a most secure single length compression function. This construction is more efficient than the construction given in [8]. Also the three computations of underlying compression functions can be done in parallel. The proof idea uses a concept of computable message which can be helpful to study security of other constructions like [8], [14], [16] etc.

1 Introduction

A hash function is a function from an arbitrary domain to a fixed domain. Hash functions have been popularly used in digital signatures schemes, public key encryption, message authentication codes etc. To have a good digital signature schemes or public key encryption, it is required that hash function should be collision resistant or preimage resistant. Intuitively, for a collision resistant hash function H it is hard to find two different inputs $X \neq Y$ such that $H(X) = H(Y)$. In case of preimage resistant hash function, given a random image it is hard to find an inverse of that image. Besides this condition, one should define hash function on an arbitrary domain. Usually, one first design a fixed domain hash function $f : \{0,1\}^{n+m} \rightarrow \{0,1\}^n$ (also known as a compression function) and then extend the domain to an arbitrary domain by iterating the compression function several times. The most popular method is known as MD-method [2], [15] with the classical iterations. We first pad the input message by some strings and

H. Gilbert and H. Handschuh (Eds.): FSE 2005, LNCS 3557, pp. 243–254, 2005.

the string representing the length so that the length of the padded message becomes multiple of m and it avoids some trivial attacks. Now for some fixed initial value $h_0 \in \{0,1\}^n$ and a padded input $M = m_1||\cdots||m_l \in (\{0,1\}^m)^*$, where $|m_i| = m$, the hash function $H^f(h_0,\cdot) : (\{0,1\}^m)^* \to \{0,1\}^n$ is defined as follow :

Algorithm $H^f(h_0, m_1||\ldots||m_l)$
For $i = 1$ **to** l
$h_i = f(h_{i-1}, m_i)$
Return h_l

There are many constructions of the underlying compression functions e.g. SHA-family i.e. SHA-0, SHA-1, SHA-256 [17], MD-family i.e. MD-4, MD-5, RIPEMD [5] [19] etc. There are several collision attacks [3] [4] [10] [21] on some of these compression functions. Also people tried to design a compression function from a block cipher known as PGV hash functions [18]. In [1], [13], the security of the PGV hash functions were studied in the black box model of the underlying block cipher.

Nowadays, people are also interested in designing a bigger size hash function to make the birthday attack infeasible. One can do it by just constructing a compression function like SHA-512. The other way to construct it from a smaller size compression function. In the later case one can study the security level of the bigger hash function assuming some security level of underlying compression functions. People also try to use block ciphers to extend the output size. There are many literatures where the double block length hash function were studied e.g. [7], [8], [11], [12], [16], [20] etc.

1.1 Motivation and Our Results

If a single length compression function has output size n then that of double length compression function is $2n$. For the smaller size hash function the birthday attack can be feasible. Thus to make birthday attack infeasible we need to construct a compression function with larger size output. In this paper, we construct a double length compression function from a single length compression function or a block cipher. We use three invocations of independent single length compression functions or block ciphers to hash two message blocks. Thus, the rate of the compression function is 2/3. We also prove the security level is $\Omega(2^{2n/3})$ and prove the bound is tight by showing an attack on this compression function with complexity $O(2^{2n/3})$.

2 Preliminaries

2.1 Some Results on Probability Distribution

In this paper we will be interested in random variables taking values on $\{0,1\}^n$ for some integer $n > 0$. A random variable X is uniformly distributed over the set $\{0,1\}^n$ if $\Pr[X=x]=1/2^n$ for all $x \in \{0,1\}^n$. We use the notation $X \sim U_n$ to denote a uniform random variable X. We say random variables $X_1,\cdots,X_k$ are independent if the joint distribution of $(X_1,\cdots,X_k)$ is the product of marginal distributions of X_i's. So if $X_1,\cdots,X_k$ are independent and $X_i \sim U_n$ for all i, then $\Pr[X_1=x_1,\cdots,X_k=x_k]=1/2^{nk}$ for all $x_i \in \{0,1\}^n$. We describe this case by the notation $(X_1,\cdots,X_k) \models U_n$. In this case, it is easy to see that $X_1||\cdots||X_k \sim U_{nk}$ i.e. uniformly distributed over the set $\{0,1\}^{nk}$. The n-bit string $0\cdots0$ (known as a zero string) is denoted by $\mathbf{0}$. For a binary vector $l=(l_1,\cdots,l_k) \in Z_2^k$, l^T denotes the transpose vector of l. Given a set of k random variables $X=(X_1,\cdots,X_k)$, $X \cdot l^T = l_1X_1 \oplus \cdots \oplus l_kX_k$, where $0X=\mathbf{0}$ and $1X=X$. For a binary matrix $L_{k\times r}=[l_1^T,\cdots,l_r^T]$, $X\cdot L$ denotes the random vector $(X\cdot l_1^T,\cdots,X\cdot l_r^T)$. Now we state a simple fact from the probability theory.

Proposition 1. *If $X=(X_1,\cdots,X_k)\models U_n$ then for any vector $l\in Z_2^k$ with $l\neq 0$, the random variable $X\cdot l^T\sim U_n$. For any matrix $L_{k\times r}$ with rank $r(\leq k)$, the random vector $X\cdot L\models U_r$.*

Example 1. Take $r=2$ and $k=3$. Let $l_1=(0,1,1)$ and $l_2=(1,1,0)$ then $X\cdot L=(X_2\oplus X_3, X_1\oplus X_2)$, where $X=(X_1,X_2,X_3)\models U_n$. By the above Proposition 1, both $X_2\oplus X_3$ and $X_1\oplus X_2$ are independently and uniformly distributed on $\{0,1\}^n$ since the matrix $L=[l_1^T,l_2^T]$ has rank 2.

2.2 (Independent) Random Functions and Permutations

A random function $f: D\to R$ taking values as random variable satisfy the following conditions

1. for any $x\in D$, $f(x)$ has uniform distribution on R.
2. for any $k>0$ and k distinct elements $x_1,\cdots x_k\in D$, the random variables $f(x_1),\cdots,f(x_k)$ are independently distributed.

More precisely, one can not construct a single function which is a random function. Consider a class of functions $Func^{D\to R}$ which consists of all function from D to R. When one says that f is a random function it means that f is drawn randomly from $Func^{D\to R}$. However, to study some security property one assume a single function as a random function. Although, it is not theoretically possible this can be meaningful for some types of adversary who only query the function f and do not explore the internal structure of f. We say two functions f_1 and f_2 from D to R are independent random functions if they are random functions and for any $k,l>0$ and k distinct elements $x_1^1,\cdots,x_k^1\in D$ and l distinct elements $x_1^1,\cdots,x_l^1\in D$ the random variables $f_1(x_1^1),\cdots,f_1(x_k^1),f_2(x_1^2),\cdots,f_2(x_l^2)$ are independently distributed. Similarly one can define that f_1,f_2 and f_3 are independent random functions and so on.

Similarly one can define a random permutation. A permutation $E: D\to D$ is said to be a random permutation if for any $k>0$ and k distinct elements $x_1,\cdots,x_k\in D$, the random variable $f(x_k)$ condition on $f(x_1)=y_1,\cdots,f(x_{k-1})$ $=y_{k-1}$ is uniformly distributed over the set $D-\{y_1,\cdots,y_{k-1}\}$. Obviously

$f(x_1), \cdots, f(x_k)$ are not independently distributed. We say $E : \{0,1\}^k \times \{0,1\}^n \to \{0,1\}^n$ by a family of permutations if for each $K \in \{0,1\}^k$, $E(K, \cdot)$ is a permutation on n-bit strings. We say a family of permutations $E : \{0,1\}^k \times \{0,1\}^n \to \{0,1\}^n$ is a random permutation if for each $K \in \{0,1\}^k$, $E(K, \cdot)$ is a random permutation and for each $s > 0$, and s distinct elements $K_1, \cdots K_s$, $E(K_1, \cdot), \cdots, E(K_s, \cdot)$ are independent function.

2.3 Some Attacks on Hash/Compression Functions

In this paper, we mainly study the collision resistant hash function but for the sake of completeness, we want to state the preimage resistance also. Given a compression function $f : \{0,1\}^N \to \{0,1\}^n$, it is called collision resistant if it is hard to find two inputs $x \neq y$ such that $f(x) = f(y)$. It is said to be a preimage resistant compression function if given a random $y \in \{0,1\}^n$, it is hard to find x such that $f(x) = y$. In the case of a random function f, the best attack is birthday attack which takes $O(2^{n/2})$ or $O(2^n)$ queries of f for collision or preimage attack, respectively. For the hash function based on a compression function, we can similarly define collision and preimage attack. But, here the initial value of the hash function is fixed and given to the adversary before starting the attack. There are also free-start collision and preimage attack where the adversary can choose the initial value. It can be easily shown that the free start attack on hash function is equivalent to the corresponding attack on the underlying compression function.

3 A New Double Length Compression Function

Let $f_i : \{0,1\}^{2n} \to \{0,1\}^n$ be independent random functions, $i = 1, 2, 3$. Define, $F : \{0,1\}^{3n} \to \{0,1\}^{2n}$, where $F(x, y, z) = (f_1(x, y) \oplus f_2(y, z)) \mid\mid (f_2(y, z) \oplus f_3(z, x))$ with $|x| = |y| = |z| = n$. We also write $F = F_1 \mid\mid F_2$, where $F_1(x, y, z) = f_1(x, y) \oplus f_2(y, z)$ and $F_2(x, y, z) = f_2(y, z) \oplus f_3(z, x)$ (see Figure 1).

Theorem 1. *$(F(x_1, y_1, z_1), F(x_2, y_2, z_2)) \models U_{2n}$, $(x_1, y_1, z_1) \neq (x_2, y_2, z_2)$. In particular, $\forall M \neq N$ and Z, $Pr[F(M) = F(N)] = \frac{1}{2^{2n}}$ and $Pr[F(M) = Z] = \frac{1}{2^{2n}}$.*

Proof. Let $M = (x_1, y_1, z_1) \neq (x_2, y_2, z_2) = N$. Assume that $x_1 \neq x_2$, $y_1 = y_2 = y$ (say), and $z_1 = z_2 = z$ (say). For the other cases, we can prove the result similarly. To prove that $(F(M), F(N)) \models U_{2n}$, it is enough to prove that $(F_1(M), F_2(M), F_1(N), F_2(N)) \models U_n$. Since f_1, f_2 and f_3 are independent random functions, $f_1(x_1, y), f_1(x_2, y), f_2(y, z), f_3(z, x_1)$ and $f_3(z, x_2)$ are independently distributed. Thus, by Proposition 1 (in Section 2.1) we know that $f_1(x_1, y) \oplus f_2(y, z)$, $f_3(z, x_1) \oplus f_2(y, z)$, $f_1(x_2, y) \oplus f_2(y, z)$ and $f_3(z, x_2) \oplus f_2(y, z)$ are independently distributed. So we have proved the proposition. □

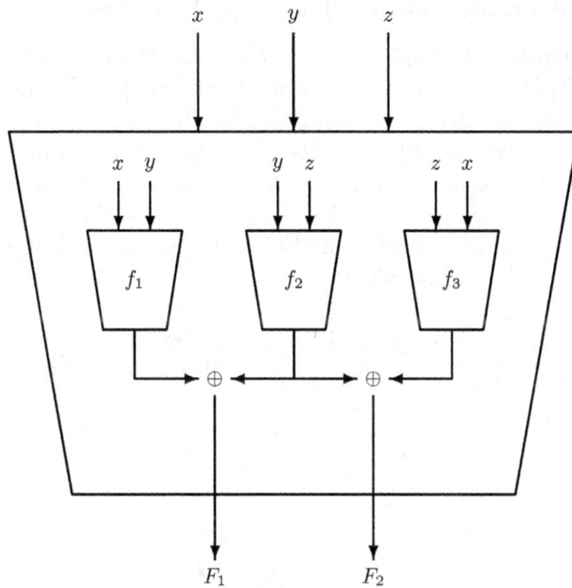

Fig. 1. A double length compression function

3.1 The Model for Adversary and Computable Message

In this subsection, we state briefly how an adversary works in the random oracle model. Adversary can ask the oracles f_1, f_2 and f_3 i.e. he can submit (a, b) to any one of the oracles f_i and he will get a response c such that $f_i(a, b) = c$. We restrict the number of queries for each f_i by at most q. Finally he outputs a pair $M \neq N$ (for collision attack of F) such that both $F(M)$ and $F(N)$ can be computed from the set of queries he made. We say adversary wins if $F(M) = F(N)$.

Definition 1. (Computable message)
Let $\mathcal{Q}_1 = \{(x_i^1, y_i^1)\}_{1 \leq i \leq q}$, $\mathcal{Q}_2 = \{(y_i^2, z_i^2)\}_{1 \leq i \leq q}$ *and* $\mathcal{Q}_3 = \{(z_i^3, x_i^3)\}_{1 \leq i \leq q}$ *be the three sets of queries for the random oracles* f_1,f_2 *and* f_3*, respectively. We say a message* $M = (x, y, z)$ *is* computable *if* $(x, y) \in \mathcal{Q}_1, (y, z) \in \mathcal{Q}_2$ *and* $(z, x) \in \mathcal{Q}_3$.

Thus it is easy to observe that a message M is *computable* if and only if $F(M)$ can be computed from the set of queries. Because of Theorem 1 of this section if we can bound the number of computable message by some number say Q then it is easy to check that the adversary will get a collision with probability at most $Q(Q-1)/2^{2n+1}$. In case of preimage attack, the probability is at most $Q/2^{2n}$. Thus the question reduces how to get an upper bound of the number of computable messages from any set of queries $\mathcal{Q}_1$, $\mathcal{Q}_2$ and $\mathcal{Q}_3$ where $|\mathcal{Q}_i| \leq q$, $1 \leq i \leq 3$. To have an upper bound we can convert our problem into a combinatorial graph theoretical problem. In the next subsection we study that problem.

3.2 A Combinatorial Graph Theoretical Problem

Tripartite Graph. A graph $G = (V, E)$ is known as a tripartite graph if $V = A \sqcup B \sqcup C$ (disjoint union) and for any edge $\{u, v\} \in E$ either $u \in A, v \in B$ or $u \in A, v \in C$ or $u \in B, v \in C$ (see Figure 2). Thus there are no edges between vertices in A or between vertices in B or between vertices in C. We use the notation $e(A, B, G)$ (or simply $e(A, B)$) for the set of edges between A and B. Similarly we can define $e(B, C)$ and $e(A, C)$. Note that for every triangle $\triangle$ in G, the vertices of $\triangle$ are from A, B and C with one vertex from each one. Now we can state the following problem.

Problem : Given an integer q, what is the maximum number of triangles of a tripartite graph G on $A \sqcup B \sqcup C$ such that $|e(A, B)|, |e(B, C)|, |e(A, C)| \leq q$.

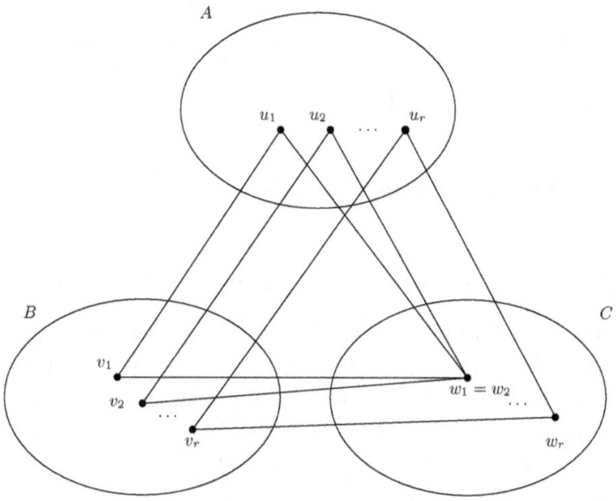

Fig. 2. A tripartite graph

We first prove a Proposition which will be useful for finding the upper bound of the problem stated above.

Proposition 2. *Let G be a tripartite graph on $A \sqcup B \sqcup C$ such that $|e(A, B)| \leq q$. For a set of edges $E_{BC} = \{v_1 w_1, \cdots, v_r w_r\} \subseteq e(B, C)$ such that v_i's are distinct vertices from B, the number of triangles in G whose one of the sides is from E_{BC} is at most q.*

Proof. Let T be the set of triangles in G one of whose side is from E_{BC}. Now we can define an injective map ρ from the set T to the set $e(A, B)$. Given a triangle $uvw \in T$ with $vw \in E_{BC}$ and $v \in B$, define $\rho(T) = uv$. Obviously the map $\rho : T \to e(A, B)$ is well defined. To see it is an injective map we just note that all v_i's are distinct (see Figure 2). So, $\rho(uvw) = \rho(u'v'w')$ with $v, v' \in B$

and $u, u' \in A$ implies that $u = u'$ and $v = v'$. Since $v = v'$ and $vw, v'w' \in E_{BC}$ implies that $w = w'$. So the two triangle uvw and $u'v'w'$ are identical. □

Thus if we can divide the set $e(B, C)$ into r sets E^i_{BC}, $1 \leq i \leq r$ such that each E^i_{BC} has the property stated in the Proposition 2 for B or C then the number of triangles in G will be at most $r \times q$. Assume $q = n^2$. We will show now that we can always divide $e(B, C)$ into $2n$ many such sets. Thus upper bound of triangles is $2n^3$. Let $G = (V, E)$ be a bipartite graph on $B \sqcup C$ with $|E| \leq n^2$. We say a set of edges $E' = \{u_1v_1, \cdots, u_rv_r\}$ in G is *good* if all $u_i \in B$ or C and u_i's are distinct.

Proposition 3. *Given a bipartite graph $G = (V, E)$ with $V = A \sqcup B$ and $|E| \leq n^2$ we can divide E into at most $2n$ good sets of edges.*

Proof. The proof is by induction on n. Assume $|E| > (n-1)^2$. Thus we can find a set B or C where number of vertices with positive degree is at least n. Without loss of generality we assume that the set B has n vertices $u_1, \cdots, u_n$ with degree at least one. Let $u_iv_i \in E$, where $v_i \in C$, $1 \leq i \leq n$. Note that v_i's are not necessarily distinct. So $E_1 = \{u_1v_1, \cdots, u_nv_n\}$ is a good set. Now consider $E - E_1$. Again, if $|E - E_1| \leq (n-1)^2$ then we can apply induction hypothesis and we will get $2(n-1)$ good sets for $E - E_1$. So the result is true. If $|E - E_1| > (n-1)^2$. Again we can find a good set E_2 of size at least n by using similar argument. Now $|E| - |E_1| - |E_2| \leq n^2 - 2n \leq (n-1)^2$. So by induction hypothesis we can get $2(n-1)$ good sets in $E - (E_1 \cup E_2)$. Thus we have $2n$ good sets whose union is the whole set E. For $n = 1$ the result is trivial. □

Theorem 2. *Given a positive integer n, the number of triangles of any tripartite graph G on $A \sqcup B \sqcup C$ such that, $|e(A, B)|, |e(B, C)|, |e(A, C)| \leq n^2$ is at most $2n^3$.*

The proof of the above theorem is immediate from Proposition 2 and 3. In fact we have better and sharp bound which is n^3. The proof is given by one of the anonymous referee. He proved a general statement as follow :

Theorem 3. *Given a positive integer n, the number of triangles of any tripartite graph G on $A \sqcup B \sqcup C$ is at most $(XYZ)^{1/2}$ such that, $|e(A, B)| \leq X, |e(A, C)| \leq Y$ and $|e(B, C)| \leq Z$. In particular, when $X = Y = Z = n^2$ the number of triangle is at most n^3.*

Proof. Let x_a be the number of edges from the vertex $a \in A$ between A and B. Similarly, y_a is the number of edges between A and C from the vertex a. Obviously,

$$\textstyle\sum_{a \in A} x_a = X \text{ and } \sum_{a \in A} y_a = Y.$$

Now the number or triangles containing the vertex a is bounded by $\min\{Z, x_ay_a\}$. Since a triangle containing the vertex a is determined by two edges containing a or determined by the opposite edge of a. But we have, $\min\{Z, x_ay_a\} \leq \sqrt{Zx_ay_a}$. Thus the number of triangles is bounded by

$\sum_a \sqrt{Zx_a y_a} = \sqrt{Z}\sum_a \sqrt{x_a y_a} \leq \sqrt{Z}.\sqrt{\sum_a x_a)(\sum_a y_a)} = \sqrt{XYZ}.$ □

Here, we use the Cauchy-Schwartz inequality. If we take $X = Y = Z = n^2$ then the number of triangle is bounded by n^3. We have an example where the number of triangles is exactly n^3 namely we take a complete tripartite graph. That is we have three disjoint set of vertices A, B and C each of size n. Consider all possible edges between A and B, between A and C and between B and C. Obviously the number of edges between A and B or B and C or A and C are exactly n^2. The number of triangles is n^3 since any vertex from A, from B and from C will contribute a triangle.

3.3 Security Study of the Double Length Compression Function

We have three disjoint vertices set each of size 2^n. In particular, take $A = \{0,1\}^n \times \{1\}, B = \{0,1\}^n \times \{2\}$ and $C = \{0,1\}^n \times \{3\}$. We can correspond each query by an edge of a tripartite graph on $A \sqcup B \sqcup C$ as follow: given a query (x, y) on f_1 we add an edge $\{(x, 1), (y, 2)\}$. The number 1,2 and 3 are used to make A, B and C disjoint. Similarly we can add edges for queries on f_2 and f_3. Now it is easy to note that a computable message corresponds to a triangle in the graph G. Thus the number of computable message is equal to the number of triangles. Also the adversary can ask at most q queries to each f_i and hence the number of edges between A and B or B and C or A and C are at most q. Thus by the Theorem 2 we have at most $2q^{3/2}$ computable inputs for F. Thus the winning probability is bounded by $2q^{3/2}(2q^{3/2} - 1)/2^{2n+1}$. So the number of queries needed to get a collision is $\Omega(2^{2n/3})$. We will show an attack which makes $O(2^{2n/3})$ queries to get a collision on F. So the security bound is tight. For preimage attack the winning probability is bounded by $q^{3/2}/2^{2n}$, thus the number of queries needed to get a preimage is $\Omega(2^{2n/3})$. This bound is also tight and one can find an attack very similar to the following collision attack.

A Collision Attack on F. The attack procedure is very much similar with the security proof. We first choose $2^{n/3}$ values of x_i, y_i and z_i independently, $1 \leq i \leq 2^{n/3}$. Now we will query $f_1(x_i, y_j)$ for all $1 \leq i, j \leq 2^{n/3}$. Thus we have to make $2^{2n/3}$ queries of f_1. Similarly, we query for f_2 and f_3. Now we have 2^n computable inputs and check whether there is any computable collision pair.

Remark 1. It is easy to note that, in the security proof of F we do not use the fact that $|x| = |y| = |z| = n$. In fact, if we have $f_i : \{0,1\}^{3n} \to \{0,1\}^n$, $1 \leq i \leq 3$ and define $F(x, y, z) = (f_1(x, y||0^n) \oplus f_2(y, z))||f_2(y, z) \oplus f_3(x, z)$, where $|x| = |y| = n$ and $|z| = 2n$ then we have same security level as in the previous definition. The proof for that is exactly same with the previous proof. Note that, $F : \{0,1\}^{4n} \to \{0,1\}^{2n}$. So we use two message block in each round function F and three parallel computations of f_i's are made. So rate of this compression function is 2/3.

Remark 2. One can define a function $F : \{0,1\}^{4n} \to \{0,1\}^{2n}$ by $F(x, y, z_1, z_2) = (f_1(x, y, z_1) \oplus f_2(y, z_1, z_2))||(f_2(y, z_1, z_2) \oplus f_3(x, z_1, z_2))$ hoping for more security.

But an attack can be shown with complexity $O(2^{2n/3})$. First, fix some z_1 and then choose $2^{n/3}$ values of x, y and z_2 independently. By the same argument like previous attack, it still has 2^n computable messages and hence we will expect to have a collision on F.

3.4 Block-Cipher Based Double Length Compression Function

Let $E : \{0,1\}^{2n} \times \{0,1\}^n \rightarrow \{0,1\}^n$ be a block cipher with $2n$-bit keys. Define a function $f : \{0,1\}^{3n} \rightarrow \{0,1\}^n$, as follow :

$$f(x, y, z) = E_{x||y}(z) \oplus z,$$

$|x| = |y| = |z| = n$, Here, we will assume $E(\cdot)$ as a family of random permutations. More precisely, given any integer $s > 0$, and s distinct keys $k_1, \cdots, k_s \in \{0,1\}^{2n}$, the functions $E_{k_1}, \cdots, E_{k_s}$ are independent random permutations. It is easy to check that if we sacrifice two bits then we can get three instances of f which will be independent to each other. That is we can define, $f_i(x, y, z) = E_{<i>||x||y}(z) \oplus z$, where $< i >$ is the two bit binary representation of i and $|x| = n - 2, |y| = |z| = n$. Then we can define similarly the double length compression function $F : \{0,1\}^{4n-2} \rightarrow \{0,1\}^{2n}$ i.e. $F(x, y, z, t) = (f_1(x, y, z) \oplus f_2(x, z, t)) \; || \; (f_2(x, z, t) \oplus f_3(x, y, t))$, where $|x| = n - 2, |y| = |z| = |t| = n$ (see Figure 3).

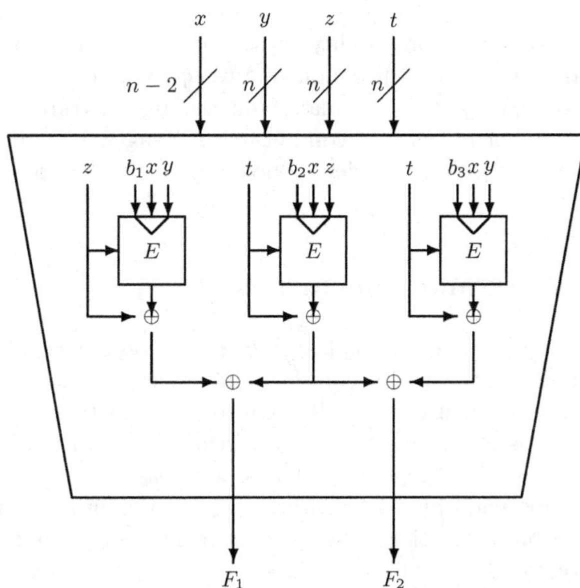

Fig. 3. A double length compression function based on a double-key length block cipher ($b_i :=< i >$)

Here an adversary can ask both E and E^{-1} query. Let $\{(k, a, b)\}$ be a query response triple (in short q-r triple), where $E_k(a) = b$. We can assume that, the first two bits of k not equal to 00 otherwise the query is useless to get a collision attack. Now, if the first two bits of k is $< i >$ with $i \neq 0$ and say k' is the remaining $2n - 2$ bits then,

$$f_i(k', a) = a \oplus b \text{ if and only if } (k, a, b) \text{ is a q-r triple.}$$

Thus given a set of q q-r triples we can have at most q computation of f_i for each i and hence we can have at most $2q^{3/2}$ computable messages. Now it is enough to find a bound of $\Pr[F(M) = F(N)]$, where $M \neq N$.

Now consider $M = (x_1, y_1, z_1, t_1) \neq (x_2, y_2, z_2, t_2) = N$. We assume that $x_1 = x_2 = x$, $y_1 = y_2 = y$, $z_1 \neq z_2$ and $t_1 \neq t_2$. For the other cases one can study similarly. Now, the event $F(M) = F(N)$ is equivalent to

$$\begin{aligned} f_1(x, y, z_1) \oplus f_2(x, z_1, t_1) &= f_1(x, y, z_2) \oplus f_2(x, z_2, t_2), \\ f_3(x, y, t_1) \oplus f_2(x, z_1, t_1) &= f_3(x, y, t_2) \oplus f_2(x, z_2, t_2). \end{aligned}$$

To compute the probability of happening above we can first condition on each term except $f_1(x, y, z_1)$ and $f_3(x, y, t_1)$. Thus the conditional event would be $f_1(x, y, z_1) = a$ and $f_3(x, y, t_1) = b$ for some string a and b. We now have,

$$\begin{aligned} \Pr[f_1(x, y, z_1) = a, f_3(x, y, t_1) = b | f_2(x, z_1, t_1) = a_1, \cdots, f_3(x, y, t_2) = a_4] \\ \leq 1/2^{n-1} \times 1/2^{n-1} \end{aligned}$$

for some $a_1, \cdots, a_4$. Thus, probability of collision for a given pair is bounded by $1/2^{2n-2}$ and hence success probability after q many queries is bounded by $2q^3/2^{2n-2}$. Note that $2q^{3/2}$ is the maximum number of computable messages and hence the number of pairs of computable messages is at most $2q^3/2^{2n-2}$. Thus we need $\Omega(2^{2n/3})$ many queries to have non-negligible success probability.

4 Future Work and Conclusion

This paper deals with a new double length compression function which can uses three parallel computations of a compression function or a double key block cipher. Although the security of this compression function is not maximum possible (i.e. there is a better attack than birthday attack) the lower bound of the number of queries is $\Omega(2^{2n/3})$. So it has better security than a most secure single length compression function. Also the security is proved for compression function. So the hash function based on the compression function has same security level for free-start collision attack. So it would be interesting to study the security level for collision attack. Also one can try to design an efficient (if possible, rate-1) double block length hash function which is maximally secure against collision attack even if the underlying compression function is not secure.

Acknowledgement. The authors would like to thank the anonymous referees of FSE-05 who gave helpful comments and provide a solution of the graph theoretical problem. This research was supported by the MIC(Ministry of Information and Communication), Korea, supervised by the IITA(Institute of Information Technology Assessment). The second author was supported by the 21st Century COE Program 'Reconstruction of Social Infrastructure Related to Information Science and Electrical Engineering' of Kyushu University.

References

1. J. Black, P. Rogaway, and T. Shrimpton. *Black-box analysis of the block-cipher-based hash function constructions from PGV*, Advances in Cryptology - Crypto'02, Lecture Notes in Computer Science, Vol. 2442, Springer-Verlag, pp. 320-335, 2002.
2. I. B. Damgård. *A design principle for hash functions*, Advances in Cryptology - Crypto'89, Lecture Notes in Computer Sciences, Vol. 435, Springer-Verlag, pp. 416-427, 1989.
3. H. Dobbertin.*Cryptanalysis of MD4*. Fast Software Encryption, Cambridge Workshop. Lecture Notes in Computer Science, vol 1039, D. Gollman ed. Springer-Verlag 1996.
4. H. Dobbertin.*Cryptanalysis of MD5* Rump Session of Eurocrypt 96, May. http//www.iacr.org/conferences/ec96/rump/index.html.
5. H. Dobbertin, A. Bosselaers and B. Preneel. *RIPEMD-160: A strengthened version of RIPEMD*, Fast Software Encryption. Lecture Notes in Computer Science 1039, D. Gollmann, ed., Springer-Verlag, 1996.
6. H. Finney. *More problems with hash functions.* The cryptographic mailing list. 24 Aug 2004. http://lists.virus.org/cryptography-0408/msg00124.html.
7. M. Hattori, S. Hirose and S. Yoshida. *Analysis of Double Block Lengh Hash Functions.* Cryptographi and Coding 2003, LNCS 2898.
8. S. Hirose. *Provably Secure Double-Block-Length Hash Functions in a Black-Box Model*, to appear in ICISC-04.
9. A. Joux. *Multicollision on Iterated Hash Function.* Advances in Cryptology, CRYPTO 2004, Lecture Notes in Computer Science 3152.
10. J. Kelsey. *A long-message attack on SHAx, MDx, Tiger, N-Hash, Whirlpool and Snefru.* Draft. Unpublished Manuscritpt.
11. L. Knudsen, X. Lai and B. Preneel. Attacks on fast double block length hash functions. *J.Cryptology, vol 11 no 1, winter 1998.*
12. L. Knudsen and B. Preneel. Construction of Secure and Fast Hash Functions Using Nonbinary Error-Correcting Codes. *IEEE transactions on information theory, VOL-48, NO. 9, Sept-2002.*
13. W. Lee, M. Nandi, P. Sarkar, D. Chang, S. Lee and K. Sakurai *A Generalization of PGV-Hash Functions and Security Analysis in Black-Box Model.* Lecture Notes in Computer Science, ACISP-2003.
14. S. Lucks. *Design principles for Iterated Hash Functions*, e-print server : http://eprint.iacr.org/2004/253.
15. R. Merkle. *One way hash functions and DES*, Advances in Cryptology - Crypto'89, Lecture Notes in Computer Sciences, Vol. 435, Springer-Verlag, pp. 428-446, 1989.
16. M. Nandi. *A Class of Secure Double Length Hash Functions.*. e-print server : http://eprint.iacr.org/2004/296.

17. NIST/NSA. *FIPS 180-2 Secure Hash Standard*, August, 2002. http://csrc.nist.gov/publications/fips/fips180-2/fips180-2.pdf
18. B. Preneel, R. Govaerts, and J. Vandewalle. *Hash functions based on block ciphers:A synthetic approach*, Advances in Cryptology-CRYPTO'93, LNCS, Springer-Verlag, pp. 368-378, 1994.
19. R. Rivest *The MD5 message digest algorithm.* http://www.ietf.org/rfc/rfc1321.txt
20. T. Satoh, M. Haga and K. Kurosawa. *Towards Secure and Fast Hash Functions.* IEICE Trans. VOL. E82-A, NO. 1 January, 1999.
21. B. Schneier. *Cryptanalysis of MD5 and SHA.* Crypto-Gram Newsletter, Sept-2004. http://www.schneier.com/crypto-gram-0409.htm#3.

Preimage and Collision Attacks on MD2

Lars R. Knudsen[1] and John E. Mathiassen[2]

[1] Department of Mathematics,
Technical University of Denmark
[2] Department of Informatics,
University of Bergen, Norway

Abstract. This paper contains several attacks on the hash function MD2 which has a hash code size of 128 bits. At Asiacrypt 2004 Muller presents the first known preimage attack on MD2. The time complexity of the attack is about 2^{104} and the preimages consist always of 128 blocks. We present a preimage attack of complexity about 2^{97} with the further advantage that the preimages are of variable lengths. Moreover we are always able to find many preimages for one given hash value. Also we introduce many new collisions for the MD2 compression function, which lead to the first known (pseudo) collisions for the full MD2 (including the checksum), but where the initial values differ. Finally we present a pseudo preimage attack of complexity 2^{95} but where the preimages can have any desired lengths.

1 Introduction

A hash function is a function that takes an arbitrary long input, and produces a fixed length output. The output is often called a fingerprint of the input. A cryptographic hash function needs to satisfy certain security criteria in order to be called a secure hash function. Let

$$H : \{0,1\}^* \rightarrow \{0,1\}^n$$

denote a hash function, whose output is of length n bits. A cryptographic hash function should be resistant against the following attacks:

- **Collision:** Find x and x' such that $x \neq x'$ and $H(x) = H(x')$.
- **2nd preimage:** Given x and $y = H(x)$ find $x' \neq x$ such that $H(x') = y$.
- **Preimage:** Given $y = H(x)$, find x' such that $H(x') = y$.

Typically one requires that there must not exist attacks of these three types which are better than brute-force methods. Thus, to find a collision should not have a lower complexity than about $2^{n/2}$ and it should not be possible to find preimages in time less than 2^n.

It is common to construct hash functions from iterating a so-called a compression function

$$h : \{0,1\}^n \times \{0,1\}^l \rightarrow \{0,1\}^n,$$

H. Gilbert and H. Handschuh (Eds.): FSE 2005, LNCS 3557, pp. 255–267, 2005.

which compresses a fixed number of bits. Here the output of one application of this function, h_i, of length n is called a chaining variable and is used as an input in the next iteration together with the next message block m_{i+1} of length l. If the design of a hash function follows the principles of Merkle and Damgård [4, 1], it can be shown that a collision for the hash function H implies a collision for the compression function h. Thus, if one can design a secure compression function, then one can also design a secure hash function. Still, the first step towards finding weaknesses in the hash function may be to find weaknesses in the compression function. The first chaining variable in an iterated hash function is often called the IV (initial value) and this is often fixed. Attacks on hash functions where the attacker is able to choose or change the IV are called pseudo attacks. Must popular hash functions are using an iterative compression function and a fixed IV. Examples are MD4, MD5, SHA-1, and RIPEMD-160.

The organisation of this paper is as follows. Section 2 presents the MD2 hash function. Section 3 presents some collision attacks on the compression function where many details are included in an appendix. Section 4 presents several attacks on MD2 (including the checksum). They are a pseudo collision attack, several preimage attacks, as well as a pseudo preimage attack. As far as we are informed the complexities of all these attacks are the lowest known today. Below is a summary of all known results on MD2, where an asterisk (*) indicates that the attack is new.

	Collision	Preimage	Comments
Compression function	2^8 [6]	2^{73} [5]	
Hash function (pseudo)	2^{16} (*)	2^{95} (*)	Arbitrary length messages
Hash function	-	2^{104} [5]	Message length 128 blocks
Hash function	-	$2^{97.6}$-2^{112} (*)	Message length 44-128 blocks

2 The MD2

The MD2 hash algorithm is designed by Ron Rivest and published in 1988[2, 3]. It is a function $H : GF(256)^* \to GF(256)^{16}$, which takes an arbitrary number of bytes $GF(256)$ and outputs a string of 16 bytes $GF(256)^{16}$. The function consists of iterations of a compression function $h : GF(256)^{16} \times GF(256)^{16} \to GF(256)^{16}$, $h_i = h(h_{i-1}, m_i)$, where the input in the ith iteration is the ith message block m_i and the chaining variable h_{i-1}. The message m to be hashed is appended with some padding bytes and a checksum c before it is processed: $m||p||c = m_1||m_2|| \cdots ||m_{t+1}$, where $|m_i| = 128$ for $i = 1, 2, \ldots, t+1$. At least one byte and at most 16 bytes of m_t are padded. Let b be the length of the message in bytes, and $i \equiv b \bmod 16$, $i \in \{0, 1, \ldots, 15\}$, then $d = 16 - i$ (represented in a byte) is added to the message d times. There is at least one byte padding, so if the length is $b \equiv 15 \bmod 16$, then $d = 1$ the byte $p = 1$ is appended the message. If the message length in bytes is 0 modulo 16 , then $d = 16$ and the byte sequence p =16$| \cdots |$16 of length 16 bytes is added to the message, so that the length of the message still is 0 modulo 16.

Algorithm 1. Algorithm to compute the checksum $c = c_0||c_1||\cdots||c_{15}$

```
for j = 0, 1, ..., 15
  c_j = 0
for i = 1 to t do
  for j = 0 to 15 do
    c_j = s(c_{j-1 mod 16} ⊕ m_{i,j}) ⊕ c_j
  end /*for i*/
end /*for j*/
```

Algorithm 2. The compression function in MD2, where the output is the 16 first bytes of $h_{i,1} \mid h_{i,2} \mid \cdots \mid h_{i,16}| \cdots |h_{i,48}$

```
for j = 1 to 16 do
  h_{i,j} = h_{i-i,j}
  h_{i,16+j} = m_{i,j}
  h_{i,32+j} = h_{i-i,j} ⊕ m_{i,j}
t=0
for r = 1 to 18 do
  for j = 1 to 48 do
    t = h_{i,j} = s(t) ⊕ h_{i,j}
  end /*for j*/
  t = r - 1modulo 256
end /*for r*/
```

Next a checksum block $m_{t+1} = c = c_0 \mid c_1 \mid \cdots \mid c_{15}$ is appended to the message. The checksum [Algorithm 1.] is generated processing every byte of the message one block at the time, starting at the first block. The checksum is initialized to 0, $c_i = 0$ for $i = 0, 1, \ldots, 15$. Then for all t message blocks, m_i for $i = 1, 2, \ldots, t$, process all 16 bytes of that block and the checksum $j = 0, 1, \ldots, 15$ by the function $c_j = s(c_{(j-1)} \oplus m_{i,j}) \oplus c_j$ where $m_{i,j}$ is the j'th byte of the i'th block of the message and where $s : \{0,1\}^8 \to \{0,1\}^8$ is a bijective mapping, which is also used in the compression function. The details of s are not important for the results in this paper. The hash function is iterated in the following way:

- $h_0 = iv = 0$
- $h_i = h(h_{i-1}, m_i)$ for $i = 1, 2, \ldots, t+1$
- $H(m) = h_{t+1}$

The compression function [Algorithm 2.] of MD2 takes two inputs of each 128 bits, cf., earlier, and consists of an 18-round iterative process, where a vector of the 48 bytes constructed from $h_{i-1}||m_i||h_{i-1} \oplus m_i$ and denoted

$$h_i = h_{i,1}||h_{i,2}|| \cdots ||h_{i,48}$$

is repeatedly processed from left to right through the use of the same round function consisting of simple byte exclusive-ors and the eight-bit bijective mapping $s()$, also used in the checksum calculation.

3 Attacks on the Compression Function

In [6] a collision attack on the compression function of MD2 is given. Recall that this function computes $h_i = h(h_{i-1}, m_i)$. Rogier and Chavaud give 141 collisions for the compression function where for all collisions h_{i-1} is fixed to the value zero. Note that the IV of MD2 as stated in [2] is zero. We found some variations of this attack. First of all we found that the collision attack extends and it is possible to find many more collisions of this form. We implemented one improvement and found 32,784 collisions, all with $h_{i-1} = 0$. This attack takes very little time. Also we found that it is possible to find so-called multi-collisions for the compression function, that is, a set of different m_is all with same output in the compression function and all with $h_{i-1} = 0$. With a complexity of about 2^{72} one expects a multiple collision of eight messages.

Another variation of Rogier and Chavauds attack is to fix m_i to zero and find different values of h_{i-1} leading to identical outputs of the compression function and yet another variation is to fix $m_i \oplus h_{i-1}$. These variants are similar to the above original one, although the complexities are slightly higher. [6] also consider cases where only a subset of the bytes of h_{i-1} are zeros. We show similar results for the variations. The details of the variant where $h_{i-1} = 0$ are descibed in Appendix B. The details of the other variants are described in an extended version of the paper available upon request.

In the next section we shall use some of the improvements and variations of the attacks on the compression function.

4 Attacks on the MD2 Hash Function

4.1 A Pseudo Collision Attack on MD2

In Section 3 we mention a collision attack on the compression function where $m_i = m'_i = 0$ and $h_{i-1} \neq h'_{i-1}$, but where $h_i = h'_i$. Using this attack we are able to find collision for MD2 (including the checksum) but using different IVs. We have found 130 such collisions in 2 seconds on a single PC, and can find $\approx 2^{15}$ such collisions in about 512 seconds (under 9 minutes) with that property. For any such collision $h_{i-1} \neq h'_{i-1}$, thus if two different IV values of MD2 are chosen to be $IV = h_{i-1}$ and $IV' = h'_{i-1}$ then one can find collisions for all of MD2 for a message using two different IVs.

- Find a pair $(h_0, m_1) \neq (h'_0, m_1)$ where $m_1 = 0$ such that $h(h_0, m_1) = h(h'_0, m_1)$.
- Set $IV = h_0$ and $IV' = h'_0$.
- Choose message blocks $m_2|m_3|, \ldots, |m_t$.
- Then clearly $H(IV, m) = H(IV', m)$, where $m = m_1|m_2|m_3|, \ldots, |m_t$.

Notice that the checksums for both hashes are identical since the message blocks are identical, and therefore we have pseudo collision for MD2.

Let us now consider a situation where such collisions could become practical. Imagine a scenario where Alice and Bob use a digital signature system using a hash function. Imagine that they are signing the same message m many times, e.g., "Alice owes Bob 100 US$". In order to avoid that the same message gives an identical signature, Alice suggests to use a time-stamp, but Bob convinces her that instead he shall send Alice a fresh random hash-IV (e.g., a nonce) to be used in every new signature. Alice agrees to this, however demands that the IV Bob chooses should be run through the hash function first. And so, they agree on the following protocol.

- Bob chooses a random IV
- Alice calculates $r = h(IV, 0)$, creates the hash as usual by $h = H(r, m)$, and signs the hash value, $sign(h)$.

Assuming that the digital signature scheme and the hash function are secure, it seems hard for Bob to cheat. In every new signature a different IV is used, so Bob cannot play the replay attack. However using MD2 in this protocol is a problem since Bob is able to find many collisions of the type $h(IV, 0) = h(IV', 0)$, and hence he is able to reuse the signature and message together with other IVs.

4.2 The Preimage Attack

In [5] F. Muller presents the first known preimage attack on MD2 faster than a brute-force attack. The attack is divided into two parts: in the first part one finds many preimages of the compression function and in the second part one finds those preimages which conform with the checksum function. Note that for most iterated hash functions a preimage attack of the compression function immediately gives at least a pseudo preimage on the hash function, but this is not true for MD2 because of the additional checksum block which is appended to the messages. [5] lists three different attacks on the compression function:

1. Given h_i and h_{i-1}, find a message m_i such that $h_i = h(h_{i-1}, m_i)$. The complexity is 2^{95}.
2. Given h_i and m_i, find a value h_{i-i} such that $h_i = h(h_{i-1}, m_i)$. The complexity is 2^{95}.
3. Given h_i, find a value h_{i-i} and a message m_i such that $h_i = h(h_{i-1}, m_i)$. The complexity is 2^{73}.

Here one unit in the complexity measures is the time to run the compression function once. All these attacks are expected to give one solution, but there might also be zero or several solutions. Assuming that the compression function is a random function, the probability that there is no solution is $(1-2^{-128})^{2^{128}}$, and the probability that there are at least w solutions is:

$$p_w \approx 1 - \sum_{i=0}^{w-1} \left[\binom{2^{128}}{i} 2^{-128i} \cdot (1-2^{-128})^{2^{128}-i} \right] \approx 1 - (\sum_{i=0}^{w-1} \frac{1}{i!}) e^{-1}.$$

The first attack above can be used to find also preimages for (all of) MD2[5]. With $h_0 = 0$ and $h = h_{128}$ the attack is as follows, where h_0 is given and i is initialised to 1:

1. Choose a random value of h_i.
2. If more than 2 solutions of m_i satisfying $h_i = h(h_{i-1}, m_i)$ is found: Increase i by 1. If $i < 128$: Goto step 1.
3. If no more than 2 solutions of m_{128} satisfying $h_{128} = h(h_{127}, m_{128})$ is found: Set i to 127 and goto step 1.

This gives 128 consecutive pairs (h_{i-1}, h_i) for which there are at least 2 different values of m_i such that $h_i = h(h_{i-1}, m_i)$. Consequently there are at least 2^{128} different messages m (of 128 blocks) such that $h = H(m)$, and therefore one of these messages is expected to conform with the checksum $m_{128} = c$. Let $c[i]$ denote the checksum after i iterations (i message blocks). Using the birthday attack on the checksum function has a complexity of about 2^{64}:

- Compute 2^{64} values of $c[64]$ by iterating the checksum function through 2^{64} possible values of the blocks $m_1, m_2, \ldots, m_{64}$.
- Compute 2^{64} values of $c[64]$ by calculating the checksum backwards through 2^{64} possible values of the blocks $m_{65}, m_{66}, \ldots, m_{128} = c$.
- Search for a collision between elements in the two lists.

The expected number of collisions in this last step is 1. The overall complexity of this attack is as follows. The probability of finding at least two solutions in the attack on the compression function is approximately $p_2 = 1 - 2e^{-1}$, and for each of the steps in the algorithm we expect p_2^{-1} repeats. So the total complexity is $128 \cdot p_2^{-1} \cdot 2^{95} \approx 2^{104}$. The padding bytes have not been considered in this attack, but it is strightforward to ensure that the preimages have correct padding without increasing the complexity of the attack[5]. One drawback of this preimage attack is that the messages always consist of 128 blocks. It is left as an open question in [5] to find preimages with fewer blocks. In the next section we give an improvement in complexity of the above attack as well as variants where the messages have fewer than 128 blocks.

4.3 Improvement of the Preimage Attack

First we give a preimage attack also with 128 blocks in the messages but with a lower complexity. We are given $h_0 = 0$ and $h = h_{128}$ and proceed as follows:

1. Given $h_0 = 0$; use the collision attack from Section 3 (see also Appendix B) to find h_1 and a collision for $u \geq 4$ different values of m_1 satisfying $h_1 = h(h_0, m_1)$.
2. Let $h_{127} = h_1$, and use the preimage attack to try to find $v \geq 1$ values of m_{128} such that $h_{128} = h(h_{127}, m_{128})$. If there are no solutions, use another collision from step 1.
3. Let $h_2 = h_1$ and find $w \geq 2$ values of m_2 such that $h_2 = h(h_1, m_2)$. If there are no solutions, repeat step 2 using another collision from step 1.
4. Set $h_i = h_1$ for $i = 3, \ldots, 126$.

This is a situation where $h_0 = 0, h_1 = h_2 = \cdots = h_{127}, h_{128} = h$, and the use of the birthday attack on the checksum is expected to give 1 solution. The first

Table 1. Complexities of the preimage attack for different message lengths, where in each case one solution is expected

$w \geq$	message length	complexity
2	128	$2^{97.6}$
3	80	$2^{99.3}$
4	64	$2^{101.4}$
5	55	$2^{103.8}$
6	50	$2^{106.4}$
7	46	$2^{109.2}$
8	43	$2^{112.2}$

step has a relative small complexity as discussed before, but we might be forced to repeat steps 2 and 3. The probability of a solution in step 2 is approximately $p_1 = 0.63$, and the probability in the third step is approximately $p_2 = 0.26$. Total complexity of the attack is then

$$p_1^{-1} \cdot p_2^{-1} \cdot 2^{95} \leq 2^{97.6}.$$

There are possible ways to shorten the number of blocks in the preimages, but at the expense of higher complexity. If we require that $w \geq 3$ in step 3, we expect a slightly higher complexity, but the number of blocks in the preimages would drop to approximately $log_3 2^{128}$. Table 1 shows the complexities and lengths of the preimages for different lower bounds of w. As an example, it is possible to lower the number of blocks in the preimages to 55 instead of 128, by requiring $w \geq 5$ in which case the complexity is $\leq 2^{104}$.

It is also possible to get more preimages without increasing the total (time) complexity. Since we use a preimage where $h_{i-1} = h_i$, the possible length of the chain in the middle can be arbitrarily long, however the length is limited by the complexity of the collision attack of the checksum. One example is an attack where the messages are of length 191 and where $w \geq 2$. This gives a memory and computational complexity of 2^{95} in the birthday attack on the checksum, and it is expected to give 2^{62} collisions and thereby 2^{62} possible preimages, but total running time of the attack is unchanged.

4.4 A Pseudo Preimage Attack on MD2

In this section we present a pseudo preimage attack on MD2 which has better complexity than the preimage attack, and where the messages can be (almost) as short or as long as we desire. This attack uses two attacks from [5] on the compression function having complexities 2^{73} and 2^{95} respectively.

Initially a hash value h is given, and we are able to find a message m and an IV which give us the desired hash value $h = H(IV, m)$. First use the method of finding pseudo preimages h_t and m_{t+1} of $h_{t+1} = h$ in the compression function. Remember that the last message block m_{t+1} is the checksum block, and we might repeat this preimage attack to find the second last message block, which also contains the padding bytes. Due to the high degree of freedom in the attack on the compression function, it is possible to choose between 1 and 16 suitable padding bytes in this message block m_t, but it is sufficient to choose the last byte of m_t equal to 1, and the attack still gives us m_t and h_{t-1} with complexity 2^{73}.

Next we need to have at least one more message block in our preimage to make the checksum consistent with the (given) initial value $c[0] = 0$, (recall that $c[i]$ denotes the checksum after i iterations (i message blocks). A potential problem with the checksum could be to fit the two fixed ends $c[0] = 0$ and $c[t] = m_{t+1}$. However it turns out to be easy to "glue" two consecutive checksum values $c[i-1]$ and $c[i]$ together by choosing an appropriate value m_i. Notice that it is also possible to calculate the checksum $c[i] = c(c[i-1], m_i)$ backwards by inverting the function, $c[i-1] = c^{-1}(c[i], m_i)$. Now suppose we have found the message values m_2 and the checksum, we compute $c[2]$ and then $c[1]$ by going backwards. We now "glue" $c[0]$ and $c[1]$ together by finding the appropriate m_1. To get a preimage of two blocks we set $h_1 = h_{t-1}$ and $m_1 = m_{t-1}$, and use another pseudo preimage attack from [5], having complexity 2^{95}, to find $IV = h_0$. Using the MD2 hash function on the IV and a message m will now give the required hash $h = H(IV, m)$. The total complexity in this situation where the message length is two, is 2^{95}.

For a required message length t, and given $h_{t+1} = h$ the algorithm is as follows:

- Find h_t and $m_{t+1}(= c)$ such that $h_{t+1} = h(h_t, m_{t+1})$.
- Find h_{t-1} and m_t (included valid padding byte), such that $h_t = h(h_{t-1}, m_t)$.
- Repeat the preimage attack $t-2$ times to find h_1 and m_2.
- Find $c[1]$ by calculating the checksum backwards by using m_i for $i = 2, 3, \ldots, t+1$
- Use special property in the checksum algorithm to find m_1 such that $c[1] = c(0, m_1)$.
- Use the other pseudo preimage attack[5] to find $IV = h_0$ given h_1 and m_1.

The complexity of three first steps of the attack is $t \cdot 2^{73}$ and the last step has complexity 2^{95}. The other parts of the algorithm have relatively small complexity and the total complexity of the attack is 2^{95} as long as $t \leq 2^{21}$. The message length could be as small as $t = 2$.

5 Conclusion

In this paper some new attacks on the hash function MD2 were presented. First some extended collision attacks on the compression function were given. Using one of these attacks it was shown to be possible to mount a pseudo collision on the MD2, which is the first known attack of its kind faster than the trivial attacks. The paper also presented the best known preimage attack on MD2 which is an improvment of a factor of 80 compared to existing attacks. Also, it was shown that the lengths of the preimages can be made smaller than in previous attacks, where the lengths were fixed and relatively high. Moreover it was shown that it is possible to extend the attack such that many preimages are found.

References

1. I.B. Damgård. A design principle for hash functions. In G. Brassard, editor, *Advances in Cryptology: CRYPTO'89, Lecture Notes in Computer Science 435*, pages 416–427. Springer Verlag, 1990.
2. B. Kaliski. The MD2 message-digest algorithm. Request for Comments (RFC) 1319, Internet Activities Board, Internet Privacy Task Force, April 1992. Available from `http://www.faqs.org/rfcs/rfc1319.html`.
3. A. J. Menezes, P. C. van Oorschot, and S. A. Vanstone. *Handbook of Applied Cryptography.* CRC Press, 1997.
4. R. Merkle. One way hash functions and DES. In G. Brassard, editor, *Advances in Cryptology - CRYPTO'89, Lecture Notes in Computer Science 435*, pages 428–446. Springer Verlag, 1990.
5. F. Muller. The MD2 hash function is not one-way. In P.J. Lee, editor, *Advances in Cryptology - ASIACRYPT 2004, LNCS 3329*, pages 214–229. Springer Verlag, 2004.
6. N. Rogier and P. Chauvaud. MD2 is not secure without the checksum byte. In *Designs, Codes and Cryptography, 12*, pages 245–251, 1997.

A Properties of the MD2 Compression Function

In order to be able to describe the attacks it is convenient to describe the compression function and its intermediate states in a 19×49-matrix

$$T = (T_{i,j})_{j=0,1,\ldots,48}^{i=0,1,\ldots,18},$$

which is also shown in Figure 1, where the first row is made from h_{i-1}, m_i and $h_{i-1} \oplus m_i$. The first element $T_{0,0}$ is never used, but $(T_{0,j})_{j=1,2,\ldots,48} = h_{i-1} \mid m_i \mid h_{i-1} \oplus m_{i-1}$.

Next the rows of the matrix is processed in an iterative manner:

- $T_{1,0} = 0$
- $T_{i,0} = T_{i-1,48} + i - 2 \bmod 256$ for $i = 2, 3, \ldots, 18$ (but not for $i = 1$)
- $T_{i,j} = T_{i-1,j} \oplus s(T_{i,j-1})$ for $i = 1, 2, \ldots, 18$ and $j = 1, 2, \ldots, 48$
- $h_i = (T_{18,j})_{j=1,2,\ldots,16}$

After this procedure the matrix contains all the states of the compression matrix. As we shall see, it is sometimes advantageous in a cryptanalytic approach to try and compute the values in the matrix in a different order than the above line by line approach. To help us do this, we have derived five computing rules directly from the algorithm. The three first rules are shown in Figure 2. The two remaining are just the dependencies between the first and last columns of T. The rules are:

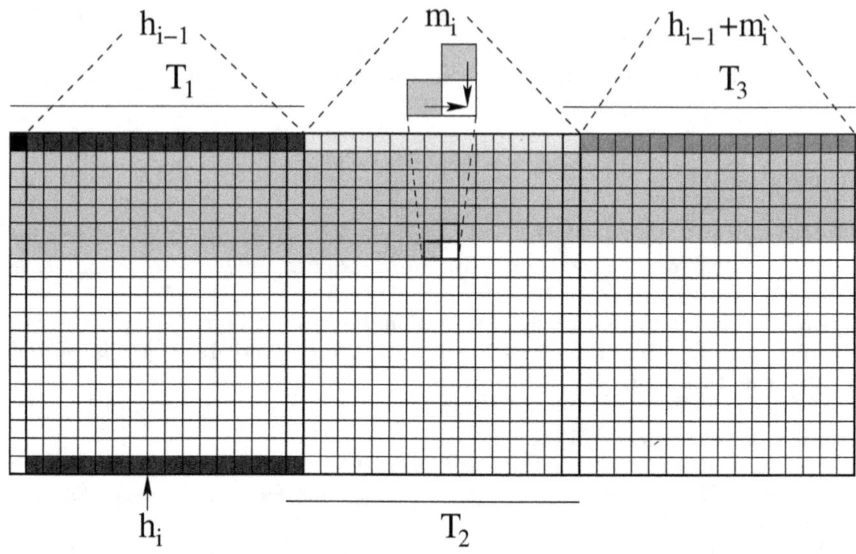

Fig. 1. The MD2 compression function calculation shown as a matrix T. It also shows how the submatrices T_1, T_2 and T_3 are defined, and one line at the time is computed from left to right. The 16 rightmost bytes of the last line of T_1 (the dark area in the last line) contains $h_i = h(h_{i-1}, m_i)$ when the matrix is completed

1. $T_{i,j} = T_{i-1,j} \oplus s(T_{i,j-1})$, where $i = 1, 2, \ldots, 18$ and $j = 1, 2, \ldots, 48$.
2. $T_{i-1,j} = T_{i,j} \oplus s(T_{i,j-1})$, where $i = 1, 2, \ldots, 18$ and $j = 1, 2, \ldots, 48$.
3. $T_{i,j-1} = s^{-1}(T_{i,j} \oplus T_{i-1,j})$, where $i = 1, 2, \ldots, 18$ and $j = 1, 2, \ldots, 48$.
4. $T_{i,0} = T_{i-1,48} + (i-2) \bmod 256$, where $i = 2, 3, \ldots, 18$.
5. $T_{i-1,48} = T_{i,0} - (i-2) \bmod 256$, where $i = 2, 3, \ldots, 18$.

The three first rules give us five properties from [6] also shown in Figure 3 and Figure 4.
Property 1: Let $k < m$ and $l < n$. If the elements $(T_{k,j})_{j=l,l+1,\ldots,n}$ from row k and $(T_{i,l})^{i=k,k+1,\ldots,m}$ from column l are known the submatrix $(T_{i,j})^{i=k,k+1,\ldots,m}_{j=l,l+1,\ldots,n}$ is uniquely determined using rule 1 (Figure 3).

Property 2: Let $k < m$ and $l < n$. If the elements $(T_{k,j})_{j=l,l+1,\ldots,n}$ from row k and $(T_{i,n})^{i=k,k+1,\ldots,m}$ from column n are known the matrix $(T_{i,j})^{i=k,k+1,\ldots,m}_{j=l,l+1,\ldots,n}$ is uniquely determined using rule 3 (Figure 3).

Property 3: Let $k < m$ and $l < n$. If the elements $(T_{m,j})_{j=l,l+1,\ldots,n}$ from row m and $(T_{i,l})^{i=k,k+1,\ldots,m}$ from column l are known the matrix $(T_{i,j})^{i=k,k+1,\ldots,m}_{j=l,l+1,\ldots,n}$ is uniquely determined using rule 2 (Figure 3).

Property 4: Let $l < n$ and $k < m$, such that $m - k = n - l$. If the elements $(T_{i,n})^{i=k,k+1,\ldots,m}$ from column n are known then half the square matrix $(T_{i,j})^{i=k,k+1,\ldots,m}_{j=l,l+1,\ldots,n}$ is uniquely determined under the diagonal $(T_{i,j})^{i=k,k+1,\ldots,m}_{j=n+k-i,(n+k-i)+1,\ldots,n}$ using rule 3 (Figure 4).

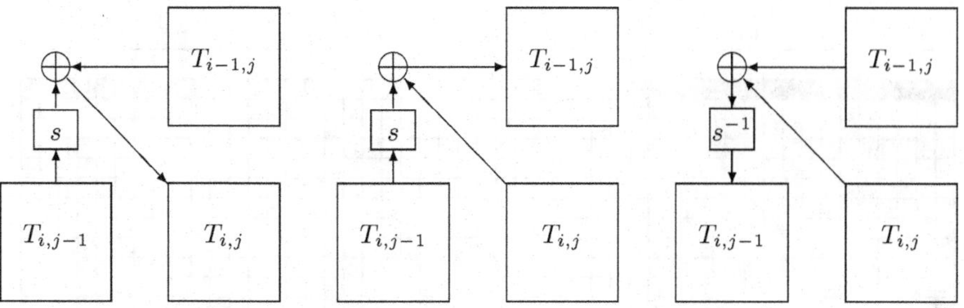

Fig. 2. The dependency of an element $T_{i,j}$ in the matrix T. These three figures show these three dependencies $T_{i,j} = T_{i-1,j} \oplus s(T_{i,j-1})$, $T_{i-1,j} = T_{i,j} \oplus s(T_{i,j-1})$ and $T_{i,j-1} = s^{-1}(T_{i,j} \oplus T_{i-1,j})$ respectively

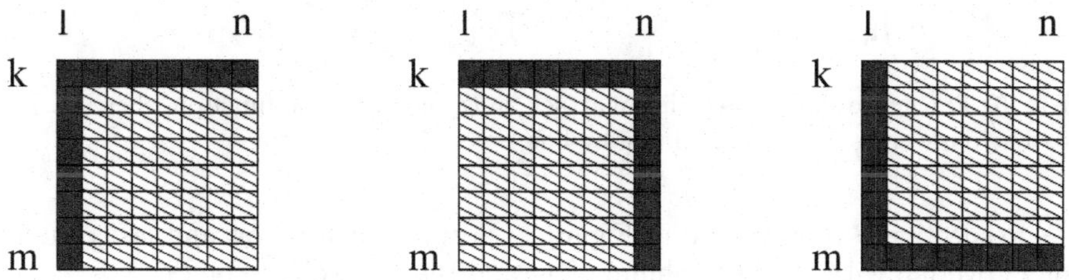

Fig. 3. The figure from left to right shows the Properties 1, 2 and 3 respectively. If the dark areas are known the rest of the matrix is uniquely defined

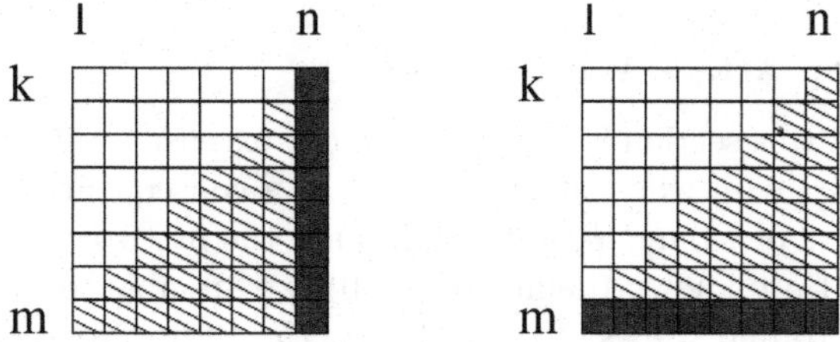

Fig. 4. Illustration of the Properties 4 and 5. If the bottom row or the rightmost column is known, the shaded triangle is uniquely defined

Property 5: Let $k < m$ and $l < n$, such that $n - l = m - k$. If the elements $(T_{m,j})_{j=l,l+1,\ldots,n}$ from row m is known then half the square matrix $(T_{i,j})_{j=l,l+1,\ldots,n}^{i=k,k+1,\ldots,m}$ is uniquely determined under the diagonal $(T_{i,j})_{j=n+k-i,(n+k-i)+1,\ldots,n}^{i=k,k+1,\ldots,m}$ using rule 2 (Figure 4).

Observe that the Properties 4 and 5 are similar and define exactly the same triangle, and that the Properties 1, 2 and 3 define the same rectangle. In the attacks of the compression function it is useful to denote the leftmost 17, the middle 17 and the rightmost 17 columns of the matrix T by (the matrices) T_1,

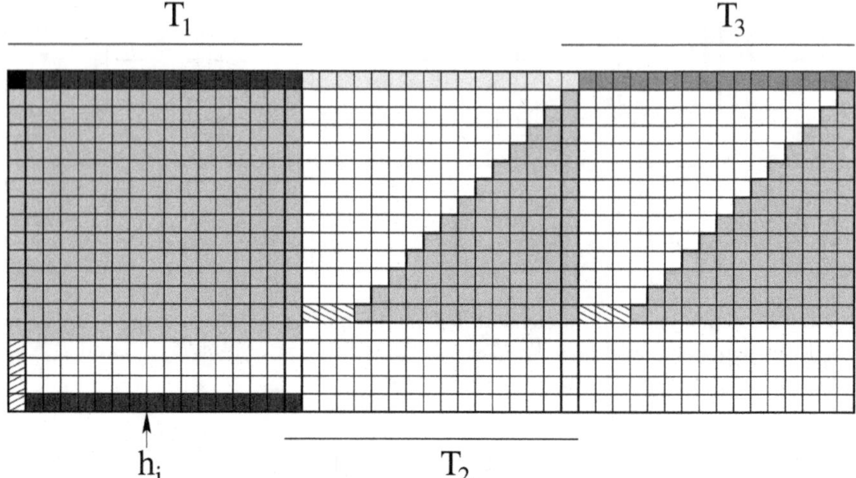

Fig. 5. The figure shows the collision attack on the compression function where $h_{i-1} = 0$. The dark areas are processed line by line

T_2, respectively T_3 as shown in Figure 1. Notice that the first and last column of T_2 overlap with the last column of T_1 and the first column of T_3.

B Collision Attacks on the Compression Function of MD2

B.1 Collision Attack Where • $_{i-1} = 0$

The first part of this section is from [6] with our extensions at the end. We shall consider a special case where $h_{i-1} = 0$ and as a consequence $m_i = h_{i-1} \oplus m_i$ and the first rows of T_2 and T_3 are equal. Since the first row of T_1 and the first element in row 1 are known (defined to be 0), we are able to calculate row 1 of T_1. Now we try to find values of m_i such that the 13 first rows of T_2 and T_3 are equal, and in order to be equal the leftmost columns of T_2 and T_3 must be equal and the rightmost columns of T_2 and T_3 must be equal. Since the rightmost column of T_2 coincide with the leftmost column of T_3, the four of them must be equal in order for the matrices to be equal. Having the rightmost element $(T_1)_{1,16}$ in the first row of T_1, we know that we must have:

$$(T_1)_{1,16} = (T_2)_{1,0} = (T_3)_{1,0} = (T_2)_{1,16} = (T_3)_{1,16} = T_{1,48}$$

and if we know $T_{1,48}$ we know that $T_{2,0} = T_{1,48} + 0 \bmod 256$, so it is simple to complete row 2 of T_1. We continue until row k:

$$(T_1)_{i,16} = (T_2)_{i,16} = (T_3)_{i,16} \text{ for } i = 1, 2, \ldots, k$$

and calculate row $k+1$ of T_1.

The k values in the right column of T_2 and T_3 are now known and we might complete a triangle in the rows $1, 2, \ldots, k$ of these two matrices according to property 2, shown in Figure 5. The figure shows the situation where 13 rows ($k = 13$) are preprocessed and the triangles are completed, and there are 3 remaining bytes to be chosen to complete row 13 of T_2 and T_3. The 2^{24} possible choices of these bytes will determine 2^{24} different first rows $m_i = h_{i-1} \oplus m_i$ (property 3) and will complete row 13 in both of these matrices, and since the first 14 rows of T_1 is already fixed we have a multi collision in:

$$((T_1)_{i,0})_{i=1,2,\ldots,14}$$

containing $(2^8)^3$ different messages m_i. It remains to find collisions among these in the last 4 rows of column 0:

$$((T_1)_{i,0})_{i=15,16,17,18}$$

and equal values in row 0 and column 0 of T_1 give an equal matrix by property 1, and we also have collisions in 16 bytes of the last row of T_1, which is the chaining variable h_i. The expected number of collisions in this case is approximately

$$(((2^8)^3)^2/2)/((2^8)^4) = 2^{15} = 32768$$

in theory, and we found 32784 collisions in practice. In [6] $k = 14$ and 2 bytes are varied, and the expected number of collisions were 128 and in practice there were 141 collisions, but to decrease k to get more collisions is not mentioned explicitly in the paper.

In general we would expect

$$(((2^8)^{16-k})^2/2)/((2^8)^{18-(k+1)}) = 2^{8(15-k)-1}$$

collisions, only depending on the choice of k. The memory and computational complexity is proportional to the number of bytes varied: $2^{8(16-k)}$.

In the preimage attack described earlier in this paper it is advantageous to use this attack when $h_0 = 0$ and to get collisions in m_1. It is possible to get more than 2 different m_1 such that all of them give the same output h_1, and if so we have a multiple collision. If we look for a d-tuple collision and we are able to vary $b = 16 - k$ bytes in the first phase of the attack, we expect

$$\binom{2^{8b}}{d} / 2^{8(b+1)(d-1)} \approx 2^{8(b+1-d)}/d!$$

d-tuple collisions. If $b = 9$ and $d = 8$ we expect $\approx 2^{0,7} \geq 1$ multiple collisions of size 8, and the complexity is approximately 2^{72}.

There are similar attacks on the compression function where $m_i = 0$ or where $h_{i-1} \oplus m_i = 0$. For these two attacks and the one where $h_{i-1} = 0$ there are generalizations which are described in detail in an extended version of the paper available at `http://www.ii.uib.no/~johnm/publications/md2-procExtended.pdf`

How to Enhance the Security of the 3GPP Confidentiality and Integrity Algorithms

Tetsu Iwata and Kaoru Kurosawa

Dept. of Computer and Information Sciences,
Ibaraki University,
4–12–1 Nakanarusawa, Hitachi, Ibaraki 316-8511, Japan
{iwata, kurosawa}@cis.ibaraki.ac.jp

Abstract. We consider the 3GPP confidentiality and integrity schemes that were adopted by Universal Mobile Telecommunication System, an emerging standard for third generation wireless communications. The schemes, known as $f8$ and $f9$, are based on the block cipher KASUMI. Although previous works claim security proofs for $f8$ and $f9'$, where $f9'$ is a generalized version of $f9$, it was shown that these proofs are incorrect; it is *impossible* to prove $f8$ and $f9'$ secure under the standard PRP assumption on the underlying block cipher. Following the results, it was shown that it is *possible* to prove $f8'$ and $f9'$ secure if we make the assumption that the underlying block cipher is a secure PRP-RKA against a certain class of related-key attacks; here $f8'$ is a generalized version of $f8$. Needless to say, the assumptions here are stronger than the standard PRP assumptions, and it is natural to seek a *practical* way to modify $f8'$ and $f9'$ to establish security proofs under the standard PRP assumption. In this paper, we propose $f8^+$ and $f9^+$, slightly modified versions of $f8'$ and $f9'$, but they allow proofs of security under the standard PRP assumption. Our results are practical in the sense that we insist on the minimal modifications; $f8^+$ is obtained from $f8'$ by setting the key modifier to all-zero, and $f9^+$ is obtained from $f9'$ by setting the key modifier to all-zero, and using the encryptions of two constants in the CBC MAC computation.

1 Introduction

Background. Within the security architecture of the 3rd Generation Partnership Project (3GPP) system there are two standardized constructions: A confidentiality scheme $f8$, and an integrity scheme $f9$ [1]. 3GPP is the body standardizing the next generation of mobile telephony. Both $f8$ and $f9$ are modes of operations based on the block cipher KASUMI [2]. $f8$ is a symmetric encryption scheme which is a variant of the Output Feedback (OFB) mode with full feedback, and $f9$ is a Message Authentication Code (MAC) which is a variant of the CBC MAC.

Provable Security. Provable security is a standard security goal for block cipher modes of operations. Indeed, many of the block cipher modes of operations are

H. Gilbert and H. Handschuh (Eds.): FSE 2005, LNCS 3557, pp. 268–283, 2005.

provably secure assuming that the underlying block cipher is a secure pseudorandom permutation, or a super-pseudorandom permutation [25]. For example, we have CTR mode [3] and CBC encryption mode [3] for symmetric encryption schemes, PMAC [9], XCBC [8] and OMAC [16] for message authentication codes, and IAPM [20], OCB mode [26], CCM mode [27, 19], EAX mode [6], CWC mode [23] and GCM mode [24] for authenticated encryption schemes.

Therefore, it is natural to ask whether $f8$ and $f9$ are provably secure if the underlying block cipher is a secure pseudorandom permutation. Making this assumption, it was claimed that $f8$ is a secure symmetric encryption scheme in the sense of left-or-right indistinguishability [21] and that $f9'$ is a secure MAC [13], where $f9'$ is a generalized version of $f9$. However, these claims were disproven [17]. One of the remarkable aspects of $f8$ and $f9$ is the use of a non-zero constant called a "key modifier," or KM. In the $f8$ and $f9$ schemes, KASUMI is keyed with K and $K \oplus \mathrm{KM}$. The paper [17] constructs a secure pseudorandom permutation F with the following property: For any key K, the encryption function with key K is the decryption function with $K \oplus \mathrm{KM}$. That is, $F_K(\cdot) = F^{-1}_{K \oplus \mathrm{KM}}(\cdot)$. Then it was shown that $f8$ and $f9'$ are insecure if F is used as the underlying block cipher. This result shows that it is *impossible* to prove the security of $f8$ and $f9'$ even if the underlying block cipher is a secure pseudorandom permutation.

Generalized Versions of $f8$ and $f9$: $f8'$ and $f9'$. Given the results in [17], it is logical to ask if there are assumptions under which $f8$ and $f9$ are actually secure and, if so, what those assumptions are. Because of the constructions' use of keys related by fixed xor differences, the natural conjecture is that if the constructions are actually secure, then the minimum assumption on the block cipher must be that the block cipher is secure against some class of xor-restricted related-key attacks, as introduced in [7] and formalized in [5].

The paper [14] proved that the above hypotheses are in fact correct and, in doing so, [14] clarifies what assumptions are actually necessary in order for the $f8$ and $f9$ modes to be secure. In more detail, [14] first considers a generalized version of $f8$, which is called $f8'$. $f8'$ is a nonce-based symmetric encryption scheme, and is the natural nonce-based extension of the original $f8$. Then it is shown that $f8'$ is a secure nonce-based deterministic symmetric encryption mode in the sense of indistinguishability from random strings if the underlying block cipher is secure against related-key attacks in which an adversary is able to obtain chosen-plaintext samples of the underlying block cipher using two keys related by a fixed known xor difference.

Then [14] next considers a generalized version of $f9$, which is called $f9'$. $f9'$ is a deterministic MAC, and is a natural extension of $f9$ that gives the user, or adversary, more liberty in controlling the input to the underlying CBC MAC core. Then it is shown that $f9'$ is a secure pseudorandom function, which provably implies a secure MAC, if the underlying block cipher resists related-key attacks in which an adversary is able to obtain chosen-plaintext samples of the underlying block cipher using two keys related by a fixed known xor difference.

Our Contribution. Because the assumptions made for $f8'$ and $f9'$ are stronger than the standard PRP assumptions (as proven necessary in [17]), in this paper, we consider the following question; What is the minimal modification on $f8'$ and $f9'$ to achieve the provable security results with the standard PRP assumptions on the underlying block cipher? We view the answer to this question gives us an important practical result. Namely, $f8$ and $f9$ can be easily replaced with minimal cost, especially to be prepared for the worst case that KASUMI is known to be vulnerable to related-key attacks.

In this paper, we propose $f8^+$ and $f9^+$, refinements of $f8'$ and $f9'$. Unlike $f8'$ and $f9'$, our $f8^+$ and $f9^+$ are provably secure with the standard PRP assumptions. Furthermore, they require very small modifications to $f8'$ and $f9'$. In particular,

- $f8^+$ is obtained from $f8'$ by setting the key modifier to all-zero, and
- $f9^+$ is obtained from $f9'$ by setting the key modifier to all-zero, using the encryption of all-zero as the initial value of the CBC chain, and xoring the encryption of all-one before the final encryption.

These small modifications *increase* the security, allowing us to prove the security of $f8^+$ and $f9^+$ under the standard PRP assumption. Intuitively, this implies that the security of $f8^+$ and $f9^+$ is irrelevant to the resistance of KASUMI against related key attacks. $f8^+$ and $f9^+$ are provably secure if KASUMI is merely secure in the sense of a PRP.

Our results are practical in the sense that we insist on the "minimal modification," and therefore we are able to switch the modes easily. Although $f8^+$ and $f9^+$ are not competitive to CTR mode and OMAC in terms of efficiency, we find that switching to CTR mode and OMAC are costly and expensive, and it is quite unreasonable to switch to them just because to reduce the security assumption on the block cipher.

We suggest the following use of our results. (1) If there is a chance to replace $f8$ and $f9$, especially to be prepared for the worst case that KASUMI is known to be vulnerable to related-key attacks, $f8^+$ and $f9^+$ are reasonable replacements since they only require small modifications and the costs for switching should not be too expensive. These modifications may be handled by "patching" $f8$ and $f9$. (2) When the whole system is updated, the future system should support more conventional modes such as CTR mode and OMAC. In this case, the cost for replacement should not be a problem.

We prove that $f8^+$ is secure in the sense of indistinguishability from random strings, and $f9^+$ is a secure pseudorandom function, which provably implies a secure MAC, if the underlying block cipher is secure in the sense of a PRP. We note that because of the "key reuse" nature in $f8^+$ and $f9^+$, their security proofs require much elaborate treatment compared to the cases for $f8'$ and $f9'$.

Related Works. Initial security evaluation of KASUMI, $f8$ and $f9$ can be found in [12]. Knudsen and Mitchell analyzed the security of $f9'$ against forgery and key recovery attacks [22]. Blunden and Escott showed related key attacks on reduced round KASUMI [10].

2 Preliminaries

Notation. If x is a string then $|x|$ denotes its length in bits. If x and y are two equal-length strings, then $x \oplus y$ denotes the xor of x and y. If x and y are strings, then $x \| y$ denotes their concatenation. Let $x \leftarrow y$ denote the assignment of y to x. If X is a set, let $x \stackrel{R}{\leftarrow} X$ denote the process of uniformly selecting at random an element from X and assigning it to x. If $F : \{0,1\}^k \times \{0,1\}^n \to \{0,1\}^m$ is a family of functions from $\{0,1\}^n$ to $\{0,1\}^m$ indexed by keys $\{0,1\}^k$, then we use the notation $F_K(D)$ as shorthand for $F(K,D)$. We say F is a family of permutations, i.e., a block cipher, if $n = m$ and $F_K(\cdot)$ is a permutation on $\{0,1\}^n$ for each $K \in \{0,1\}^k$. Let $\mathrm{Rand}(n,m)$ denote the set of all functions from $\{0,1\}^n$ to $\{0,1\}^m$. When we refer to the time of an algorithm or experiment in the provable security sections of this paper, we include the size of the code (in some fixed encoding). There is also an implicit big-$\mathcal{O}$ surrounding all such time references.

PRPs. The PRP notion was introduced in [25] and later made concrete in [4].

Let $\mathrm{Perm}(n)$ denote the set of all permutations on $\{0,1\}^n$, and let $E : \{0,1\}^k \times \{0,1\}^n \to \{0,1\}^n$ be a family of permutations, i.e., a block cipher. Let $\mathcal{A}$ be an adversary with access to an oracle and returns a bit. Then

$$\mathbf{Adv}_E^{\mathrm{prp}}(\mathcal{A}) \stackrel{\mathrm{def}}{=} \left| \Pr(K \stackrel{R}{\leftarrow} \{0,1\}^k : \mathcal{A}^{E_K(\cdot)} = 1) - \Pr(G \stackrel{R}{\leftarrow} \mathrm{Perm}(n) : \mathcal{A}^{G(\cdot)} = 1) \right|$$

is defined as the *PRP-advantage* of $\mathcal{A}$ on E. Intuitively, we say that E is a *secure PRP* if the PRP-advantage of all adversaries using reasonable resources is small.

We briefly remark that modern block ciphers, e.g., AES [11], are designed to be secure PRP.

3 Specifications of • 8, • 9, • 8′ and • 9′

3GPP Confidentiality Algorithm f8 [1]. $f8$ is a symmetric encryption scheme standardized by 3GPP [1]. It uses a block cipher $\mathrm{KASUMI} : \{0,1\}^{128} \times \{0,1\}^{64} \to \{0,1\}^{64}$ as the underlying primitive. The $f8$ key generation algorithm returns a random 128-bit key K. The $f8$ encryption algorithm takes a 128-bit key K, a 32-bit counter COUNT, a 5-bit radio bearer identifier BEARER, a 1-bit direction identifier DIRECTION, and a message $M \in \{0,1\}^*$ to return a ciphertext C, which is the same length as M. Also, it uses a 128-bit constant $\mathrm{KM} = (01)^{64}$ (or `0x55...55` in hexadecimal) called the key modifier. In more detail, the encryption algorithm is defined in Fig. 1. In Fig. 1, $[i-1]_{64}$ denotes the 64-bit binary representation of $i-1$. The decryption algorithm, which takes COUNT, BEARER, DIRECTION, and a ciphertext C as input and returns a plaintext M, is defined in the natural way.

[1] The original specification [1] refers $f8$ as a symmetric synchronous stream cipher. The specification presented here is fully compatible with the original one.

Algorithm $\text{f8-Encrypt}_K(\text{COUNT}, \text{BEARER}, \text{DIRECTION}, M)$
$m \leftarrow \lceil |M|/64 \rceil$
$Y[0] \leftarrow 0^{64}$
$A \leftarrow \text{COUNT} \| \text{BEARER} \| \text{DIRECTION} \| 0^{26}$
$A \leftarrow \text{KASUMI}_{K \oplus \text{KM}}(A)$
For $i \leftarrow 1$ to m do:
$X[i] \leftarrow A \oplus [i-1]_{64} \oplus Y[i-1]$
$Y[i] \leftarrow \text{KASUMI}_K(X[i])$
$C \leftarrow M \oplus$ (the leftmost $|M|$ bits of $Y[1] \| \cdots \| Y[m]$)
Return C

Fig. 1. Algorithm $\text{f8-Encrypt}_K(\text{COUNT}, \text{BEARER}, \text{DIRECTION}, M)$

Since we analyze and prove results about a variant of $f8$ whose encryption algorithm takes a nonce as input in lieu of COUNT, BEARER, and DIRECTION, we do not describe the specifics of how COUNT, BEARER, and DIRECTION are used in real 3GPP applications. We do note that 3GPP applications will never invoke the $f8$ encryption algorithm twice with the same (COUNT, BEARER, DIRECTION) triple, which means that our nonce-based variant is appropriate.

3GPP Integrity Algorithm f9 [1]. $f9$ is a message authentication code standardized by 3GPP. It uses KASUMI as the underlying primitive. The $f9$ key generation algorithm returns a random 128-bit key K. The $f9$ tagging algorithm takes a 128-bit key K, a 32-bit counter COUNT, a 32-bit random number FRESH, a 1-bit direction identifier DIRECTION, and a message $M \in \{0,1\}^*$ and returns a 32-bit tag T. It uses a 128-bit constant $\text{KM} = (10)^{64}$ (or `0xAA...AA` in hexadecimal), called the key modifier.

Let $M = M[1] \| \cdots \| M[m]$ be a message, where each $M[i]$ $(1 \le i \le m-1)$ is 64 bits. The last block $M[m]$ may have fewer than 64 bits. We define $\text{pad}_{64}(\text{COUNT}, \text{FRESH}, \text{DIRECTION}, M)$ as follows: It concatenates COUNT, FRESH, M and DIRECTION, and then appends a single "1" bit, followed by between 0 and 63 "0" bits so that the total length is a multiple of 64 bits. More precisely,

$$\text{pad}_{64}(\text{COUNT}, \text{FRESH}, \text{DIRECTION}, M) \\ = \text{COUNT} \| \text{FRESH} \| M \| \text{DIRECTION} \| 1 \| 0^{63-(|M|+1 \bmod 64)} .$$

Then the tagging algorithm is defined in Fig. 2. In Fig. 2, "$M[1] \| \cdots \| M[m] \leftarrow M$" is a shorthand for "break M into 64-bit blocks $M[1] \| \cdots \| M[m]$." The $f9$ verification algorithm is defined in the natural way by tag recomputation.

As with $f8$, since we analyze and prove the security of a generalized version of $f9$, we do not describe how COUNT, FRESH, and DIRECTION are used in real 3GPP applications.

A Generalized Version of f8: f8′ [17, 14]. $f8'$ is a nonce-based deterministic symmetric encryption scheme, which is a generalized (and weakened) version

```
Algorithm f9-Tag_K(COUNT, FRESH, DIRECTION, M)
    M ← pad_64(COUNT, FRESH, DIRECTION, M)
    M[1]||···||M[m] ← M
    Y[0] ← 0^64
    For i ← 1 to m do:
        X[i] ← M[i] ⊕ Y[i − 1]
        Y[i] ← KASUMI_K(X[i])
    T ← KASUMI_{K⊕KM}(Y[1] ⊕ ··· ⊕ Y[m])
    T ← the leftmost 32 bits of T
    Return T
```

Fig. 2. Algorithm $\mathsf{f9\text{-}Tag}_K(\mathrm{COUNT}, \mathrm{FRESH}, \mathrm{DIRECTION}, M)$

of $f8$. It uses a block cipher $E : \{0,1\}^k \times \{0,1\}^n \to \{0,1\}^n$ as the underlying primitive. Let $f8'[E, \Delta]$ be $f8'$, where E is used as the underlying primitive and Δ is a non-zero k-bit key modifier. The $f8'[E, \Delta]$ key generation algorithm returns a random k-bit key K. The $f8'[E, \Delta]$ encryption algorithm, which we call f8′-Encrypt, takes an n-bit nonce N instead of COUNT, BEARER and DIRECTION. That is, the encryption algorithm takes a k-bit key K, an n-bit nonce N, and a message $M \in \{0,1\}^*$ to return a ciphertext C, which is the same length as M. Then the encryption algorithm is in Fig. 3. In Fig. 3, $[i-1]_n$ denotes n-bit binary representation of $i - 1$. Decryption is done in an obvious way.

Notice that we treat COUNT, BEARER and DIRECTION as a nonce. That is, we allow the adversary to choose these values. Consequently, $f8'$ can be considered as a weakened version of $f8$ since it gives the adversary the ability to control the entire initial value of A, rather than only a subset of the bits as would be the case for an adversary attacking $f8$.

A Generalized Version of f9: f9′ [13, 22, 17, 14]. The message authentication code $f9'$ is a generalized (and weakened) version of $f9$ that gives the user (or adversary) almost complete control over the input the underlying CBC MAC core. It uses a block cipher $E : \{0,1\}^k \times \{0,1\}^n \to \{0,1\}^n$ as the underlying primitive. Let $f9'[E, \Delta, l]$ be $f9'$, where E is used as the underlying block cipher, Δ is a non-zero k-bit key modifier, and the tag length is l, where $1 \le l \le n$. The key generation algorithm returns a random k-bit key K. The tagging algorithm, which we call f9′-Tag, takes a k-bit key K and a message $M \in \{0,1\}^*$ as input and returns an l-bit tag T.

Let $M = M[1]\|\cdots\|M[m]$ be a message, where each $M[i]$ $(1 \le i \le m-1)$ is n bits. The last block $M[m]$ may have fewer than n bits. In $f9'$, we use pad'_n instead of pad_{64}. $\mathsf{pad}'_n(M)$ works as follows: It simply appends a single "1" bit, followed by between 0 and $n-1$ "0" bits so that the total length is a multiple of n bits. More precisely,

$$\mathsf{pad}'_n(M) = M\|1\|0^{n-1-(|M| \bmod n)} \ . \qquad (1)$$

Thus, we simply ignore COUNT, FRESH and DIRECTION. Equivalently, we consider them as a part of the message. The rest of the tagging algorithm is the same as $f9$. The pseudocode is given in Fig. 5. In Fig. 5, "$M[1]\|\cdots\|M[m] \leftarrow M$" is a shorthand for "break M into n-bit blocks $M[1]\|\cdots\|M[m]$."

Note that the adversary is allowed to choose COUNT, FRESH, and DIRECTION since $f9'$ treats them as a part of the message. In this sense, $f9'$ can be considered as a weakened version of $f9$.

4 Proposed Schemes: Specifications of • 8^+ and • 9^+

4.1 Proposed Refinement of • $8'$: • 8^+

$f8^+$ is a nonce-based deterministic symmetric encryption scheme, which is a refinement of $f8'$. Definening $f8^+$ is simple: we set $\Delta \leftarrow 0^n$ in $f8'$.

For full specification, $f8^+$ uses a block cipher $E : \{0,1\}^k \times \{0,1\}^n \to \{0,1\}^n$ as the underlying primitive. Let $f8^+[E]$ be $f8^+$, where E is used as the underlying primitive. The $f8^+[E]$ key generation algorithm returns a random k-bit key K. The $f8^+[E]$ encryption algorithm, which we call $\mathsf{f8}^+$-$\mathsf{Encrypt}$, takes an n-bit nonce N. That is, the encryption algorithm takes a k-bit key K, an n-bit nonce N, and a message $M \in \{0,1\}^*$ to return a ciphertext C, which is the same length as M. Then the encryption algorithm is given in Fig. 4.

Algorithm $\mathsf{f8}'$-$\mathsf{Encrypt}_K(N, M)$
 $m \leftarrow \lceil |M|/n \rceil$
 $Y[0] \leftarrow 0^n$
 $A \leftarrow N$
 $A \leftarrow E_{K\oplus\Delta}(A)$
 For $i \leftarrow 1$ to m do:
  $X[i] \leftarrow A \oplus [i-1]_n \oplus Y[i-1]$
  $Y[i] \leftarrow E_K(X[i])$
 $C \leftarrow M \oplus$ (the leftmost $|M|$ bits of $Y[1]\|\cdots\|Y[m]$)
 Return C

Fig. 3. Algorithm $\mathsf{f8}'$-$\mathsf{Encrypt}_K(N, M)$

Algorithm $\mathsf{f8}^+$-$\mathsf{Encrypt}_K(N, M)$
 $m \leftarrow \lceil |M|/n \rceil$
 $Y[0] \leftarrow 0^n$
 $A \leftarrow N$
 $A \leftarrow E_K(A)$
 For $i \leftarrow 1$ to m do:
  $X[i] \leftarrow A \oplus [i-1]_n \oplus Y[i-1]$
  $Y[i] \leftarrow E_K(X[i])$
 $C \leftarrow M \oplus$ (the leftmost $|M|$ bits of $Y[1]\|\cdots\|Y[m]$)
 Return C

Fig. 4. Algorithm $\mathsf{f8}^+$-$\mathsf{Encrypt}_K(N, M)$

Decryption is done in an obvious way.

Note that the only difference between $\mathsf{f8}^+$-$\mathsf{Encrypt}_K(\cdot,\cdot)$ and $\mathsf{f8}'$-$\mathsf{Encrypt}_K(\cdot,\cdot)$ is in the 4-th line. With this small modification, as we will show shortly, $f8^+$ has much higher security assurance than that of $f8'$.

4.2 Proposed Refinement of • $9'$: • 9^+

$f9^+$ is a message authentication code, which is a refinement of $f9'$. As with $f8^+$, definening $f9^+$ is simple:

- We set $\Delta \leftarrow 0^n$ in $f9'$,
- We set the initial value of the CBC chain to be $E_K(0^n)$ instead of 0^n, and
- We xor $E_K(1^n)$ before the last encryption.

For full specification, $f9^+$ uses a block cipher $E : \{0,1\}^k \times \{0,1\}^n \to \{0,1\}^n$ as the underlying primitive. Let $f9^+[E,l]$ be $f9^+$, where E is used as the underlying block cipher, and the tag length is l, where $1 \le l \le n$. The key generation algorithm returns a random k-bit key K. The tagging algorithm, which we call $\mathsf{f9^+\text{-}Tag}$, takes a k-bit key K and a message $M \in \{0,1\}^*$ as input and returns an l-bit tag T. $f9^+$ uses $\mathsf{pad}'_n(\cdot)$ defined in (1). The pseudocode is given in Fig. 6. In Fig. 6, "$M[1]\|\cdots\|M[m] \leftarrow M$" is a shorthand for "break M into n-bit blocks $M[1]\|\cdots\|M[m]$."

Algorithm $\mathsf{f9'\text{-}Tag}_K(M)$
- $M \leftarrow \mathsf{pad}'_n(M)$
- $M[1]\|\cdots\|M[m] \leftarrow M$
- $Y[0] \leftarrow 0^n$
- For $i \leftarrow 1$ to m do:
  - $X[i] \leftarrow M[i] \oplus Y[i-1]$
  - $Y[i] \leftarrow E_K(X[i])$
- $T \leftarrow E_{K\oplus\Delta}(Y[1] \oplus \cdots \oplus Y[m])$
- $T \leftarrow$ the leftmost l bits of T
- Return T

Fig. 5. Algorithm $\mathsf{f9'\text{-}Tag}_K(M)$

Algorithm $\mathsf{f9^+\text{-}Tag}_K(M)$
- $M \leftarrow \mathsf{pad}'_n(M)$
- $M[1]\|\cdots\|M[m] \leftarrow M$
- $Y[0] \leftarrow E_K(0^n)$
- For $i \leftarrow 1$ to m do:
  - $X[i] \leftarrow M[i] \oplus Y[i-1]$
  - $Y[i] \leftarrow E_K(X[i])$
- $T \leftarrow E_K(Y[1] \oplus \cdots \oplus Y[m] \oplus E_K(1^n))$
- $T \leftarrow$ the leftmost l bits of T
- Return T

Fig. 6. Algorithm $\mathsf{f9^+\text{-}Tag}_K(M)$

The verification algorithm is defined in the natural way.

Notice that the only difference between $\mathsf{f9^+\text{-}Tag}_K(\cdot)$ and $\mathsf{f9'\text{-}Tag}_K(\cdot)$ is in the 3-rd and 7-th lines. With these small modifications, as we will show shortly, $f9^+$ has much higher security assurance than that of $f9'$.

4.3 Design Rational on • 9^+

One might try to set $\Delta \leftarrow 0^n$ in $f9'$ (which eliminates the needs of related keys), and preserve the rest of $f9'$. Unlike the case for $f8^+$, this does not work. In fact, the above mentioned MAC is easily forgeable. The attack proceeds as follows.

1. The adversary $\mathcal{A}$ first queries a message M_1 such that $1 \le |M_1| < n$, to obtain the tag $T_1 = E_K(E_K(\mathsf{pad}'_n(M_1)))$.
2. Then $\mathcal{A}$ queries a message $M_2 \leftarrow \mathsf{pad}'_n(M_1)\|0^n\|M'_2$, where M'_2 is a string such that $|M'_2| < n$ and $\mathsf{pad}'_n(M'_2) = T_1 \oplus \mathsf{pad}'_n(M_1)$, to obtain the tag T_2. By a simple calculation, one can verify that $T_2 = E_K(T_1)$.
3. Next $\mathcal{A}$ queries a message $M_3 \leftarrow T_1\|M'_3$, where M'_3 is a string such that $|M'_3| < n$ and $\mathsf{pad}'_n(M'_3) = T_1 \oplus T_2$, to obtain the tag T_3. By a simple calculation, one can verify that $T_3 = E_K(0^n)$.

4. Finally, $\mathcal{A}$ outputs a forgery attempt (M^*, T^*), where $M^* \leftarrow 0^n \| T_3'$, T_3' is a string such that $|T_3'| < n$ and $\mathsf{pad}_n'(T_3') = T_3$, and $T^* \leftarrow T_3$.

There are several cases where this attack fails. These cases are $T_1 \oplus \mathsf{pad}_n'(M_1) = 0^n$ (since M_2' does not exist), $T_1 \oplus T_2 = 0^n$ (since M_3' does not exist), and $M^* = M_3$ (since M^* should be a new massage). Assuming that the underlying block cipher is a random permutation, these cases occur with only negligible probabilities. Therefore, the above attack succeeds in forgery with overwhelming probability even if the underlying block cipher is a random permutation [2].

Therefore, merely setting $\Delta \leftarrow 0^n$ does not work. This motivates us to "mask" the input to the first block cipher invocation, as well as the final invocation. We used $Y[0] \leftarrow E_K(0^n)$ and $E_K(1^n)$ as masks, however, any other constant is fine.

5 Security of • 8+

Definitions. Before proving the security of $f8^+$, we first formally define what we mean by a nonce-based encryption scheme, and what it means for such an encryption scheme to be secure.

A nonce-based symmetric encryption scheme $\mathcal{SE} = (\mathcal{K}, \mathcal{E}, \mathcal{D})$ consists of three algorithms and is defined for some nonce length n. The randomized key generation algorithm $\mathcal{K}$ takes no input and returns a random key K. The stateless and deterministic encryption algorithm takes a key K, a nonce $N \in \{0,1\}^n$, and a message $M \in \{0,1\}^*$ as input and returns a ciphertext C such that $|C| = |M|$; we write $C \leftarrow \mathcal{E}_K(N, M)$. The stateless and deterministic decryption algorithm takes a key K, a nonce $N \in \{0,1\}^n$, and a ciphertext $C \in \{0,1\}^*$ as input and returns a message M such that $|M| = |C|$; we write $M \leftarrow \mathcal{D}_K(N, C)$. For consistency, we require that for all keys K, nonces N, and messages M, $\mathcal{D}_K(N, \mathcal{E}_K(N, M)) = M$.

We adopt the strong notion of privacy for nonce-based encryption schemes from [26]. This notion, which we call indistinguishability from random strings, provably implies the more standard notions given in [3]. Let $\$(\cdot,\cdot)$ denote an oracle that on input a pair of strings (N, M) returns a random string of length $|M|$. If $\mathcal{A}$ is an adversary with access to an oracle, then

$$\mathbf{Adv}^{\mathrm{priv}}_{\mathcal{SE}}(\mathcal{A}) \stackrel{\mathrm{def}}{=} \left| \Pr(K \stackrel{R}{\leftarrow} \mathcal{K} : \mathcal{A}^{\mathcal{E}_K(\cdot,\cdot)} = 1) - \Pr(\mathcal{A}^{\$(\cdot,\cdot)} = 1) \right|$$

is defined as the *PRIV-advantage* of $\mathcal{A}$ in distinguishing the outputs of the encryption algorithm with a randomly selected key from random strings. We say that $\mathcal{A}$ is nonce-respecting if it never queries its oracle twice with the same nonce value. Intuitively, we say that an encryption scheme *preserves privacy under chosen-plaintext attacks* if the PRIV-advantage of all nonce-respecting adversaries $\mathcal{A}$ using reasonable resources is small.

[2] We note that [13–p. 157, Section 2.2] shows similar attack. But the attack in [13] cannot be applied here since padding is not considered.

Provable Security Results. Let $p8^+[n]$ be a variant of $f8^+$ that uses a random function on n bits instead of E_K. Specifically, the key generation algorithm for $p8^+[n]$ returns a randomly selected function R from $\mathrm{Rand}(n, n)$. The encryption algorithm for $p8^+[n]$, $\mathsf{p8^+\text{-}Encrypt}$, takes R as a "key" and uses it instead of E_K. The decryption algorithm is defined in the natural way.

We first upper-bound the advantage of an adversary in breaking the privacy of $p8^+[n]$. Let (N_i, M_i) denote a privacy adversary's i-th oracle query. If the adversary makes exactly q oracle queries, then we define the total number of blocks for the adversary's queries as $\sigma = \sum_{1 \le i \le q} \lceil |M_i|/n \rceil$.

Lemma 1. *Let $p8^+[n]$ be as described above and let $\mathcal{A}$ be a nonce-respecting privacy adversary which asks at most q queries totaling at most σ blocks. Then*

$$\mathbf{Adv}^{\mathrm{priv}}_{p8^+[n]}(\mathcal{A}) \le \frac{2\sigma^2}{2^n} \ . \tag{2}$$

A proof sketch is given in Appendix A, and a proof is given in the full version of this paper [18].

We now present our main result for $f8^+$ (Theorem 1 below). At a high level, our theorem shows that if a block cipher E is a secure PRP, then the construction $f8^+[E]$ based on E will be a provably secure encryption scheme. In more detail, our theorem states that given any adversary $\mathcal{A}$ attacking the privacy of $f8^+[E]$ and making at most q oracle queries totaling at most σ blocks, we can construct a PRP adversary $\mathcal{B}$ attacking E such that $\mathcal{B}$ uses similar resources as $\mathcal{A}$ and $\mathcal{B}$ has advantage $\mathbf{Adv}^{\mathrm{prp}}_{E}(\mathcal{B}) \ge \mathbf{Adv}^{\mathrm{priv}}_{f8^+[E]}(\mathcal{A}) - 4\sigma^2/2^n$. If we assume that E is a secure PRP and that $\mathcal{A}$ (and therefore $\mathcal{B}$) uses reasonable resources, then $\mathbf{Adv}^{\mathrm{prp}}_{E}(\mathcal{B})$ must be small by definition, and thus $\mathbf{Adv}^{\mathrm{priv}}_{f8^+[E]}(\mathcal{A})$ must also be small. This means that under the assumptions on E being a secure PRP, $f8^+[E]$ is provably secure.

Since many block ciphers, including AES and KASUMI, are believed to be a secure PRP, this theorem means that $f8^+$ constructions built from these block ciphers will be provably secure.

Our main theorem statement for $f8^+$ is given below.

Theorem 1 (Main Theorem for $f8^+$). *Let $E : \{0,1\}^k \times \{0,1\}^n \to \{0,1\}^n$ be a block cipher. Let $f8^+[E]$ be as described in Section 4.1. If $\mathcal{A}$ is a nonce-respecting privacy adversary which asks at most q queries totaling at most σ blocks, then we can construct a PRP adversary $\mathcal{B}$ against E such that*

$$\mathbf{Adv}^{\mathrm{priv}}_{f8^+[E]}(\mathcal{A}) \le \frac{4\sigma^2}{2^n} + \mathbf{Adv}^{\mathrm{prp}}_{E}(\mathcal{B}) \ . \tag{3}$$

Furthermore, $\mathcal{B}$ makes at most $\sigma + q$ oracle queries and uses the same time as $\mathcal{A}$.

A proof is done by applying a well known PRF/PRP switching lemma (see [4–Proposition 2.5]), and we add $(q+\sigma)^2/2^{n+1} \le 2\sigma^2/2^n$ to the bound in Lemma 1, and the rest of the proof of Theorem 1 is completely standard.

Notice the difference between Theorem 1 and the result for $f8'$ in [14–p. 435, Theorem 4.1]. $\mathbf{Adv}^{\mathrm{priv}}_{f8'[E,\Delta]}(\mathcal{A})$ is upper bounded by $(3\sigma^2 + q^2)/2^{n+1} +$

$\mathbf{Adv}^{\text{prp-rka}}_{\Phi,E}(\mathcal{B})$. Intuitively, $f8'$ preserves privacy under chosen-plaintext attacks if the pair $(E_K(\cdot), E_{K\oplus\Delta}(\cdot))$ and a pair of two independent random permutations are indistinguishable, while $f8^+$ achieves the same security goal if $E_K(\cdot)$ is a secure PRP.

6 Security of • 9^+

Definitions. Before proving the security of $f9^+$, we formally define what we mean by a MAC, and what it means for a MAC to be secure.

A message authentication scheme or MAC $\mathcal{MA} = (\mathcal{K}, \mathcal{T}, \mathcal{V})$ consists of three algorithms and is defined for some tag length l. The randomized key generation algorithm $\mathcal{K}$ takes no input and returns a random key K. The stateless and deterministic tagging algorithm takes a key K and a message $M \in \{0,1\}^*$ as input and returns a tag $T \in \{0,1\}^l$; we write $T \leftarrow \mathcal{T}_K(M)$. The stateless and deterministic verification algorithm takes a key K, a message $M \in \{0,1\}^*$, and a candidate tag $T \in \{0,1\}^l$ as input and returns a bit b; we write $b \leftarrow \mathcal{V}_K(M,T)$. For consistency, we require that for all keys K and messages M, $\mathcal{V}_K(M, \mathcal{T}_K(M)) = 1$.

For security, we adopt a strong notion of security for MACs, namely pseudorandomness (PRF). In [4] it was proven that if a MAC is secure PRF, then it is also unforgeable. If $\mathcal{A}$ is an adversary with access to an oracle, then

$$\mathbf{Adv}^{\text{prf}}_{\mathcal{MA}}(\mathcal{A}) \stackrel{\text{def}}{=} \left|\Pr(K \stackrel{R}{\leftarrow} \mathcal{K} : \mathcal{A}^{\mathcal{T}_K(\cdot)} = 1) - \Pr(g \stackrel{R}{\leftarrow} \text{Rand}(*, l) : \mathcal{A}^{g(\cdot)} = 1)\right|$$

is defined as the *PRF-advantage* of $\mathcal{A}$ in distinguishing the outputs of the tagging algorithm with a randomly selected key from the outputs of a random function with the same domain and range. Intuitively, we say that a message authentication code is *pseudorandom* or secure if the PRF-advantage of all adversaries $\mathcal{A}$ using reasonable resources is small.

Provable Security Results. Let $p9^+[n]$ be a variant of $f9^+$ that always outputs a full n-bit tag and that uses a random function on n bits instead of E_K. Specifically, the key generation algorithm for $p9^+[n]$ returns a randomly selected functions R from $\text{Rand}(n,n)$. The tagging algorithm for $p9^+[n]$, $\mathsf{p9^+\text{-}Tag}$, takes R as a "key" and uses it instead of E_K. The verification algorithm is defined in the natural way.

We first upper-bound the advantage of an adversary in attacking the pseudorandomness of $p9^+[n]$. Let M_i denote an adversary's i-th oracle query. If an adversary makes exactly q oracle queries, then we define the total number of blocks for the adversary's queries as $\sigma = \sum_{1\le i\le q} |\mathsf{pad}'_n(M_i)|/n$.

Lemma 2. *Let $p9^+[n]$ be as described above and let $\mathcal{A}$ be an adversary which asks at most q queries totaling at most σ blocks. Then*

$$\mathbf{Adv}^{\text{prf}}_{p9^+[n]}(\mathcal{A}) \le \frac{5\sigma^2}{2^n} \ . \tag{4}$$

A proof sketch is given in Appendix B, and a proof is given in the full version of this paper [18].

We now present our main result for $f9^+$ (Theorem 2), which we interpret as follows: our theorem shows that if a block cipher E is a secure PRP, then the construction $f9^+[E,l]$ based on E will be a provably secure message authentication code. In more detail, we show that given any adversary $\mathcal{A}$ attacking $f9^+[E,l]$ and making at most q oracle queries totaling at most σ blocks, we can construct a PRP adversary $\mathcal{B}$ against E such that $\mathcal{B}$ uses similar resources as $\mathcal{A}$ and $\mathcal{B}$ has advantage $\mathbf{Adv}_E^{\mathrm{prp}}(\mathcal{B}) \geq \mathbf{Adv}_{f9^+[E,l]}^{\mathrm{prf}}(\mathcal{A}) - 10\sigma^2/2^n$. If we assume that E is a secure PRP and that $\mathcal{A}$ (and therefore $\mathcal{B}$) uses reasonable resources, then $\mathbf{Adv}_E^{\mathrm{prp}}(\mathcal{B})$ must be small by definition. Therefore $\mathbf{Adv}_{f9^+[E,l]}^{\mathrm{prf}}(\mathcal{A})$ must be small as well, proving that under the assumption on E being a secure PRP, $f9^+[E,l]$ is provably secure.

Since many block ciphers, including AES and KASUMI, are believed to be a secure PRP, this theorem means that $f9^+$ constructions built from these block ciphers will be provably secure.

The precise theorem statement is as follows:

Theorem 2 (Main Theorem for $f9^+$). *Let $E : \{0,1\}^k \times \{0,1\}^n \to \{0,1\}^n$ be a block cipher, and let l, $1 \leq l \leq n$, be a constant. Let $f9^+[E,l]$ be as described in Section 4.2. If $\mathcal{A}$ is a PRF adversary which asks at most q queries totaling at most σ blocks, then we can construct a PRP adversary $\mathcal{B}$ against E such that*

$$\mathbf{Adv}_{f9^+[E,l]}^{\mathrm{prf}}(\mathcal{A}) \leq \frac{10\sigma^2}{2^n} + \mathbf{Adv}_E^{\mathrm{prp}}(\mathcal{B}) \ . \tag{5}$$

Furthermore, $\mathcal{B}$ makes at most $\sigma + q + 2$ oracle queries and uses the same time as $\mathcal{A}$.

As is the case in Theorem 1, a proof is done by applying the PRF/PRP switching lemma, and $(q+\sigma+2)^2/2^{n+1} \leq 9\sigma^2/2^{n+1}$ is added to the bound in Lemma 2, and the rest of the proof is standard.

Notice the difference between Theorem 2 and the result for $f9'$ in [14–p. 438, Theorem 5.1]. $\mathbf{Adv}_{f9'[E,\Delta,l]}^{\mathrm{prf}}(\mathcal{A})$ is upper bounded by $(3q^2+2\sigma^2+2\sigma q)/2^{n+1} + \mathbf{Adv}_{\Phi,E}^{\mathrm{prp\text{-}rka}}(\mathcal{B})$. Intuitively, $f9'$ is pseudorandom if the pair $(E_K(\cdot), E_{K\oplus\Delta}(\cdot))$ and a pair of two independent random permutations are indistinguishable, while $f9^+$ is pseudorandom if $E_K(\cdot)$ is a secure PRP.

7 Conclusion

In this paper, we proposed $f8^+$ and $f9^+$, which are refinements of the original $f8'$ and $f9'$. $f8^+$ and $f9^+$ are designed with two goals; (1) minimal modifications to $f8'$ and $f9'$, and (2) provable security results with the standard PRP assumption on the underlying block cipher. Since we make only "small" modifications, these modes can be practical candidates for future replacement of $f8$ and $f9$. Especially, we believe that $f8^+$ is simple enough to be replaced easily.

References

1. 3GPP TS 35.201 v 3.1.1. Specification of the 3GPP confidentiality and integrity algorithms, Document 1: $f8$ and $f9$ specification. Available at `http://www.3gpp.org/tb/other/algorithms.htm`.
2. 3GPP TS 35.202 v 3.1.1. Specification of the 3GPP confidentiality and integrity algorithms, Document 2: KASUMI specification. Available at `http://www.3gpp.org/tb/other/algorithms.htm`.
3. M. Bellare, A. Desai, E. Jokipii, and P. Rogaway. A concrete security treatment of symmetric encryption. Proceedings of *The 38th Annual Symposium on Foundations of Computer Science, FOCS '97,* pages 394–405, IEEE, 1997.
4. M. Bellare, J. Kilian, and P. Rogaway. The security of the cipher block chaining message authentication code. *JCSS,* vol. 61, no. 3, pages 362–399, 2000. Earlier version in Y. Desmedt, editor, *Advances in Cryptology – CRYPTO '94*, volume 839 of *Lecture Notes in Computer Science*, pages 341–358. Springer-Verlag, Berlin Germany, 1994.
5. M. Bellare, and T. Kohno. A theoretical treatment of related-key attacks: RKA-PRPs, RKA-PRFs, and applications. In E. Biham, editor, *Advances in Cryptology – EUROCRYPT 2003*, volume 2656 of *Lecture Notes in Computer Science*, pages 491–506. Springer-Verlag, Berlin Germany, 2003.
6. M. Bellare, P. Rogaway, and D. Wagner. The EAX mode of operation. In B. Roy and W. Meier, editors, *Fast Software Encryption, FSE 2004*, volume 3017 of *Lecture Notes in Computer Science*, pages 389–407. Springer-Verlag, Berlin Germany, 2004.
7. E. Biham. New types of cryptanalytic attacks using related keys. In T. Helleseth, editor, *Advances in Cryptology – EUROCRYPT '93*, volume 765 of *Lecture Notes in Computer Science*, pages 398–409. Springer-Verlag, Berlin Germany, 1993.
8. J. Black and P. Rogaway. CBC MACs for arbitrary-length messages: The three key constructions. In M. Bellare, editor, *Advances in Cryptology – CRYPTO 2000*, volume 1880 of *Lecture Notes in Computer Science*, pages 197–215. Springer-Verlag, Berlin Germany, 2000.
9. J. Black and P. Rogaway. A block-cipher mode of operation for parallelizable message authentication. In L.R. Knudsen, editor, *Advances in Cryptology – EUROCRYPT 2002*, volume 2332 of *Lecture Notes in Computer Science*, pages 384–397. Springer-Verlag, Berlin Germany, 2002.
10. M. Blunden and A. Escott. Related key attacks on reduced round KASUMI. In M. Matsui, editor, *Fast Software Encryption, FSE 2001*, volume 2355 of *Lecture Notes in Computer Science*, pages 277–285. Springer-Verlag, Berlin Germany, 2002.
11. J. Daemen and V. Rijmen. *The Design of Rijndael.* Springer-Verlag, Berlin Germany, 2002.
12. Evaluation report (version 2.0). Specification of the 3GPP confidentiality and integrity algorithms, Report on the evaluation of 3GPP confidentiality and integrity algorithms. Available at `http://www.3gpp.org/tb/other/algorithms.htm`.
13. D. Hong, J-S. Kang, B. Preneel and H. Ryu. A concrete security analysis for 3GPP-MAC. In T. Johansson, editor, *Fast Software Encryption, FSE 2003*, volume 2887 of *Lecture Notes in Computer Science*, pages 154–169. Springer-Verlag, Berlin Germany, 2003.
14. T. Iwata and T. Kohno. New security proofs for the 3GPP confidentiality and integrity algorithms. In B. Roy and W. Meier, editors, *Fast Software Encryption, FSE 2004*, volume 3017 of *Lecture Notes in Computer Science*, pages 427–445. Springer-Verlag, Berlin Germany, 2004.

15. T. Iwata and T. Kohno. New security proofs for the 3GPP confidentiality and integrity algorithms. Full version of [14], available at IACR Cryptology ePrint Archive, Report 2004/019, `http://eprint.iacr.org/`, 2004.
16. T. Iwata and K. Kurosawa. OMAC: One-Key CBC MAC. In T. Johansson, editor, *Fast Software Encryption, FSE 2003*, volume 2887 of *Lecture Notes in Computer Science*, pages 129–153. Springer-Verlag, Berlin Germany, 2003.
17. T. Iwata and K. Kurosawa. On the correctness of security proofs for the 3GPP confidentiality and integrity algorithms. In K.G. Paterson, editor, *Cryptography and Coding, Ninth IMA International Conference*, volume 2898 of *Lecture Notes in Computer Science*, pages 306–318. Springer-Verlag, Berlin Germany, 2003.
18. T. Iwata and K. Kurosawa. How to enhance the security of the 3GPP confidentiality and integrity algorithms. Full version of this paper, available from the authors, 2005.
19. J. Jonsson. On the Security of CTR + CBC-MAC. In K. Nyberg and H.M. Heys, editors, *Selected Areas in Cryptography, 9th Annual Workshop (SAC 2002)*, volume 2595 of *Lecture Notes in Computer Science*, pages 76–93. Springer-Verlag, Berlin Germany, 2002.
20. C.S. Jutla. Encryption modes with almost free message integrity. In B. Pfitzmann, editor, *Advances in Cryptology – EUROCRYPT 2001*, volume 2045 of *Lecture Notes in Computer Science*, pages 529–544. Springer-Verlag, Berlin Germany, 2001.
21. J-S. Kang, S-U. Shin, D. Hong and O. Yi. Provable security of KASUMI and 3GPP encryption mode $f8$. In C. Boyd, editor, *Advances in Cryptology – ASIACRYPT 2001*, volume 2248 of *Lecture Notes in Computer Science*, pages 255–271. Springer-Verlag, Berlin Germany, 2001.
22. L.R. Knudsen and C.J. Mitchell. Analysis of 3gpp-MAC and two-key 3gpp-MAC. *Discrete Applied Mathematics*, vol. 128, no. 1, pages 181–191, 2003.
23. T. Kohno, J. Viega, and D. Whiting. CWC: A high-performance conventional authenticated encryption mode. In B. Roy and W. Meier, editors, *Fast Software Encryption, FSE 2004*, volume 3017 of *Lecture Notes in Computer Science*, pages 408–426. Springer-Verlag, Berlin Germany, 2004.
24. D.A. McGrew, and J. Viega. The security and performance of the Galois/Counter Mode of operation. In IACR Cryptology ePrint Archive, Report 2004/193, `http://eprint.iacr.org/`, 2004.
25. M. Luby and C. Rackoff. How to construct pseudorandom permutations from pseudorandom functions. *SIAM J. Comput.*, vol. 17, no. 2, pages 373–386, April 1988.
26. P. Rogaway, M. Bellare, J. Black, and T. Krovetz. OCB: a block-cipher mode of operation for efficient authenticated encryption. *Proceedings of ACM Conference on Computer and Communications Security, ACM CCS 2001,* ACM, 2001.
27. D. Whiting, R. Housley, and N. Ferguson. Counter with CBC-MAC (CCM). Submission to NIST. Available at `http://csrc.nist.gov/CryptoToolkit/modes/`.

A Proof Sketch of Lemma 5.1

We sketch the proof of Lemma 1 here, leaving the details to [18]. The adversary has an oracle which is either $\mathsf{p8}^{+}\text{-}\mathsf{Encrypt}_R(\cdot,\cdot)$ or $\$(\cdot,\cdot)$. We fix some notation. For q and σ in Lemma 1, let $m_1, \ldots, m_q$ be integers such that $m_i \geq 1$ and

$\sigma \geq m_1 + \cdots + m_q$. Let $N_1, \ldots, N_q$ be fixed and distinct bit strings such that $|N_i| = n$. Let $M_1, \ldots, M_q$ be arbitrarily fixed bit strings such that $|M_i| = m_i n$, and let $M_i = M_i[1] \| \cdots \| M_i[m_i]$, where $M_i[j] \in \{0,1\}^n$. Also, let $C_1, \ldots, C_q$ be fixed bit strings such that $|C_i| = m_i n$ and, let $C_i = C_i[1] \| \cdots \| C_i[m_i]$, where $C_i[j] \in \{0,1\}^n$. Assume $C_1, \ldots, C_q$ satisfy the following condition:

For any i $(1 \leq i \leq q)$, the multiset

$$\{0^n, M_i[1] \oplus C_i[1] \oplus [1]_n, \ldots, M_i[m_i-1] \oplus C_i[m_i-1] \oplus [m_i-1]_n\} \quad (6)$$

has m_i distinct points

(there is no condition on $C_1[m_1], \ldots, C_q[m_q]$).

For (N_i, M_i) and the function R, let $A_i = R(N_i)$, and $M_i[0] \oplus C_i[0] = 0^n$. For $1 \leq j \leq m_i$, let $X_i[j] = A_i \oplus M_i[j-1] \oplus C_i[j-1] \oplus [j-1]_n$ and $Y_i[j] = R(X_i[j])$. Further, for $1 \leq i \leq q$, let $\bullet_i \stackrel{\text{def}}{=} \{X_i[j] \mid 1 \leq j \leq m_i\}$, and $\bullet \stackrel{\text{def}}{=} \{N_i \mid 1 \leq i \leq q\}$.

Then for randomly chosen A_t (this will fix $\bullet_t$), define the following $(t-1)+1 = t$ conditions: Cond. A-s $(1 \leq s \leq t-1)$ and Cond. B.

Cond. A-s $(1 \leq s \leq t-1)$: $\bullet_s \cap \bullet_t \neq \emptyset$.
Cond. B: $\bullet \cap \bullet_t \neq \emptyset$.

We say that BAD$[t]$ occurs if at least one of the above t events occurs.

Intuitively, we show that if all the query-answer pairs satisfy (6) and BAD$[t]$ does not occur, then the adversary cannot distinguish between p8$^+$-Encrypt$_R(\cdot,\cdot)$ and $\$(\cdot,\cdot)$. The proof is completed by upper bounding the probability that some query-answer pair fails to satisfy (6), or some BAD$[t]$ occurs.

B Proof Sketch of Lemma

To prove Lemma 2, we define $p9^+$-E$[n]$, a variant of $p9^+[n]$. The tagging algorithm for $p9^+$-E$[n]$ takes only messages of length multiple of n. That is, we consider that messages have already padded. Also, it does not perform the final encryption and it does not mask with $R(1^n)$. Specifically, the key generation algorithm for $p9^+$-E$[n]$ returns a randomly selected function R from Rand(n, n). The tagging algorithm for $p9^+$-E$[n]$, p9$^+$-E-Tag, takes R as a "key" and a message M such that $|M| = mn$ for some $m \geq 1$. The pseudocode is given in Fig. 7. "$M[1] \| \cdots \| M[m] \leftarrow M$" is a shorthand for "break M into n-bit blocks $M[1] \| \cdots \| M[m]$." The verification algorithm is defined in the natural way.

We next fix some notation. For q and σ in Lemma 2, let $m_1, \ldots, m_q$ be integers such that $m_i \geq 1$ and $\sigma \geq m_1 + \cdots + m_q$. Let $M_1, \ldots, M_q$ be fixed and distinct bit strings such that $|M_i| = m_i n$.

Then we have the following lemma.

Lemma 3. *Let q, $m_1, \ldots, m_q$, σ, $M_1, \ldots, M_q$ be as described above. Then the probablity of*

```
Algorithm p9+-E-Tag_R(M)
    M[1]||···||M[m] ← M
    Y[0] ← R(0^n)
    For i ← 1 to m do:
        X[i] ← M[i] ⊕ Y[i − 1]
        Y[i] ← R(X[i])
    Return Y[1] ⊕ ··· ⊕ Y[m]
```

Fig. 7. Algorithm $\mathsf{p9^+\text{-}E\text{-}Tag}_R(M)$

- $1 \leq {}^{\exists}i < {}^{\exists}j \leq q, \mathsf{p9^+\text{-}E\text{-}Tag}_R(M_i) = \mathsf{p9^+\text{-}E\text{-}Tag}_R(M_j)$, *or*
- $1 \leq {}^{\exists}i \leq q$, $R(1^n)$ *is used in the computation of* $\mathsf{p9^+\text{-}E\text{-}Tag}_R(M_i)$

is at most $3\sigma^2/2^n$ *where the probability is taken over the random choice of* $R \stackrel{R}{\leftarrow} Rand(n,n)$.

Given the above Lemma 3, it is easy to prove the following lemma.

Lemma 4. *Let* q *and* σ *be as in Lemma 2. Also, let* $M_1, \ldots, M_q$ *be arbitrarily fixed and distinct bit strings, and let* $T_1, \ldots, T_q$ *be arbitrarily fixed* n*-bit strings. Then*

$$\Pr(R \stackrel{R}{\leftarrow} Rand(n,n) : 1 \leq {}^{\forall}i \leq q, \mathsf{p9^+\text{-}Tag}_R(M_i) = T_i) \geq \frac{1}{2^{qn}}\left(1 - \frac{5\sigma^2}{2^n}\right) .$$

Given the above lemma, the proof of Lemma 2 is standard.

Two-Pass Authenticated Encryption Faster Than Generic Composition

Stefan Lucks

University of Mannheim, Germany
http://th.informatik.uni-mannheim.de/people/lucks/

Abstract. This paper introduces CCFB and CCFB+H, two patent-free authenticated encryption schemes. CCFB+H also supports the authentication of associated data. Our schemes can employ any block cipher and are provably secure under standard assumptions. The schemes and their proofs of security are simple and straightforward. CCFB and CCFB+H restrict the sizes of nonce and authentication tags and can, depending on these sizes, perform significantly better than both generic composition and other two-pass schemes for authenticated encryption, such as the EAX mode.

Keywords: authenticated encryption, associated data, provable security, OMAC.

1 Introduction

An *Authenticated Encryption (AE)* scheme is a secret-key cryptosystem designed for simultaneously protecting *both* a message's privacy *and* its authenticity. Traditionally, these two security goals had been handled separately by the means of encryption schemes and message authentication codes (MACs). In practice, however, the same message often needs to be kept both private and authentic, and gluing together encryption and message authentication is surprisingly tricky and error-prone. Hence, a couple of block cipher based AE schemes have been developed recently.

Even more recently, people discovered that AE is not quite sufficient. Often, some header *(associated data, AD)* is not confidential, but vital for authentication. *Authenticated Encryption with Associated Data (AEAD)* schemes authenticate both the message and the associated data, but only encrypt the message. Most of today's AE and AEAD schemes are either "two-pass" schemes and thus as slow as encrypting and authenticating independently, or "one-pass" schemes whose usage is hindered by the patent situation. This paper proposes a new two-pass scheme. Depending on the size of the authentication tag, our solution can run significantly faster than generic composition or other non-patented two-pass AE(AD) schemes. Another advantage is simplicity: compared to other AE(AD) schemes, our solution and its proof of security is very simple and straightforward.

H. Gilbert and H. Handschuh (Eds.): FSE 2005, LNCS 3557, pp. 284–298, 2005.

1.1 The Development of Authenticated Encryption

In 2000, Bellare and Namprempre proposed *generic composition*: a privacy-protecting encryption scheme and a MAC are used jointly (but securely) under independent keys [3]. This is not very efficient – it takes the time to encrypt plus the time to authenticate and makes block cipher based authenticated encryption twice as slow as either encryption or authentication. The generic approach can provide AEAD as well as AE. Generic composition can be *minimal-expanding*[1], i.e. the size of a ciphertext is the *plaintext size* plus τ *bit for the authentication tag*, where τ is a plaintext-size-independent constant, and the forgery probability is close to $1/2^\tau$.

In the same year, Katz and Yung presented the RPC block cipher mode for authenticated encryption [8]. It is a *single-pass* AE scheme, but the message expansion is not minimal – it is linear in the plaintext size. Depending on the size of the authentication tag, RPC can run significantly faster than generic AE, but always less than twice as fast[2]. For historical reasons, the authors of RPC did not consider AEAD.

In 2001, several *single-pass* minimal-expanding AE schemes have been proposed: IAPM, OCB and XCBC [7, 13, 4]. These combine minimal expansion with a close-to-optimal running time: for large messages, these schemes are almost as fast as conventional encryption (without authenticity), i.e. twice as fast as the generic approach. In 2002, a single-pass AEAD scheme based on OCB has been proposed [12].

Unfortunately, several patents cover the usage of the fast single-pass schemes. The patent situation has turned out to be a significant deterrence. To avoid patents, new *two-pass* AEAD schemes have been developed, with one pass for encryption and another one for authentication. The first was CCM [15], followed by EAX, CWC, and GCM [1, 2, 9, 10, 11], which addressed some shortcomings [14] of CCM. All these modes are minimal expanding, but as (in)efficient as generic composition. Their main advantage over generic composition is that a single block cipher key suffices for the entire scheme.

1.2 Contributions and Outline of This Paper

This paper proposes CCFB (Counter-CipherFeedBack) – another two-pass AE mode for block ciphers, but with a different separation of duties between the passes. It has been developed with low-end devices in the mind, such as smart-cards, small embedded systems, sensor network motes, and RFID tags. CCFB is related to RPC, which has been published *before* the patented single-pass schemes. The first pass of CCFB is for privacy and "local" authentication, while the second computes a single "global" authentication tag from the local ones. CCFB+H is

1 ... depending on the underlying encryption and MAC scheme.

2 E.g., AES-RPC with 32-bit authentication tags is 50 % faster than AES-based generic composition.

- a new *minimal-expanding* and *two-pass* AEAD scheme (avoiding the patents on single-pass schemes [3], similarly to EAX, CWC, and GCM),
- which can run significantly faster than previously published two-pass schemes[4], especially on low-end devices.

Like EAX, CWC, and GCM,

- CCFB+H can use any block cipher and even a pseudorandom function (PRF) as the underlying primitive,
- it uses a single block cipher (or PRF) key for all its work, and a block cipher is only used in encryption mode,
- CCFB+H allows the (pre-)processing of the header, independently from the message,
- CCFB+H is provably secure under standard assumptions on the security of the underlying block cipher or PRF,
- and we analyse our schemes' concrete security.

A drawback, inherited from RPC, is that the sizes for nonces and authentication tags are limited (in contrast to EAX, CWC, and GCM). More specifically, if n is the block size of the underlying block cipher or PRF, then

$$\overbrace{\text{maximum size of nonce}}^{\delta} = \overbrace{\text{block size}}^{n} - \overbrace{\text{size of authentication tag}}^{\tau}.$$

Section 2 describes CCFB, Section 4 analyses it with respect to the notions of security defined in Section 3. Section 5 extends CCFB to an AEAD scheme CCFB+H (CCFB with Header). Using OMAC [5,6], a block cipher based message authentication code, Sections 6 and 7 develop a block cipher based instantiation of CCFB. Section 8 compares CCFB+H and EAX security-wise and performance-wise. The proof of Theorem 2 and some figures are deferred to the appendix.

2 CCFB Authenticated Encryption

We define *CCFB authenticated encryption* under a function $F : \{0,1\}^n \rightarrow \{0,1\}^n$. Fix the tag size $\tau \leq n/2$. Set $\delta = n - \tau$. The notation "$(d,t) := F(\cdot)$" implies $d \in \{0,1\}^\delta$ and $t \in \{0,1\}^\tau$. For $i \in \{1,\ldots,2^\tau - 1\}$, we write $\langle i \rangle_\tau$ for the corresponding τ-bit string. We write "$\|$" for the concatenation of bit-strings.

---

[3] We neither have, nor are aware of any patents or pending patents relevant to CCFB+H. We do not intend to apply for such patents.

[4] EAX and our instantiation of CCFB+H are dominated by the block cipher operations, and can run on any low-end device capable of running block cipher operations. This enables a "platform-independent" performance evaluation by counting the number of block cipher calls, see Section 8. In the same section, we also explain why CWC and GCM appear to be poor choices for low-end devices.

If X is a bit-string of length $\geq \lambda$, we write $\text{MSB}_\lambda(X)$ for the first λ bits of X. The input for CCFB encryption consists of a nonce $N \in \{0,1\}^\delta$ (shorter nonces can be padded), and a message M of any length $|M|$ between 1 bit and $(2^\tau - 3)\delta$ bit. The algorithm is described in Figure 1. See also Figures 2 and 3 for an illustration of CCFB encryption.

Algorithm: CCFB encryption.
Input: nonce $N \in \{0,1\}^\delta$ and $M \in \{0,1\}^*$, $1 \leq |M| \leq (2^\tau - 3)\delta$;
First pass:

1. parse M as $(M_1, \ldots, M_m)$ with
   $|M_1| = \cdots = |M_{m-1}| = \delta$, $|M_m| \in \{1, \ldots, \delta\}$;
2. $C_0 := N$;
3. for $1 \leq i \leq m-1$: $(\text{tmp}, A_i) := F(C_{i-1}, \langle i \rangle_\tau)$;
   $C_i := \text{tmp} \oplus M_i$;
4. $(\text{tmp}, A_m) := F(C_{m-1}, \langle m \rangle_\tau)$;
5. if $|M_m| = \delta$ then $d := 1; \text{pad} := ()$(*empty string*);
   else $d := 2; \text{pad} := (1||0^{\delta-|M_m|-1})$;
6. $C_m := \text{MSB}_{|M_m|}(\text{tmp}) \oplus M_m$;
7. $C' := \text{tmp} \oplus (M_m||\text{pad})$;
8. $(\text{dummy}, A_{m+1}) := F(C', \langle m+d \rangle_\tau)$;

Second pass:

9. $T := A_1 \oplus \cdots \oplus A_{m+1}$;

Output: ciphertext $(C_1, \ldots, C_m, T)$ with
$C_1, \ldots, C_{m-1} \in \{0,1\}^\delta$, and $C_m \in \{0,1\}^{|M_m|}$.

Fig. 1. CCFB encryption under $F : \{0,1\}^\delta \times \{0,1\}^\tau \rightarrow \{0,1\}^\delta \times \{0,1\}^\tau$

Observe that if the length $|M|$ of M is a multiple of δ, i.e., $|M_m| = \delta$, steps 3 to 8 simplify to the following short algorithm:

- for $1 \leq i \leq m$: $(\text{tmp}, A_i) := F(C_{i-1}, \langle i \rangle_\tau)$;
  $C_i := \text{tmp} \oplus M_i$;
- $(\text{dummy}, A_{m+1}) := F(C_m, \langle m+1 \rangle_\tau)$;

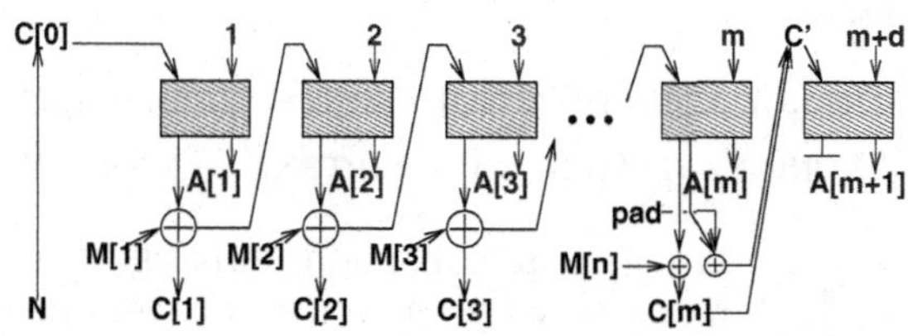

Fig. 2. 1st phase of CCFB encryption: compute the C_i and the local tags A_i; $d \in \{1,2\}$

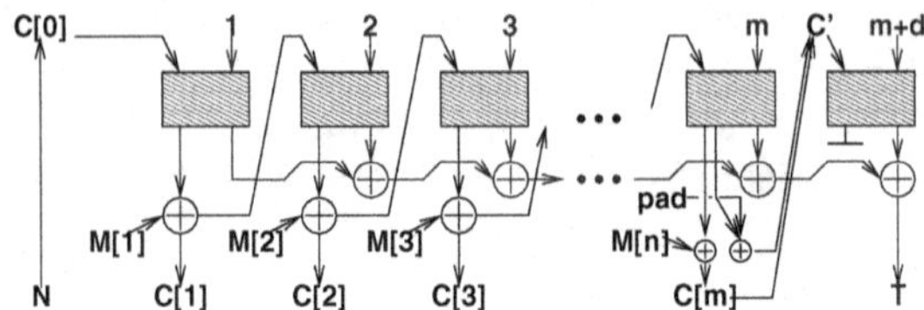

Fig. 3. Full CCFB encryption: The global tag T is computed in the second phase

An $|M|$-bit message M is split into $m = \lceil |M|/\delta \rceil + 1$ blocks M_i, and encrypting M requires $2m$ XORs and $m + 1$ random function (or block cipher) calls. Thus, CCFB runs at essentially the same speed as RPC [8]. The most important differences between CCFB and RPC, cf. Figure 7 in the appendix, are:

- CCFB employs CipherFeedBack, where RCB uses the ECB mode. Accordingly, RPC assumes F to be a permutation.
- The output of RPC consists of the encryption blocks C_i and the local authentication tags A_i. CCFB extends RPC by the second pass, which makes CCFB minimal-expanding. The output of CCFB is a single "global" authentication tag $T = \bigoplus A_i$.
- To protect against cut-and-paste attacks, RPC requires a message encoding with reserved "start" and "stop" blocks. CCFB does not need a message encoding.

Given a nonce $N \in \{0,1\}^\delta$ and a ciphertext $C = (C_1, \ldots, C_m, T)$, **CCFB decryption** is straightforward and needs as much computation as the encryption, see Figure 6 in the appendix.

As ususal for modes of operations, nonces must not be re-used. E.g., if we encrypt two messages $(M_1, \ldots, M_m)$ and $(M'_1, \ldots, M'_{m'})$ under the same nonce, the corresponding first ciphertext blocks satisfy $C_1 \oplus C'_1 = M_1 \oplus M'_1$.

3 Notions of Security for Authenticated Encryption

Before we analyse the security of CCFB (and later CCFB+H), we have to specify what we mean by "secure". Our notions of security are standard, see e.g. [1, 2]. An **AEAD scheme** is a pair (E, D) of deterministic algorithms E for encryption and D for decryption:

$$E: \text{KEY} \times \text{NONCE} \times \text{HEADER} \times \text{MESSAGE} \rightarrow \text{CIPHERTEXT},$$
$$D: \text{KEY} \times \text{NONCE} \times \text{HEADER} \times \text{CIPHERTEXT} \rightarrow \text{MESSAGE} \cup \text{(none)}.$$

The sets KEY, NONCE, HEADER, MESSAGE, and CIPHERTEXT are bit-strings, i.e., subsets of $\{0,1\}^*$. For simplicity, we assume KEY to be finite. An **adversary** with access to an **encryption oracle** $E(K, \cdot, \cdot, \cdot)$ chooses triples (N^1, H^1, M^1), $\ldots$, $(N^q, H^q, M^q) \in$ NONCE$\times$HEADER$\times$MESSAGE and receives the corresponding ciphertexts $C^i = E(K, N^i, H^i, M^i)$. The adversary is **nonce-respecting**, if for all $i \neq j$, $N^i \neq N^j$. If NONCE is finite, a **nonce-randomising** adversary chooses a fresh uniformly distributed random $N^i \in$ NONCE for each query (N^i, H^i, M^i).

In a privacy attack, the adversary is either given access to the real encryption oracle, or to a fake oracle $F(K, \cdot, \cdot, \cdot)$, which on input (N^i, H^i, M^i) returns a random ciphertext $F(N^i, H^i, M^i)$ of the same length as the real ciphertext $C^i = E(K, N^i, H^i, M^i)$. The adversay has to distinguish between both oracles. Let K be a random key. An AEAD scheme is p-**private** against a class of adversaries, if for all adversaries A of that class, the advantage in distinguishing E from F is

$$\left|\Pr\left[A^{E(K,\cdot,\cdot,\cdot)} = 1\right] - \Pr\left[A^{F(\cdot,\cdot,\cdot)} = 1\right]\right| \leq p.$$

A forger asks queries (N^1, H^1, M^1), ..., (N^q, H^q, M^q), receives the corresponding ciphertexts C^1, ..., C^q, and finally chooses a ciphertext C, a nonce N, and a header H. The forger succeeds, if $(C, H) \notin \{(C^1, H^1) \ldots, (C^q, H^q)\}$[5] and $D(K, N, H, C) \neq$ (none).

An AEAD scheme is p-**authentic** against a class of forgers, if for all forgers A_F of that class and a random key K

$$\Pr\left[A_F \text{ succeeds}\right] \leq p.$$

An **AE scheme** is an AEAD scheme without a choice for the headers: HEADER $= \{0\}$.

4 Analysis of CCFB Authenticated Encryption

Consider a chosen plaintext scenario where the adversary $\mathcal{A}$ selects q messages $M^1 = (M_1^1, \ldots, M_{m_1}^1)$, ..., $M^q = (M_1^q, \ldots, M_{m_q}^q)$ with $r = \sum_{1 \leq i \leq q} m_i$ blocks in total. We write $N^1 = C_0^1$, ..., $N^q = C_0^q$ for the corresponding nonces chosen by $\mathcal{A}$, and $C^1 = (C_1^1, \ldots, C_{m_1}^1, T^1)$, ..., $C^q = (C_1^q, \ldots, C_{m_q}^1, T^q)$ for the ciphertexts. Consider the inputs for F:

$$D_k^i = \begin{cases} (C_k^i, k+1) & \text{if } k < m_i \\ ((C')^i, m_i + d) & \text{if } k = m_i \ (d = 1 \text{ if } |M_m| = \delta, \text{ else } d = 2). \end{cases} \tag{1}$$

Here $(C')^i$ corresponds to the "internal" value C' from Figure 1. An "input-collision" is an input-pair (D_k^i, D_k^j) with

$$D_k^i = D_k^j \text{ with } 1 \leq i < j \leq q \text{ and } k \in \{0, \ldots, \min\{m_i, m_j\}\}. \tag{2}$$

We assume the adversaries to ask q queries to the encryption oracle with, in total, r message blocks, i.e. $r = \sum_{1 \leq i \leq q} m_i$.

[5] Even if the forger is nonce-respecting $N \in \{N^1, \ldots, N^q\}$ is permissable.

Lemma 1. *For CCFB encryption under a random function F, the probability for any nonce-respecting adversary to generate an input-collision is at most*

$$\frac{qr}{2^{\delta+1}}.$$

Similarly, the probability for any nonce-randomising adversary to generate an input-collision is at most

$$\frac{q(r+q)}{2^{\delta+1}}.$$

Proof. First, consider a nonce-respecting adversary. There is no input-collision with $k=0$. Thus, we can concentrate on $k \geq 1$.

A collision $D_k^i = D_k^j$ implies $F(D_{k-1}^i) = F(D_{k-1}^j)$, and if $D_{k-1}^i \neq D_{k-1}^j$, then $\Pr[D_k^i = D_k^j] \leq 1/2^\delta$. The number of triples (i,j,k) with $1 \leq i < j \leq q$ and $1 < k \leq \min\{m_i, m_j\}$, is at most $(q-1)r/2$. The probability that at least one of these triples collides is thus at most $\frac{(q-1)r}{2} * \frac{1}{2^\delta} = \frac{(q-1)r}{2^{\delta+1}}$.

Second, consider a nonce-randomising adversary. If there is no input-collision with $k=0$, then the adversary happens to be nonce-respecting. The additional chance to generate an input-collision at the level $k=0$ – which is in fact a nonce-collision – is a most $(q(q-1)/2)/2^\delta \leq q^2/2^{\delta+1}$. The second claim follows from $qr + q^2 = q(r+q)$. □

Theorem 1 (Information-Theoretic Privacy of CCFB).
CCFB encryption using a random F is

$\frac{qr}{2^{\delta+1}}$*-private against nonce-respecting adversaries and*

$\frac{q(r+q)}{2^{\delta+1}}$*-private against nonce-randomising adversaries.*

Proof. Without any input-collision D_k^i $(k \geq 0)$, all the inputs to the random function F are different, all of its outputs are distributed uniformly at random. Thus, the outputs from the "real" ecnryption oracle and the fake oracle are distributed equally. To distinguish the oracles, the adversary would need an input-collision. The claims follow from the bounds given in Lemma 1. □

Theorem 2 (Information-Theoretic Authenticity of CCFB).
CCFB encryption, using a random F, is

$\left(\frac{qr}{2^{\delta+1}} + \frac{1}{2^\tau}\right)$*-authentic with respect to nonce-respecting adversaries and*

$\left(\frac{q(r+q)}{2^{\delta+1}} + \frac{1}{2^\tau}\right)$*-authentic with respect to nonce-randomising adversaries.*

The proof will be given in the appendix.

5 The CCFB+H AEAD Mode and Its Analysis

Let $F' : \{0,1\}^* \rightarrow \{0,1\}^\delta$ be an additional random function, chosen independently from F. Note that F' is defined for a variable input length, in contrast to F. We write $H \in \{0,1\}^*$ for the associate ("header") data and tweak both the CCFB encryption algorithm and its decryption counterpart by changing instruction 2 in Figure 1 and in Figure 6: replace $\boxed{C_0 := N;}$ by: $\boxed{C_0 := N \oplus F'(H);}$ see Figure 4 for an illustration of the modified encryption. Since CCFB+H is a tweaked CCFB, we conveniently inherit most the analysis from CCFB.

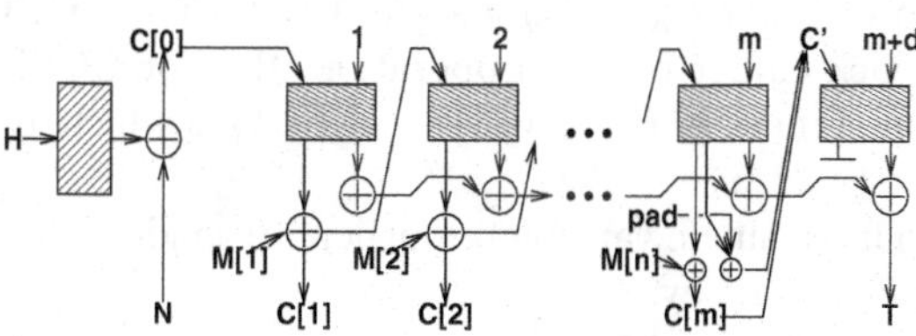

Fig. 4. CCFB Authenticated Encryption of Message $M[\cdot]$ with Associated Data H

Recall the definition of q and r from the previous section.

Lemma 2. *For CCFB encryption under a random F, the probability for a nonce-respecting or nonce-randomising adversary to generate an input-collision is at most*

$$\frac{q(r+q)}{2^{\delta+1}}.$$

Proof. For a nonce-randomising adversary, the result follows immediately from the second claim of Lemma 1. We will show that for a nonce-respecting adversary, the probability for an input-collision at level $k = 0$ is no more than $q^2/2^{\delta+1}$. We write H^i for the header of the i-th chosen ciphertext query. If $H^i = H^j$, then $D^i \neq D^j$, since the adversary is nonce-respecting.

Consider $H^i \neq H^j$ and $\Delta(i,j) := F'(H^i) \oplus F'(H^j) \in \{0,1\}^\delta$. We get $C_0^i = C_0^j$ if and only if $N^i \oplus N^j = \Delta(i,j)$, i.e. with at most the probability $1/2^\delta$. There are $q(q-1)/2$ pairs (i,j) with $1 \leq i < j \leq q$, so the probability for an input-collision at level $k = 0$ is $q(q-1)/2^{\delta+1}$. □

The proofs for privacy and authenticity of CCFB+H are the same as their counterparts in Section 4. We consider adversaries, who are either nonce-respec ting or nonce-randomising.

Theorem 3 (Information-Theoretic Privacy of CCFB+H). *CCFB encryption using a random F is*

$$\frac{qr+q^2}{2^{\delta+1}}\textit{-private.}$$

Theorem 4 (Information-Theoretic Authenticity of CCFB+H). *CCFB encryption using a random F is*

$$\left(\frac{qr+q^2}{2^{\delta+1}}+\frac{1}{2^\tau}\right)\text{-authentic.}$$

6 Using a Single Random Function •

For CCFB, we can instantiate the random function $F:\{0,1\}^n \to \{0,1\}^n$ by a PRF – or by a block cipher E_K under a secret key K. Thus, CCFB can obviously be viewed as a block cipher mode of operation. But for CCFB+H, we need an additional random function $F':\{0,1\}^* \to \{0,1\}^\delta$, which is supposed to be independent from F.

We propose to use a single variable-input-length random function $f:\{0,1\}^*\to\{0,1\}^n$, defining F and F' by

$$\begin{aligned} F(x) &= f(x) && \text{for } x \in \{0,1\}^n \\ F'(y) &= \mathrm{MSB}_{|M_m|}(f(0^n||y)) && \text{for } y \in \{0,1\}^* \end{aligned}$$

By the definition of CCFB and CCFB+H, the first τ bits of any input for F represent a number between 1 and $2^{\tau-1}$, i.e. are never zero. Thus, inputs x and $0^n||y$ for f are never the same,[6] and F and F' behave exactly like two independent random functions.

7 Instantiating • by OMAC

OMAC [5, 6], described in Figure 5,[7] is a message authentication code under a function $E_K:\{0,1\}^n \to \{0,1\}^n$. It

- can use any block cipher or PRF as the underlying primitive,
- uses a block cipher E only in encryption mode,
- uses a single block cipher (or PRF key) K,
- and is provably secure in the standard model, see Theorem 5 for OMAC's **information-theoretical security as a variable-input-size PRF** in a concrete security setting.

---

[6] In fact, we could replace $0^n||y$ by $0^\tau||y$. The only reason why we propose the longer 0^n-prefix is the improved efficiency for our OMAC based instantiation of f.

[7] [6] describes two flavours of OMAC, OMAC1 and OMAC2. In this paper, we set OMAC=OMAC1, but we could use OMAC2 just as well.

For the definition of u and "$*$" in GF(2^n) see [5, 6]. We stress that computing $L*u$ and $L*u^2$ can be done very efficiently by shifting and conditional XORing.

Algorithm: OMAC.
Init: $L_0 := E_K(0)$; $L_1 := L_0 * u$; $L_2 := (L_0 * u^2)$; (* in GF(2^n) *)
Input: $X \in \{0,1\}^n$.
1. parse X as $X_1, \ldots X_m$;
2. $Z := 0^n$;
3. for i in $1 \le i \le m-1$: $Y := X_i \oplus Z$;
   $Z := E_K(Y)$;
4. if $|M_m| = n$ then $Y := Y \oplus L_1$;
   else $Y := Y \oplus L_2$;

Output: authentication tag $E_K(Y)$.

Fig. 5. OMAC

Theorem 5 (Lemma 5.2 of [6]). *Consider OMAC under a random permutation* $E_K : \{0,1\}^n \rightarrow \{0,1\}^n$. *An adversary asking at most* q' *queries, each at most* $\mu < 2^n/4$ *blocks long, cannot distinguish OMAC from a random function with an advantage exceeding*

$$\frac{(5\mu^2+1)q'}{2^n}.$$

Thus, we propose to instantiate f by OMAC under a block cipher E (e.g., E=AES) and a secret block cipher key K. The performance figures are:

- Computing $F'(H) = \mathrm{OMAC}_K(0^n||H)$ can be done by calling the block cipher E_K only $\lceil |H|/n \rceil$ times. The first iteration of the loop in Figure 5 can easily be optimised away, since it produces $E_K(0) = L_0$, which has been computed before, in the initialisation phase.
- Each computation of a value $F(C_{i-1}, i)$ or $F(C', m+d)$ boils down to a single block cipher call.

⇒ Computing CCFB+H(H,M) thus needs

$$\left\lceil \frac{|H|}{n} \right\rceil + \left\lceil \frac{|M|}{\delta} + 1 \right\rceil \text{ block cipher calls.} \quad (3)$$

8 A Comparison: EAX ↔ CCFB+H

In this section, we extrapolate the performance of CCFB+H from EAX' performance. Based on these results, one can compare the performance of CCFB+H with other modes, such as CWC and GCM, and one can verify these findings by benchmarking CCFB+H directly.

This has not been done in the current paper, which's focus is on low-end systems. As stressed in [9], CWC has not been developed for low-end devices. CWC combines counter-mode encryption with a Carter-Wegman hash function over GF($2^{127}-1$). Due to the heavy use of large-scale integer multiplications, CWC actually appears to be very unattractive for low-end devices. Similarly to

CWC, GCM combines counter-mode encryption with a universal hash function, namely a polynomial hash over some binary field GF(2^w). Efficient software implementations would need large tables, i.e. more storage space than available on usual low-end systems. It thus seems natural to consider EAX as the main "competitor" for CCFB+H.

To the security architect, CCFB offers a trade-off between the size τ of the authentication tag and the size δ of the message blocks. This has an obvious impact on the performance, but also determines the security level. Table 1 highlights this. Apart from the bound on τ, what is the impact of replacing a term $\Theta(r^2/2^n)$ (for EAX) by $\Theta(qr/2^{n-\tau})$ (for CCFB+H)?

Table 1. Asymptotical Security of EAX and CCFB+H

	provable privacy	provable authenticity	limit for τ
EAX	$\Theta\left(\frac{r^2}{2^n}\right)$	$\Theta\left(\min\left\{\frac{r^2}{2^n}, \frac{1}{2^\tau}\right\}\right)$	$\tau \le n$
CCFB+H	$\Theta\left(\frac{qr}{2^{n-\tau}}\right)$	$\Theta\left(\min\left\{\frac{qr}{2^{n-\tau}}, \frac{1}{2^\tau}\right\}\right)$	$\tau \le n - \delta$

$r = \sum m_i$: accumulated number of message blocks q: number of messages

The maximum message length for CCFB+H is $(n-\tau)(2^\tau - 3)$ bit, i.e. appxoximately 2^τ blocks. Thus, if the average message size is large, CCFB+H can be about as secure as EAX. On the other hand, CCFB+H has been designed with low-end devices in mind. Typical applications for low-end devices mostly transmit small messages. So let us consider a concrete example with small messages:

block cipher: E=AES, and thus $n = 128$,
tag size: $\tau = 32$, and thus $\delta = 128 - 32 = 96$,
number of messages in the lifetime of a secret key: $q \le 2^{28}$
average message size: ≤ 16 blocks $(16 * \delta = 1536\,\text{bit}) \Rightarrow r \le 2^{32}$.

While EAX would provide better security than CCFB+H, we still get good privacy and almost the authenticity we would expect from an ideal MAC with 32-bit authentication tags:

$$\text{privacy (Thm. 3):} \quad \frac{qr+q^2}{2^{\delta+1}} \approx 2^{60}/2^{97} \approx 2^{-37}$$
$$\text{authenticity (Thm. 4):} \quad \frac{qr+q^2}{2^{\delta+1}} + \frac{1}{2^\tau} \approx 2^{-37} + 2^{-32} \approx 2^{-32}$$

The above results apply in an information-theoretic setting. Since we propose to use OMAC as a pseudorandom function, Theorem 5 comes into play. Note that each header H and each message block M_i in the CCFB+H setting is, from OMAC's point of view, a message of its own right – OMAC thus authenticates $q' = q+r$ messages. Theorem 5 also considers the length μ of the largest message (in blocks). By the specification of CCFB, we have $\mu \le 2^\tau - 3$. Even if we assume $\mu \approx 2^\tau$, the advantage is bounded by

$$\text{the pseudorandomness of OMAC (Thm. 5): } \frac{(5\mu^2+1)q'}{2^n} \le 2^{-64}.$$

This is negligible, compared to the 2^{-37} and 2^{-32} from above.

Finally, we also compare the performance of the concrete CCFB+H example with the security of AES-based EAX. CCFB+H allows the precomputation of a header checksum $F'(H)$ in advance, before knowing M. EAX offers a similar feature. Thus, in Table 2, we consider authenticated encryption with and without header precomputation. It turns out that

- The header-dependent work is the exactly same for EAX and CCFB+H: Computing OMAC(H) by making $\lceil |H|/128 \rceil$ AES calls.
- Apart from the header-dependent work, we see the following:
  - For short messages ($|M| \le 96$), CCFB+H makes two AES calls, while EAX makes three. E.e., CCFB+H is 50 % faster than EAX.
  - With $|M|$ increasing, CCFB+H is at least as fast as EAX ($97 \le |M| \le 128$), and at most 66.7 % faster ($128 \le |M| \le 192$).
  - In the long run, EAX makes about $|M|/64$ calls. CCFB+H with $|M|/96$ calls is 50 % faster.

Table 2. Performance of AES-based EAX and CCFB+H in # of AES calls

	full computation	header has been preprocessed
EAX	$\lceil \frac{\lvert M\rvert}{128} \rceil + \lceil \frac{\lvert M\rvert}{128} \rceil + 1 + \lceil \frac{\lvert H\rvert}{128} \rceil$	$\lceil \frac{\lvert M\rvert}{128} \rceil + \lceil \frac{\lvert M\rvert}{128} \rceil + 1$
CCFB+H	$\lceil \frac{\lvert M\rvert}{96} \rceil + 1 + \lceil \frac{\lvert H\rvert}{128} \rceil$	$\lceil \frac{\lvert M\rvert}{96} \rceil + 1$

$|M|$ = message length $|H|$ = header length nonce-length $\le$ 128 bit

Acknowledgement

The author thanks Ulrich Kühn for suggesting the name "Counter CFB", Nico Schmoigl for his survey on EAX and CWC, and the referees for their support in improving this presentation.

References

1. M. Bellare, P. Rogaway, D. Wagner. EAX: a conventional authenticated encryption mode. FSE 2004.
2. M. Bellare, P. Rogaway, D. Wagner. EAX: a conventional authenticated encryption mode. Extended version of [1]. http://www.cs.berkeley.edu/~daw/papers/eax-fse04.ps

3. M. Bellare, C. Namprempre. Authenticated Encryption: relations among notions and analysis of the generic composition paradigm. Asiacrypt 00.
4. V. Gligor, P. Donescu. Fast encryption and authentication: XCBC encryption and XECB authentication modes. FSE 01.
5. T. Iwata and K. Kurosawa. OMAC: One-Key CBC MAC. Fast Software Encryption, FSE 03.
6. T. Iwata and K. Kurosawa. OMAC: One-Key CBC MAC. Extended Version of [5]. `http://crypt.cis.ibaraki.ac.jp/omac/docs/omac.pdf`
7. C Jutla. Encryption modes with almost free message integrity. Eurocrypt 01.
8. J. Katz, M. Yung. Unforgeable encryption and adaptively secure modes of operation. FSE 00.
9. T. Kohno, J. Viega, D. Whiting. CWC: a high performance conventional authenticated encryption mode. FSE 04.
10. T. Kohno, J. Viega, D. Whiting. CWC: a high performance conventional authenticated encryption mode. Extended version of [9]. `http://eprint.iacr.org/2003/106.ps.gz`
11. D. McGrew, J. Viega. The Security and Performance of the Galois/Counter Mode of Operation (Full Version). `http://eprint.iacr.org/2004/193`
12. P. Rogaway. Authenticated encryption with associated data. Computer and Communications Security, ACM, 2002.
13. P. Rogaway, M. Bellare, J. Black, T. Krovetz. OCB: A block-cipher mode of operation for efficient authenticated encryption. Computer and Communications Security, ACM, 2001.
14. P. Rogaway, D. Wagner. A critique of CCM. Unpublished manuscript. February 2, 2003. `http://www.cs.berkeley.edu/~{}daw/papers/ccm.html`
15. D. Whiting, R. Hously, N. Ferguson. Counter with CBC-MAC (CCM). Submission to NIST.

Appendix: Deferred Proof and Figures

Theorem 2 (Information-Theoretic Authenticity of CCFB)
CCFB encryption, using a random F, is

$$\left(\frac{qr}{2^{\delta+1}} + \frac{1}{2^\tau}\right)\text{-authentic with respect to nonce-respecting adversaries and}$$

$$\left(\frac{q(r+q)}{2^{\delta+1}} + \frac{1}{2^\tau}\right)\text{-authentic with respect to nonce-randomising adversaries.}$$

Proof. We will show that the chance to succeed in forging a message *without* having found an input-collision is at most $1/2^\tau$. The claimed theorem then follows from Lemma 1.

The adversary's knowledge about the local authentication tags A^i_j can be described by

$$q \text{ linear equations} \qquad T^i = A^i_1 \oplus \bigoplus_{2 \le j \le m_i+1} A^i_j \qquad \text{with } 1 \le i \le q$$

Algorithm: CCFB decryption.
Input: nonce $N \in \{0,1\}^\delta$ and $C \in \{0,1\}^*$, $\tau + 1 \le |C| \le (2^\tau - 3)\delta + \tau$;
First pass:

1. parse C as $(C_1, \dots, C_m, T)$ with
   $|C_1| = \dots = |C_{m-1}| = \delta$, $|C_m| \in \{1, \dots, \delta\}$, $|T| = \delta$;
2. $C_0 := N$;
3. for $1 \le i \le m-1$: $(\text{tmp}, A_i) := F(C_{i-1}, \langle i \rangle_\tau)$;
   $M_i := \text{tmp} \oplus C_i$;
4. $(\text{tmp}, A_m) := F(C_{m-1}, \langle m \rangle_\tau)$;
5. if $|C_m| = \delta$ then $d := 1; \text{pad} := ()$(*empty string*);
   else $d := 2; \text{pad} := (1 || 0^{\delta - |C_m| - 1})$;
6. $M_m := \text{MSB}_{|C_m|}(\text{tmp}) \oplus C_m$;
7. $C' := \text{tmp} \oplus (M_m || \text{pad})$;
8. $(\text{dummy}, A_{m+1}) := F(C', \langle m + d \rangle_\tau)$;

Second pass:

9. $T' := A_1 \oplus \dots \oplus A_{m+1}$;

Output: If $T = T'$
then output plaintext $(M_1, \dots, M_m)$ with
$M_1, \dots, M_{m-1} \in \{0,1\}^\delta$, and $M_m \in \{0,1\}^{|M_m|}$
else output (none).

Fig. 6. CCFB decryption under $F : \{0,1\}^\delta \times \{0,1\}^\tau \to \{0,1\}^\delta \times \{0,1\}^\tau$

over GF(2^τ). We stress that only the T^i are known – the unknowns A_j^i ($j \ge 1$) are uniformly distributed independent random values from GF(2^τ) (since we assumed no input-collision). Due to the statistical independence of the A_1^i, all q linear equations are linearly independent.

A forgery (C_0, C) with $C = (C_1, \dots, C_{m-1}, C_m, T)$ succeeds if and only if C is different from all the other ciphertexts C^i and the linear equation

$$T = A_1 \oplus \bigoplus_{2 \le j \le m+1} A_j \tag{4}$$

holds. We claim that Equation 4 is linearly independent from the q equations above. I.e., we show that the sum of Equation 4 with any subset of equations $T^i = \dots$ is the sum of some non-dissappearing unknowns A_j^i or A_j with $j \ge 1$ and $1 \le i \le q$.

If $D_0 \notin \{D_0^1, \dots, D_0^q\}$,[8] the term A_1 cannot dissappear. So assuming w.l.o.g. $D_0 = D_0^1$, this is equivalent to $C_0 = C_0^1$, from which $A_1^1 = A_1$ follows. By adding T and T^1, we get

$$T^1 \oplus T = \bigoplus_{2 \le j \le m_1+1} A_j^1 \oplus \bigoplus_{2 \le j \le m+1} A_j.$$

[8] The inputs D_j for F are defined similarly to the D_j^i in Equation 1.

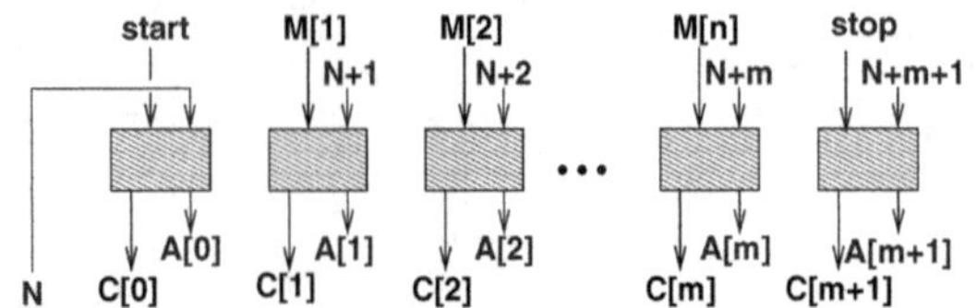

plaintext: $(M_1, \ldots M_m) \in \{0,1\}^{\delta m}$, $M_i \notin \{\text{start}, \text{stop}\}$;
nonce: $N \in \{0,1\}^\tau$; ciphertext: $((C_0, A_0), \ldots, (C_{m+1}, A_{m+1})) \in \{0,1\}^{n(m+2)}$

Fig. 7. RPC encryption under a permutation

Any terms $A_j^1 = A_j$ cancel out if $D_j = D_j^k$. We define the set

$$A^* = \{A_j^1 | 2 \leq j \leq m_1 + 1, D_j^1 \neq D_j\} \cup \{A_j | 2 \leq j \leq m + 1, D_j \neq D_j^1\}$$

of terms which don't cancel out and rewrite $T^1 \oplus T$ as

$$T^1 \oplus T = \bigoplus_{A \in A^*} A.$$

Since $C \neq C^i$, the set A^* is not empty.[9] For $i > 1$, each equation $T^i = \ldots$ with $i > 1$ added to $T^1 \oplus T$ introduces a non-disappearing term A_1^i to the sum. Thus, equation 4 is linearly independent from the equations for the T^i, as claimed.

Since Equation 4 is linearly independent from the q equations for the T^i, the sum $T = A_1 \oplus \bigoplus_{2 \leq j \leq m+1} A_j$ can take any value in $T \in \mathrm{GF}(2^\tau)$, and the number of solutions for each T is the same. All $T \in \mathrm{GF}(2^\tau)$ are equally likely to be the "correct" solution, which finally yields the claimed probability $1/2^\tau$. □

[9] Technically, $C \neq C^i$ could mean $T \neq T^i$, $m = m_i$, and $C_j = C_j^i$ for $1 \leq j \leq m$. But this type of forgery would fail: $A^* = \{\}$ and thus $T^1 \oplus T = 0$, contradicting $T \neq T^i$.

Padding Oracle Attacks on CBC-Mode Encryption with Secret and Random IVs

Arnold K. L. Yau*, Kenneth G. Paterson, and Chris J. Mitchell

Information Security Group,
Royal Holloway, University of London,
Egham, Surrey, TW20 0EX, UK
{a.yau, kenny.paterson, c.mitchell}@rhul.ac.uk

Abstract. In [8], Paterson and Yau presented padding oracle attacks against a committee draft version of a revision of the ISO CBC-mode encryption standard [3]. Some of the attacks in [8] require knowledge and manipulation of the initialisation vector (IV). The latest draft of the revision of the standard [4] recommends the use of IVs that are secret and random. This obviates most of the attacks of [8]. In this paper we consider the security of CBC-mode encryption against padding oracle attacks in this secret, random IV setting. We present new attacks showing that several ISO padding methods are still weak in this situation.

Keywords: padding oracle; CBC-mode; ISO standards; side channel.

1 Introduction

Vaudenay [9] introduced the notion of padding oracle attacks on CBC-mode encryption. His work showed that several uses of CBC-mode encryption in well-known products and standards are potentially vulnerable to attack whenever the attacker can submit ciphertexts for decryption and has access to a side-channel which tells him only whether or not the corresponding plaintext is correctly padded. Canvel *et al.* [7] applied and extended the ideas of [9] to show that a particular implementation of SSL used to protect e-mail passwords could be attacked and the passwords extracted. Further padding methods were examined in [6].

In [8], Paterson and Yau examined the security of the ISO standard for CBC-mode encryption with respect to padding oracle attacks. The draft revision of the standard [3] analyzed in [8] proposes the use of padding methods from ISO/IEC 9797-1 [1] and ISO/IEC 10118-1 [2]. Paterson and Yau showed that several of these padding methods, when used with CBC-mode encryption, are vulnerable

* Supported by EPSRC and Hewlett-Packard Laboratories Bristol through CASE award 01301027.

H. Gilbert and H. Handschuh (Eds.): FSE 2005, LNCS 3557, pp. 299–319, 2005.

to padding oracle attacks[1]. The work of [8] highlights the dangers of "cutting-and-pasting" methods from one set of standards into another.

Partly as a consequence of the work of [8], a later draft of the revised ISO standard [4] omits all mention of padding methods. Additionally, it recommends that "integrity-protected secret" and "randomly chosen statistically unique" IVs be used. The motivation for using secret IVs given in [4] is "to prevent information leakage". The recommendation for random IVs is in-line with the formal security analysis of [5] which shows (in a sense that can be made precise) that CBC-mode is secure provided that the underlying block cipher is strong and that the IV is random. We also note that [4] allows the use of multiple IVs (called starting variables, or SVs in [4]) and interleaving of multiple cipher block chains; this allows for parallelism in encryption. We expect that in most applications, a single IV will be used, and this is the situation we focus on here.

The attack model in [8] assumes that the IV can be chosen by the attacker and is submitted to the padding oracle along with the ciphertext. To be successful, most of the attacks in [8] do in fact require the attacker to have knowledge of the IV and the ability to manipulate it. For this reason, the attacks in [8] would not apply to CBC-mode as defined in [4] if the padding methods of [1] and [2] were used and if the new recommendations to use secret, random IVs were followed. More specifically, the only attack in [8] that remains practicable is Attack 2 against padding method 3 of [2]. This attack on its own arguably has a small impact on the confidentiality of data because it works only against the last one or two blocks of a target ciphertext and recovers relatively few useful data bits.

Despite their omission from the draft ISO standard [4], padding methods are needed in order to fully specify the CBC-mode of operation. It is not unreasonable to assume that, in the absence of any other guidance, an implementer of CBC-mode according to [4] might borrow techniques from other ISO standards, as was indeed proposed in [3]. Here, we demonstrate that padding oracle attacks can still be effective against CBC-mode encryption even when IVs are secret and random. In particular, we show that several padding methods from [1, 2] are still weak even in this situation.

1.1 Attack Models

Before giving details of our attacks, we clarify the attack models under which these attacks will take place.

When IVs are secret and random, a variety of practical methods could be used to ensure the IVs are available to both encrypting and decrypting parties. For example, the IV could be encrypted using ECB-mode and prefixed to the ciphertext. Alternatively, a value V could be prefixed to the ciphertext and the IV generated by encrypting V using ECB mode. Or, as a third possibility, a pre-

---

[1] In fact, [8] claims padding oracle attacks against the second edition of ISO/IEC 10116, though this edition of the standard makes no mention of padding methods. Padding methods did not appear in draft revisions of the standard until the committee draft stage in the proposed 3rd edition of ISO/IEC 10116.

agreed list of IVs could be used and an index sent with ciphertexts to indicate which entry in the list was used as the IV. Because these approaches include information determining IVs along with ciphertexts, they allow the adversary to influence which IV is used by the padding oracle when decrypting, without the adversary necessarily knowing the *actual* value of the IV. In particular, they allow the adversary to force the oracle to re-use an old IV. We can model this kind of attack by assuming that, when submitting a ciphertext to the padding oracle, the attacker specifies an additional string I which in some way determines the IV used by the padding oracle. The contents of I will depend on the particular method used for establishing IVs: for example, in the case of encrypted IVs, I will simply be the encrypted IV, while in the case of a pre-established list, I would be an index in the list.

We expect that the above kind of approach for establishing secret, random IVs is most likely to be used in practice. But it is also conceivable that a second approach, in which no information at all about the IV is transmitted as part of ciphertexts, might be used. For example, the communicating parties may be able to maintain a synchronised counter and then obtain IVs by applying a keyed pseudo-random function to the counter. We also want to model attacks in this scenario, which presents a tougher attack environment to the adversary. We can do this by assuming that the padding oracle simply selects a fresh, random IV before every decryption and that no IV-related information is included in ciphertexts.

Thus in this paper, we will consider two slightly different attack models. In the first model, IVs are secret and random but are determined by additional information I available to the attacker and submitted to the oracle. In the second model, IVs are secret and random and the attacker has no control over the IV used by the padding oracle. Obviously, attacks in the second model are more powerful, but attacks in the first model already capture many likely practical situations.

1.2 Our Results

In Section 3.2, we introduce a new padding oracle attack against CBC-mode when used with padding method 3 of [1]. Our new attack applies for secret, random IVs in the first attack model. The new attack uses a set of auxiliary ciphertexts corresponding to plaintexts of different lengths as an aid to recovering the plaintext corresponding to a target ciphertext block. The complexity of the attack depends on the spread of lengths of the auxiliary ciphertexts; it can be as low as n queries to the padding oracle, where n is the block size.

We have been able to adapt the attacks of [8] against CBC-mode when used with padding method 3 of [2] to the secret and random IV setting without significant penalties on complexity or generality. These attacks are applicable in our second, tougher attack scenario. An attack applicable to any ciphertext block is presented in Section 4.2. This attack first constructs a valid ciphertext with the target block as the final block and then uses the attack of Section 4.3 to decrypt that block. The first phase requires, on average, roughly 2^{r-1} calls to the padding oracle. Here r is a parameter associated with the padding method.

The attack of Section 4.3 is applicable to the final block of any ciphertext and is always efficient, requiring only $O(n)$ oracle queries to recover all the plaintext bits in the last block.

We note that our results do not contradict the results of [5], since the security model of [5] does not cater for the kind of side-channel information that a padding oracle provides to an attacker. We also note that all of our attacks are independent of the particular block cipher used.

Our attacks can be further developed to handle the situation where multiple IVs are in use. Again, we can obtain attacks against method 3 of [1] for multiple secret, random IVs in the first attack model. We can also find attacks against method 3 of [2] for multiple secret, random IVs in the second attack model. Since the modifications to our existing attacks are quite straightforward, we do not include the details in this paper. Nor have we analyzed the other padding methods from [1,2] in the secret and random IV setting. Padding method 1 in both standards does not de-pad uniquely and is only useful when plaintexts have fixed or known lengths. We expect that padding oracle attacks may be possible against this method. As was noted in [6,8], padding method 2 in the two standards seems to be largely immune to such a side-channel analysis and indeed makes a good candidate for recommendation as a padding method in the ISO standard for CBC-mode encryption.

2 Symbols and Notation

We largely use the same notation as in [8], with only one major difference. In [8], the first block of the ciphertext C_0 submitted to the padding oracle was taken to be the IV. Here, the attacker no longer submits the IV (since he does not know it), but he may or may not submit additional information I, depending on whether the attack is in the first or second attack model. Therefore in our new notation, the first block of the ciphertext will be the first encrypted block C_1, and, in making padding oracle queries, we will prepend additional information I to ciphertexts whenever appropriate. The context will make clear when this is being done.

For a detailed description of CBC-mode encryption, see [8–Section 2.2]. We summarise our other frequently used notation here for ease of reference.

C : ciphertext output after CBC-mode encryption; target ciphertext the attacker is trying to decrypt.
C' : ciphertext to be submitted to the padding oracle during an attack.
$d_K(Y)$: decryption of ciphertext block Y under key K.
D : unpadded data string to be CBC-mode encrypted.
$e_K(X)$: encryption of plaintext block X under key K.
I : information determining the IV in our first attack model.
IV : the initialisation vector used in CBC-mode.
L_D : the length (in bits) of the data string D.
n : the block size (in bits) of the block cipher.

P : the result of applying a given padding method to D.
q : the number of blocks in data string P after padding.
VALID and INVALID: padding oracle responses to, respectively, correct and incorrect padding after receipt and decryption of ciphertext.
$X||Y$: the result of concatenation of strings X and Y.
$X \oplus Y$: the result of exclusive-or (XOR) of strings X and Y.
$(X)_2$: the binary representation of the value X.
X_j : the jth block of the plaintext or ciphertext X $(1 \leq j \leq q)$.
$X_{j,k}$: the kth bit of the plaintext or ciphertext block X_j, $0 \leq k < n$.

3 Analysis of Padding Method 3 of ISO/IEC 9797-1

3.1 Review of Padding Method and Previous Attack

We reproduce the original text of the padding method from [1]:

> "The data string D to be input to the [...] algorithm shall be right-padded with as few (possibly none) '0' bits as necessary to obtain a data string whose length (in bits) is a positive integer multiple of n. The resulting string shall then be left-padded with a block L. The block L consists of the binary representation of the length (in bits) L_D of the unpadded data string D, left-padded with as few (possibly none) '0' bits as necessary to obtain an n-bit block. The right-most bit of the block L corresponds to the least significant bit of the binary representation of L_D."

The attack in [8–Section 3.4] decrypts, one block at a time, arbitrary ciphertexts $C_1||C_2||\ldots||C_q$ that are padded using the above method. The attack makes repeated use of a padding oracle and has two phases.

The general case of the first phase applies to ciphertexts consisting of three or more blocks and was presented as Algorithm 9797-1-m3-get-L_D-general in [8]. The algorithm, when given a q-block valid ciphertext as input, finds L_D by manipulating the padding bits. The procedure requires the re-use of old IVs. Since we will use it in our new attack, we reproduce this algorithm here as Algorithm 1., with notation modified to reflect the use of additional information I to determine IVs. In the algorithm (which, in common with all the algorithms presented here, can be found in the Appendix), I denotes the IV-determining information that accompanied the target ciphertext.

The special case of the first phase applies to two-block ciphertexts and was presented as Algorithm 9797-1-m3-get-L_D-special in [8]. This algorithm does require the ability to directly manipulate bits in the IV and so does not apply in either of our attack models.

The second phase of the attack on Method 3 of ISO/IEC 9797-1 in [8–Section 3.4] is the actual decryption. Algorithm 9797-1-m3-decrypt in [8] returns the rightmost $n-1$ bits of a plaintext block but in so doing makes repeated updates to the IV. It is therefore unusable in our attack models. Algorithm 9797-1-m3-decrypt-last-bit in [8] returns the leftmost bit of a plaintext block. It is also unusable, since it requires a customised setting of the IV and a successful run of Algorithm 9797-1-m3-decrypt.

3.2 An Attack with Secret and Random IVs

We require some further mild assumptions in order to obtain an attack against padding method 3 of [1] with secret and random IVs. The attack is in our first attack model. We assume that, in addition to having a target ciphertext C which he wishes to decrypt, the attacker has also gathered a set of m auxiliary ciphertexts labelled $C^1, C^2, \ldots, C^m$, and associated IV-determining information $I^1, \ldots, I^m$. We write q_j for the number of blocks in ciphertext C^j and require that $q_j \geq 3$ for each j. The attacker can immediately use Algorithm 1. and the padding oracle to find the length L_j of each ciphertext C^j. We write $F_j = L_j \bmod n$. We require that the F_j be distinct and that no F_j is equal to zero. Without loss of generality, we can then write $1 \leq F_1 < F_2 < \ldots < F_m \leq n - 1$. We also set $F_{m+1} = n$.

Notice that auxiliary ciphertexts with the required properties can easily be selected from a larger pool of ciphertexts. The auxiliary ciphertexts are not themselves decrypted in the course of the attack (though they can individually be used as target ciphertexts if their decryption is desired).

Our attack is presented in Algorithm 2. and described in words below.

The attack attempts to recover the plaintext block P_k matching the block C_k of the q-block ciphertext C. In fact, we are only able to extract the rightmost $n - F_1$ bits of P_k for each $k \geq 2$. The attack attempts to construct, for decreasing values of j, a valid q_j-block ciphertext whose last block is the target block C_k and whose first block is C_1^j. Because of the padding rule, such a ciphertext must correspond to a plaintext in which the last block P'_{q_j} consists entirely of '0's in the rightmost $n - F_j$ positions. By carefully controlling the values in the penultimate ciphertext block, we can ensure that only a relatively small number of trials is needed in order to achieve this for each successive value of j. Eventually, when $j = 1$, we have a ciphertext with last block C_k where the matching plaintext block P'_{q_1} has '0's in the rightmost $n - F_1$ positions. From this information and C_{k-1} it is easy to recover the rightmost $n - F_1$ positions of the original plaintext block P_k.

We now explain in more detail the operation of the attack. We begin by considering the rightmost $n - F_m$ positions. Consider submitting to the padding oracle a ciphertext of the form:

$$I^m, C_1^m || \underbrace{00\ldots0|| \ldots ||00\ldots0}_{q_m-3 \text{ blocks}} ||S||C_k$$

where S is a block taking on a random value in the rightmost $n - F_m$ positions. Because I^m determines the original IV used in obtaining C^m, block C_1^m indicates that $n - F_m$ '0' padding bits should be found in the last plaintext block, and hence the oracle will return VALID with a probability of $2^{F_m - n}$. An INVALID response indicates that another value of S should be tested. In the algorithm we simply use an increasing $(n - F_m)$-bit counter for this purpose. After an average of around 2^{n-F_m-1} and at most 2^{n-F_m} trials, we will obtain a VALID response. In this case, we learn that $S \oplus d_K(C_k)$ is equal to '0' in the rightmost $n - F_m$ positions.

Notice that from this information and knowledge of C_{k-1}, we could immediately recover the rightmost $n - F_m$ bits of P_k. However, we now preserve the successful value of S by setting $R = S$, and proceed to examine the rightmost $n - F_{m-1}$ bits. Now consider submitting to the padding oracle a ciphertext of the form:

$$I^{m-1}, C_1^{m-1}||\underbrace{00\ldots0||\ldots||00\ldots0}_{q_{m-1}-3 \text{ blocks}}||S||C_k$$

where now S is a block taking on a random $(F_m - F_{m-1})$-bit value in positions $F_{m-1}, F_{m-1}+1, \ldots, F_m - 1$, and equalling R in the rightmost $n - F_m$ positions. Now block C_1^{m-1} indicates that $n - F_{m-1}$ '0' padding bits should be found in the last plaintext block. By using R to set the rightmost $n - F_m$ bits of S, we have already arranged '0' bits in the rightmost $n - F_m$ positions of the last plaintext block. So the oracle returns a VALID response with probability $2^{-(F_m - F_{m-1})}$. Again, we use a counter to test the $2^{F_m - F_{m-1}}$ values in positions $F_{m-1}, F_{m-1}+1, \ldots, F_m - 1$. After an average of about $2^{F_m - F_{m-1}-1}$ and at most $2^{F_m - F_{m-1}}$ trials, we will obtain a VALID response. In this case, we learn that $S \oplus d_K(C_k)$ is equal to '0' in the rightmost $n - F_{m-1}$ positions.

It is now straightforward to see how Algorithm 2. proceeds in this manner to eventually construct a valid ciphertext of the form:

$$I^1, C_1^1||\underbrace{00\ldots0||\ldots||00\ldots0}_{q_1-3 \text{ blocks}}||R||C_k$$

so that the corresponding last plaintext block contains '0' padding bits in the rightmost $n - F_1$ positions. Then a simple calculation shows that the rightmost $n - F_1$ bits of P_k are equal to the rightmost $n - F_1$ bits of the block $R \oplus C_{k-1}$.

3.3 Complexity and Impact

It takes an average of just over $2^{F_{j+1}-F_j-1}$ oracle queries to obtain a VALID response and recover the bits at positions F_j to $F_{j+1} - 1$ of P_k. So the average number of oracle queries needed to recover $n - F_1$ bits of plaintext is $\sum_{j=1}^{m} 2^{F_{j+1}-F_j-1}$. The worst-case complexity is twice this. Notice that when $F_1 = 1$ and $F_{j+1} - F_j = 1$ for each j, the average number of oracle queries needed to decrypt all but the leftmost bit of an n-bit block is just $n - 1$. In this case, at most two oracle queries are made for each j. In fact, since the outcome of the second oracle query is determined by the first, it is trivial to modify the attack so that $n - 1$ queries also represents the worst-case performance.

As an example, suppose the block size $n = 64$ and the data is byte-oriented. Suppose we can obtain 7 auxiliary ciphertexts whose lengths modulo 64 are 8,

16, 24, ..., 56. Then we have $m = 7$ and the average number of oracle queries needed to obtain 56 out of 64 plaintext bits is roughly 900. If the plaintext has some sort of predictability (e.g. ASCII characters making up an English text, or certain positions in a message within some known protocol), then the remaining byte might be easily guessed.

3.4 Limitations

Unfortunately, we have not succeeded in finding a method to extract the leftmost $F_1 \geq 1$ bits of the plaintext block P_k. The underlying reason is that, when the original data fits exactly within blocks, the default padding rule is to add no padding bits at all. This makes it difficult to set up a padding oracle test giving plaintext information.

Algorithm 1. can only find the contents of the length block for ciphertexts with at least 3 blocks. Whilst we are usually more interested in plaintext bits than length information, it would be convenient if Algorithm 2. could be applied to block C_1 of a two-block target ciphertext to extract the length information L_D. However, this would require knowledge of the IV (since block C_{k-1} is used at the last stage of our attack to recover the original plaintext bits). A lower bound on this length can be found by running Algorithm 2. on target block C_2 and finding the position of the rightmost one in P_2.

3.5 Comparison

The secret and random conditions on IVs have forced us to develop a completely new attack strategy against padding method 3 of [1]. The corresponding attack in [8] makes near-optimal use of the padding oracle and extracts all plaintext bits. To be efficient, our new attack requires the collection of auxiliary ciphertexts with a good spread of data lengths. There might be scenarios where this is unrealistic. Our new attack can never extract the leftmost data bits in each block. In the best case, it can recover all but the leftmost bit of plaintext using an optimal number of oracle queries (if we ignore the cost of finding the lengths of the auxiliary ciphertexts). Our attack cannot be extended to yield efficient attacks in the second attack scenario in which the adversary has no information about IVs at all. The reason is that the length information is placed in the first plaintext block – as a result, a random setting of the IV is almost certain to produce an INVALID response from the padding oracle.

In summary, in comparison to [8], the secret IV restriction has succeeded in increasing the complexity and decreasing the effectiveness of an attack. However, the attack is still feasible in many circumstances.

4 Analysis of Padding Method 3 of ISO/IEC 10118-1

4.1 Review of Padding Method and Previous Attacks

We reproduce below the original description of the padding method from [2], except that here, and throughout, we use n in place of L_1 to denote the block size:

"This padding method requires the selection of a parameter r (where $r \leq n$), e.g. $r = 64$, and a method of encoding the bit length of the data D, i.e. L_D as a bit string of length r. The choice for r will limit the length of D, in that $L_D < 2^r$.

"The data D [...] is padded using the following procedure.

1. D is concatenated with a single '1' bit.
2. The result of the previous step is concatenated with between zero and $n-1$ '0' bits, such that the length of the resultant string is congruent to $n-r$ modulo n. The result will be a bit string whose length will be r bits short of an integer multiple of n bits (in the case $r = n$, the result will be a bit string whose length is an exact multiple of n bits).
3. Append an r-bit encoding of L_D using the selected encoding method, yielding the padded version of D."

No encoding method (for L_D) is specified in the standard. We assume that base 2 encoding is used. Our attacks here work no matter which encoding method is used, though the attacker needs to know this method.

Using this padding method, the padding bits for data string D are appended in one of two ways:

Same-block Here $(L_D \bmod n) \leq (n - r - 1)$. The last block of D has enough space after the last data bit to contain at least a single '1' bit and the r bits encoding L_D. The number of padding bits (including the length information) is between $r+1$ and $n-1$.

Cross-block Here $(L_D \bmod n) \geq (n - r)$. The last block of D does not have enough space to contain a '1' bit and the r bits encoding L_D. The number of bits padded is between n and $n+r$ and the padding either fits exactly into an extra block or extends over two blocks. Note that this will always be the case when $r = n$ or $r = n - 1$.

In [8], the authors presented two inter-dependent attacks against this padding method. The first attack creates a valid ciphertext with the target ciphertext block as the last block, while the second attack decrypts the last block of any ciphertext.

In more detail, Attack 1 of [8] (named "directed IV search") takes a ciphertext block C_k as input, and outputs a valid ciphertext of the form $IV'||C_k$. It operates by searching for an IV setting that produces a valid ciphertext. This ciphertext is then fed into Attack 2 for decryption. The need to vary the IV in a controlled manner means that the attack does not work when IVs are secret.

Attack 2 of [8] (named "attacking the last block(s)") takes as input a whole ciphertext and operates in two phases. In the first phase, it finds L_D; in some cases (including those resulting from Attack 1 of [8]) this involves changing bits in the IV. So this phase does not work in general for secret IVs. In the second phase plaintext bits are extracted. In the case of a same-block padded ciphertext, this second phase does not require any control over the IV. So it will continue to function with only minor modifications in the new setting. In the case of a cross-block padded ciphertext, the second phase can be used to speed up Attack 1 of [8]. This will fail with secret IVs, since Attack 1 of [8] requires their controlled modification.

Despite the failure of Attacks 1 and 2 of [8], a similar strategy can be followed and the original attacks can be modified to work in the tougher of our two attack scenarios. Analogues of Attacks 1 and 2 of [8] are presented in Sections 4.2 and 4.3.

4.2 Attacking an Arbitrary Ciphertext Block

The attack we present in this section attempts to decrypt an arbitrary block C_k of a ciphertext $C_1||C_2||\dots||C_q$. In fact, our attacks only work for $k \geq 2$. It proceeds in two phases. In the first phase, a valid ciphertext is constructed having C_k as the final block. In the second phase, the attack of Section 4.3 is used to decrypt that final block. From this, P_k is easily found. Note that if C_q is the target block, then one should proceed directly to the attack of Section 4.3.

Phase 1: Constructing a Valid Ciphertext. In this phase, we construct a valid three-block or four-block ciphertext having target block C_k as the last block. We aim for ciphertexts of these lengths because they simplify the second phase of the attack: we will see in Section 4.3 that ciphertexts containing $q \geq 3$ blocks are the easiest ones to deal with.

This phase splits into two cases, dependent on the value of r.

In the first case, we have $r < n$. The algorithm for this case is given in Algorithm 3. and is next described in words. The algorithm essentially submits three-block ciphertexts of the form:

$$\underbrace{00\dots0}_{n}||R_2||C_k$$

to the padding oracle, for various values of R_2 chosen in such a way that at least one choice is guaranteed to produce a valid ciphertext. Our algorithm works no matter what IVs are used by the padding oracle. Note that we suppress any information I in submissions to the padding oracle here, and throughout this section, because we are operating in the second attack model.

In more detail, a counter i is used to determine the rightmost $r+1$ bits of R_2, while the leftmost $n-r-1$ bits are set to '0'. This effectively means that ciphertexts with all possible values of the length field in plaintext block P_3' are submitted to the oracle as i runs between 0 and $2^r - 1$, the first half of the search space. At least one choice of i in this range is guaranteed to result in a VALID response from the oracle unless C_k and the selection of R_2 mean that the leftmost $n-r$ bits of P_3' are all '0'. If this last case occurs, then considering all i between 2^r and $2^{r+1}-1$ ensures that one of the leftmost $n-r$ bits of P_3' is a '1' and that at least one choice of i results in a VALID response. We will evaluate the average and worst-case complexity of this case of Phase 1 below.

In the second case, where $r = n$, a similar attack applies. We now submit four-block ciphertexts of the form:

$$\underbrace{00\ldots0}_{n}||R_1||R_2||C_k$$

to the padding oracle, where we try all possible settings of R_2 and the rightmost bit of R_1. We are then guaranteed to encounter a valid ciphertext after a maximum of 2^{n+1} oracle calls. The algorithm for this case is given in Algorithm 4.; we analyse its complexity in detail below.

Phase 2: Decrypting • $_k$. Once we have a valid three or four-block ciphertext, the attack of Section 4.3 can be applied to obtain the plaintext block P'_3 (or P'_4 in the four-block case) corresponding to the final block of C'. From P'_3, the original plaintext block P_k can be recovered using the relation $P_k = P'_3 \oplus R_2 \oplus C_{k-1}$. (A similar procedure applies for the four-block case.) As we shall see below, the attack of Section 4.3 is always efficient when attacking the last block of a three-block (or four-block) ciphertext. So this approach allows efficient extraction of P_k.

A little more detail is appropriate at this stage. We focus on the three-block case. The first phase of the attack in Section 4.3 finds the length L_D of the data encrypted in C'. If $L_D > 2n$, then the data is same-block padded, while if $L_D \leq 2n$ it is cross-block padded. If it happens that the data is cross-block padded, then all the bits in P'_3 (or P'_4 in the four-block case) are already determined and are of the form:

$$\underbrace{00\ldots0}_{n-r}\underbrace{(L_D)_2}_{r} \quad \text{or} \quad \underbrace{10\ldots0}_{n-r}\underbrace{(L_D)_2}_{r}.$$

So in this case no actual decryption step is needed to recover P_k. Notice that this case will always apply when $r = n$ or $r = n - 1$. When the data is same-block padded, we must proceed to the second phase of the attack in Section 4.3. In the three-block case, this phase will efficiently recover the entire plaintext block P'_3 consisting of (in general) data bits, padding bits and length information. From P'_3, we can recover P_k using the relation $P_k = P'_3 \oplus R_2 \oplus C_{k-1}$. A similar procedure applies for the four-block case.

Complexity. We begin by analyzing Phase 1 of the attack in the case where $r < n$. The analysis is complicated by the fact that Algorithm 3. might output a valid three-block ciphertext C' for which the corresponding plaintext $P' = P'_1||P'_2||P'_3$ is cross-block padded. This will have the effect of slightly lowering the average-case complexity when compared to the corresponding attack in [8]. Such a cross-block padded plaintext requires that blocks $P'_2||P'_3$ take the form:

$$P'_{2,0}P'_{2,1}\cdots P'_{2,L_D-n-1}\underbrace{10\ldots0}_{2n-L_D}||\underbrace{00\ldots0}_{n-r}\underbrace{(L_D)_2}_{r}$$

where each $P'_{2,i}$ can be either a '0' or '1' bit and $(2n - r) \leq L_D \leq (2n - 1)$. There are r n-bit patterns (corresponding to the r possible values of L_D) for P'_3 that have the correct form. So the probability that Phase 1 produces cross-block padding is at most $r2^{r-n}$ as we vary the rightmost r bits of R_2 in Algorithm 3.. Of course, such cross-block padding may never occur during the execution of Algorithm 3.: given that R_1 and the decryption key K are fixed, there may be no choice of R_2 that produces the required bit pattern in $P'_2 = d_K(R_2) \oplus R_1$.

In any case, we see that there is a probability of at least $1 - 2^{r-n}$ that either there is a '1' somewhere in the leftmost $n - r$ bits of P'_3, or we obtain a cross-block padded ciphertext. In these cases, Algorithm 3. takes on average 2^{r-1} oracle calls. On the other hand, there is a probability of at most 2^{r-n} that the leftmost $n - r$ bits of P'_3 are all '0' and Algorithm 3. tries all 2^r possible settings for the rightmost bits of P'_3 without a VALID response. Algorithm 3. will then take on average a further 2^{r-1} oracle calls before obtaining a VALID response. A simple calculation now shows that the average number of oracle calls needed by Algorithm 3. is at most $2^{r-1} + 2^{2r-n}$, while in the worst-case it is 2^{r+1}. When r is small relative to n, the average-case complexity is dominated by the term 2^{r-1}.

Phase 1 of the attack in the case $r = n$ uses Algorithm 4.. This algorithm uses on average 2^n oracle calls to obtain a VALID response and 2^{n+1} in the worst case.

Phase 2 uses the attack in Section 4.3 for the same-block padded case, which has a complexity of $O(n)$ oracle calls. So Phase 2 does not contribute significantly to the overall complexity required to decrypt a single block (unless r is very small).

Impact. This attack applies to any ciphertext block C_k of a ciphertext $C_1||C_2||\ldots||C_q$, except for the first block C_1. It is not possible to decrypt C_1 because of the use of the relation $P_k = P'_3 \oplus C_{k-1} \oplus R_2$ at the end of the attack: this would necessitate an XOR with the secret IV. The attack recovers all n bits within the block and does so many orders faster than exhaustive search for many choices or r. When $r = n$ our attack is still better than exhaustive key search for block ciphers whose key size is greater than the block length. We restate the observation from [8] that the seemingly innocuous parameter r has unexpected implications for security.

Comparison. This attack is an adaptation of Attack 1 in [8] to the second of our attack models, where IVs are secret, random and completely hidden from the adversary. These extra restrictions do not seem to be a major hindrance to the effectiveness of the attack. Specifically, the complexity of the attack has remained practically the same as the corresponding attack in [8], and, except for the first ciphertext block, the impact remains unchanged. The attack uses three-block or four-block ciphertexts instead of two-block ones when $r < n$; this is not expected to be of any practical significance.

4.3 Attacking the Last Block(s)

The attack we present in this section attempts to decrypt the last block C_q of a ciphertext $C_1||C_2||\ldots||C_q$. It is an adaptation of Attack 2 in Section 4.3 of [8] to the secret and random IV setting, and, like that attack, proceeds in two phases. Phase 1 determines the length L_D of the ciphertext, while Phase 2 will recover plaintext bits in the mixed block containing both padding and data bits. (If there is such a block, then it is unique.) Recall that, as well as being directly applicable to the last block C_q, our attack can also be used in conjunction with the attack in Section 4.2 to decrypt arbitrary ciphertext blocks.

Phase 1: Finding • $_D$. This phase of our attack is derived from the corresponding phase in [8]. The case $q = 2$ requires special treatment and our methods fail completely when $q = 1$. We first examine the general case $q \geq 3$.

For ease of presentation we take $r \leq n - 2$, but Algorithm 5. handles all values of r. Here, in the same-block padded case, the last plaintext block P_q has the following format:

$$\underbrace{[\mathsf{DATA}]}_{t}\,\underbrace{10\ldots0}_{p}\,\underbrace{(L_D)_2}_{r}$$

where $t+p+r = n$ and $p \geq 1$. In the cross-block padded case, the above format spans the last two blocks P_{q-1} and P_q and we put $t+p+r = 2n$. We note that the attacker does not, at first, know which of the cases he is faced with.

Given our q-block ciphertext, the rightmost position at which a data bit could ever reside is at $P_{q,n-r-2}$. Consider then submitting to the padding oracle the ciphertext:

$$C_1||C_2||\ldots||C_{q-1} \oplus \underbrace{00\ldots0}_{n-r-2}1\underbrace{00\ldots0}_{r+1}||C_q.$$

The oracle will return either:

- VALID, meaning the padding has not been disturbed so the bit flipped in P'_q by modifying C_{q-1} is a data bit. Since this bit is at the rightmost possible data bit position, we can deduce that the data length L_D equals $(q-1)n + n - r - 1 = qn - r - 1$.
- or INVALID, meaning a padding bit has been flipped so the padding is no longer valid. Therefore the padding boundary is somewhere to the left of this bit.

We can generalise the above observation about $P_{q,n-r-2}$ to produce Algorithm 5., a binary search algorithm to find L_D. In this algorithm, we initialise two pointers l and u at the extremities of the possible padding range and modify the ciphertext so as to invert the plaintext bit that lies in the middle position $h := \lfloor (l+u)/2 \rfloor$ of the range. We then submit the ciphertext to the oracle. A VALID response means the start of the padding is to the right of this test bit so we set the lower pointer l to the position $h+1$, whereas INVALID indicates it is to the left and we set the upper pointer u to h. We must then reset the test bit before proceeding to the next test. This process is repeated until the upper and lower pointers coincide, at which point they indicate the rightmost data bit. It is then easy to determine L_D. Clearly, the algorithm makes roughly $\log_2 n$ calls to the padding oracle and so is efficient.

This completes our discussion of the general case where $q \geq 3$. Next we focus on the case $q = 2$. This case requires special treatment because setting up a binary search as above requires the ability to modifiy plaintext bits in the whole range of padding positions, which in this case includes those in the rightmost r positions of the plaintext block P_1. This in turns necessitates the ability to modify bits in the corresponding positions in the IV, which is not possible in the setting of secret and random IVs.

Our solution, presented in Algorithm 6., is to perform a binary search over the restricted range of those padding positions in the second (and last) plaintext block P_2. This is done by initializing the lower and upper pointers to n and $2n + r - 1$ respectively. If the search finishes pointing to any position between $P_{2,1}$ and $P_{2,n-r-1}$ then this indicates the actual leftmost padding position from which L_D can be determined. On the other hand, if the search ends pointing at $P_{2,0}$, then we can deduce that the bit at that position is a padding bit and hence the boundary is somewhere to the left of that position. From this we can deduce that the plaintext block P_2 consists only of padding bits and encoded length information, and that $L_d \leq n$. We could go further and deduce most of the contents of block P_2, but these bits are not usually of much interest to the attacker. In this case, we cannot continue with the attack.

We note that this $q = 2$ version of the length-finding algorithm is never invoked by the attack in Section 4.2 (unless C_2 is the last block and happens to be the initial target).

Finally we consider the case $q = 1$. Here we are not able to find L_D by performing any kind of search for the data/padding boundary since this would require manipulating the IV. Thus our methods fail in this case.

Phase 2: Decrypting. We assume that $q \geq 2$ and that L_D has been successfully obtained from Phase 1. This will always be the case for $q \geq 3$ and often the case for $q = 2$. Same-block and cross-block padded messages are treated differently; recall that knowledge of L_D indicates with which case the attacker is faced.

Decrypting: Same-block Recall the structure of the last plaintext block P_q: t unknown data bits, followed by p padding bits in the form $10\ldots0$ and finally r bits encoding the data length L_D. The only bits remaining to be found are the t data bits. We can assume that $t \geq 1$ and recover these as follows. Consider submitting to the oracle the ciphertext $C' = R||C_q$ where:

$$R = C_{q-1} \oplus \underbrace{00\ldots0}_{n-r}\underbrace{(L_D)_2}_{r} \oplus \underbrace{00\ldots0}_{t}\underbrace{10\ldots0}_{p}\underbrace{(n+t-1)_2}_{r}.$$

This ciphertext is constructed in such a way that, after decryption to obtain plaintext $P'_1||P'_2$, the length block in P'_2 encodes the length $n + t - 1$, while the

p padding bits are modified to be all '0's. Moreover, data bits are copied intact from P_q to P'_2, so that $P_{q,i} = P'_{2,i}$ for $0 \leq i < t$. From the construction of C', we see that the oracle will output VALID if and only if $P'_{2,t-1} = 1$. Since we have $P_{q,t-1} = P'_{2,t-1}$, we can obtain the last data bit of block P_q.

This idea can be extended to recover all t data bits in P_q in a similar manner: we reduce the length field in P'_2 one step at a time whilst fixing the data in all recovered bit positions to be '0' so that they become part of a valid padding. A single bit of P'_2 and hence of P_q is revealed at each iteration, until all the data bits in P_q are recovered. This procedure is given in detail in Algorithm 7.. Note that the algorithm makes use of the function $\bar{\Omega}$ defined by:

$$\bar{\Omega}(C) = \begin{cases} 1 & \text{if the padding oracle returns } \mathsf{VALID} \text{ for input } C, \\ 0 & \text{if the padding oracle returns } \mathsf{INVALID} \text{ for input } C. \end{cases}$$

Note that $\bar{\Omega}$ is the complement of the function Ω in [8].

Decrypting: Cross-block For cross-block padded plaintexts with $q \geq 3$ blocks, P_q is determined completely by L_D and the padding. However, the padding often extends into the penultimate plaintext block P_{q-1} and we can exploit this fact when decrypting block C_{q-1}.

Suppose $t = L_D \bmod n$ and $t \neq 0$. Then $u = n - t$ bits of padding of the form $\underbrace{10\ldots0}_{u}$ are present in P_{q-1}. We show how to decrypt C_{q-1} using the attack in Section 4.2, but with a speed-up factor of 2^{u-1}. Consider ciphertexts of the form $C' = 00\ldots0||R_2||C_{q-1}$ where:

$$R_2 = C_{q-2} \oplus \underbrace{00\ldots0}_{t}\underbrace{10\ldots0}_{u} \oplus \underbrace{00\ldots0}_{n-r}\underbrace{(3n-r-1)_2}_{r}.$$

Upon decryption, this ciphertext will produce a plaintext block P'_3 of the form:

$$P'_{3,0}P'_{3,1}\cdots P'_{3,t-1}y_0y_1\cdots y_{u-1}$$

where $y_0y_1\cdots y_{u-1}$ are the u least significant bits of the binary encoding of the length field $3n-r-1$. Now it is straightforward to see that running through all 2^{r-u+1} settings of the $r-u+1$ bits immediately to the left of the rightmost u bits (by varying the relevant bits of R_2) will ensure that at least one valid three-block ciphertext C' is obtained. Naturally, after obtaining such a valid C', we can apply the attack of this section again, now using C' as the input ciphertext. Eventually, that attack will output a candidate P'_3 for the decryption of block C_{q-1} in ciphertext C'; from this we can deduce the decryption P_{q-1} of C_{q-1} in the original ciphertext C using the relation $P_{q-1} = P'_3 \oplus R_2 \oplus C_{q-2}$.

This strategy takes on average about 2^{r-u} oracle calls which is roughly a fraction $2^{-(u-1)}$ of the number of oracle calls needed on average for the corresponding attack in Algorithm 3. without the knowledge of the u padding bits.

Unfortunately this strategy does not work for two-block cross-block padded ciphertexts in our attack model, because the very last step would need to use IV in place of C_{q-2}.

Complexity. For $q \geq 3$, Phase 1 of the attack takes roughly $\log_2 n$ oracle calls to find the data length L_D. For same-block padded plaintexts, Phase 2 then takes one call per bit for decrypting. So to recover the t data bits in the last block, $t + \log_2 n$ oracle calls are required. For cross-block padded plaintexts, the block P_q is completely determined by L_D. Then Phase 2 needs on average around 2^{r-u} oracle calls to recover the whole of the penultimate plaintext block P_{q-1}. Here u is the number of known padding bits in P_{q-1} and we have ignored the comparatively small cost of running the length-finding and last-block decryption algorithms of this section.

For two-block ciphertexts, Phase 1 will take on average $\log_2(n - r)$ oracle calls to find either the actual value of L_D or to find that $L_D \leq n$. In the former case, the complexity of Phase 2 is exactly as above. In the latter case, the data is cross-block padded but we are not able to recover the penultimate plaintext block. Phase 1 of the attack is not successful for single-block ciphertexts and no data bits can be extracted using our attack in this case.

It is important to note that, even though the two attacks presented here and in Section 4.2 are inter-dependent, there is no possibility of the attack entering an infinite loop. This is not difficult to show.

Impact. The attack is highly efficient (in terms of oracle access) at extracting plaintext bits in the last plaintext block P_q. A maximum of $n - r - 1$ bits of data can be recovered in this way and the attack is therefore significant for short messages, especially in combination with a small r. One might argue that $r = n$ is a natural choice for the implementor. In this case, the padding is always cross-block and the attacker must resort to the speeded-up version of the attack in Section 4.2.

Comparison. One impact of assuming that IVs are secret and random on the attack in this section is that Phase 1 of the attack is prevented from determining the exact data length of single-block ciphertexts, and two-block ones when the plaintext is cross-block padded. This, in turn, stops us from extracting any data bits in these cases. This is in contrast to the corresponding cases in [8], where the ability to manipulate the IV can be used to advantage.

The complexity of the two phases remains unchanged when compared to the corresponding attack in [8] ($\log_2 n$ oracle calls to find L_D and one oracle call per data bit extracted for same-block padding). Short ciphertexts, typically two or three blocks long, are used throughout, so there is little or no message expansion.

5 Conclusions

We have shown that the use of IVs that are secret and random does not prevent padding oracle attacks on CBC-mode encryption. We have shown this to be the case in the context of two padding methods previously analyzed in [8]. The use of secret, random IVs required us to develop new ideas and to extend the analysis of [8]. The new attacks are, at best, of roughly equal complexity to those of [8] and the assumptions we have made to obtain attacks seem reasonable. The attacks recover most, if not all, plaintext bits many orders of magnitude faster than exhaustive key search.

The 2004 FCD text for the 3rd edition of ISO/IEC 10116 [4], which supersedes [3], contains new text regarding padding methods in Clause 5 (Requirements). It now reads

> *...Padding techniques...are not within the scope of this International Standard, and throughout this standard it is assumed that any padding, as necessary, has already been applied.*

This effectively off-loads the responsibility of choosing a padding method to the implementor of this standard (if it is published with the text as it stands). In our view, not specifying a padding method at all has the potential to be even more dangerous than specifying a method that is known to be weak against certain attack types. After all, there is no guarantee that an implementor will not choose a method that falls to some even more realistic form of attack. Methods that appear to resist padding oracle attacks have been analysed [6]. For example, padding method 2 of [1], in which the plaintext is padded with a single '1' and as many '0's as are necessary to complete a block, seems like a good candidate. We currently know of no reason not to recommend it for use. We argue that the more complete and unambiguous a specification is, the smaller the chance for insecure approaches to be taken by an implementor.

Finally, we wish to repeat the point made in [6, 8] that padding oracle attacks can be easily thwarted by the proper use of strong integrity checks. It is now widely held that encryption should be accompanied by a data integrity mechanism whenever feasible and appropriate. Of course there are situations (for example, constrained environments) where the use of a MAC algorithm in addition to encryption is not possible. In these scenarios, the careful selection of a padding method and the avoidance of padding oracles in implementations is of paramount importance.

References

1. *ISO/IEC 9797-1: Information technology — Security techniques — Message Authentication Codes (MACs) — Part 1: Mechanisms using a block cipher*, 1999.
2. *ISO/IEC 10118-1 (2nd edition): Information technology — Security techniques — Hash-functions — Part 1: General*, 2000.
3. *ISO/IEC 2nd CD 10116 (revision): Information technology — Security techniques — Modes of operation for an n-bit block cipher*, 2002. (Second committee draft of proposed 3rd edition of the standard).
4. *ISO/IEC FCD 10116 (2nd edition): Information technology — Security techniques — Modes of operation for an n-bit block cipher*, 2004. (Final committee draft of proposed 3rd edition of the standard).

5. M. Bellare, A. Desai, E. Jokipii, and P. Rogaway. A Concrete Analysis of Symmetric Encryption: Analysis of the DES Modes of Operations. In *38th IEEE Symposium on Foundations of Computer Science*, pages 394–409. IEEE, 1997.
6. J. Black and H. Urtubia. Side-Channel Attacks on Symmetric Encryption Schemes: The Case for Authenticated Encryption. In *Proceedings of the 11th USENIX Security Symposium, San Francisco, CA, USA, August 5-9, 2002*, pages 327–338. USENIX, 2002.
7. B. Canvel, A. Hiltgen, S. Vaudenay, and M. Vuagnoux. Password Interception in a SSL/TLS Channel. In D. Boneh, editor, *Advances in Cryptology — CRYPTO 2003*, volume 2729 of *Lecture Notes in Computer Science*, pages 583–599. Springer-Verlag, 2003.
8. K.G. Paterson and A. Yau. Padding Oracle Attacks on the ISO CBC Mode Padding Standard. In T. Okamoto, editor, *Topics in Cryptology — CT-RSA 2004*, volume 2964 of *Lecture Notes in Computer Science*, pages 305–323. Springer-Verlag, 2004.
9. S. Vaudenay. Security Flaws Induced by CBC Padding — Applications to SSL, IPSEC, WTLS .... In L. Knudsen, editor, *Advances in Cryptology — EUROCRYPT 2002*, volume 2332 of *Lecture Notes in Computer Science*, pages 534–545. Springer-Verlag, 2002.

Appendix

We present here pseudo-code for the various algorithms developed in the text.

Algorithm 1.

Input: $I, C_1||C_2||\ldots||C_q$
Output: L_D

function 9797-1-M3-GET-L_D-GENERAL
  $l := 0$
  $u := n-1$
  repeat
    $h := \lceil (l+u)/2 \rceil$
    $C_{q-1,h} := C_{q-1,h} \oplus 1$
    if ORACLE$(I, C_1||C_2||\ldots||C_q)$ = VALID **then**
      $l := h$
    else
      $u := h-1$
    end if
    $C_{q-1,h} := C_{q-1,h} \oplus 1$
  until $l = u$
  return $L_D := (q-1)n + l + 1$
end function

Algorithm 2.

Input: auxiliary ciphertexts C^1, C^2, ..., C^m, IV-determining information I^1, I^2, ..., I^m, length information $q_1, \ldots, q_m$ and $F_1, \ldots, F_m$, target ciphertext blocks C_{k-1}, C_k

Output: rightmost $n - F_1$ bits of P_k

function 9797-1-M3-DECRYPT

  $R := \underbrace{00\ldots0}_{n}$

  $F_{m+1} := n$

  for $j := m$ to 1 **do**

    $i := -1$

    repeat

      $i := i + 1$

      $S := R \oplus \underbrace{00\ldots0}_{F_j} \; \underbrace{(i)_2}_{F_{j+1}-F_j} \; \underbrace{00\ldots0}_{n-F_{j+1}}$

    until ORACLE$(I^j, C_1^j || \underbrace{00\ldots0||\ldots||00\ldots0}_{q_j-3 \text{ blocks}} ||S||C_k) = \mathsf{VALID}$

    $R := R \oplus \underbrace{00\ldots0}_{F_j} \; \underbrace{(i)_2}_{F_{j+1}-F_j} \; \underbrace{00\ldots0}_{n-F_{j+1}}$

  end for

  return rightmost $n - F_1$ bits of $R \oplus C_{k-1}$

end function

Algorithm 3.

Input: C_k, r, n

Output: A valid three-block ciphertext, the last block of which is C_k

Require: $1 \leq r < n$

function 10118-1-M3-GENERAL(C_k, r, n)

  $R_1 := \underbrace{00\ldots0}_{n}$

  $R_2 := \underbrace{00\ldots0}_{n}$

  $i := 0$

  while ORACLE$(R_1||R_2||C_k) = \mathsf{INVALID}$ **do**

    $i := i + 1$

    $R_2 := \underbrace{00\ldots0}_{n-r-1} \underbrace{(i)_2}_{r+1}$

  end while

  return $R_1||R_2||C_k$

end function

Algorithm 4.

Input: C_k, r, n
Output: A valid four-block ciphertext, the last block of which is C_k
Require: $r = n$

```
function 10118-1-M3-SPECIAL(C_k, r, n)
    R_1 := 00...0  (n bits)
    R_2 := 00...0  (n bits)
    i := 0
    while ORACLE(00...0 (n bits) || R_1 || R_2 || C_k) = INVALID do
        i := i + 1
        if i = 2^r then
            i := 0
            R_1 := 00...01  (n bits)
        end if
        R_2 := (i)_2  (n bits)
    end while
    return 00...0 (n bits) || R_1 || R_2 || C_k
end function
```

Algorithm 5.

Input: $C_1||C_2||\ldots||C_q, n, r$
Output: L_D
Require: $q \geq 3$

```
function 10118-1-M3-FIND-L_D-GENERAL(C_1||C_2||...||C_q, n, r)
    C := C_1||C_2||...||C_q
    l := (q - 2)n + n - r
    u := (q - 1)n + n - r - 1
    repeat
        h := ⌊(l + u)/2⌋
        C_{⌊h/n⌋, h mod n} := C_{⌊h/n⌋, h mod n} ⊕ 1
        if ORACLE(C) = VALID then
            l := h + 1
        else
            u := h
        end if
        C_{⌊h/n⌋, h mod n} := C_{⌊h/n⌋, h mod n} ⊕ 1
    until l = u
    return L_D := l
end function
```

Algorithm 6.

Input: $C_1||C_2, n, r$
Output: L_D or "Plaintext length at most n"

function 10118-1-M3-FIND-L_D-SPECIAL($C_1||C_2$, n, r)
  $C := C_1||C_2$
  $l := n$
  $u := 2n - r - 1$
  repeat
    $h := \lfloor (l + u)/2 \rfloor$
    $C_{\lfloor h/n \rfloor, h \bmod n} := C_{\lfloor h/n \rfloor, h \bmod n} \oplus 1$
    if ORACLE(C) = VALID **then**
      $l := h + 1$
    else
      $u := h$
    end if
    $C_{\lfloor h/n \rfloor, h \bmod n} := C_{\lfloor h/n \rfloor, h \bmod n} \oplus 1$
  until $l = u$
  if $l > n$ **then**
    return $L_D := l$
  else
    return "Plaintext length at most n"
  end if
end function

Algorithm 7.

Input: L_D, C_{q-1}, C_q, r, n
Output: $P_q := P_{q,0}P_{q,1} \ldots P_{q,t-1} \underbrace{10\ldots0}_{p} \underbrace{(L_D)_2}_{r}$

Require: L_D indicates that the plaintext is same-block padded

function 10118-1-M3-DECRYPT(L_D, C_{q-1}, C_q, r, n)
  $t := L_D \bmod n$
  $p := n - r - t$
  $R := C_{q-1} \oplus \underbrace{00\ldots0}_{t} \underbrace{10\ldots0}_{p} \underbrace{(L_D)_2}_{r} \oplus \underbrace{00\ldots0}_{n-r} \underbrace{(n+t)_2}_{r}$
  for $j := t - 1$ to 0 **do**
    $R := R \oplus \underbrace{00\ldots0}_{n-r} \underbrace{(n+j+1)_2}_{r} \oplus \underbrace{00\ldots0}_{n-r} \underbrace{(n+j)_2}_{r}$
    $P_{q,j} := \bar{\Omega}(R||C_q)$
    $R_j := R_j \oplus P_{q,j}$
  end for
  return $P_q := P_{q,0}P_{q,1} \ldots P_{q,t-1} \underbrace{10\ldots0}_{p} \underbrace{(L_D)_2}_{r}$
end function

Analysis of the Non-linear Part of Mugi

Alex Biryukov[1,*] and Adi Shamir[2]

[1] Katholieke Universiteit Leuven, Dept. ESAT/SCD-COSIC, Kasteelpark Arenberg 10, B–3001 Heverlee, Belgium
http://www.esat.kuleuven.ac.be/~abiryuko/

[2] Department of Applied Mathematics and Computer Science, Weizmann Institute of Science, Rehovot 76100, Israel
shamir@wisdom.weizmann.ac.il

Abstract. This paper presents the results of a preliminary analysis of the stream cipher MUGI. We study the nonlinear component of this cipher and identify several potential weaknesses in its design. While we can not break the full MUGI design, we show that it is extremely sensitive to small variations. For example, it is possible to recover the full 1216-bit state of the cipher and the original 128-bit secret key using just 56 words of known stream and in 2^{14} steps of analysis if the cipher outputs any state word which is different than the one used in the actual design. If the linear part is eliminated from the design, then the secret non-linear 192-bit state can be recovered given only three output words and in just 2^{32} steps. If it is kept in the design but in a simplified form, then the scheme can be broken by an attack which is slightly faster than exhaustive search.

Keywords: Cryptanalysis, Stream ciphers, MUGI.

1 Introduction

MUGI is a fast 128-bit key stream cipher [4] designed for efficient software and hardware implementations (achieves speeds which are 2-3 times faster than RIJNDAEL in hardware and slightly faster in software). The cipher was selected for standardization by the Japanese government project CRYPTREC and is also one of the two proposed ISO stream cipher standards.

Previous analysis of MUGI given by its designers in [1] concentrated on linear cryptanalysis and resynchronization attacks. A recent work by Golic [3] studied

* This author's work was supported in part by the Concerted Research Action (GOA) Mefisto 2000/06 and Ambiorix 2005/11 of the Flemish Government and in part by the European Commission through the IST Programme under Contract IST-2002-507932 ECRYPT. The information in this document reflects only the author's views, is provided as is and no guarantee or warranty is given that the information is fit for any particular purpose. The user thereof uses the information at its sole risk and liability.

H. Gilbert and H. Handschuh (Eds.): FSE 2005, LNCS 3557, pp. 320–329, 2005.

only the linear component of the cipher. No security flaw of the full cipher has been reported so far.

In this paper we study the non-linear component of this cipher, and identify several potential weaknesses in its design. However we do not make any claim as to the security of the full MUGI which remains unbroken.

This paper is organized as follows: in Sect. 2 we describe the cipher MUGI, in Sect. 3 we describe two attacks on its non-linear component, and Sect. 4 concludes the paper.

2 Description of Mugi

The design of MUGI is based on the design philosophy proposed by J. Daemen and C. Clapp in their stream cipher PANAMA [2]. The internal state of the cipher at time t consists of two parts: a linearly changed large buffer $b^{(t)}$, and a non-linearly evolving shorter state $a^{(t)}$. See Fig. 1 for a schematic description. The

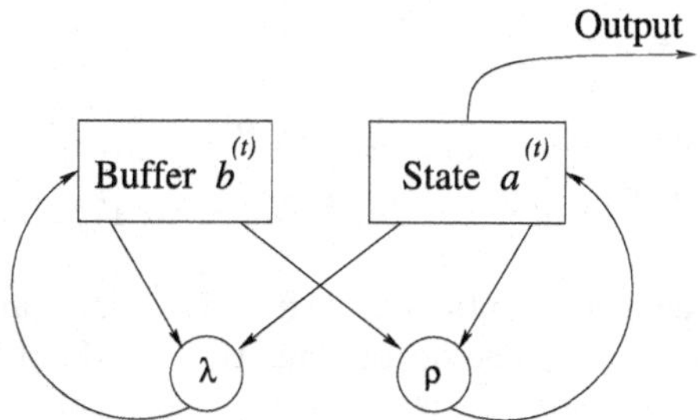

Fig. 1. The evolution of the MUGI state

cipher works in steps called rounds. At each round the internal state is updated and one of its three 64-bit words is produced as output. The evolution of the buffer which consists of sixteen 64-bit words happens in a LFSR-like slow fashion, together with a 64-bit feedback from the nonlinear state at each round:

$$b^{(t+1)} = \lambda(b^{(t)}, a^{(t)}),$$

or more explicitly:

$$\begin{aligned} b_j^{(t+1)} &= b_{j-1}^{(t)} (j \neq 0, 4, 10) \\ b_0^{(t+1)} &= b_{15}^{(t)} \oplus a_0^{(t)} \\ b_4^{(t+1)} &= b_3^{(t)} \oplus b_7^{(t)} \\ b_{10}^{(t+1)} &= b_9^{(t)} \oplus (b_{13}^{(t)} \lll 32) \end{aligned}$$

The evolution of the state $a^{(t)}$ is essentially a block cipher-like invertible process, in which two 64-bit words coming from the buffer are used as subkeys:

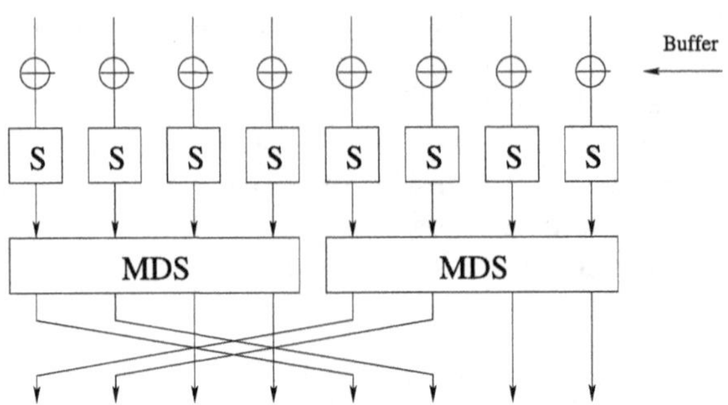

Fig. 2. The F-function

$$a^{(t+1)} = \rho(a^{(t)}, b^{(t)}),$$

or more explicitly:

$$\begin{aligned} {a_0}^{(t+1)} &= {a_1}^{(t)} \\ {a_1}^{(t+1)} &= {a_2}^{(t)} \oplus F({a_1}^{(t)}, {b_4}^{(t)}) \oplus C_1 \\ {a_2}^{(t+1)} &= {a_0}^{(t)} \oplus F({a_1}^{(t)}, {b_{10}}^{(t)} \lll 17) \oplus C_2\,, \end{aligned}$$

where C_1, C_2 are known constants. The F function reuses an S-box and MDS of the AES as shown in Fig. 2. The output function of MUGI is very simple:

$$Output[t] = a_2^{(t)},$$

i.e. one of the three 64-bit words in the internal state is given as output *prior* to the evaluation of the round t.

3 Cryptanalysis of the Two Variants of Mugi

In this section we cryptanalyse several variants of MUGI.

3.1 Change in the Output Function

The output function of MUGI outputs the 64-bit word $a_2^{(t)}$ before the tth round. In this section we show that MUGI is very sensitive to small changes in the output function. For example, if the output would consist of the same word $a_2^{(t)}$ but *after* the evaluation of the round t (or equivalently the word $a_1^{(t+1)}$), then the full cipher could be easily broken with practical complexity. In general, for any choice of a output word other than the one used in the actual design (i.e., either $a_0^{(t)}$ or $a_1^{(t)}$) the cipher can be easily broken.

We denote by the Greek letters $\alpha, \beta, \gamma, \delta, \epsilon$ an intermediate value $a_1^{(t)}$ which is used as the input to the two F-functions in each round. Suppose that $Output[t] = a_1^{(t)}$. Let us make several important observations. The first observation is that $a_1^{(t)} = a_0^{(t+1)}$ and thus we know the word that updates the buffer at each round. This would allow us to run the buffer update function $\lambda(\cdot)$ symbolically and write linear equations for the buffer bits, including those that enter the non-linear ρ function as "sub-keys". The second observation is that the word $a_1^{(t)}$ is used as the input to both F-functions of the round. The final observation is that we can write the following equation:

$$\alpha \oplus \delta = F(\beta, b_{10}^{(t+1)} \lll 17) \oplus F(\gamma, b_4^{(t+2)}). \tag{1}$$

In this equation α, β, γ and δ are words from the output stream and are thus known to the attacker (see also Fig. 3). Let us denote the inverse of the MDS

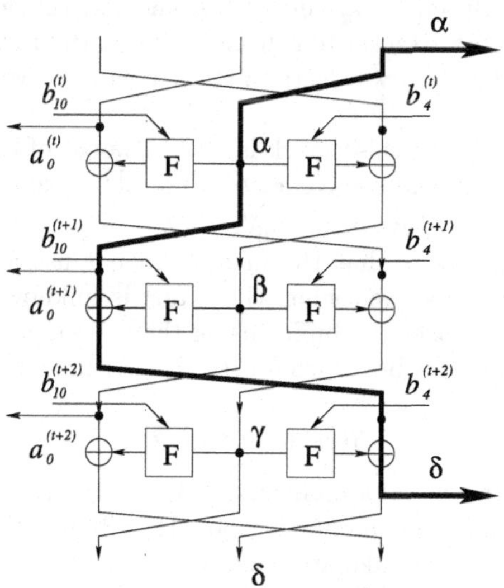

Fig. 3. The attack path

and the byte swap layers by M^{-1} and the eight S-box layer by S. By applying the inverse of the linear layer of F to both sides of the equation we can get the simplified equation:

$$M^{-1}(\alpha \oplus \delta) = S(\beta \oplus (b_{10}^{(t+1)} \lll 17)) \oplus S(\gamma \oplus b_4^{(t+2)}). \tag{2}$$

which decomposes into 8 independent equations in the 8 bit values of each S-box. One of the possible ways to use this equation is to look for points in the output stream in which at least one of the bytes of $M^{-1}(\alpha \oplus \delta)$ is zero. For a particular S-box this would lead to a simpler equation:

$$S(\beta \oplus (b_{10}^{(t+1)} \lll 17)) = S(\gamma \oplus b_4^{(t+2)}), \tag{3}$$

which further simplifies to:

$$(b_{10}^{(t+1)} \lll 17) \oplus b_4^{(t+2)} = \beta \oplus \gamma. \tag{4}$$

This is an 8-bit linear constraint on the bits of the buffer. It is sufficient to gather about $1024/8 = 128$ such equations in order to get a system of linear equations for all the buffer bits. The probability that at least one zero value occurs in each 64-bit block is $1 - (1 - 1/256)^8 \approx 2^{-5}$. Thus given about $2^5 \cdot 128 = 2^{12}$ output words the attacker would expect to obtain a solvable system of equations and reconstruct the full 1024-bit buffer. Knowing the buffer at some round t and the output words $a_1^{(t-1)}, a_1^{(t)}, a_1^{(t+1)}$ the attacker can recover the state $a^{(t)}$ as follows:

$$(a_0^{(t)}, a_1^{(t)}, a_2^{(t)}) = (a_1^{(t-1)}, a_1^{(t)}, F(a_1^{(t)}, b_4^{(t)}) \oplus a_1^{(t+1)}).$$

Since all the steps in the cipher are invertible, knowing the buffer and the state at some point t enables the attacker to run the cipher both forwards and backwards. By running the cipher backwards the attacker can recover the initial 128-bit secret key.

The total complexity of this attack is $O(2^{12})$ words of known stream (about 2^{15} bytes) and $O(2^{30})$ steps to solve a system of 2^{10} equations. The same attack would work for the case $Output[t] = a_0^{(t)}$.

An important remark is that the attack did not use any properties of the S-box, and thus can work even when the S-box is unknown or key-dependent. The attack will first recover the buffer using the technique described above and then derive the unknown S-boxes from Eq. 2 by writing a set of equations of the form

$$S(c_1) \oplus S(c_2) = c_3, \tag{5}$$

where c_1, c_2, c_3 are known values. Given $k+3$ output stream words the attacker can write $8k$ equation for the unknown S-box. He will need $O(256)$ 64-bit outputs due to a "coupon collector"-like argument in order to write and solve a system of linear equations in terms of the entries of the unknown S-box. Note that in this attack it is important that the S-box is invertible, since otherwise we can write Eq. (4) only probabilistically.

Another observation is that we can use the "buffer elimination" equations from [3] to attack this variant of MUGI with even smaller data and time complexity, since we will not need to spend $O(2^{30})$ operations in order to solve the system of linear equations.

The equations are as follows ($t \geq 48$, and $^{<32}$ stands for cyclic 32-bit rotation of a 64-bit word to the left):

$$\begin{aligned} &a_1^{(t)} \oplus a_1^{(t-48)} \oplus F^{-1}(a_1^{(t+1)} \oplus a_2^{(t)} \oplus C_1) \oplus F^{-1}(a_1^{(t-47)} \oplus a_2^{(t-48)} \oplus C_1) = \\ &a_1^{(t-6)} \oplus a_1^{(t-10)} \oplus a_1^{(t-14)} \oplus a_1^{(t-18)} \oplus a_1^{(t-26)} \oplus a_1^{(t-30)} \oplus a_1^{(t-34)} \oplus a_1^{(t-38)} \\ &\qquad \oplus^{<32} a_1^{(t-26)} \oplus^{<32} a_1^{(t-42)} \end{aligned}$$

$$a_1^{(t)} \oplus a_1^{(t-48)} \oplus F^{-1}(a_1^{(t-1)} \oplus a_2^{(t+1)} \oplus C_2) \oplus F^{-1}(a_1^{(t-49)} \oplus a_2^{(t-47)} \oplus C_2) = \\ {}^{<17}a_1^{(t-12)} \oplus^{<17} a_1^{(t-16)} \oplus^{<17} a_1^{(t-32)} \oplus^{<17} a_1^{(t-44)} \oplus^{<49} a_1^{(t-16)} \\ \oplus^{<49} a_1^{(t-20)} \oplus^{<49} a_1^{(t-32)} \oplus^{<49} a_1^{(t-36)}$$

One notices two interesting properties of these two equations: there are only two instances of a_2 in each equation and they are 48 time steps appart in both equations. Thus if we assume that the sequence $a_1^{(t)}$ is known, those equations reduce to simpler equations:

$$F^{-1}(a_2^{(t)} \oplus C_1') \oplus F^{-1}(a_2^{(t-48)} \oplus C_1'') = const_1 \tag{6}$$

$$F^{-1}(a_2^{(t)} \oplus C_2') \oplus F^{-1}(a_2^{(t-48)} \oplus C_2'') = const_2 \tag{7}$$

in which all the quantities are known except for the two unknowns $a_2^{(t)}$ and $a_2^{(t-48)}$. If we denote $x = M^{-1}(a_2^{(t)})$ and $y = M^{-1}(a_2^{(t-48)})$, then we can take care of the linear mapping, and write further simplified equations:

$$S_1(x) \oplus S_2(y) = const_1 \tag{8}$$

$$S_3(x) \oplus S_4(y) = const_2 \tag{9}$$

where S-boxes S_1, S_2, S_3, S_4 are known and are derived from S^{-1} and the known constants. By solving the system we derive the full words $a_2^{(t)}$ and $a_2^{(t-48)}$ in just $8 \cdot 2^8 = 2^{11}$ steps. Repeating this procedure about eight times for $t, t+1, \ldots, t+7$ we find eight consecutive words $a_2^{(t)}, a_2^{(t+1)}, \ldots, a_2^{(t+7)}$, which allows us to directly derive the bits of the buffer without solving a system of linear equations. The complexity of this approach is about 2^{14} very simple steps and it uses about 56 output words and negligible memory[1]. This second attack uses the knowledge of the S-boxes and the special properties of the buffer update function, which were not important in our first attack.

3.2 Attacking the Non-linear Part of Mugi

In this section we present an attack that efficiently recovers the 192-bit non-linear state of the cipher when part of the buffer is known to the attacker. It turns out that it is sufficient to know only two words of the buffer $b_4^{(t)}$ and $b_{10}^{(t)}$ in order to mount this attack. The attack uses only 3 output words and has a complexity of $O(2^{32})$ steps. Note that the cipher is unchanged, including the original output function of MUGI: $Output[t] = a_2^{(t)}$.

[1] The attack can be done even faster in only 2^6 steps, if we are allowed to do a single precomputation of 2^{32} steps in order to produce a table that would occupy 2^{32} bytes and would store the solution for the system of equations (8) and (9).

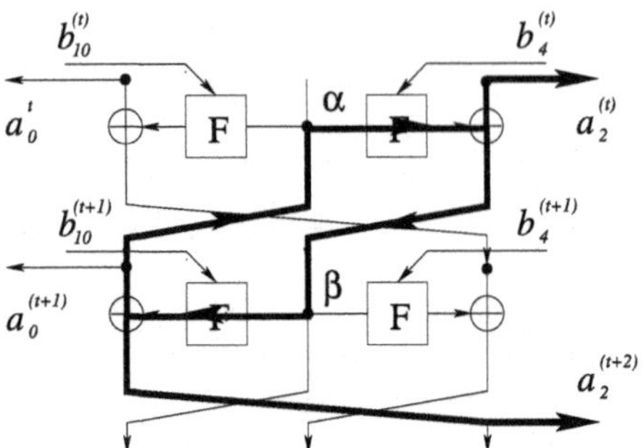

Fig. 4. The loop in the ρ function

The attacker can consider the following loop in the ρ function (see also Fig. 4):

$$F(\alpha, b_4^{(t)}) \oplus a_2^{(t)} = \beta \tag{10}$$

$$F(\beta, b_{10}^{(t+1)}) \oplus a_2^{(t+2)} = \alpha. \tag{11}$$

Here the $a_2^{(t)}, a_2^{(t+2)}$ and the keys $b_4^{(t)}$, $b_{10}^{(t+1)}$ are known. Then we have a system of two non-linear but simple one-round equations with two variables α and β. In order to find the solution of the system the attacker may proceed as follows:

1. Guess the first four bytes of β (denoted 0,1,2,3, with 2^{32} possibilities). Here we enumerate the bytes of Fig. 2 from left to right, starting from 0.
2. For each possibility, find the middle four bytes (2,3,4,5) of α.
3. At this stage, we know two bytes (number 2,3) at the input and at the output of the first MDS at round t. The first two bytes (number 0,1) of the input can be found from a system of linear equations:

$$\begin{cases} x_0 + x_1 = c_0 \\ 3 \cdot x_0 + x_1 = c_1, \end{cases}$$

   where c_0, c_1 are known values computed from bytes 2,3 of input and output. Knowing the new input bytes x_0, x_1 we can calculate the missing output bytes y_0, y_1. As a result we will find two additional bytes of α and β and thus we know bytes number 0,1,2,3,4,5 of these words. Due to the byte swap after the MDS being an involution, we can get in a similar fashion a similar set of constraints knowing bytes 4,5 at the input and the output of the second MDS and solving a system of equations for the bytes 6,7 at the same round t. As a result we completely recover α and β.
4. At this point we know 128 bits of the state a by knowing $a_1^{(t)} = \alpha$, and the known output $a_2^{(t)}$. The remaining 64 bits can be found as follows:

$$a_0^{(t)} = F(\alpha, b_4^{(t)}) \oplus a_2^{(t+1)}.$$

We see that given two buffer words and 3 output words, we can completely recover the 192-bit secret state $a^{(t)}$ in $O(2^{32})$ simple steps. If this attack has to be performed multiple times for example as part of another attack on the full cipher – it can be sped up by precomputation.

3.3 Mixing One Buffer Word per Round

For MUGI it is crucial that two 64-bit buffer words are mixed at each round, while only a single 64-bit word is given as output. In this section we present an attack that recovers the full secret state of the cipher (both the state $a^{(t)}$ and the buffer $b^{(t)}$) when only one word is mixed per round. The rest of the cipher structure is kept intact. The complexity of this attack is equivalent to $O(2^{126.5})$ key searching steps steps, but the similarity between this complexity and exhaustive key search is accidental since the complexity of our attack would remain the same even if MUGI had much larger keys[2]. Note that this is much faster than the alternative type of exhaustive search over the 192-bit state and the 1024-bit buffer.

Consider a variant of MUGI in which both F-functions of each round share a common key[3], for example and without loss of generality, $b_4^{(t)}$. We could reuse the previous attack by guessing the two buffer words $b_4^{(t)}, b_4^{(t+1)}$ and recovering the state $a^{(t)}$ in $O(2^{160})$ steps. However there is a much faster direct attack:

1. At round t we know word $a_2^{(t)}$ and we guess the remaining parts of the non-linear state: $a_0^{(t)}, a_1^{(t)}$ (a 128-bit guess).
2. We find the buffer word $b_4^{(t)}$ used as the key in round t from the equation:

$$F(a_1^{(t)}, b_4^{(t)}) \oplus a_0^{(t)} = a_2^{(t+1)},$$

in which all values except for the key are either guessed or known from the output stream.
3. We use the newly recovered key to compute the unknown word $a_1^{(t+1)}$ of the next state:

$$a_1^{(t+1)} = F(a_1^{(t)}, b_4^{(t)}) \oplus a_2^{(t)}.$$

4. Go to step 2, and repeat it 16 times for the following rounds till we know the full 16 word buffer. Once we start reusing buffer words we can check them for possible contradictions.

Note that since at each step we know the full non-linear state, we know the update function of the buffer and we can thus run the buffer update-function

[2] This complexity is about 3 times faster than exhaustive search of the actual 128-bit key of MUGI ; to verify each guessed key, MUGI has to perform 48 initial key setup rounds, whereas our attack verifies or discards a guessed pair of buffer words after only 16 rounds.

[3] The attack can also tolerate a construction in which given a buffer word b the keys of the round are derived via known and easy to invert permutations: $k_1 = f(b), k_2 = g(b)$. For example, a rotation of the buffer word by 17 as used in MUGI would not make the attack any harder.

$\lambda(\cdot)$ symbolically. The attack uses 18 output stream words, and its complexity is equivalent to $\frac{16}{48} \cdot 2^{128}$ key searching steps . The attack completely recovers the 1024+192-bit internal state of the cipher, and by using the invertibility of the operations we can also recover the 128-bit original secret key. The Table 1 summarizes the results of our attacks on the non-linear component of MUGI.

Table 1. Our attacks on the non-linear component of MUGI

Cipher Variant	Known Stream[a]	Time Complexity	The Attack Recovers
Change in the output function[b]	$O(2^{12})$	$O(2^{30})$	full 192+1024 state
Change in the output function[c]	56	$O(2^{14})$	full 192+1024 state
Attack on non-linear component	3	$O(2^{32})$	192 bits of $a^{(t)}$
One buffer word per round	18	$O(2^{126.4})$	full 192+1024 state
Exhaustive search	2	$O(2^{128})$	full 192+1024 state

[a] Expressed in 64-bit words.
[b] The same complexities even if S-boxes are unknown or key-dependent.
[c] Using the buffer elimination observation from [3] and the knowledge of the S-boxes.

4 Conclusions

In this paper we have identified several potential weaknesses of MUGI:

- The output function reveals 1/3 of the state without any masking. It would be better to reveal fewer bits which are a more complex function of the current state.
- The mixing affects only 4 bytes per round. It would be better to mix all the 8 bytes simultaneously.
- Knowledge of the middle word allows for very powerful attacks to be mounted. However, this attack is avoided by the actual design.
- The feedback from the nonlinear state is added to the buffer in a linear way. It would be better to avoid this additional linearity.

In spite of these potential weaknesses, the full MUGI is still unbroken. The two factors that make our attacks impractical on the real MUGI is the large size of the buffer and the fact that the cipher mixes two key-words per-round while it outputs only a single word.

References

[1] "MUGI pseudorandom number generator, self-evaluation report," Technical report, Hitachi, 18.12.2001.

[2] J. Daemen and C. S. K. Clapp, "Fast hashing and stream encryption with PANAMA," in *Fast Software Encryption, FSE'98* (S. Vaudenay, ed.), vol. 1372 of *Lecture Notes in Computer Science*, pp. 60–74, Springer-Verlag, 1998.

[3] J. D. Golic, "A weakness of the linear part of stream cipher MUGI," in *Fast Software Encryption, FSE 2004* (B. K. Roy and W. Meier, eds.), vol. 3017 of *Lecture Notes in Computer Science*, pp. 178–192, Springer-Verlag, 2004.
[4] D. Watanabe, S. Furuya, H. Yoshida, K. Takaragi, and B. Preneel, "A new keystream generator MUGI," in *Fast Software Encryption, FSE 2002* (J. Daemen and V. Rijmen, eds.), vol. 2365 of *Lecture Notes in Computer Science*, pp. 179–194, Springer-Verlag, 2002.

Two Attacks Against the HBB Stream Cipher

Antoine Joux[1] and Frédéric Muller[2]

[1] DGA and Univ. Versailles St-Quentin
Antoine.Joux@m4x.org
[2] DCSSI Crypto Lab
Frederic.Muller@sgdn.pm.gouv.fr

Abstract. Hiji-Bij-Bij (HBB) is a new stream cipher proposed by Sarkar at Indocrypt'03. In this algorithm, classical LFSRs are replaced by cellular automata (CA). This idea of using CAs in such constructions was initially proposed by Sarkar at Crypto'02, in order to instantiate its new Filter-Combiner model.

In this paper, we show two attacks against HBB. First we apply differential cryptanalysis to the self-synchronizing mode. The resulting attack is very efficient since it recovers the secret key by processing a chosen message of length only 2 Kbytes. Then we describe an algebraic attack against the basic mode of HBB. This attack is much faster than exhaustive search for secret keys of length 256 bits.

1 Introduction

Stream ciphers are an important class of secret key cryptosystems. Unlike block ciphers which view the plaintext as blocks of bits (typically 64 bits for the DES or 128 bits for the AES), stream ciphers handle each bit of plaintext separately. Basically, a stream cipher generates a long pseudo-random sequence (or keystream) from a seed (usually the secret key). This sequence is XORed with the plaintext to produce the ciphertext. It is widely believed that secure stream ciphers can be much faster than block ciphers.

Yet, over the last years, few stream cipher proposals have resisted cryptanalysis efforts. For instance, none of the stream ciphers candidate for the NESSIE project has been selected in the final portfolio [24], since all schemes revealed various degrees of weakness. Many of these attacks originate from the mathematical structure of Linear Feedback Shift Registers (LFSR) [5, 8], which are used as a building block by many stream ciphers. To avoid these security concerns, alternative solutions have been recently proposed. For instance, Klimov and Shamir have suggested to replace LFSRs by software-efficient nonlinear mappings based on T-functions [19].

Another contribution came from Sarkar at Crypto'02 [26]. He showed that some classical models for LFSR-based stream ciphers (Nonlinear Filter and Nonlinear Combiner) do not provide optimal security against Correlation Attacks [30]. He proposed to mix these two concepts, using a new paradigm, the Filter-Combiner Model. Unfortunately, he also showed that such a construction

H. Gilbert and H. Handschuh (Eds.): FSE 2005, LNCS 3557, pp. 330–341, 2005.

cannot be instantiated with LFSRs since they do not fulfill some of the basic requirements. Instead, the author showed that cellular automata are good candidates to replace LFSRs in this model. Moreover, they seem to improve the resistance against some classical attacks such as Inversion Attacks [15, 16] or the Anderson Leakage [1]. However it was recently shown that this new construction did not necessarily increase the level of security [17].

In this paper we focus on the HBB stream cipher. This new algorithm [27] was proposed at Indocrypt'03 by Sarkar. HBB is not exactly an instantiation of the Filter-Combiner model (since the non-linear component has a memory) although its linear map is based on cellular automata. The outputs of the cellular automata are combined with a nonlinear map achieved using some of the primitives from Rijndael [12]. In addition, HBB has the particularity of offering a Self-Synchronizing (SS) mode of operation, in addition to the basic (B) mode of operation. Self-synchronizing stream ciphers are a rare primitive which can be useful in specific contexts [21]. However few dedicated designs have been proposed and many published proposals (such as [11]) did not resist cryptanalysis [18]. In fact, it is an open problem to design a secure dedicated self-synchronizing stream cipher.

In this paper, we show that both modes of operation of HBB suffer from important flaws. Against the SS mode, we use a differential attack which recovers the secret key by processing 2^{14} bits of chosen ciphertext. We also describe an algebraic attack against the B mode of operation, faster than exhaustive search for key size of 256 bits. In a first section, we give a brief overview of cellular automata and of the HBB cipher. Then, we describe our attack against the Self-Synchronizing mode of operation. Finally, we focus on the Basic mode of operation and apply algebraic cryptanalysis.

2 Stream Ciphers Based on Cellular Automata

2.1 Cellular Automata Preliminaries

In general, an automaton consists in a set of l memory cells, represented at time t by $S^{(t)} = (s_1^{(t)}, \ldots, s_l^{(t)})$, with a rule of evolution for each cell depending on the content of neighboring cells. Details of the theory of cellular automata are not relevant here, refer to [27] for more information. Basically, the only property we really take advantage of is their linear behavior. More precisely, a cellular automaton can be associated with a matrix M that characterizes its evolution. $S^{(t+1)}$ can then be computed by multiplying $S^{(t)}$ with M. This matrix has the additional properties of being tridiagonal and having a primitive characteristic polynomial. This guarantees that the linear recurrence has maximal period $2^l - 1$.

2.2 Overview of the HBB Cipher

HBB is a classical keystream generator, which contains a linear finite state machine and a nonlinear part. It is not a basic instantiation of the Filter-Combiner model [26] since its nonlinear part has memory (128 bits of internal state), however it belongs to the same family. According to [27], the use of cellular automata should improve the security of the cipher against some attacks using the specific properties of LFSRs.

General Structure of the Cipher. An overview of HBB is given in Figure 1. LC represents the Linear Component (which contains 512 bits of internal state), and NLC represents the Non Linear Component (which contains 128 bits of internal state).

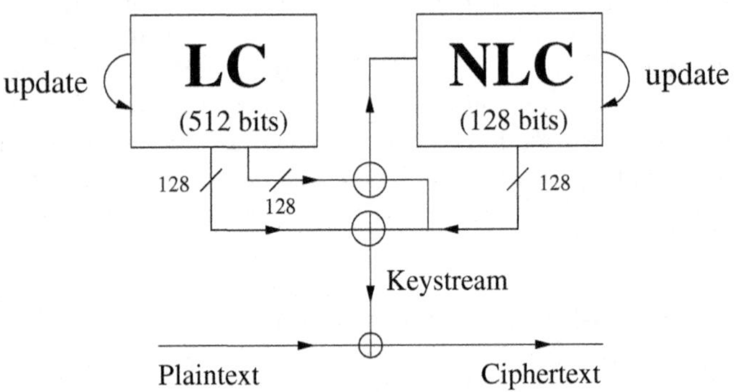

Fig. 1. One round of Hiji-Bij-Bij

Both LC and NLC have an update function which is applied at each round to their internal state. Then, 128 bits of keystream are extracted using a linear transform. To summarize, one round of encryption can be expressed as

1. Update the internal state of NLC
2. Update the internal state of LC
3. Extract 128 bits from LC and XOR it with NLC. The result is the keystream.
4. Extract 128 other bits from LC and XOR it with NLC. The result is the next state of NLC.

In general, this technique for producing keystream bits with a whitening layer after the nonlinear operations is called "linear masking" (see [5]). The initial states of LC and NLC are derived linearly from the secret key K. Then 16 rounds are applied for divergence, without using the output keystream for encryption. However the last 512 bits of keystream are XORed to the internal state of LC just before the beginning of encryption. Two key sizes are suggested for HBB : 128 and 256 bits.

The Round Function. The updating function of LC is based on the cellular automata theory. Details can easily be obtained from [27]. In fact, there are two cellular automata of size 256 bits each. Both matrices contains entries of the form $(c_1, \ldots, c_{256})$ on the main diagonal, and all entries equal to 1 on the upper and lower subdiagonals.

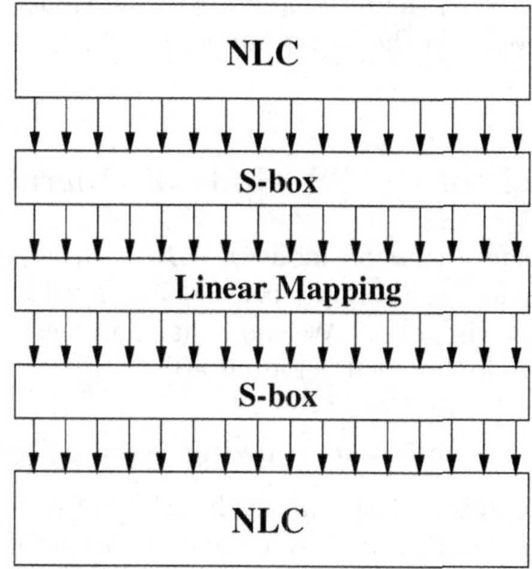

Fig. 2. The NLC updating function

The updating function of NLC can be seen as one round of a block cipher (see Figure 2). It consists of three consecutive layers. The first and third layers apply the Rijndael S-box [12] to each byte of NLC. The intermediate layer is a simple linear application over $\mathbb{F}_2$. The author of HBB shows that the global function has full diffusion, thus any output bit depends on all input bits. Moreover, linear approximation have been analyzed and it is shown in [27] that none can have a bias better than 2^{-12} (actually, it says 2^{-13}, but bias are represented as $0.5 \pm \varepsilon$, while we prefer the convention $0.5(1 \pm \varepsilon)$). Thus, according to the author, the cipher should resist attacks based on linear approximations, such as linear cryptanalysis [20] and correlation attacks [30]. Finally, the previous elements give an implicit description of the Basic (B) mode of operation of HBB.

The Self-synchronizing (SS) Mode of Operation. A Self-Synchronizing mode can be easily derived from the above description by making the keystream dependent on the previous bits of ciphertext. To do this, one additional step is added to the previous description.

At the end of round i, the last four blocks of ciphertext (128 bits each), represented as C_i, C_{i-1}, C_{i-2} and C_{i-3}, are XORed together with the secret key K (if K is 256 bits long, both halves of K are first XORed together). The resulting value is the new internal state of NLC :

$$NLC = Fold(K) \oplus C_i \oplus C_{i-1} \oplus C_{i-2} \oplus C_{i-3}$$

On the linear side of the generator, the secret key K is linearly expanded into a 512 bits value, then XORed with the concatenation of the last four ciphertext blocks, to produce the new value of LC :

$$LC = Expand(K) \oplus (C_i || C_{i-1} || C_{i-2} || C_{i-3})$$

Therefore, in the SS mode, the internal state of the cipher at the beginning of each round depends only on the secret key and the last 512 bits of ciphertext, in a linear manner.

3 Cryptanalysis of the SS Mode of Operation

In this section, we focus on the SS mode of HBB with secret key K of 128 bits long. Each round of encryption simply consists in a fixed function applied to K and the last 4 blocks of ciphertext. We show that this construction actually does not resist to a chosen-ciphertext differential attack.

3.1 Background on Self-synchronizing Stream Ciphers

Generally a Self-Synchronizing Stream Cipher (SSSC) is one in which the keystream bit is a function of the key and a fixed number m of previous ciphertext bits.

This parameter m is called the *memory* of the cipher.

In order to describe formally a SSSC, let x_t denote plaintext bit number t, y_t the corresponding ciphertext bit and z_t the corresponding keystream bit. The encryption is generally described

$$y_t = x_t \oplus z_t$$

where the keystream bit is computed as

$$z_t = F(y_{t-1}, \ldots, y_{t-m}, K)$$

F denotes the keystream function and K the secret key. The basic idea is to encrypt each plaintext bit with a function depending only on the secret key and the previous m ciphertext bits. Therefore each ciphertext bit can be correctly deciphered as long as the previous m ciphertext bits have been successfully received.

General properties and design criteria for SSSCs have been studied by Maurer [21]. It pointed out that a self-synchronization mechanism has several advantages from an engineering point of view. For instance, it may be helpful in contexts where no lower layer of protocol is present to assure error-correction. In particular, it prevents long bursts of error when a bit insertion or deletion occurs during the transmission of the ciphertext.

However in terms of security the analysis of SSSCs requires a completely different approach from conventional stream ciphers. Indeed, since the pseudo-random generator does not behave autonomously, the cipher might be subject to chosen message attacks. Therefore it is not straightforward to turn a conventional stream cipher into a SSSC and few dedicated designs have been proposed. Using a block cipher in 1-bit Cipher FeedBack (CFB) mode [14] is usually quite inefficient in terms of encryption speed, and reduced-round optimizations may be subject to attacks [25]. KNOT, one of the few dedicated designs [11] was broken at FSE'03 [18]. An improved version of KNOT, described in Daemen's thesis [10], appears to resist this attack, but currently there is no other secure dedicated SSSC in the literature.

3.2 The Attack Against HBB in SS Mode

This section describes a differential attack using chosen ciphertexts. Accordingly, we suppose that an attacker gains access to a decryption oracle and introduces chosen blocks of ciphertext.

The goal of the attacker is to obtain two inputs of the NLC round function that differ only on one byte, with difference δ. This can be achieved by introducing the blocks of ciphertext C_1, C_2, C_3, C_4 and then $C_1, C_2, C_3, C_4 \oplus \delta$. Let j denote the position of this one byte difference and K_j the corresponding byte of secret key.

At first sight, it seems that the attacker has no access to the output difference of the NLC updating function because of the linear masking. However at the beginning of each round, LC is set to the value (see Section 2.2)

$$LC = Expand(K) \oplus (C_i || C_{i-1} || C_{i-2} || C_{i-3})$$

thus the difference of linear masking is a purely linear function of the introduced ciphertexts only. Thus an attacker able to observe the keystreams can cancel out the difference of linear masking and observe directly the output difference of the NLC updating function.

Initially, the difference δ is confined to one byte. In addition, after the first layer of S-box in this computation, the difference still only concerns the byte number j in the internal state. This difference is of the form

$$\delta' = S(K_j \oplus x) \oplus S(K_j \oplus x \oplus \delta) \qquad (1)$$

where x is a known byte depending on the introduced ciphertext. If δ' has hamming weight equal to 1, it is true by construction that the hamming weight of the difference after the linear layer will be exactly 3 (see [27]). Thus, only 3 bytes will differ after the last layer of S-boxes. This difference trail on NLC is described in Figure 3 where dashed areas represent the differences.

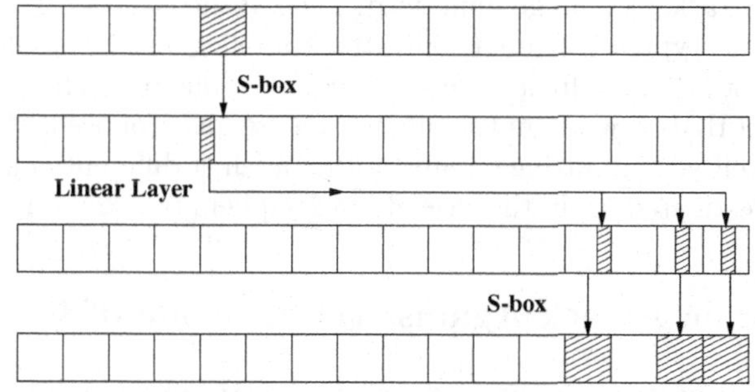

Fig. 3. An illustration of the Differential Trail on NLC

We observe that, when $\mathtt{hamming}(\delta') = 1$, the difference on the output of the NLC updating function is null for 13 bytes of 16. Otherwise this property is very unlikely. In average the event that the hamming weight of δ' is 1 happens with probability $\frac{8}{256} = 2^{-5}$. Thus, according to the birthday paradox, testing 2^3 values of δ should be sufficient to detect a collision on 13 out of 16 output bytes. This event provides a condition on K_j using relation (1), of the form

$$\mathtt{hamming}(S(K_j \oplus x) \oplus S(K_j \oplus x \oplus \delta)) = 1$$

In general, only a few K_j values will verify it. Furthermore it is straightforward to eliminate false candidates for K_j with a few extra decryption queries. To summarize, one byte of secret key can be recovered by processing about $2^3 = 8$ blocks of keystream in average. Repeating it 16 times, the full secret key can be easily obtained.

Moreover this attack can be mounted by processing just a single message. Indeed an attacker can just concatenate all the chosen ciphertexts he needs for successive applications of the attack and submit the resulting message to the decryption oracle. In practice this can be done by just adapting the choice of ciphertext block number i to the previous blocks of ciphertext. Therefore, to process $2^3 \times 16$ blocks of keystream, a message of length $3 + (2^3 \times 16)$ blocks is sufficient. This corresponds to about 2 Kbytes of chosen ciphertext. Besides, when δ' has hamming weight equal to $2, 3$ or 4, we also obtain detectable collisions on the keystream. So, we think it is even possible to lower a little our data requirement, using a precise analysis of all these events. We have implemented our attack using only the case of hamming weight 1 and were able to recover successfully the bytes of secret key as expected, with about 8 blocks/byte. An attack against the 256 bits secret keys would work by first recovering the 128 bits used in NLC, and then obtaining the remaining key bits by other means.

This attack represents a real threat in practical applications where the SS mode of the HBB cipher is used. Indeed, an active attacker can easily introduce a chosen ciphertext sequence of a few Kbytes in the communication layer. Since self-synchronizing modes of operation are usually implemented for fast streaming communications on faulty channels, it is even likely that the error caused by the action of the attacker would go unnoticed. Then, if the attacker is able to observe the resulting decrypted plaintext, our attack applies and he would recover the complete secret key with little offline processing. Thus, we believe the SS mode of operation of HBB is weak and should not be used as proposed. More generally, resistance of self-synchronizing stream ciphers against differential attacks should always be investigated, as in the case of block ciphers (see [22] for more details).

4 Algebraic Attack Against the B Mode of Operation

In the previous section, we have described a very efficient differential attack against the SS mode of HBB. Obviously, this attack does not apply to the Basic mode of operation since the attacker cannot choose the inputs of NLC (or LC) at each round. However, other cryptanalysis techniques may be envisaged here. In particular we propose an algebraic attack against HBB which is faster than exhaustive search for a key size of 256 bits.

4.1 Algebraic Attacks and Stream Ciphers

Algebraic attacks form a class of cryptanalysis techniques which has received a huge interest in the last years. Indeed new applications have been described in various fields including block ciphers [9, 23], stream ciphers [7, 8] and even public key cryptography [13]. Algebraic attacks exploit polynomial equations describing exactly an algorithm. There is a contrast with classical cryptanalysis techniques which are often based on approximations of the behavior of the algorithm. In the recent years many stream ciphers [2, 3, 4, 8] have been broken using algebraic attacks and it has now become important to investigate the security of new algorithms regarding these techniques.

The general idea in algebraic attacks on stream ciphers is to write keystream bits as a polynomial of low enough degree in the bits of the secret key, and then apply an appropriate algorithm for solving this polynomial system. Many strategies exist like the simple linearization attack or the refined relinearization attack [28]. In these basic attacks, all monomials are replaced by new variables, then the resulting linear system is solved by usual linear algebra. In some cases, better strategies may also apply, such as sparse linear algebra or other dedicated algorithms. For instance, Gröbner base techniques (for an illustration of these techniques, see [13]) are a well-known mathematical tool helpful in the case of algebraic attacks. Besides an alternative solution, the XL algorithm has also been proposed [29] to resolve polynomial systems.

In general, for a recent cipher, it should be impossible to write low degree equations involving the secret key bits and the plaintext bits. In the next section, we show that HBB fails to meet these requirements.

4.2 The Case of HBB

It is not straightforward to express directly the keystream in function of the secret key because of the divergence steps executed before the beginning of encryption. However, an attacker can focus on recovering the initial state of LC (of length 512 bits). Then in a second phase, it might be possible to recover the key from the initial state, if necessary. Therefore we first focus on writing polynomial equations between the keystream and the initial state.

In Figure 1, we can see that the internal state of NLC at the beginning of any round i is a linear function of the initial state of LC and keystream bits. Looking at two consecutive states of NLC, we can express both the input and the output of the NLC updating function as linear functions of the initial state and the keystream bits.

Besides, it is easy to write equations of degree 7 relating inputs and outputs of the NLC updating function (referred to as Φ). Let $(a_i)_{0\leq i\leq 127}$ and $(b_i)_{0\leq i\leq 127}$ respectively denote the input and output bits of Φ. Notations c_i and d_i are used for the intermediate states after the first and second layer of Φ. It is well known that the Rijndael S-box (and its inverse) has algebraic degree 7, i.e. the output bits of the S-box (and its inverse) can be expressed as degree 7 polynomials in its input bits. Consequently,

$$b_i = P_i(a_0, \ldots, a_{127})$$

where P_i is a polynomial of degree 7. Similarly,

$$c_i = Q_i(d_0, \ldots, d_{127})$$

where Q_i is a polynomial of degree 7. Furthermore, b_i and c_i are related by a linear transform. Thus, 128 relations of degree 7 of the form

$$\sum_{i=0}^{127} \lambda_i P_i(a_0, \ldots, a_{127}) \oplus \sum_{i=0}^{127} \mu_i Q_i(d_0, \ldots, d_{127}) = 0$$

can be written. As we argued previously the a_i's and d_i's depend linearly on the initial state and the keystream, thus the previous relation can be rewritten as

$$R_t(v_0, \ldots, v_{511}, Z_t) = 0$$

where R_t is a degree 7 polynomial, the v_i's are the bits of initial state and Z_t is the keystream block number t (which is known).

The number of monomial of degree 7 on 512 unknowns is

$$\binom{512}{7} \simeq 2^{50.6}$$

which is not sufficient to provide a security level of 256 bits. Indeed, a simple linearization attack would proceed by linear algebra on all monomials of degree 7. Using Gaussian elimination, the corresponding complexity would be about

$$\binom{512}{7}^3 \simeq 2^{151.8}$$

basic binary instructions for inverting the matrix ($2^{146.8}$ on a 32 bit processor). Besides, about $2^{50.6}$ bits of known plaintext would be needed. The solution of this system reveals the initial state of the cipher from which it might be possible to retrieve the actual secret key. This algebraic attack is faster than exhaustive search only for keys of 256 bits.

However, with better linear algebra such as algorithms[1] with exponent $w = \log_2(7)$, the complexity can be lowered to $2^{142.2}$ binary instructions, and we even expect that better techniques exist, for instance if we exploit the sparsity of the system. According to Courtois [7], this kind of attacks can also be further improved by using optimized resolution algorithms and a precomputation step. On the whole, it is likely that an improved version of this attack could break the 128 bit version of HBB.

This attack illustrates the fact that linear masking techniques and cellular automata by themselves do not provide protection against the class of algebraic attacks. Basically there is too much linearity in this algorithm. This type

[1] We did not consider other algorithms with smaller exponent such as the Coppersmith-Winograd algorithm, because they are not practical enough.

of problem is also encountered when using LFSRs as a building block. Sound countermeasures are irregularly clocked stream ciphers, like the Shrinking Generator [6] or the use of nonlinear mappings as a building block instead of LFSRs or cellular automata [19].

4.3 Recovering the Key from the Initial State

After the previous attack, we know the initial state of the linear component LC. This is not fully satisfying in practice. It is often expected to go further and recover the secret key. We refer to the initial state of LC as $\mathcal{S}$. Our goal is to derive from $\mathcal{S}$ the secret key K. We consider first the key size of 256 bits, since the previous attack has a complexity larger than 2^{128}.

By construction, $\mathcal{S}$ is a linear function of K and 4 keystream blocks (of 128 bits each) produced during the key schedule and discarded immediately after. We call these blocks $T_0, \ldots, T_3$ as in [27]. The first block of keystream (called Z_0) can be expressed by (see also Figure 1) :

$$Z_0 = \lambda_1(\mathcal{S}) \oplus \phi(T_3 \oplus \lambda_2(K)) \tag{2}$$

where λ_1, λ_2 are linear functions and ϕ is the NLC updating function. From (2), we retrieve $T_3 \oplus \lambda_2(K)$, since ϕ is invertible. With this additional equation, we get a total of

$$512 + 128 = 640$$

binary linear equations involving

$$256 + 4 \times 128 = 768$$

unknowns (corresponding to K, T_0, T_1, T_2 and T_3). In average this linear system should contain 2^{128} solutions, one of them corresponding to the correct K. These solutions can be obtained with the usual Gaussian reduction algorithm. The complexity required here is about 2^{128} steps of computation. Besides, for keys of 128 bits, there are 128 unknowns less in the system, hence only one solution is expected.

To summarize, the extra work required to find K from the initial state of LC depends on the key size. For keys of length 256 bits, this complexity is about 2^{128}, and for keys of 128 bits, the complexity is negligble. These results allow to complete the attacks against HBB without increasing the overall complexity.

5 Conclusion

In this paper, we have described two attacks against the new HBB stream cipher. First a very efficient differential attack breaks the self-synchronizing (SS) mode of operation by processing a message of length about 2 Kbytes. Then we described an algebraic attack which breaks HBB in B mode with workload of about 2^{142} steps of computation. This is faster than exhaustive search for secret keys of length 256 bits and we believe optimized versions could threaten 128 bit keys as well.

These attacks enlighten some important weaknesses in the design of the new HBB stream cipher. In addition, by breaking the SS mode, we have also shown that the challenge of designing a secure dedicated self-synchronizing stream cipher, initially proposed by [21] is still an open problem.

References

1. R. Anderson. Searching for the Optimum Correlation Attack. In B. Preneel, editor, *Fast Software Encryption – 1994*, volume 1008 of *Lectures Notes in Computer Science*, pages 137–143. Springer, 1995.
2. F. Armknecht and M. Krause. Algebraic Attacks on Combiners with Memory. In D. Boneh, editor, *Advances in Cryptology – Crypto'03*, volume 2729 of *Lectures Notes in Computer Science*, pages 162–175. Springer, 2003.
3. E. Barkan, E. Biham, and N. Keller. Instant Ciphertext-Only Cryptanalysis of GSM Encrypted Communication. In D. Boneh, editor, *Advances in Cryptology – Crypto'03*, volume 2729 of *Lectures Notes in Computer Science*, pages 600–616. Springer, 2003.
4. J.Y. Cho and J. Pieprzyk. Algebraic Attacks on SOBER-t32 and SOBER-t16 without Stuttering. In B. Roy and W. Meier, editors, *Fast Software Encryption – 2004*, volume 3017 of *Lectures Notes in Computer Science*, pages 49–64. Springer, 2004.
5. D. Coppersmith, S. Halevi, and C. Jutla. Cryptanalysis of Stream Ciphers with Linear Masking. In M. Yung, editor, *Advances in Cryptology – Crypto'02*, volume 2442 of *Lectures Notes in Computer Science*, pages 515–532. Springer, 2002.
6. D. Coppersmith, H. Krawczyk, and Y. Mansour. The Shrinking Generator. In D. Stinson, editor, *Advances in Cryptology – Crypto'93*, volume 773 of *Lectures Notes in Computer Science*, pages 22–39. Springer, 1994.
7. N. Courtois. Fast Algebraic Attacks on Stream Ciphers with Linear Feedback. In D. Boneh, editor, *Advances in Cryptology – CRYPTO'03*, volume 2729 of *Lectures Notes in Computer Science*, pages 176–194. Springer, 2003.
8. N. Courtois and W. Meier. Algebraic Attacks on Stream Ciphers with Linear Feedback. In E. Biham, editor, *Advances in Cryptology – Eurocrypt'03*, volume 2656 of *Lectures Notes in Computer Science*, pages 345–359. Springer, 2003.
9. N. Courtois and J. Pieprzyk. Cryptanalysis of Block Ciphers with Overdefined Systems of Equations. In Y. Zheng, editor, *Advances in Cryptology – Asiacrypt'02*, volume 2501 of *Lectures Notes in Computer Science*, pages 267–287. Springer, 2002.
10. J. Daemen. *Cipher and hash function design. Strategies based on linear and differential cryptanalysis.* PhD thesis, march 1995. Chapter 9.
11. J. Daemen, R. Govaerts, and J. Vandewalle. A Practical Approach to the Design of High Speed Self-Synchronizing Stream Ciphers. In *Singapore ICCS/ISITA '92*, pages 279–283. IEEE, 1992.
12. J. Daemen and V. Rijmen. AES Proposal: Rijndael (Version 2). 1999. NIST AES website `http://csrc.nist.gov/encryption/aes`.
13. J-C. Faugère and A. Joux. Algebraic Cryptanalysis of Hidden Field Equations (HFE) Cryptosystems Using Gröbner Bases. In D. Boneh, editor, *Advances in Cryptology – CRYPTO'03*, volume 2729 of *Lectures Notes in Computer Science*, pages 44–60. Springer, 2003.
14. FIPS PUB 81. *DES Modes of Operation*, 1980.

15. J. Golić. On the Security of Nonlinear Filter Generators. In D. Gollman, editor, *Fast Software Encryption – 1996*, volume 1039 of *Lectures Notes in Computer Science*, pages 173–188. Springer, 1996.
16. J. Golić, A. Clark, and E. Dawson. Generalized Inversion Attack on Nonlinear Filter Generators. In *IEEE Transactions on Computers 49(10)*, pages 1100–1109, 2000.
17. J. Hong, D.H. Lee, S. Chee, and P. Sarkar. Vulnerability of Nonlinear Filter Generators Based on Linear Finite State Machines. In B. Roy and W. Meier, editors, *Fast Software Encryption – 2004*, volume 3017 of *Lectures Notes in Computer Science*, pages 193–209. Springer, 2004.
18. A. Joux and F. Muller. Loosening the KNOT. In T. Johansson, editor, *Fast Software Encryption – 2003*, volume 2887 of *Lectures Notes in Computer Science*, pages 87–99. Springer, 2003.
19. A. Klimov and A. Shamir. Cryptographic Applications of T-functions. In M. Matsui and R. Zuccherato, editors, *Selected Areas in Cryptography – 2003*, volume 3006 of *Lectures Notes in Computer Science*, pages 248–261. Springer, 2004.
20. M. Matsui. Linear Cryptanalysis Method for DES Cipher. In T. Helleseth, editor, *Advances in Cryptology – Eurocrypt'93*, volume 765 of *Lectures Notes in Computer Science*, pages 386–397. Springer, 1993.
21. U. Maurer. New Approaches to the Design of Self-Synchronizing Stream Ciphers. In D.W. Davies, editor, *Advances in Cryptology – Eurocrypt'91*, volume 547 of *Lectures Notes in Computer Science*, pages 458–471. Springer, 1991.
22. F. Muller. Differential Attacks and Stream Ciphers. In *State of the Art in Stream Ciphers*. ECRYPT Network of Excellence in Cryptology, 2004. Workshop Record.
23. S. Murphy and M. Robshaw. Essential Algebraic Structure within the AES. In M. Yung, editor, *Advances in Cryptology - CRYPTO'02*, volume 2442 of *Lectures Notes in Computer Science*, pages 1–16. Springer, 2002.
24. NESSIE - New European Schemes for Signature, Integrity and Encryption. `http://www.cryptonessie.org`.
25. B. Preneel, M. Nuttin, R. Rijmen, and J. Buelens. Cryptanalysis of the CFB Mode of the DES with a Reduced Number of Rounds. In D.R. Stinson, editor, *Advances in Cryptology – Crypto'93*, volume 773 of *Lectures Notes in Computer Science*. Springer, 1993.
26. P. Sarkar. The Filter-Combiner Model for Memoryless Synchronous Stream Ciphers. In M. Yung, editor, *Advances in Cryptology – Crypto'02*, volume 2442 of *Lectures Notes in Computer Science*, pages 533–548. Springer, 2002.
27. P. Sarkar. Hiji-Bij-Bij : A New Stream Cipher with a Self-Synchronizing Mode of Operation. In T. Johansson and S. Maitra, editors, *Progress in Cryptology – INDOCRYPT'03*, volume 2904 of *Lectures Notes in Computer Science*, pages 36–51. Springer, 2003.
28. A. Shamir and A. Kipnis. Cryptanalysis of the HFE Public Key Cryptosystem. In M. Wiener, editor, *Advances in Cryptology – Crypto'99*, volume 1666 of *Lectures Notes in Computer Science*, pages 19–30. Springer, 1999.
29. A. Shamir, J. Patarin, N. Courtois, and A. Klimov. Efficient Algorithms for solving Overdefined Systems of Multivariate Polynomial Equations. In B. Preneel, editor, *Advances in Cryptology – Eurocrypt'00*, volume 1807 of *Lectures Notes in Computer Science*, pages 392–407. Springer, 2000.
30. T. Siegenthaler. Correlation-immunity of Nonlinear Combining Functions for Cryptographic Applications. In *IEEE Transactions on Information Theory*, volume 30, pages 776–780, 1984.

Two Linear Distinguishing Attacks on VMPC and RC4A and Weakness of RC4 Family of Stream Ciphers

Alexander Maximov

Dept. of Information Technology, Lund University, Sweden
P.O. Box 118, 221 00 Lund, Sweden
movax@it.lth.se

Abstract. At FSE 2004 two new stream ciphers VMPC and RC4A have been proposed. VMPC is a generalisation of the stream cipher RC4, whereas RC4A is an attempt to increase the security of RC4 by introducing an additional permuter in the design. This paper is the first work presenting attacks on VMPC and RC4A. We propose two linear distinguishing attacks, one on VMPC of complexity 2^{54}, and one on RC4A of complexity 2^{58}. We investigate the RC4 family of stream ciphers and show some theoretical weaknesses of such constructions.

Keywords: RC4, VMPC, RC4A, cryptanalysis, linear distinguishing attack.

1 Introduction

Stream ciphers are very important cryptographic primitives. Many new designs appear at different conferences and proceedings every year. In 1987, Ron Rivest from RSA Data Security, Inc. made a design of a byte oriented stream cipher called RC4 [1]. This cipher found its application in many Internet and security protocols. The design was kept secret up to 1994, when the alleged specification of RC4 was leaked for the first time [2]. Since that time many cryptanalysis attempts were done on RC4 [3, 4, 5, 6, 7].

At FSE 2004, a new stream cipher VMPC [8] was proposed by Bartosz Zoltak, which appeared to be a modification of the RC4 stream cipher. In cryptanalysis, a linear distinguishing attack is one of the most common attacks on stream ciphers. In the paper [8] it was claimed that VMPC is designed especially to resist distinguishing attacks.

At the same conference, FSE 2004, another cipher RC4A [9] was proposed by Souradyuti Paul and Bart Preneel. This cipher is another modification of RC4.

In our paper we point out a general theoretical weakness of such ciphers, which, in some cases, can tell us without additional calculations whether a new construction is weak against distinguishing attacks. We also investigate VMPC and RC4A in particular and find two linear distinguishing attacks on them. VMPC can be distinguished from random using around 2^{54} bytes of the

H. Gilbert and H. Handschuh (Eds.): FSE 2005, LNCS 3557, pp. 342–358, 2005.

keystream, whereas the attack on RC4A needs only 2^{58} bytes. This is the first paper that proposes attacks on VMPC and RC4A.

This paper is organized as follows. In Section 2 we describe RC4, RC4A, and the VMPC ciphers. In Section 3 we study digraphs on an instance of VMPC, and then we demonstrate a theoretical weakness of the RC4 family of stream ciphers in general. We propose our distinguishers for both VMPC and RC4A in Sections 4 and 5. Finally, we summarize the results and make our conclusions in Section 6.

1.1 Notations

The algorithms VMPC, RC4A and RC4 are byte oriented stream ciphers. For notation purposes we consider VMPC-n, RC4A-n, and RC4-n to be n-bit oriented ciphers, i.e., the originals are when $n = 8$. Therefore, in the design of these ciphers, $+$ means addition modulo 2^n. For simplicity in formulas, let q be the size of permuters used in these ciphers, i.e.

$$q = 2^n. \tag{1}$$

The ciphers have an internal state consisting of one or two permuters of length q, and a few iterators. The idea of these designs is derived from the RC4 stream cipher, therefore, we call ciphers with a structure similar to RC4 as *the RC4 family of stream ciphers.* We denote by O_t the n-bit output symbol at time t. When a permuter $P[\cdot]$ is applied k times, e.g., $P[P[\ldots P[x]\ldots]]$, then, for simplicity, we sometimes denote it as $P^k[x]$.

1.2 Preliminaries: A Linear Distinguishing Attack

In a *linear distinguishing attack* one can observe a keystream of some length (known plaintext attack), and give an answer: whether the stream comes from the considered cipher, or from a truly random source. Distinguishers are usually based on statistical analysis of the given stream. At any point t in the stream we observe b linear combinations, the joint value of which is called a *sample* at time t. If the stream is completely random, then the sample is from the *random distribution* denoted as D_{Random}. If the stream is the keystream from the considered cipher, then the sample is from the *cipher distribution* denoted as D_{Cipher}.

To give an answer whether the given stream is from D_{Random} or D_{Cipher} one has to collect N samples from the stream at different points. These N samples form an *empirical distribution*, named also *type* and denoted as D_{Type}. If the distance from D_{Type} to D_{Cipher} is less than the distance to D_{Random}, then we conclude that the stream is from the cipher, otherwise it is decided to be from a random source.

The distance between two distributions is given as

$$\delta = |D_{\text{A}} - D_{\text{B}}| = \sum_{\text{all } x} |\Pr\{x|x \in D_{\text{A}}\} - \Pr\{x|x \in D_{\text{B}}\}|. \tag{2}$$

From statistical analysis the following fact is well known. The closer the distributions D_{Cipher} and D_{Random} are to each other, the larger the number of samples N should be, in order to distinguish with a negligible probability of error. The distance $\epsilon = |D_{\text{Cipher}} - D_{\text{Random}}|$ is then called the *bias*. The bias and the number of required samples N, from which we form our type D_{Type}, are related by the formula $N = \frac{\text{const}}{\epsilon^2}$, where the constant influences on the probability of the decision error. For more details we refer to [10]. However, the following relation is enough to have a rather negligible probability of error, and we use this formula in our paper.

$$N = \frac{1}{\epsilon^2} \tag{3}$$

1.3 Cryptanalysis Assumptions

We start our analysis of the RC4 family of stream ciphers by making a few reasonable assumptions.

(1) We assume that the initialisation procedure is perfect, i.e., all internal variables (except known iterators) are from the uniform distribution. In practice this is not true, but we make this assumption as long as we do not investigate the initialisation procedures;
(2) In our distinguishers we construct a type D_{Type} by collecting samples from the given keystream. Each derived sample at time t is from some *local distribution* of the keystream. We assume that at any time the internal state of a cipher is uniformly distributed and we don't have any knowledge about it. This assumption will be used to investigate different local distributions in the next sections. In our simulations we checked that the internal state of VMPC is roughly uniformly distributed. But for RC4A the internal state is not uniformly distributed;
(3) We consider that adjacent samples are independent. In the real life it is not true, because between two consecutive samples the internal states of a cipher are dependent. It means that samples might have a strong dependency, which may influence on the resulting type D_{Type}. To reduce these dependencies we suggest to skip few samples before accept one, then the consecutive adjacent samples will be much less dependent on each other.

2 Descriptions of VMPC-• , RC4-• , and RC4A-•

The stream cipher RC4-n [1] was designed by Ron Rivest in 1987. It produces an infinite pseudo-random sequence of n-bit symbols, which is, actually, the keystream. Encryption is then performed in a typical way for stream ciphers: `Ciphertext = Plaintext` $\oplus$ `Keystream`. The structure of RC4-n is shown in Figure 1(left).

The stream cipher VMPC-n [8] was proposed at FSE 2004 by Bartosz Zoltak. This cipher is also byte oriented ($n = 8$), and is a generalised version of RC4-n. The structure of VMPC-n is shown in Figure 1(right).

Internal variables:
i, j – integers $\in [0 \dots q-1]$
$P[0 \dots q-1]$ – a permuter of integers $0 \dots q-1$
The RC4-n cipher
1. $P[\cdot]$ – are initialised with the secret key
   $i = j = 0$
2. Loop until get enough n-bit symbols
   | $i++$
   | $j+ = P[b]$
   | swap$(P[i], P[j])$
   | output $\leftarrow P[P[i] + P[j]]$

Internal variables:
i, j – integers $\in [0 \dots q-1]$
$P[0 \dots q-1]$ – a permuter of integers $0 \dots q-1$
The VMPC-n cipher
1. $j, P[\cdot]$ – are initialised with the secret key
   $i = 0$
2. Loop until get enough n-bit symbols
   | $j = P[j + P[i]]$
   | output $\leftarrow P[P[P[j]] + 1]$
   | swap$(P[i], P[j])$
   | $i++$

Internal variables:
i, j_1, j_2 – integers $\in [0 \dots q-1]$
$P_1[0 \dots q-1], P_2[0 \dots q-1]$ – two permuters of integers $0 \dots q-1$
The RC4A-n cipher
1. $P_1[\cdot], P_2[\cdot]$ – are initialised with the secret key
   $i = j_1 = j_2 = 0$
2. Loop until get enough n-bit symbols
   | $i++$
   | $j_1+ = P_1[i]$
   | swap$(P_1[i],\ P_1[j_1])$
   | output $\leftarrow P_2[P_1[i] + P_1[j_1]]$
   | $j_2+ = P_2[i]$
   | swap$(P_2[i],\ P_2[j_2])$
   | output $\leftarrow P_1[P_2[i] + P_2[j_2]]$

Fig. 1. The structures of RC4-n (left), VMPC-n (right), and RC4A-n (bottom) ciphers

The stream cipher RC4A-n[9] was proposed at FSE 2004 by Souradyuti Paul and Bart Preneel. This cipher is an attempt to hide the correlation between the internal states and the keystream. The authors suggested to introduce a second permuter in the design. The structure of RC4A-n is shown in Figure 1(bottom).

3 Investigation of the RC4 Family of Stream Ciphers

In this section we approximate different *local distributions* of the accessible keystream in the RC4 family of stream ciphers, with the assumptions that were made in Section 1.3. Since in the real cipher the internal state is not from the uniform distribution, the real local distribution differs from our approximation.

However, in practice we will show that this does not make our distinguishers worse.

3.1 Digraphs Approach, on the Instance of VMPC-•

In this subsection we give the idea of how a distinguisher for VMPC can be built. In the previous work [5] the cipher RC4-n was analysed. The authors suggested to observe two consecutive output symbols O_t, O_{t+1}, and the *known* variable i jointly. For RC4-5 they could calculate theoretical probabilities $\Pr\{(i, O_t = x, O_{t+1} = y)\}$, for all possible n^3 values of the triple (i, x, y) (let us denote such distribution as $D_{(i,O_t,O_{t+1})}$). But for RC4-8 they could only approximate the bias for the distribution above due to the high complexity of calculations, and show that a distinguisher needs around $2^{30.6}$ samples (the required length of the plaintext to know).

We use a similar idea to create a distinguisher for VMPC-n. For this purpose we investigate the pair (O_t, O_{t+1}) in the following scheme.

i – known value at time t
$j, P[\cdot]$ – are from a random source

1. $O_t = P[P^2[j] + 1]$
2. $\texttt{swap}(P[i], P[j])$
3. $j' = j + P[i + 1]$
4. $O_{t+1} = P[P^3[j'] + 1]$

Below we give the explicit algorithm to calculate the approximated distribution table $D_{(i,O_t,O_{t+1})}$. For each value i, in each cell of a table T we want to store an integer number $T[i, x, y]$ of possible pairs $(i, P[\cdot])$, which cause the corresponding output pair $(O_t = x, O_{t+1} = y)$. It means, that the probability of any triple (i, O_t, O_{t+1}) is then calculated as:

$$\Pr\{(i, O_t = x, O_{t+1} = y)\} = \frac{T[i, x, y]}{q \cdot q!}. \tag{4}$$

As we can see from the algorithm, its complexity is $O(2^{11n})$ [1]. In our simulations we could manage to calculate the approximation of $D_{(i,O_t,O_{t+1})}$ only for the reduced version VMPC-4. The bias of such table appeared to be around $\epsilon \approx 2^{-8.7}$. It means that we can distinguish VMPC-4 from random having plaintext of length around 2^{18} 4-bits symbols. For notation purposes, let $D^{\text{VMPC}-n}_{(i,O_t,O_{t+1})}$ be the distribution $D_{(i,O_t,O_{t+1})}$ for VMPC-n, and similar for $D^{\text{RC4}-n}_{(i,O_t,O_{t+1})}$.

[1] The complexity to construct such a table with a similar algorithm for RC4-n is $O(2^{6n})$ [5].

Algorithm 1. *Recursive construction of the approximated distribution table* $D_{(i,O_t,O_{t+1})}$

Prepare the permuter: $P[i] = \infty$ at all positions, i.e., all cells of the permuter are undefined. In the algorithm the operation *define* $P[i]$ means that for the cell i in the permuter $P[\cdot]$ we need to try all possible values $0 \ldots (q-1)$. Note, we cannot select a value which has been already used in another cell of the permuter in a previous step. Before making a step back by the recursion, restore the value $P[i] = \infty$. In the case when the cell $P[i]$ was already defined (is not ∞) due to previous steps, then we just go to the next step directly. Do the following steps recursively:

- for all $i = 0 \ldots q-1$;
- for all $j = 0 \ldots q-1$;
- define $P[j]$;
- define $P^2[j]$;
- define $P[P^2[j]+1] \Rightarrow$ remember $x = P[P^2[j]+1]$;
- define $P[i]$;
- swap$(P[i], P[j])$;
- define $P[i+1] \Rightarrow$ calculate $j' = j + P[i+1]$;
- define $P[j']$, then $P^2[j']$, then $P^3[j']$;
- define $P[P^3[j]+1] \Rightarrow$ remember $y = P[P^3[j]+1]$;
- $T[i,x,y] += (q-r)!$, where r is the actual number of currently defined cells in the permuter $P[\cdot]$.

The calculation of a similar distribution table for VMPC-8 meets computational difficulties, as well as for RC4-8 in [5]. One of the ideas in [5] was to approximate the biases from small n's to a larger n, but we decided not to go this way. Instead, in the next sections we will present only precise theoretical results on VMPC-8, and on the RC4 family of stream ciphers in general.

3.2 Theoretical Weakness of the RC4 Family of Stream Ciphers

The recursive Algorithm 1 is trivial and slow, but we use it to show the further theoretical results. We prove that the approximated distribution table $D_{(i,O_t,O_{t+1})}$ cannot be the uniform distribution when n is larger than some threshold n_0. Moreover, we prove that *each* probability of the approximated distribution $D_{(i,O_t,O_{t+1})}$ differs from the corresponding probability in the case of a random source. In other words, the approximated distribution $D_{(i,O_t,O_{t+1})}$ is biased and we find the lower bound of the bias $\epsilon_{\min}$.

Theorem 1. *For VMPC-n, where $n \geq 8$, under the assumptions made in Section 1.3, the following hold.*

1. *Each probability* $\Pr\{(i, O_t = x, O_{t+1} = y)\} \neq 1/q^3$ *(in a random case it should be $1/q^3$).*

2. *The bias* $|D_{\text{Random}} - D^{\text{VMPC}-n}_{(i,O_t,O_{t+1})}|$ *is bounded by*

$$q^{-8n} \le \epsilon_{\min} = \frac{|\delta_{\min}| \cdot q \cdot (q-9)!}{q!} \le \epsilon = |D_{\text{Random}} - D^{\text{VMPC}-n}_{(i,O_t,O_{t+1})}|, \quad (5)$$

where $|\delta_{\min}|$ *is the minimum value, such that*

$$(q-1)(q-2)\cdot\ldots\cdot(q-8) + \delta_{\min} \equiv 0 \pmod{q}.$$

3. *For VMPC-8, we have* $\epsilon_{\min} \approx 2^{-56.8}$.

Proof:
1) Consider Algorithm 1. In the last step the value of r, the number of currently placed positions in the permuter, can be at most 9. It means that when the algorithm is finished, each cell in $D^{\text{VMPC}-n}_{(i,O_t,O_{t+1})}$ *can be written in the form* $k \cdot (q-9)!$, *for some integer number k.*

On the other hand, for a truly random sequence, the probability must be $\Pr\{(i,O_t,O_{t+1})\} = 1/q^3$. *From (4) it follows that* $\frac{k\cdot(q-9)!}{q\cdot q!}$ *must be equal to* $\frac{1}{q^3}$, *i.e.,*

$$k \quad \text{must be equal to} \quad \frac{q\cdot(q-1)\cdot\ldots\cdot(q-8)}{q^2}. \quad (6)$$

Since k is an integer, then q must divide $(q-1)\cdot\ldots\cdot(q-8)$. *It is easy to show that starting from* $n \ge 8$ *this is not true.*
2) We now try to choose k such that $\Pr\{(i,O_t,O_{t+1})\}$ *is as close to* $1/q^3$ *as possible. Let* $|\delta_{\min}|$ *be the smallest value such that* $(q-1)\cdot\ldots\cdot(q-8)+\delta_{\min}$ *is divisible by q. Then* $\Pr\{(i,O_t,O_{t+1})\} = \frac{1}{q^3} \pm \frac{q\cdot|\delta_{\min}|\cdot(q-9)!}{q^3\cdot q!}$. *The minimum value of* $|D_{\text{Random}} - D^{\text{VMPC}-n}_{(i,O_t,O_{t+1})}|$ *is then derived as*

$$\epsilon_{\min} = q^3 \cdot \frac{q\cdot|\delta_{\min}|\cdot(q-9)!}{q^3\cdot q!} = \frac{|\delta_{\min}|\cdot q\cdot(q-9)!}{q!}. \quad (7)$$

3) for VMPC-8, the minimum $\delta_{\min}$ *is 128. Hence, the lower bound for the bias is* $\epsilon_{\min} \approx 2^{-56.8}$. □

For RC4-n a maximum of 6 positions can be fixed, if we use a similar algorithm. Hence, all cells of the distribution table $D^{\text{RC4}-n}_{(i,O_t,O_{t+1})}$ can be written in the form $k\cdot(q-6)!$. By similar arguments as above, we conclude:

Corollary 1. *For RC4-n,* $n \ge 4$, *under the assumptions made in Section 1.3, the following hold.*

1. *Each probability in* $D^{\text{RC4}-n}_{(i,O_t,O_{t+1})} \ne 1/q^3$;
2. *The minimum value* $|D_{\text{Random}} - D^{\text{RC4}-n}_{(i,O_t,O_{t+1})}|$ *is bounded by*

$$q^{-5n} \le \epsilon_{\min} = \frac{|\delta_{\min}|\cdot q\cdot(q-6)!}{q!} \le \epsilon = |D_{\text{Random}} - D^{\text{RC4}-n}_{(i,O_t,O_{t+1})}|, \quad (8)$$

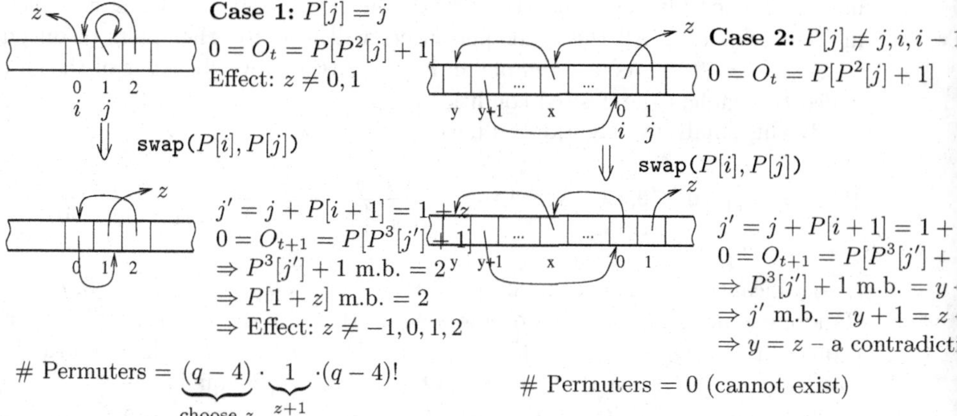

Fig. 2. Condition: $O_t = O_{t+1} = 0$, $i = 0$, $j = 1$. The only case when the condition is satisfied (left), and one of the cases when it is not (right)

where $|\delta_{\min}|$ *is the minimum value, such that*

$$(q-1)(q-2)\cdot\ldots\cdot(q-5) + \delta_{\min} \equiv 0 \pmod q;$$

3. For RC4-n, $n = 4, \ldots, 8$, *we have the following lower bounds.*

	n=4	n=5	n=6	n=7	n=8
$\delta_{\min}$	+8	−8	−8	−8	−120
$\epsilon_{\min}$	$2^{-15.46}$	$2^{-21.28}$	$2^{-26.65}$	$2^{-31.83}$	$2^{-33.01}$

□

The above theorem shows us the way how one can think when designing a new cipher from the RC4 family of stream ciphers to avoid these weaknesses. For the case of VMPC-8, for instance, we can say that the structure seem to be weak in advance, without deep additional investigations of the cipher.

On the contrary, for RC4A-8 our theorem gave us a very small lower bound, so that a hypothetical distinguisher would be slower than an exhaustive search. It means that this cipher would resist distinguishing attacks better than, for example, VMPC-8 or RC4-8. Note, these conclusions were made with the assumptions from Section 1.3. However, in the next sections we investigate digraphs for both ciphers VMPC-n and RC4A-n in detail.

4 Our Distinguisher for VMPC-•

4.1 What the Probability That • $_t$ = • $_{t+1}$ = 0, When • = 0 and • = 1, Should Be?

If VMPC-n would be a truly random generator, then the answer to the question of this section would be $1/q^2$, because when i and j are fixed, then $\Pr\{O_t =$

$0, O_{t+1} = 0|i = 0, j = 1, \text{Random source}\} = 1/q^2$. In the case of VMPC-$n$ this is not true. The only case when the desired outputs can be produced is depicted in Figure 2 (left). All the other permuters will lead to other pairs of outputs $(O_t, O_{t+1}) \neq (0, 0)$. As an example, in Figure 2 (right) we show one of the cases, which contradicts the desired conditions.

By this small investigation we have shown that

$$\Pr\{O_t = O_{t+1} = 0|i = 0, j = 1, \text{VMPC-}n\} = \frac{(q-4)(q-4)!}{q!} = \frac{q-4}{q(q-1)(q-2)(q-3)} \approx 1/q^3$$

is significantly smaller compared to $\Pr\{O_t = O_{t+1} = 0|i = 0, j = 1, \text{Random source}\} = 1/q^2$. If we now assume that for the other values of j the probability $\Pr\{O_t = O_{t+1} = 0|i = 0, j \neq 1, \text{VMPC-}n\} \approx 1/q^2$ – like in a random case, then we can derive that $\Pr\{O_t = O_{t+1} = 0|i = 0\}$ is equal to $(\frac{1}{q} \cdot \frac{1}{q^3} + \frac{q-1}{q} \cdot \frac{1}{q^2})$ (in a random case it should be $1/q^2$). Then, we have a bias $\epsilon \approx 2^{-3n}$, and our hypothetical distinguisher needs to observe the event $O_t = O_{t+1} = i = 0$ from around 2^{6n} samples (2^{7n} bytes of the keystream). It means that VMPC-8 can be distinguished from random having around 2^{56} bytes of keystream. But this estimated bias is still too rough for VMPC-8, and in the next section we show how to compute the exact probability $\Pr\{O_t = O_{t+1} = 0|i = 0\}$ for VMPC-8.

4.2 Calculating Pr• $_t$ = • $_{t+1}$ = 0|• = 0, When • and • [·] Are Random

We could calculate the complete distribution table $D_{(i, O_t = x, O_{t+1} = y)}$ for VMPC-4, and the bias appeared to be $\epsilon \approx 2^{-8.7}$. Unfortunately, we could not apply Algorithm 1 for VMPC-8, because the complexity is 2^{88} – infeasible for a common PC. Instead, we propose to consider only two events $\{O_t = O_{t+1} = 0\}$ and its complement for $i = 0$. We distinguish between the following two binary distributions:

$$D_{\text{VMPC}-n} = \left(\begin{array}{c} \Pr\{O_t = O_{t+1} = 0\} \\ 1 - \Pr\{O_t = O_{t+1} = 0\} \end{array} \right) \Bigg|_{i=0}, \quad \text{and} \quad D_{\text{Random}} = \left(\begin{array}{c} 1/q^2 \\ 1 - 1/q^2 \end{array} \right) \Bigg|_{i=0} \tag{9}$$

Here we give the algorithm to calculate the probability $\Pr\{O_t = O_{t+1} = 0|i = 0\}$. The Algorithm 2 has complexity $O(2^{5n})$, i.e., to calculate $\Pr\{O_t = O_{t+1} = 0|i = 0\}$ for VMPC-8 we need to make only 2^{40} operations. After simulation we got the following result.

Theorem 2. *For VMPC-8, under the assumptions made in Section 1.3,*

$$\Pr\{O_t = O_{t+1} = 0|i = 0\} = \frac{15938227062862998000}{256 \cdot 4096374767995023500000},$$

and the bias is $\epsilon \approx 2 \cdot 2^{-23.98322} \approx 2^{-23}$. I.e., we can distinguish VMPC-8 from random having around 2^{46} samples, or $2^8 \cdot 2^{46} = 2^{54}$ bytes of the keystream, when the two events from the equation (9) are considered. The cipher and random *distributions are the following,*

$$D_{\text{Random}} = \begin{pmatrix} 2^{-16} \\ 1 - 2^{-16} \end{pmatrix}\Bigg|_{i=0}, \quad D_{\text{VMPC}-8} = \begin{pmatrix} 2^{-16} - 2^{-23.98} \\ 1 - 2^{-16} + 2^{-23.98} \end{pmatrix}\Bigg|_{i=0}. \quad (10)$$

□

Algorithm 2. *Recursive computation of the probability* $\Pr\{O_t = O_{t+1} = 0|i = 0\}$

We use the same operation *define* $P[i]$ as in Algorithm 1.

Do the following steps recursively:

· for all $j = 0 \ldots q - 1$;

· define $P[j]$, then $P^2[j]$;

· Since $O_t = 0$, then *fix* the position $P[P^2[j] + 1] = 0$. If this position is already defined ($\neq \infty$), and the value is not 0, or pointer to 0 is already used, then track back by the recursion;

· define $P[i = 0]$;

· swap$(P[i], P[j])$;

· set $P[i + 1] = P[1]$, if possible, otherwise return by recursion;

· calculate $j' = j + P[i + 1]$ which is the same as $j + P[1]$;

· Since $O_{t+1} = 0$, and 0 is already placed in the permuter $P[\cdot]$, then we know the value $P^3[j'] + 1$, hence, we also know the value $P^3[j'] = c$. We can calculate the number of permuters of size q, where $P^3[j'] = c$, and r positions are fixed from the previous steps, by the subalgorithm of complexity $O(q)$, given in Appendix A.

4.3 Simulations of the Attack on VMPC-•

Our theoretical distinguisher from the previous subsection is based on a few assumptions from Section 1.3. First of all, by simulations we have checked the distribution of the internal state of VMPC-n for different values of n, and we did not find any noticeable anomalies. From this we conclude that the internal state in real is distributed close to the uniform distribution, and our theoretical distinguisher should work. Secondly, we can argue that the samples are quite independent. It happens because each sample is connected to the known variable i, and the distance between two samples (for a fixed i) is q rounds of the internal loop.

Theorem 2 says that the complexity of the attack on VMPC-8 is $O(2^{54})$, and, due to such a high complexity, we could not perform simulations of our attack on this cipher. Instead, we could perform simulations on the reduced version VMPC-4, and show the attack in practice.

VMPC-4 has one permuter of size 16, and the internal indices i and j are taken modulo 16. In our simulations we made $N = 2^{34}$ iterations and from 2^{34} received samples we have constructed the type (empirical distribution) with probabilities $\Pr\{O_t = x, O_{t+1} = y|i\}$. Below we show this table (type) partly.

$N = 2^{34}$	i=0				i=1				...
$x \Rightarrow$	0	1	2	...	0	1	2	...	
	To get the probability of the event $(O_t = x, O_{t+1} = y)|i$ the corresponding cell should be divided by 16^2. In the case of a random source each such event has the probability $1/16^2$.								
$y \Rightarrow$ 0	**0.92474**	0.99866	1.00432		0.99287	0.99086	0.99890		
1	1.00085	0.98815	1.01204		0.99309	0.99656	0.99068		
2	1.00519	1.00569	1.00343	...	0.99496	1.06880	1.06524	...	...
3	1.00631	0.99999	0.99562		1.00080	0.99260	0.99767		
⋮	⋮	⋮	⋮		⋮	⋮	⋮		
15	0.99744	0.98926	1.00845		1.00052	0.99124	0.99495		

This table represents the type D_{Type} and we can see that many probabilities are far away from $1/16^2$, and the most biased probability is in the cell $(0,0)$, which corresponds to $\Pr\{O_t = O_{t+1} = 0|i = 0\} = \frac{0.924744}{16^2}$. When the type (the table with probabilities) is derived, one can analyze two possible distinguishers for VMPC-4.

(1) In the first scenario we consider the whole distribution table, i.e., we consider all events of the form $(i, O_t = x, O_{t+1} = y)$. The probability of each event in this case is $1/16^3$. I.e., each cell of the table (type) should be divided by $1/16^3$.
The bias of the received type is $\epsilon_0 = 2^{-8.679648}$, which is close to the theoretical value calculated in the previous section $\epsilon = 2^{-8.7}$. However, we could not calculate a theoretical bias for VMPC-8, therefore, we consider the second scenario;

(2) In this scenario we observe only two events $\{O_t = O_{t+1} = 0|i = 0, \textit{ the others}\}$ – as in (9). As we have mentioned, the probability of the event $(O_t = O_{t+1} = 0)|i = 0$ is much lower than the corresponding probability in the case of a random source. In this example, the received bias appears to be $\epsilon_0 = 2 \cdot \frac{1.0-0.924744}{16^2} \approx 2^{-10.73205}$, which, again, is close to the theoretical value $\epsilon = 2^{-10.755716}$ (calculated in a similar way as for VMPC-8 in Theorem 2). For other values of n the simulation results are presented in the following table.

	n=3	n=4	n=5	n=6	n=7	n=8
Theoretical bias, ϵ	$2^{-7.551}$	$2^{-10.756}$	$2^{-13.871}$	$2^{-16.934}$	$2^{-19.967}$	$2^{-22.98}$
	Simulations of the Attack on VMPC-n					
Number of rounds made, N_0	2^{30}	2^{30}	2^{30}	2^{35}	—	—
The real bias, ϵ_0	$2^{-7.558}$	$2^{-10.732}$	$2^{-13.931}$	$2^{-16.912}$	—	—

Our simulations show that the attack on VMPC-n works in practice. We have also shown that the dependency of the adjacent samples does not influence much on the type.

5 Our Distinguisher for RC4A-•

5.1 Building a Distinguisher

In this section we investigate the cipher RC4A-n (see Figure 1(bottom)), and propose our distinguisher for RC4A-8. We again idealize the situation by the preliminary assumptions from Section 1.3, i.e., at any time t the values $j_1, j_2, P_1[\cdot]$, and $P_2[\cdot]$ are considered from the uniform distribution, and unknown for us. We would like to investigate the following scheme.

i – known value at time t-even
$j_1, j_2, P_1[\cdot], P_2[\cdot]$ – are from a random source
1. $O_t = P_2[P_1[i] + P_1[j_1]]$
2. `swap(`$P_2[i], P_2[j_2]$`)`
3. $O_{t+1} = \ldots$
4. $O_{t+2} = P_2[P_1[i+1] + P_1[j_1 + P_1[i+1]]]$

For cryptanalysis of RC4A-n, we use ideas as before. Our methodology of finding anomalies for both VMPC-n and RC4A-n was just to consider the distribution tables like $D_{(i,O_t,O_{t+2})}$ for small values of n, using an Algorithm 1-like procedure. If some anomaly is found then we concentrate on them in particular for larger values of n, and try to understand why anomalies exist.

For RC4A-n we have noticed that $\Pr\{O_t = O_{t+2} | \ i \text{ is even}\} \neq 1/q$, i.e., does not correspond to the random distribution, whereas the other probabilities $\Pr\{O_t \neq O_{t+2} | \ i \text{ is even}\}$ are equal to each other, but not equal to $1/q$. From the other hand, all probabilities $\Pr\{O_t = O_{t+2} | \ i \text{ is odd}\} = 1/q$ – correspond to the random distribution. So, our target is to calculate the probabilities $\Pr\{O_t = O_{t+2} | \ i \text{ is even}\}$ for RC4A-8. We have used a similar idea as in the Algorithm 2, but much simpler. Our optimized search algorithm to find all such probabilities has complexity $O(2^{6n})$. The result of this work is the following.

Theorem 3. *For RC4A-n,*

under the assumptions made in Section 1.3, consider the following vector of events, and its random distribution,

$$\text{Events} = \begin{pmatrix} O_t = O_{t+2} | i = 0 \\ O_t = O_{t+2} | i = 2 \\ \vdots \\ O_t = O_{t+2} | i = q-2 \\ \text{other cases} \end{pmatrix}, \quad D_{\text{Random}} = \begin{pmatrix} 1/q^2 \\ 1/q^2 \\ \vdots \\ 1/q^2 \\ 1 - 1/(2q) \end{pmatrix}. \qquad (11)$$

For RC4A-8, the bias $D_{\mathrm{RC4A-8}}$ is $\epsilon \approx 2 \cdot 2^{-30.05}$. Hence, our distinguisher needs around 2^{58} bytes of the keystream. □

5.2 Checking the Assumptions

By simulations we found that the internal state of RC4A-n is not close to the uniform distribution. We could clearly see these anomalies running simulations many times for different n each time sampling from at least $N = 2^{30}$ rounds of the loop. To begin counting anomalies, we would like to note that the internal variables $j_1, P_1[\cdot]$ are updated independently from $j_2, P_2[\cdot]$ as follows.

One-Round-Update for $j_*, P_*[\cdot]$, where $*$ is 1 or 2
1. $i++$;
2. $j_* += P_*[i]$
3. swap$(P_*[i], P_*[j_*])$

It means that all anomalies found for $j_1, P_1[\cdot]$ are true for $j_2, P_2[\cdot]$ as well.

We found an event for which the probability is far from the probability of this event in the case of a random source. In particular, $\Pr\{j_1 = i+1\} \approx \frac{q-1}{q^2}$, when in the random case it should be $1/q$. Other probabilities are $\Pr\{j_1|i, j_1 \neq i+1\} \approx \frac{q^2-q+1}{q^2(q-1)}$. For example, for RC4A-4, it appeared that $\Pr\{j_1 = i+1\} \approx 0.05859375$, and the others are $\Pr\{j_1|i, j_1 \neq i\} \approx 0.06276042$ – the difference is noticeable. Some other less notable non-uniformities in the internal state also were found.

5.3 Simulations of the Attack on RC4A-•

Despite finding the non-uniformity of the internal state of RC4A-n we make a set of simulations to see how our distinguisher behaves itself. We will consider the attack scenario as in Theorem 3.

	n=3	n=4	n=5	n=6	n=7	n=8
Theoretical bias, ϵ	$2^{-10.014}$	$2^{-14.005}$	$2^{-18.001}$	$2^{-22.00}$	$2^{-26.00}$	$2^{-29.05}$
	Simulations of the Attack on RC4A-n					
Number of rounds made, N_0	2^{30}	2^{30}	2^{34}	2^{40}	2^{40}	—
The real bias, ϵ_0	$2^{-8.9181}$	$2^{-12.2703}$	$2^{-15.073}$	$2^{-18.042}$	$2^{-20.025}$	—

Note that the number of actual samples N_0 in our simulations is larger than $1/\epsilon_0^2$. From (3) it means that we have distinguished the cipher with a very small probability of error, and the real theoretical bias without pre-assumptions should be close to what we get in our simulations. From the table above we see that the bias in practice (when the internal state is not from the unoform distribution) is larger than the approximated value of the bias (the uniformly distributed internal state), for $n = 3, \ldots, 7$. The same behaviour of the distinguisher we expect for $n = 8$ as well. Since we could not perform simulations for $n = 8$, we decided to leave theoretical bias as the lower bound of the attack, i.e., $\epsilon = 2^{-29.05}$ for $n = 8$, the complexity is $O(2^{58})$. However, in the real life we expect this bias to be even larger, and complexity of the attack lower.

6 Results and Conclusions

In this paper we have shown some theoretical weaknesses of the RC4 family of stream ciphers. We have also investigated recently suggested stream ciphers VMPC-n and RC4A-n, and found linear distinguishing attacks on them. They are regarded as academic attacks which show weak places in these ciphers. The summarizing table of our results is below:

Cipher	Theoretical Lower Bound for ϵ, $n = 8$	Our Distinguishers Complexity (# of symbols)					
		$n = 3$	$n = 4$	$n = 5$	$n = 6$	$n = 7$	$n = 8$
RC4-n (1987)	2^{-33} (Corr.1)	—	—	—	—	—	$2^{30.6}$ (from [5])
VMPC-n (2004)	$2^{-56.8}$ (Thr.1)	2^{23}	* 2^{29}	2^{35}	2^{41}	2^{48}	$\mathbf{2^{54}}$
RC4A-n (2004)	—	2^{18}	2^{28}	2^{36}	2^{44}	2^{52}	$\mathbf{2^{58}}$

The distinguisher for VMPC-8 that we propose is the following [2]:

Distinguisher for VMPC-8:

1. Observe $N = 2^{54}$ output bytes. Calculate the number L of occurences such that $a = O_t = O_{t+1} = 0$.
2. Calculate two distances:
   $\lambda_{\text{Random}} = |2^{-16} - 2^8 \cdot L/N|$
   $\lambda_{\text{VMPC}} = |(2^{-16} - 2^{-23.98322}) - 2^8 \cdot L/N|$
3. If $\lambda_{\text{Random}} > \lambda_{\text{VMPC}}$ then **keystream of VMPC-8**, else **a random sequence**.

If the internal state of a cipher from the RC4 family is uniformly distributed, then, based on our discussions in Section 3, we conclude that such ciphers are not very secure. When the internal state is non-uniformly distributed then the real bias would more likely be larger rather than smaller, and the complexity of the attack would be lower, in most cases. That effect we could observe on the example of RC4A-n. It seems that the security level of such constructions

[2] The distinguisher for RC4A-8 is in a similar fashion as for VMPC-8.

* In the first scenario from Subsection 4.3 the attack complexity for VMPC-4 is $O(2^{18})$.

depends more on the degree of the recursive relations between output symbols and internal states, rather than on the length of the permuter(s).

One of the solutions to protect against of such distinguishing attacks is to increase the number of accesses to the permuter(s) in the loop. This solution will increase the relation complexity between adjacent outputs. Another solution is to discard some output symbols before to accept one. Unfortunately, both the suggestions significantly decrease the speed of these ciphers – the main purpose of such designs (speed) is then destroyed.

Acknowledgements

We thank Willi Meier for his useful suggestions on this research direction that made this paper possible. We also thank Thomas Johansson and anonymous reviewers for their editing advises and critical comments.

References

1. N. Smart. Cryptography: An Introduction, 2003.
2. B. Schneier. *Applied Cryptography: Protocols, Algorithms, and Source Code in C.* John Wiley and Sons, New York, 2nd edition, 1996.
3. J.D. Golić. Linear statistical weakness of alleged RC4 keystream generator. In W. Fumy, editor, *Advances in Cryptology—EUROCRYPT'97*, volume 1233 of *Lecture Notes in Computer Science*, pages 226–238. Springer-Verlag, 1997.
4. L.R. Knudsen, W. Meier, B. Preneel, V. Rijmen, and S. Verdoolaege. Analysis methods for (alleged) RC4. In K. Ohta and D. Pei, editors, *Advances in Cryptology—ASIACRYPT'98*, volume 1998 of *Lecture Notes in Computer Science*, pages 327–341. Springer-Verlag, 1998.
5. S. R. Fluhrer and D. A. McGrew. Statistical analysis of the alleged RC4 keystream generator. In B. Schneier, editor, *Fast Software Encryption 2000*, volume 1978 of *Lecture Notes in Computer Science*, pages 19–30. Springer-Verlag, 2000.
6. I. Mantin and A. Shamir. Practical attack on broadcast RC4. In M. Matsui, editor, *Fast Software Encryption 2001*, volume 2355 of *Lecture Notes in Computer Science*, pages 152–164. Springer-Verlag, 2001.
7. S. Paul and B. Preneel. Analysis of non-fortuitous predictive states of the RC4 keystream generator. In T. Johansson and S. Maitra, editors, *Progress in Cryptology—INDOCRYPT 2003*, volume 2904 of *Lecture Notes in Computer Science*, pages 52–67. Springer-Verlag, 2003.

---

[0] The work described in this paper has been supported in part by Grant VR 621-2001-2149, in part by the Graduate School in Personal Computing and Communication PCC++, and in part by the European Commission through the IST Program under Contract IST-2002-507932 ECRYPT.

The information in this document reflects only the author's views, is provided as is and no guarantee or warranty is given that the information is fit for any particular purpose. The user thereof uses the information at its sole risk and liability.

8. B. Zoltak. VMPC one-way function and stream cipher. In B. Roy and W. Meier, editors, *Fast Software Encryption 2004*, volume 3017 of *Lecture Notes in Computer Science*, pages 210–225. Springer-Verlag, 2004.
9. S. Paul and B. Preneel. A new weakness in the RC4 keystream generator and an approach to improve the security of the cipher. In B. Roy and W. Meier, editors, *Fast Software Encryption 2004*, volume 3017 of *Lecture Notes in Computer Science*, pages 245–259. Springer-Verlag, 2004.
10. D. Coppersmith, S. Halevi, and C.S. Jutla. Cryptanalysis of stream ciphers with linear masking. In M. Yung, editor, *Advances in Cryptology—CRYPTO 2002*, volume 2442 of *Lecture Notes in Computer Science*, pages 515–532. Springer-Verlag, 2002.

Appendix A: Subalgorithm for Algorithm 2

Problem statement: We are given a permuter template of size q, where r positions are already placed, whereas the rest are undefined. We want to calculate the number of permuters satisfying the given template, such that $P^3[j'] = c$, where j' and c are some known positions in the permuter.

Sub-Algorithm:[a]

1. Go forward by the path $j' \to P[j'] \to P^2[j'] \to P^3[j']$, as much as possible, but not more then 3 steps. Let g be the point in this path where we have stopped, and l_g be the number of steps we made (from 0 to 3).
2. Go backward by the path $c \to P^{-1}[c] \to P^{-2}[c] \to P^{-3}[c]$, as much as possible, but not more then 3 steps. Let h be the point in the path where we have stoped, and l_h be the number of steps we made (from 0 to 3).
3. if ($l_g = 3$ and $g \neq c$) or ($l_h = 3$ and $h \neq j'$) then return 0;
   if ($l_g = 3$ and $g = c$) or ($l_h = 3$ and $h = j'$) then return $(q-r)!$;
   if ($l_g + l_h \geq 3$) return 0;
4. Count the number t_1 of positions $x \neq g, h$ in the permuter $P[\cdot]$ for which $P[x] = P^{-1}[x] = \infty$ (see Fig. 3(1)).
   Count the number t_2 of positions $x \neq g, h$, for which $P[x] \neq \infty, g, h$, and $P^{-1}[x] = P^2[x] = \infty$ (see Fig. 3(2)).
5. Now there could be 7 possibilities to connect positions g and h, and they are depicted in Figure 3(a–g):
   a) $g = h,\ l_g + l_h = 0 \Rightarrow$ add $(q-r-1)!$ combinations;
   b) $g = h,\ l_g + l_h = 0, t_1 \geq 2 \Rightarrow$ add $t_1(t_1-1)(q-r-3)!$ combinations;
   c) $g = h,\ l_g + l_h = 0 \Rightarrow$ add $t_2(q-r-2)!$ combinations;
   d) $g \neq h,\ l_g + l_h = 2 \Rightarrow$ add $(q-r-1)!$ combinations;
   e) $g \neq h,\ l_g + l_h = 1 \Rightarrow$ add $t_1(q-r-2)!$ combinations;
   f) $g \neq h,\ l_g + l_h = 0, t_1 \geq 2 \Rightarrow$ add $t_1(t_1-1)(q-r-3)!$ combinations;
   g) $g \neq h,\ l_g + l_h = 0 \Rightarrow$ add $t_2(q-r-2)!$ combinations;

[a] The complexity of the subalgorithm is $O(q)$

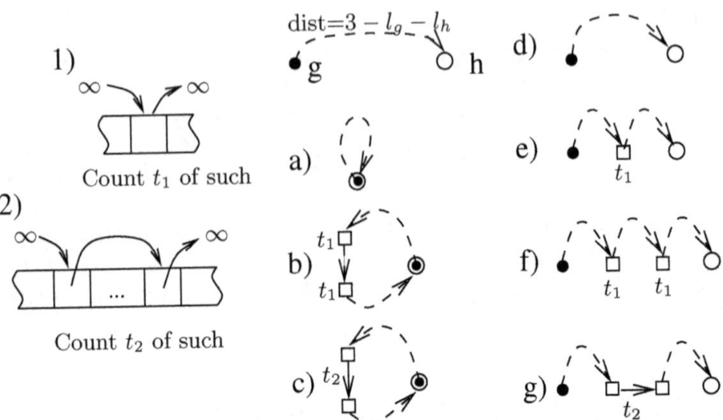

Fig. 3. Possibilities to connect g and h, used in subalgorithm

Impossible Fault Analysis of RC4 and Differential Fault Analysis of RC4

Eli Biham[1, ⋆], Louis Granboulan[2,⋆⋆], and Phong Q. Nguyễn[3,⋆⋆]

[1] Computer Science Department,
Technion – Israel Institute of Technology,
Haifa 32000, Israel
biham@cs.technion.ac.il
http://www.cs.technion.ac.il/$\sim$biham/

[2] École normale supérieure, DI, 45 rue d'Ulm, 75005 Paris, France
Louis.Granboulan@di.ens.fr
http://www.di.ens.fr/~{}granboul/

[3] CNRS/École normale supérieure, DI, 45 rue d'Ulm, 75005 Paris, France
Phong.Nguyen@di.ens.fr
http://www.di.ens.fr/~{}pnguyen/

Abstract. In this paper we introduce the notion of *impossible fault analysis*, and present an impossible fault analysis of RC4, whose complexity 2^{21} is smaller than the previously best known attack of Hoch and Shamir (2^{26}), along with an even faster fault analysis of RC4, based on different ideas, with complexity smaller than 2^{16}.

1 Introduction

RC4 is a stream cipher designed by Ron Rivest in 1987, and used by RSADSI in their products. It was never officially published, but a reverse-engineered copy of the code appeared anonymously in the sci.crypt newsgroup in 1994. Nowadays, RC4 is one of the most widely used stream ciphers in a wide range of applications.

Fault analysis was introduced in 1996, when an attack on implementations of RSA and other public key algorithms was described [3]. Shortly after, differential fault analysis of secret key cryptosystems such as DES has followed [2]. These attacks can be made in practice, and various techniques have been described that induce faults during cryptographic computations [1, 9, 14, 16].

Various observations were made on the design and properties of RC4 since its publication, e.g., in [5, 6, 7, 10, 12, 13, 15], however till recently no fault analysis was performed. In a recent paper [8] Hoch and Shamir study fault analytic attacks against various stream ciphers, including RC4. Their results show that

⋆ Part of this work was done while visiting the École normale supérieure, Paris, France.

⋆⋆ The work described in this article is supported in part by the French government through X-Crypt, and in part by the Commission of the European Communities through the IST program under contract IST-2002-507932 ECRYPT.

H. Gilbert and H. Handschuh (Eds.): FSE 2005, LNCS 3557, pp. 359–367, 2005.

Table 1. Summary of the Results for Finding the Initial States

Attack	#Faults	Data	#Rekey	Time	Space
Hoch and Shamir [8]	2^{16}	2^{26}	2^{16}	2^{26}	2^{16}
This paper:					
Impossible Fault Analysis	2^{16}	2^{21}	–	online (2^{21})	2^{8}
Differential Fault Analysis	2^{10}	2^{16}	2^{10}	2^{16}	2^{10}

the initial state of RC4 can be recovered given 2^{26} bytes of stream with about 2^{16} key setups, and 2^{26} steps of analysis.

In this paper we present the notion of *impossible fault analysis*, which uses faults to force the internal state of the stream cipher to become an impossible state, i.e., a state that can never occur in a regular use of the cipher. We use this notion to force RC4 to enter impossible states that were first observed by Hal Finney in [4], and were later described with additional properties in [11]. Once the internal state falls into Finney's impossible states, the output stream become a copy (actually 255 interleaved copies) of the internal state — a phenomena which is very unexpected. In order to fall to these state we need only 2^{21} bytes of stream (without any additional key setups), and once we fall into Finney's states, the recovery of the internal state is one of the simplest ever described for any cipher, as the internal state is simply copied into the output stream.

We then describe another, more standard, fault analytic attack against RC4, which require less than 2^{16} stream bytes with less than a thousand (or a few thousands) key setups, and whose analysis time complexity is also lower than 2^{16}. However, because the assumptions used by the two attacks are different, it is perhaps not clear which attack would be more efficient in practice.

A summary of our results, and of the previously published results is given in Table 1.

The paper is organized as follows: In Section 2 we give a short description of RC4. In Section 3 we describe the notion of impossible fault analysis and the application to RC4. In Section 4 we describe the best known fault analysis of RC4, and in Section 5 we summarize the paper.

2 A Brief Description of RC4

For the discussion of this paper it suffices to describe the step function of RC4. The key setup is very similar, but is not necessary for our analysis.

RC4's internal state consists of two indices i and j, and an array S of size 256. In a real computation of RC4, this array is always a permutation of all the values from 0 to 255. The content of the S array is the result of the key schedule, while i and j are always set to 0 by the key schedule. Each step of RC4 is then as follows

```
i ++
j += S[i]
swap the values of S[i] and S[j]
Output S[S[i] + S[j]]
```

3 Impossible Fault Analysis of RC4

In [4] Hal Finney presented the following family of internal states of RC4, where

$$j = i + 1, \qquad \text{and} \qquad S[j] = 1. \tag{1}$$

He observed that if an internal state is a member of this family at some step, then all the next and prior states are also in this family, as if the internal state is in the family, then the value 1 is swapped with the next value, and i and j are incremented by 1. In his thesis, Itsik Mantin [11] also observes that due to these swaps, every 255 steps the array S is rotated by one byte, but remains with the same circular order of byte values, and that every 255 states, the index of the output byte repeats (as it is just the sum of the two entries pointed by i and j, i.e., the first repeats every 255 states, and the other is always 1). As a result of these two observations, the output stream takes the same entry from the array every 255 output bytes, but the byte values are already rotated, making the new value of the entry after 255 states being the next value (that was in the next entry), and any further 255th byte in the output stream being the next value of the internal state in a cyclic order. We thus receive 255 interleaved streams, each of them is a copy of the internal state with a (possibly) different starting value. It is easy to see that once we fall into these states, it is very easy to identify that we fell into these states, and also very easy to deduce the content of the internal state (due to an additional property it is even easy to know which is the first value in the S array, thus to know the exact location of each value in the array, not only the relative order).

However, the key schedule of RC4 sets $i = j = 0$ causing these states to be impossible, thus these states can never appear in real RC4 streams, and we cannot expect to fall into these states in our analysis.[1]

Fortunately, for the purposes of this paper, fault analysis can make the impossible possible, by modifying the indices i or j, or even the content of the S array. Therefore, if a fault occurs during the computation of RC4, it may occur that the resulting internal state becomes a member of this impossible family, and once the internal state is a member of the family, it is very easy to deduce the internal state by looking at the output stream.

[1] Some RC4 variants with modified key schedule may allow this family of states to occurs for some fraction (usually about 2^{-16}) of the keys. In such cases, this fraction of the keys form a class of weak keys, in which the initial state is being copied to the output stream.

For the attack we assume that faults are injected into either register i or register j at any time the attacker wishes. Once a fault modifies the internal state to become one of Finney's states, it is very easy to identify this fact after several hundred bytes (after two or three bytes from each interleaved cycle are given, i.e., after a total of about 500 or 700 bytes). However, as we show below, the identification of non-Finney states can be made much earlier.

We first observe that the probability of falling into Finney's states by a fault in i or j is 2^{-16}, because the two equalities defining Finney's states in Eq. (1) compare a total of 16 bits, where

1. If the fault is in i, the probability that $S[j] = 1$ holds is 2^{-8}, and the probability that the new i is set to $j-1$ is also 2^{-8}.
2. If the fault is in j, the probability that $S[i+1] = 1$ holds is 2^{-8}, and the probability that the new j is set to $i+1$ is also 2^{-8}.

Therefore, we expect that the internal states will fall to Finney's states after about 2^{16} faults are induced.

Once a fault is induced, we would like to identify as soon as possible if the internal state is a Finney's state or not. We observe that Finney's states have the property that the consecutive (or close) output states fetch the output bytes from different locations in the S array (because a different value is added to the 1 to form the index of the output byte), and that the array is a permutation of all the 256 values from 0 to 255. So usually close bytes of the output stream should be distinct, though, bytes that were swapped already with the prior location in the array may appear a second time. Therefore, we expect that a collision of two bytes of the output stream should occur after about 80 bytes of the output stream, while in the case of non-Finney's states a collision should occur after less than 20 bytes due to the birthday paradox (we verified this behavior by a simulation).[2]

Based on these observations, we use the following procedure for identifying whether a state is a Finney's state or not:

1. For each new output byte check whether the byte value already appeared since the last fault.

[2] A short approximate calculation of the case of Finney's states is as follows. We ignore the case where the colliding output byte has value 1, since it is rare, and affects the result only marginally. Then, (when the output is not 1) a collision of a byte a states after the fault and another byte b states after the fault can occur only if $S[i_a]+S[j_a] = S[i_b]+S[j_b]+1$ and $i_a \leq S[i_a]+S[j_a] \leq i_b$ (in a cyclic manner, where i_x and j_x are the values of the registers x states after the fault). The probability of the first equation to hold is 1/256, while the probability of the second equation is $(b-a)/256$. Therefore, under some independence assumptions, the probability that a collision occurs within the first L bytes after the fault is approximated by $\sum_{0 \leq a < b < L}(1/256)((b-a)/256)$ which is about $L^3/(3 \cdot 2^{17})$. By forcing this probability to be about 1 we get that $L \approx 73$, and by forcing it to be about 1/2 we get that $L \approx 58$. We expect that accurate computation will show slightly higher values.

2. If not, process the next byte (goto step 1).
3. If this byte value already appeared at least once.
   Denote the number of bytes since the last fault by n.
   (a) If $n < 30$, the state is most probably not a Finney's state.
   (b) If $30 \leq n < 40$ and this is the second collision, or if $40 \leq n < 50$ and this is the third collision, (etc., with a few additional similar tests designed to distinguish between the two cases), then the state is most probably not a Finney's state.[3]
   (c) If $n > 255$, check the relation to the 255th preceding byte, and (unless one of these two values is 1) verify that they are different, and that no other earlier pair of this kind appeared with a combination of the first value with another second value, or the second value with another first value. If such a pair appeared, the state is not a Finney's state.
   (d) If $n > 600$, the state is most probably a Finney's state.
   (e) Otherwise, process the next byte (goto step 1).
4. Once the state is identified as either (most probably) a Finney's state, or (most probably) not a Finney's state, the algorithm stops and outputs this information (for the rest of the attack, we will not distinguish between a most probable identification or an absolute identification).

The attack is thus as follows: start encryption with RC4 with some unknown key, and apply the procedure repeatedly, as many times as required, till a Finney's state is identified. Each time the procedure identifies a non-Finney's state, inject a new fault to either i or j before the next application of the procedure. Once the procedure identifies a Finney's state, there are a few options

1. The simplest but slow method is to process extra $255 \cdot 256$ bytes, and select one of the interleaved cycles as the internal state.
2. A much faster method is to use the information we already got about consecutive values in the internal state to recover the full state.
3. In both cases the assignment of the cycle to the exact location in the S array can be done using information obtained at time that the output byte is taken from either location i or location j (i.e., $S[i]+S[j]$ is either i or j), in which the outputs are easily identified, e.g., when $S[i]+S[j]=i$ then the output byte is necessarily 1.

Once the internal state is obtained at some point in time, it is easy to find the current values of i and j, and then to process the states backwards till the prior fault. It is also possible to identify the prior value of the faulty register by careful analysis of the prior bytes of the output stream (i.e., given the internal state, one of the two registers, and the output byte, it is easy to recover the value of the other register), thus enabling the attacker to obtain the key-dependent initial state.

---

[3] Our actual simulation used a slightly different set of thresholds, but for simplicity of the description we present a rounded version.

The attacker can deduce the key from this initial state, e.g., with the technique described in section A.2 of [11].

We programmed most parts of this attack, and verified that it works and that the expected number of required bytes of the output stream is about 2^{21} (i.e., the average number of bytes required for recognition of a non-Finney's state is not far from 32). In all our tests, our simulation of the attack always identified Finney's states correctly, and was not sensitive to whether the faults were induced in register i or register j (in fact, even if the attacker have no control on the selection of the register in which the fault is induced, and even if each fault is induced on one of the registers at random, still the attacker will be able to recover the initial state without additional complexity).

This attack have various variants: The faults can also be injected to the S array, but then if the fault is injected at a random location in the array, the probability to get a Finney's state is only 2^{-24} (i.e., $i = j - 1$ occurs with probability 2^{-8}, the fault occurs at the new j with probability 2^{-8}, and the fault modifies this location to a value 1 with probability 2^{-8}). An additional problem in this case is that the fault change the content of the S array quite rapidly, making it difficult to follow the changes backwards to the initial state once a Finney's state occurs (a solution to this latter problem is to make a new key setup once in a while, like is made is [8] and in the next section).

Once we make fault in the array, it is possible to replace usage of the value 1 in the attack by two consecutive 2's in the following way

$$j = i + 2, \qquad \text{and} \qquad S[j-1] = S[j] = 2,$$

or with 3's

$$j = i + 3, \qquad \text{and} \qquad S[j-2] = S[j-1] = S[j] = 3,$$

or similarly with larger values. The probability to reach such states is smaller, but once they appear, the internal state is copied to the output in the same way as in Finney's states.

If we assume a stronger fault model than we did, in which the attacker can induce the faults and select the exact new value of the register (i or j or an entry of the S array), the attack would be much faster (but then we assume that there exists an even more efficient attack designed specifically for this stronger model).

Finally, as in Finney's states the output stream repeats with known period, it may be possible to mount a ciphertext only attack where the attacker is given the ciphertext rather than the output stream. Given a known statistics of the plaintext (e.g., the knowledge that the plaintext is written in English), these attacks would required much longer streams for the identification of Finney's internal states, but once these states are identified, the combined knowledge on the statistics of the plaintext's language and the properties of the internal state may leak full information on the plaintext and on the initial state (this attack may be applicable in cases where a smartcard encrypts using RC4 while being at the possession of the attacker).

4 A Differential Fault Analysis of RC4

This attack makes use of the fact that each step of RC4 accesses only three bytes of the S array. In this section we present the main ideas behind this attack, as there are some delicate cases where a more complicated handling should be performed, which we will not describe here. We first collect the following data

1. Process a non-faulty stream
   (a) Perform a key setup with the unknown key
   (b) Run RC4 256 steps, keeping the output stream for later analysis. Call this output stream R.
2. Process the following 256 times with l being set from 0 to 255, giving 256 output streams of length about 30
   (a) Perform a key setup with the unknown key
   (b) Make a fault in $S[l]$
   (c) Run RC4 30 steps, keeping the output stream for later analysis. Call this output stream O_l.
3. Repeat again but with the fault being injected after 30 steps of RC4, calling the output streams T_l, and keeping longer output streams in the T_l's.
4. Possibly repeat again with a few additional such sets.

Evidently, the first byte of all the O_l's, except for three of them, are the same as in the real stream R. The identification of these three streams leak the values of i, j, and $S[i]+S[j]$, but not which is which. Evidently, the value of i is always known, thus it remains only to identify which is j and which is $S[i]+S[j]$. We continue doing so for the following bytes of the output streams (ignoring streams O_l which already had earlier faults). This technique can be called *cascade guessing*.

Now, in any location in the stream, whichever of the two values is j, we can subtract the previous j from the new j and get $S[i]$, then testing whether we already know that this value appears in another location in the array, allowing us to discard many wrong cases. Similarly, the other value is $S[i]+S[j]$, and by knowing the output byte (of the real stream R) we know the content of the location pointed by this value $S[i]+S[j]$, which also allows us to make a test for a contradiction. In addition, also $S[j]$ is observed, and can also be tested for a contradiction. Due to the birthday paradox, we expect to get a contradiction after an average of less than 20 observed bytes, i.e., after processing about 7 bytes of the output stream. Therefore, the analysis for finding the internal state should find a considerable number of bytes of the internal state with a small complexity.

However, as the faults (made before the first step of RC4) become more and more distant from the analyzed bytes (computed after several steps of RC4), the identification of the three values become less and less successful, due to earlier effects of these faults on the streams. This problem have several solutions: one of them is to have several sets with faults made at different times (like the T_l's mentioned above). Another solution is to make a more delicate analysis and use the faulty streams to recover more information on the key (as the faults changed the state slightly, different bytes may be reached from these streams). Or alternatively, it is possible to allow the algorithm to ignore some unknown values during the computation, and continue to the next stream byte. A combination of these methods can recover a large majority of the bytes of the internal state with a relatively small complexity, after which much simpler analysis can complete the full content of the internal state.

The complexity of this attack is bounded by 2^{16} using a total of less than 2^{16} stream bytes (with less than a thousand, only a few thousands, key setups).

5 Summary

In this paper we presented two fault attacks on RC4. The first introduces the notion of impossible fault analysis, and shows how to apply it with complexity smaller than of previously published fault attacks against RC4. The second attack uses more key setups and analyzes the difference behavior of the key stream once the internal states is modified due to faults. This later attack is currently the fastest known fault analysis of RC4.

Acknowledgments

We would like to thank Adi Shamir, Jonathan Hoch, and Itsik Mantin for various valuable comments and discussions which improved the results of this paper.

References

1. R. J. Anderson and S. Skorobogatov, "Optical fault induction attacks." in *Proceedings of CHES'02* (B. S. Kaliski, Çetin Kaya Koç, and C. Paar, eds.), no. 2535 in Lecture Notes in Computer Science, Springer-Verlag, 2002.
2. E. Biham and A. Shamir, "Differential fault analysis of secret key cryptosystems." in *Proceedings of Crypto'97* (B. S. Kaliski, Jr, ed.), no. 1294 in Lecture Notes in Computer Science, pp. 513–525, Springer-Verlag, 1997.
3. D. Boneh, R. A. DeMillo, and R. J. Lipton, "On the importance of checking cryptographic protocols for faults." in *Proceedings of Eurocrypt'97* (W. Fumy, ed.), no. 1233 in Lecture Notes in Computer Science, pp. 37–51, Springer-Verlag, 1997.
4. H. Finney, *An RC4 Cycle that Can't Happen*, September 1994.
5. S. R. Fluhrer, I. Mantin, and A. Shamir, "Weaknesses in the key scheduling algorithm of RC4." in *Proceedings of Selected Areas in Cryptography – SAC'01* (S. Vaudenay and A. M. Youssef, eds.), no. 2259 in Lecture Notes in Computer Science, pp. 1–24, Springer-Verlag, 2001.
6. S. R. Fluhrer and D. A. McGrew, "Statistical analysis of the alleged RC4 stream cipher." in *Proceedings of Fast Software Encryption – FSE'00* (B. Schneier, ed.), no. 1978 in Lecture Notes in Computer Science, pp. 19–30, Springer-Verlag, 2000.
7. J.D. Golic, "Linear Statistical Weakness of Alleged RC4 Keystream Generator." in *Proceedings of Eurocrypt'97* (W. Fumy, ed.), no. 1233 in Lecture Notes in Computer Science, pp. 226–238, Springer-Verlag, 1997.

8. J. J. Hoch and A. Shamir, "Fault Analysis of Stream Ciphers." in *Proceedings of CHES'04* (M. Joye and J.-J. Quisquater, eds.), no. 3156 in Lecture Notes in Computer Science, pp. 240–253, Springer-Verlag, 2004.
9. O. Kömmerling and M.G. Kuhn, "Design Principles for Tamper-Resistant Smartcard Processors." in *Proceedings of the USENIX Workshop on Smartcard Technology (Smartcard '99)*, pp. 9–20, USENIX Association, 1999.
10. L.R. Knudsen, W. Meier, B. Preneel, V. Rijmen, and S. Verdoolaege, "Analysis Methods for (Alleged) RC4." in *Proceedings of Asiacrypt'98* (K. Ohta and D. Pei, eds.), no. 1514 in Lecture Notes in Computer Science, pp. 327–341, Springer-Verlag, 1998.
11. I. Mantin, *Analysis of the Stream Cipher RC4*, M.Sc. thesis, The Weizmann Institute of Science, 2001. Available at `http://www.wisdom.weizmann.ac.il/~itsik/RC4/rc4.html`.
12. I. Mantin and A. Shamir, "A practical attack on broadcast RC4." in *Proceedings of Fast Software Encryption – FSE'01* (M. Matsui, ed.), no. 2355 in Lecture Notes in Computer Science, pp. 152–164, Springer-Verlag, 2001.
13. I. Mironov, "(Not So) Random Shuffles of RC4." in *Proceedings of Crypto'02* (M. Yung, ed.), no. 2442 in Lecture Notes in Computer Science, pp. 304–319, Springer-Verlag, 2002.
14. D. Naccache, P.Q. Nguyễn, M. Tunstall and C. Whelan, "Experimenting with Faults, Lattices and the DSA" in *Proceedings of PKC '05* (S. Vaudenay, ed.), no. 3386 in Lecture Notes in Computer Science, Springer-Verlag, 2005.
15. Souradyuti Paul, Bart Preneel, "A New Weakness in the RC4 Keystream Generator and an Approach to Improve the Security of the Cipher.", proceedings of Fast Software Encryption, 11th International Workshop, FSE 2004, Delhi, India, Lecture Notes in Computer Science 3017, pp. 245–259, Springer-Verlag, 2004.
16. J.-J. Quisquater and D. Samyde, "ElectroMagnetic Analysis (EMA): Measures and counter-measures for smart cards." in *Proceedings of E-smart 2001* (I. Attali and T. P. Jensen, eds.), no. 2140 in Lecture Notes in Computer Science, pp. 200–210, Springer-Verlag, 2001.

Related-Key Rectangle Attacks on Reduced Versions of SHACAL-1 and AES-192*

Seokhie Hong[1], Jongsung Kim[2], Sangjin Lee[2], and Bart Preneel[1]

[1] Katholieke Universiteit Leuven, ESAT/SCD-COSIC,
Kasteelpark Arenberg 10, B-3001 Leuven-Heverlee, Belgium
{Seokhie.Hong, Bart.Preneel}@esat.kuleuven.ac.be
[2] Center for Information Security Technologies(CIST),
Korea University, Seoul, Korea
{joshep, sangjin}@cist.korea.ac.kr

Abstract. In this paper we propose a notion of related-key rectangle attack using 4 related keys. It is based on two consecutive related-key differentials which are independent of each other. Using this attack we can break SHACAL-1 with 512-bit keys up to 70 rounds out of 80 rounds and AES with 192-bit keys up to 8 rounds out of 12 rounds, which are faster than exhaustive search.

1 Introduction

Differential cryptanalysis [1] introduced by E. Biham and A. Shamir is one of the most powerful known attacks on block ciphers. After this attack was introduced, various variants of the attack have been proposed, such as the truncated differential attack [18], the higher order differential attack [18], the differential-linear attack [20], the impossible differential attack [3], the boomerang attack [23], the rectangle attack [4] and so on.

In 1993, E. Biham introduced the related-key attack [2] in which the attacker can choose the relationship between two unknown keys. It is based on a key scheduling algorithm and shows that a block cipher with a weak key scheduling algorithm may be vulnerable to this kind of attack. Several cryptanalytic results of this attack were reported in [6, 12, 13, 22].

In [10], P. Hawkes showed that the related-key attack can be combined with the differential-linear attack and that this combined attack can find a relatively large weak-key class of block cipher IDEA. After this, G. Jakimoski and Y. Desmedt [11] exploited a combination of the related-key and the impossible differential attacks to analyze 8-round AES with 192-bit keys. Recently, J. Kim et al. [15] introduced a combination of the related-key and the rectangle attacks,

* The first author was supported by the Post-doctoral Fellowship Program of Korea Science & Engineering Foundation (KOSEF). The second and the third authors were supported by the MIC(Ministry of Information and Communication), Korea, supervised by the IITA(Institute of Information Technology Assessment.)

H. Gilbert and H. Handschuh (Eds.): FSE 2005, LNCS 3557, pp. 368–383, 2005.

Table 1. Comparison of our attacks with the previous ones

Block Cipher	Type of Attack	Number of Rounds	Complexity Data / Time
SHACAL-1 (80 rounds)	Differential	30(0-29)	2^{110}CP / $2^{75.1}$[17]
		41(0-40)	2^{141}CP / 2^{491}[17]
	Amplified Boomerang	47(0-46)	$2^{158.5}$CP / $2^{508.4}$[17]
	Rectangle	47(0-46)	$2^{151.9}$CP / $2^{482.6}$[5]
		49(22-70)	$2^{151.9}$CP / $2^{508.5}$[5]
		49(29-77)	$2^{151.9}$CC / $2^{508.5}$[5]
	Related-Key Rectangle	57(0-56)	$2^{154.75}$RK-CP / $2^{503.38}$ [15]
		59(0-58)	$2^{149.72}$RK-CP / $2^{498.30}$ [15]
		70(0-69)	$2^{151.75}$RK-CP / $2^{500.08}$ (New)
AES-192 (12 rounds)	Square	7(0-6)	2^{32}CP / 2^{184} [21]
	Partial Sums	7(0-6)	$19 \cdot 2^{32}$CP / 2^{155} [8]
		7(0-6)	$2^{128} - 2^{119}$CP / 2^{120}[8]
		8(0-7)	$2^{128} - 2^{119}$CP / 2^{188}[8]
	Related-Key Impossible	7(0-6)	2^{111}RK-CP / 2^{116} [11]
		8(0-7)	2^{88}RK-CP / 2^{183} [11]
	Related-Key Rectangle	8(0-7)	$2^{86.5}$RK-CP / $2^{86.5}$ (New)

CP: Chosen Plaintexts, RK-CP: Related-Key Chosen Plaintexts,
CC: Chosen Ciphertexts, Time: Encryption units

called the related-key rectangle attack, in which the attacker can use consecutive two differentials; one is a related-key differential and the other one is a differential.

Until now, a relation of two keys has been considered in almost all attacks relevant to related-key attacks but in this paper we consider 4 related keys. Our basic idea is similar to the related-rectangle attack presented in [15] except that our attack uses 4 related keys. In our attack we use two consecutive related-key (truncated) differentials which are independent of each other. Our attack allows us to break SHACAL-1 with 512-bit keys up to 70 rounds out of 80 rounds and AES with 192-bit keys up to 8 rounds out of 12 rounds. See Table 1 for a summary of our results and their comparison with the previous attacks.

Our paper is organized as follows. In Sect. 2, we introduce the related-key rectangle attack using 4 related keys. Two applications on SHACAL-1 and AES are presented in Sect. 3 and Sect. 4, respectively. We conclude our paper in Sect. 5.

2 The Related-Key Rectangle Attack

The related-key rectangle attack introduced in [15] is a combination of the related-key and the rectangle attacks. It exploits two types of related-key rectangle distinguishers to retrieve the related keys of the underlying block cipher. Each of these two types of distinguishers uses two consecutive differentials; one is a related-key differential and the other one is a differential. However, we can extend the range of distinguishers by considering two consecutive related-key differentials. The distinguishers presented in [15] can be useful in analyzing block ciphers which have a good related-key differential followed by a good differential, or which have a good differential followed by a good related-key differential, while our distinguishers can be efficiently used in analyzing block ciphers which have a good related-key differential followed by another good related-key differential.

We now describe two related-key rectangle distinguishers based on two consecutive related-key differentials. Let $E : \{0,1\}^k \times \{0,1\}^n \rightarrow \{0,1\}^n$ be a block cipher that uses $\{0,1\}^k$ and $\{0,1\}^n$ as a key space and a plaintext/ciphertext space, respectively, and that is composed of a cascade $E = E^1 \circ E^0$, i.e., $E_K(P) = E^1_K \circ E^0_K(P)$.

A related-key rectangle distinguisher can be formed by building quartets of plaintexts $(P_i, P^*_i, P'_j, P'^*_j)$ that satisfy the below four differential conditions. Assume that P_i, P^*_i, P'_j and P'^*_j are encrypted by using keys K, K^*, K' and K'^*, respectively, where K, K^*, K' and K'^* are related to each other. Let I_i, I^*_i, I'_j and I'^*_j denote the intermediate encrypted values of P_i, P^*_i, P'_j and P'^*_j under E^0, respectively, and C_i, C^*_i, C'_j and C'^*_j denote the encrypted values of I_i, I^*_i, I'_j and I'^*_j under E^1, respectively. If the following four differential conditions are satisfied, we call such a quartet $(P_i, P^*_i, P'_j, P'^*_j)$ *a right quartet.*

- Differential Condition 1 : $P_i \oplus P^*_i = P'_j \oplus P'^*_j = \alpha$
- Differential Condition 2 : $I_i \oplus I^*_i = I'_j \oplus I'^*_j = \beta$
- Differential Condition 3 : $I_i \oplus I'_j = \gamma$
- Differential Condition 4 : $C_i \oplus C'_j = C^*_i \oplus C'^*_j = \delta$

In these four differential conditions, the α and the δ represent specific differences and the β and the γ represent arbitrary differences. Note that the differential conditions 2 and 3 allow us to get $I^*_i \oplus I'^*_j = \gamma$ with probability 1. See Fig. 1 for

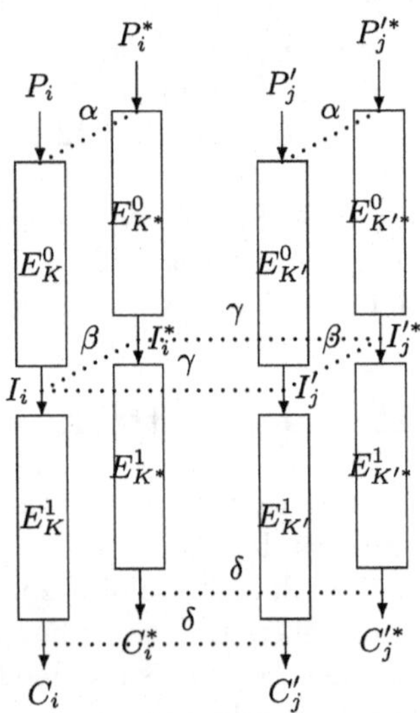

Fig. 1. A Related-Key Rectangle Distinguisher (A Right Quartet)

a description of such a right quartet. In Fig. 1, we set relations of K, K^*, K' and K'^* as follows: $K \oplus K^* = K' \oplus K'^* = \Delta K^*$ and $K \oplus K' = K^* \oplus K'^* = \Delta K'$, where the ΔK^* and the $\Delta K'$ represent specific key differences.

When does a right quartet described in Fig. 1 form a distinguisher? In order to answer this question, we first assume the following two related-key differentials of the E^0 and the E^1; for E^0 there exists a related-key differential $\alpha \rightarrow \beta$ with probability $p^*_{\alpha,\beta}$ and for E^1 there exists a related-key differential $\gamma \rightarrow \delta$ with probability $q^*_{\gamma,\delta}$. These assumptions mean that $p^*_{\alpha,\beta} = Pr_{X,K}[E^0_K(X) \oplus E^0_{K^*}(X^*) = \beta | X \oplus X^* = \alpha, K \oplus K^* = \Delta K^*]$, $q^*_{\gamma,\delta} = Pr_{X,K}[E^1_K(X) \oplus E^1_{K'}(X') = \delta | X \oplus X' = \gamma, K \oplus K' = \Delta K']$.

Assume that we have m_1 pairs of (P_i, P_i^*) and m_2 pairs of $(P_j', P_j'^*)$ with difference α. Then about $m_1 \cdot p^*_{\alpha,\beta}$ and $m_2 \cdot p^*_{\alpha,\beta}$ pairs satisfy the related-key differential $\alpha \rightarrow \beta$ for E^0. Thus we have about $m_1 \cdot m_2 \cdot (p^*_{\alpha,\beta})^2$ quartets satisfying the differential conditions 1 and 2. If we assume that the intermediate encryption values are distributed uniformly over all possible values, we get $I_i \oplus I_j' = \gamma$ with a probability 2^{-n}. This assumption enables us to obtain about $m_1 \cdot m_2 \cdot 2^{-n} \cdot (p^*_{\alpha,\beta})^2$ quartets satisfying the differential conditions 1, 2 and 3. As stated above, the differential conditions 2 and 3 allow us to get $I_i^* \oplus I_j'^* = \gamma$ with probability 1. Moreover, each of the pairs (I_i, I_j') and $(I_i^*, I_j'^*)$ satisfies the related-key differential $\gamma \rightarrow \delta$ for E^1 with probability $q^*_{\gamma,\delta}$. Therefore, the expected number of right quartets is

$$\sum_{\beta,\gamma} m_1 \cdot m_2 \cdot 2^{-n} \cdot (p^*_{\alpha,\beta})^2 \cdot (q^*_{\gamma,\delta})^2 = m_1 \cdot m_2 \cdot 2^{-n} \cdot (\hat{p^*_\alpha})^2 \cdot (\hat{q^*_\delta})^2 ,$$

where $\hat{p^*_\alpha} = (\sum_\beta (p^*_{\alpha,\beta})^2)^{\frac{1}{2}}$ and $\hat{q^*_\delta} = (\sum_\gamma (q^*_{\gamma,\delta})^2)^{\frac{1}{2}}$.

For a random permutation the expected number of right quartets is $m_1 \cdot m_2 \cdot 2^{-2n}$, since there are $m_1 \cdot m_2$ possible quartets and each of the pairs (C_i, C_j') and $(C_i^*, C_j'^*)$ satisfies the δ difference with probability 2^{-n}. Thus, $\hat{p^*_\alpha} \cdot \hat{q^*_\delta} > 2^{-n/2}$ must hold for the related-key rectangle distinguisher to work. This kind of distinguisher will be used in attaking 70-round SHACAL with 512-bit keys.

The above related-key rectangle distinguisher can be extended by considering a number of output differences for E^1. That is, we can use a related-key truncated differential for E^1 whose input difference is of γ and output difference is in a set $D \neq \emptyset$. $q^*_{\gamma,D}$ denotes the probability of this related-key truncated differential. In this case, the expected number of right quartets is

$$\sum_{\beta,\gamma} m_1 \cdot m_2 \cdot 2^{-n} \cdot (p^*_{\alpha,\beta})^2 \cdot (q^*_{\gamma,D})^2 = m_1 \cdot m_2 \cdot 2^{-n} \cdot (\hat{p^*_\alpha})^2 \cdot (\hat{q^*_D})^2 ,$$

where $\hat{q^*_D} = (\sum_\gamma (q^*_{\gamma,D})^2)^{\frac{1}{2}}$. In the case of a random permutation, the expected number of right quartets is $m_1 \cdot m_2 \cdot 2^{-2n} \cdot |D|^2$, since there are $m_1 \cdot m_2$ possible quartets and each of the pairs (C_i, C_j') and $(C_i^*, C_j'^*)$ satisfies one of the differences in a set D with probability $2^{-n} \cdot |D|$ where $|D|$ is the number of elements in D. Thus, $\hat{p^*_\alpha} \cdot \hat{q^*_D} > 2^{-n/2} \cdot |D|$ must hold for the related-key rectangle distinguisher to work. This kind of distinguisher will be used in attacking 8-round AES with 192-bit keys.

3 Related-Key Rectangle Attack on Reduced Rounds of SHACAL-1

Firstly, we briefly describe SHACAL-1. Secondly, we describe a 59-round related-key rectangle distinguisher of SHACAL-1 and use it to attack 70-round SHACAL-1.

3.1 A Description of SHACAL-1

The SHACAL-1 cipher [9] is a 160-bit block cipher based on the compression function of the hash standard SHA-1 [19]. It consists of 80 rounds and uses a variable key length up to 512 bits.

A 160-bit plaintext P is composed of five 32-bit words A,B,C,D and E. X_r denotes the value of 32-bit word X before the r-th round. According to this notation, the plaintext P is divided into A_0,B_0,C_0,D_0 and E_0, and the corresponding ciphertext C is divided into A_{80},B_{80},C_{80},D_{80} and E_{80}. The r-th round of encryption is performed as follows:

$$\begin{aligned}
A_{r+1} &= K_r + ROTL_5(A_r) + f_r(B_r, C_r, D_r) + E_r + Cst_r \\
B_{r+1} &= A_r \\
C_{r+1} &= ROTL_{30}(B_r) \\
D_{r+1} &= C_r \\
E_{r+1} &= D_r
\end{aligned}$$

for $r = 0, \cdots, 79$, where $ROTL_j(X)$ represents rotation of the 32-bit word X to the left over j bits, K_r is the round subkey, Cst_r is the round constant, and

$$\begin{aligned}
f_r(B_r, C_r, D_r) &= (B_r \& C_r) | (\neg B_r \& D_r), && (0 \le r \le 19) \\
f_r(B_r, C_r, D_r) &= B_r \oplus C_r \oplus D_r, && (20 \le r \le 39,\ 60 \le r \le 79) \\
f_r(B_r, C_r, D_r) &= (B_r \& C_r) | (B_r \& D_r) | (C_r \& D_r), && (40 \le r \le 59).
\end{aligned}$$

As stated above, SHACAL-1 supports a variable key length up to 512 bits. However, SHACAL-1 is not intended to be used with a key shorter than 128 bits. In case a shorter key than 512 bits is inserted in the cipher, the key is padded with zeros to a 512-bit string. Let the 512-bit key string be denoted $K = [K_0 || K_1 || \cdots || K_{15}]$, where each K_i is a 32-bit word. The key expansion of 512 bits K to 2560 bits is defined by

$$K_i = ROTL_1(K_{i-3} \oplus K_{i-8} \oplus K_{i-14} \oplus K_{i-16}), \quad (16 \le i \le 79) .$$

3.2 Attack on 70-Round SHACAL-1 with 512-Bit Keys

In the key schedule of SHACAL-1, fixing differences of any consecutive 16 round keys determines differences of the remaining 64 round keys. Indeed the key schedule of SHACAL-1 corresponds to a linear feedback shift register (LFSR) with left rotation. Moreover, the key schedule of SHACAL-1 has relatively low difference propagations. These weaknesses of the key schedule allow us to get two consecutive good related-key differential characteristics of SHACAL-1. That is, we can construct a 33-round related-key differential characteristic $\alpha \to \beta$ for rounds 0-32 (E^0) with probability 2^{-45} ($\approx p^*_{\alpha,\beta}$) and a 26-round related-key differential characteristic $\gamma \to \delta$ for rounds 33-58 (E^1) with probability 2^{-25} ($\approx q^*_{\gamma,\delta}$), where $\alpha = (0, e_{8,22,1}, e_{1,15}, e_{10}, e_{5,31})$, $\beta = (e_{1,5,15,30}, e_{10}, e_3, e_{30}, 0)$, $\gamma = (e_{1,8}, 0, e_{3,6,31}, e_{1,3,31}, e_{3,13,31})$ and $\delta = (e_{1,15}, e_{10}, e_3, e_{30}, 0)$. Here, e_i denotes a 32-bit word that has $0's$ in all bit positions except for bit i and $e_{i_1,\cdots,i_k}$ denotes $e_{i_1} \oplus \cdots \oplus e_{i_k}$. These two consecutive related-key differential characteristics are combined to construct our 59-round related-key rectangle distinguisher of SHACAL-1.

The first 33-round related-key differential characteristic is same as that of [15] (Sect. 4) except for the condition of plaintext pairs. The 33-round related-key differential characteristic presented in [15] exploits plaintext pairs for which 6 bits are fixed, while our related-key differential characteristic has plaintext pairs for which 10 bits are fixed as follows:

$$\begin{aligned} a_1 = a_1^* = 1,\ b_3 = b_3^* = 0,\ b_{10} = b_{10}^* = 1,\ b_{15} = b_{15}^* = 0,\ c_8 = c_8^* = 0\ , \\ c_{10} = c_{10}^* = 0,\ c_{22} = c_{22}^* = 0,\ d_8 = d_8^* = 0,\ d_{15} = d_{15}^* = 0,\ d_{22} = d_{22}^* = 0\ , \end{aligned} \tag{1}$$

where $P = (A, \cdots, E)$, $P^* = (A^*, \cdots, E^*)$ and x_i is the i-th bit of 32-bit word X. This stronger condition increases the probability of [15] by a factor of four. See Tables 2 and 3 in Appendix A for the details of this related-key differential characteristic and the associated key differences. As shown in Table 2, the difference of the master keys is $(e_{31}, e_{31}, e_{31}, e_{31}, 0, e_{31}, 0, e_{31}, 0, 0, 0, 0, 0, 0, 0, e_{31})$. Let ΔK^* denote this difference of the master keys and Δk^* denote the difference of keys for rounds 59 $\sim$ 69 depicted in Table 2. These two notations will be used in our attack algorithm.

The second 26-round related-key differential characteristic is very similar to that of [15] (Sect. 5). The related-key differential characteristic presented in [15] works through rounds 21-47, while our related-key differential characteristic works through rounds 33-58. Since the SHACAL-1 cipher uses a different f function every 20 rounds, the probability of our related-key differential characteristic is slightly different from that of [15]. See Tables 4 and 5 for the details of this related-key differential characteristic and the associated key differences. As shown in Table 4, the difference of the master keys is $(0, e_{31}, e_{31}, e_{30}, 0, e_{29,30,31}, e_{31}, 0, e_{31}, e_{29}, 0, e_{30}, 0, e_{30}, e_{31}, e_{30,31})$. Let $\Delta K'$ be this difference of the master keys and $\Delta k'$ be the difference for rounds 59 $\sim$ 69 depicted in Table 4.

According to [15] we can increase the lower bounds for $\hat{p^*_\alpha}$ and $\hat{q^*_\delta}$ to $2^{-44.17}$ and $2^{-24.08}$. These lower bounds are derived from taking into account as many related-key differential characteristics associated with $\hat{p^*_\alpha}$ or $\hat{q^*_\delta}$ as possible. Since the value $\hat{p^*_\alpha} \cdot \hat{q^*_\beta}$ ($\approx 2^{-68.25}$) is greater than 2^{-80}, our related-key differential characteristics can form a 59-round related-key rectangle distinguisher of SHACAL-1.

We are now ready to show how to exploit the above 59-round distinguisher to attack 70-round SHACAL-1. We assume that the 70-round SHACAL-1 cipher uses the master key K as well as the related keys K^*, K', K'^* with differences $K \oplus K^* = K' \oplus K'^* = \Delta K^*$ and $K \oplus K' = K^* \oplus K'^* = \Delta K'$. The following is an attack procedure for 70-round SHACAL-1.

Input: Two pools of $2^{149.75}$ plaintext pairs.
Output: Master key quartet (K, K^*, K', K'^*)

---

1. Choose two pools of $2^{149.75}$ plaintext pairs (P_i, P_i^*) and $(P_j', P_j'^*)$ with the difference α and 10-bit fixed values of (1). With a chosen plaintext attack, the P_i, P_i^*, P_j' and $P_j'^*$ are encrypted using the keys K, K^*, K' and K'^*, respectively, relating in the ciphertexts C_i, C_i^*, C_j' and $C_j'^*$. We keep all these ciphertexts in a table.

2. Guess a 352-bit key quartet (k, k^*, k', k'^*) for rounds 59-69 where $k^* = k \oplus \Delta k^*$, $k' = k \oplus \Delta k'$ and $k'^* = k^* \oplus \Delta k'$. For (k, k^*, k', k'^*) do the following:

2.1 For each i, decrypt C_i and C_i^* through rounds 69-59 using k and k^*, and denote the decrypted values by T_i and T_i^*. Let $T' = T_i \oplus \delta$ and $T'^* = T_i^* \oplus \delta$ and encrypt them through rounds 59-69 using k' and k'^* and denote the encrypted values by C' and C'^*. Find a j such that $(C_j', C_j'^*) = (C', C'^*)$.

2.2 If the number of (i, j) satisfying Step 2.1 is greater than or equal to 6, go to Step 3. Otherwise, go to Step 2.

3 For the suggested key k, do an exhaustive search for the remaining 160 key bits using trial encryption. During this procedure, if a 512-bit key satisfies three known plaintext and ciphertext pairs, output this 512-bit key, denoted by $\mathcal{K}$, as the master key K of 70-round SHACAL-1. We also output $\mathcal{K} \oplus \Delta K^*, \mathcal{K} \oplus \Delta K'$ and $\mathcal{K} \oplus \Delta K^* \oplus \Delta K'$ as the related keys K^*, K' and K'^*. Otherwise, go to Step 2.

This attack requires two pools of $2^{149.75}$ plaintext pairs and thus the data complexity of this attack is $2^{151.75}$ related-key chosen plaintexts. This attack also requires about $2^{156.08}$ ($=2^{151.75} \cdot 20$) memory bytes since the memory complexity of this attack is dominated by Step 1.

We now analyze the time complexity of this attack. The time complexity of Step 1 is $2^{151.75}$ 70-round SHACAL-2 encryptions. In Step 2.1, this attack seeks colliding quartets for all i, j which seems to require a great amount of time complexity. However, this procedure can be done efficiently by sorting the ciphertext pairs, $(C_j', C_j'^*)$'s by C_j''s. Hence the time complexity of Step 2.1 is dominated by the partial decryption/encryption procedure and thus the time complexity of Step 2 is about $2^{500.08}$ ($\approx 2^{352} \cdot 2^{151.75} \cdot \frac{1}{2} \cdot \frac{11}{70}$) on average. In order to estimate the time complexity of Step 3 we should check the expected number of wrong key quartets suggested in Step 2. In Step 2.1, the probability that for each wrong key quartet there exist at least 6 colliding quartets is about $2^{-132.49}$ ($\approx \sum_{i=6}^{t} \left(\binom{t}{i} \cdot (2^{-160 \cdot 2})^i \cdot (1 - 2^{-160 \cdot 2})^{t-i}\right)$) where $t = 2^{299.50}$ and t represents the number of all possible quartets generated by the two pools of $2^{149.75}$ plaintext pairs. From the above analysis we expect about $2^{218.51}$ ($\approx 2^{352} \cdot 2^{-132.49} \cdot \frac{1}{2}$) wrong key quartets on average which are suggested in Step 2 and thus Step 3 requires about $2^{378.51}$ ($\approx 2^{218.51} \cdot 2^{160}$) 70-round SHACAL-1 encryptions. Therefore, the time complexity of this attack is about $2^{500.08}$ 70-round SHACAL-1 encryptions.

In Step 3, the probability that each 512-bit wrong key is suggested is about 2^{-480} ($\approx 2^{-160 \cdot 3}$). It follows that the expected number of 512-bit wrong keys which are suggested in Step 3 is about $2^{-101.49}$ ($= 2^{-480} \cdot 2^{378.51}$). Thus, the possibility that the output of the above attack algorithm is a wrong key quartet is very low. Moreover, the expected number of right quartets is about 8 ($= (2^{149.75})^2 \cdot 2^{-160} \cdot (2^{-68.25})^2$) and thus the expected number of colliding quartets for the right key quartet is about 8. This is due to the fact $\hat{p_\alpha} \cdot \hat{q_\delta} \approx 2^{-68.25}$. Since the probability that for the right key quartet there exist at least 6 colliding quartets is about 0.80 ($\approx \sum_{i=6}^{t} (\binom{t}{i} \cdot (2^{-160} \cdot 2^{-68.25 \cdot 2})^i \cdot (1 - 2^{-160} \cdot 2^{-68.25 \cdot 2})^{t-i})$) where $t = 2^{299.50}$, the success rate of this attack is about 0.80.

4 Related-Key Rectangle Attack on Reduced Rounds of AES

Firstly, we briefly describe AES [7]. Secondly, we describe a 7-round related-key rectangle distinguisher of AES and use it to attack 8-round AES.

4.1 A Description of AES

AES encrypts data blocks of 128 bits with 128, 192 or 256-bit key. A round function of AES consists of four basic transformations as follows:

- ByteSub (BS): 8×8 S-box transformation
- ShiftRow (SR): Left rotation of each row
- MixColumn (MC): Matrix multiplication in each column
- AddRoundKey (KA): Key exclusive-or

Each round function of AES applies the BS, SR, MC and KA steps in order, but the MC is omitted in the last round. Before the first round, an extra KA step is applied. We call the key used in this step a whitening key. In this paper we concentrate on the 192-bit key version of the AES which is composed of 12 rounds. For more details of the above four transformations, refer to [7].

The 192-bit key schedule is described in Fig. 2. In Fig. 2, the whitening key is (W_0, W_1, W_2, W_3), the subkey of round 0 is (W_4, W_5, W_6, W_7), the subkey of round 1 is $(W_8, W_9, W_{10}, W_{11})$, $\cdots$, the subkey of round 11 is $(W_{48}, W_{49}, W_{50}, W_{51})$, where the 192-bit master key is $W_0||W_1|| \cdots ||W_5$ and W_i is a 32-bit word. The *Rcon* denotes fixed constants and the SubByte is a byte-wise S-box transformation and the RotByte represents one byte left rotation.

4.2 Attack on 8-Round AES -192

We describe two related-key truncated differentials on which our 7-round related-key rectangle distinguisher is based and then we present our related-key rectangle

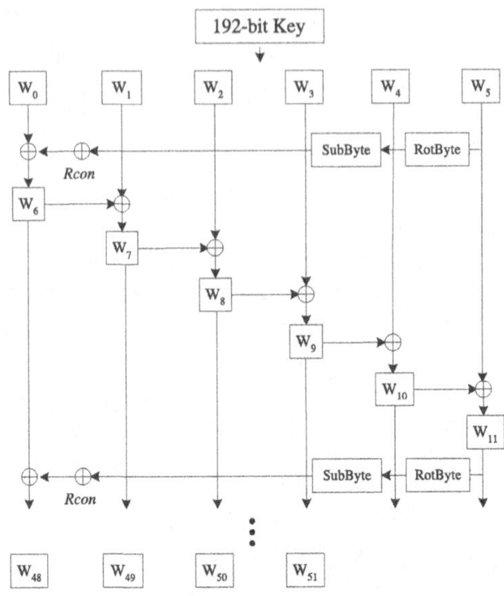

Fig. 2. AES Key schedule(KS) for 192-bit keys

attack on 8-round AES with 192-bit keys. Before describing the two related-key truncated differentials, we define some notations.

- $K_w, K_w^*, K_w', K_w'^*$: whitening keys generated from master keys K, K^*, K', K'^*, respectively.
- $K_i, K_i^*, K_i', K_i'^*$: subkeys of round i generated from master keys K, K^*, K', K'^*, respectively.
- a: a fixed nonzero byte value.
- b: output difference of S-box for fixed input difference a.
- $*$: a variable and unknown byte.
- $\Delta K^*, \Delta K', \Delta P^*, \Delta I'$: particular differences described in Figs. 3 and 4.
- $\Delta T, \Delta O$: particular difference set described in Fig. 4.
- $E_K(\cdot)$: 8-round AES encryption with key K.
- $E_K^0(\cdot)$: 4-round AES encryption from round 0 to round 3 with key K but excluding the exclusive-or with K_3.
- $E_K^1(\cdot)$: 3-round AES encryption from round 4 to round 6 with key K including the exclusive-or with K_3

Figs. 3 and 4 show our two related-key truncated differentials with probability 1. If the master key difference is ΔK^* (resp., $\Delta K'$), then the subkey difference in rounds 0-2 (resp., 3-6) is $\Delta K_w^*, \Delta K_0, \Delta K_1^*$ and ΔK_2^* (resp., $\Delta K_3', \Delta K_4', \Delta K_5'$ and $\Delta K_6'$) described in Fig. 3 (resp., Fig. 4).

Let K and K^* be two keys with difference ΔK^* and P and P^* be two plaintexts with difference ΔP^*. If the plaintexts P and P^* are encrypted under E_K^0 and $E_{K^*}^0$, respectively, then P and P^* satisfy the 4-round related-key truncated

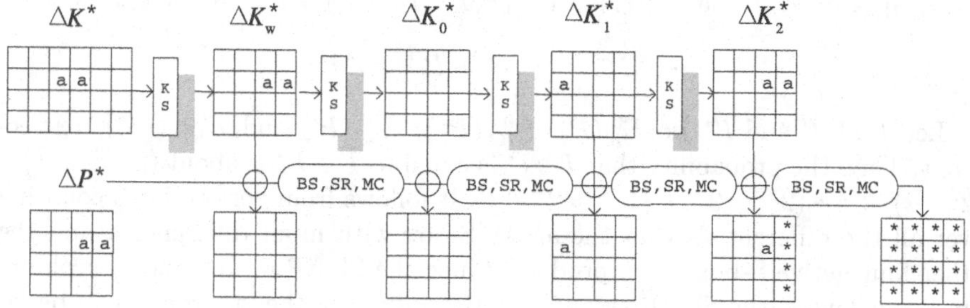

Fig. 3. The first related-key truncated differential for rounds 0-3 (E^0) of AES

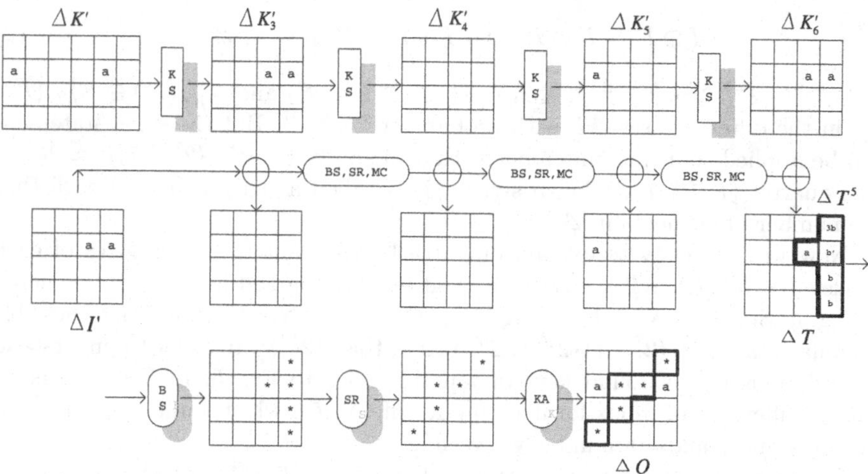

Fig. 4. The second related key truncated differential for rounds 4-6 (E^1) of AES

differential, described in Fig. 3. A similar argument can be applied to two keys, K and K', and two intermediate values, I and I'. Let K and K' be two keys with difference $\Delta K'$ and P and P' be two plaintexts. If $E^0_K(P) \oplus E^0_{K'}(P') = \Delta I'$, i.e., $I \oplus I' = \Delta I'$, then I and I' satisfy the 3-round related-key truncated differential described in Fig. 4. Note that the output difference of this 3-round differential is one of the elements in ΔT. In Fig. 4, b is an unknown variable which can be one of $2^7 - 1$ elements since the b is the output difference of the S-box for a given input difference a, and $b' = 2b \oplus a$.

As stated above, these two related-key truncated differentials can form a 7-round related-key rectangle distinguisher which has a relatively high probability. In order to compute the probability of this distinguisher we need the following two assumptions.

Assumption 1. The key quartet (K, K^*, K', K'^*) is related as follows;

$$K \oplus K^* = K' \oplus K'^* = \Delta K^*, \ K \oplus K' = K^* \oplus K'^* = \Delta K' .$$

Assumption 2. A plaintext quartet (P, P^*, P', P'^*) is related as follows;

$$P \oplus P^* = P' \oplus P'^* = \Delta P^* .$$

Let I, I^*, I' and I'^* be $E^0_K(P)$, $E^0_{K^*}(P^*)$, $E^0_{K'}(P')$ and $E^0_{K'^*}(P'^*)$, respectively. Then the probability that $I \oplus I^*$ is equal to $I' \oplus I'^*$ is about $(2^{-32} \cdot 2^{-7})^2 \cdot (2^7 - 2) \cdot 2^{32} + (2^{-32} \cdot 2^{-6})^2 \cdot 2^{32} \approx 2^{-38.97}$. It follows from performing a counting over all the differentials that the active S-box with input difference a and the other four active S-boxes can produce. Since the ShiftRow and the Mixcolumn are linear layers, the ShiftRow and the Mixcolumn of the last round can be ignored in computing the probability (See Fig. 3). Moreover the probability that $I \oplus I' = I^* \oplus I'^* = \Delta I'$ is 2^{-128} under the condition that $I \oplus I^* = I' \oplus I'^*$. So the probability that

$$I \oplus I^* = I' \oplus I'^* \text{ and } I \oplus I' = I^* \oplus I'^* = \Delta I' \qquad (2)$$

is $2^{-38.97} \cdot 2^{-128} = 2^{-166.97}$. Hence $E^1_K(I) \oplus E^1_{K'}(I')$ and $E^1_{K^*}(I^*) \oplus E^1_{K'^*}(I'^*)$ are in the difference set ΔT with probability $2^{-166.97}$. But the same statement can be applied to a random cipher with probability $(2^{-128} \cdot (2^7 - 1))^2 \approx 2^{-242}$. The quartet (P, P^*, P', P'^*) satisfying (2) is called a right quartet. Recall that the number of elements in ΔT is $2^7 - 1$.

Now we are ready to explain our attack. We want to find 5 bytes of each subkey K_7, K^*_7, K'_7, K'^*_7 whose byte positions are marked as $*$ on ΔO depicted in Fig. 4. Since the keys K, K^*, K' and K'^* are related, the number of possible key quartets is $2^{40} \cdot (2^7 - 1) \cdot 2^{16} \approx 2^{63}$ rather than $(2^{40})^4$. In order to understand the relations of the round keys of round 7 refer to Fig. 5. In Fig. 5, b is an output difference of S-box for fixed input difference a which can be one of $2^7 - 1$ elements and c and d are unknown varibles.

The basic idea of our attack is simple. Let (P, P^*, P', P'^*) be right quartet and (C, C^*, C', C'^*) be the corresponding ciphertext quartet. Define $D_k(\cdot)$ as a partial one round decryption with k, where k is a 5-byte key candidate of round 7. Then we guess a 5-byte key quartet (k, k^*, k', k'^*) and check that $D_k(C) \oplus D_{k'}(C') \in$

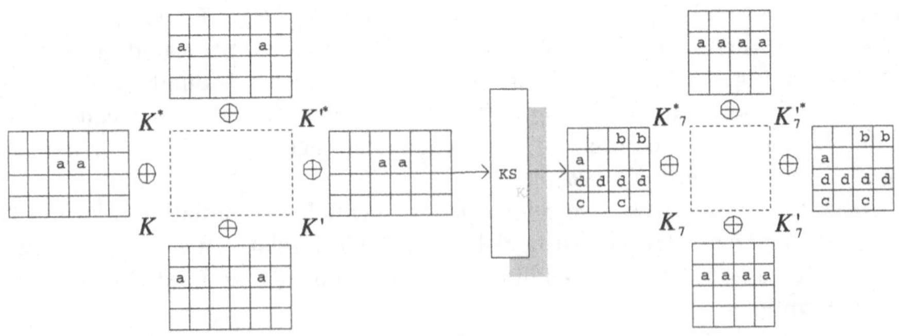

Fig. 5. Differential Property of 4 related keys for rounds 0-7 of AES

ΔT^5 and $D_{k^*}(C^*) \oplus D_{k'^*}(C'^*) \in \Delta T^5$ where ΔT^5 is a set described in Fig. 4. If the number of ciphertext quartets passing the above test is more than an appropriate threshold, we consider the guessed key quartet as the right one.

Input: Two pools of $2^{84.5}$ plaintext pairs.
Output: 5-byte key quartet of round 7.

---

1. Choose $2^{84.5}$ plaintext pairs (P_i, P_i^*) and $2^{84.5}$ plaintext pairs (P'_j, P'^*_j) with $P_i \oplus P_i^* = P'_j \oplus P'^*_j = \Delta P^*$. With a chosen plaintext attack, the P_i, P_i^*, P'_j, P'^*_j are encrypted using the keys K, K^*, K' and K'^*, respectively, relating in the ciphertexts C_i, C_i^*, C'_j and C'^*_j. We keep all these ciphertexts in a table.
2. Check that $C_i \oplus C'_j \in \Delta O$ and $C_i^* \oplus C'^*_j \in \Delta O$ for all i, j.
3. Guess a 5-byte key quartet (k, k^*, k', k'^*) for round 7.
  3.1 For all ciphertext quartets $(C_i, C_i^*, C'_j, C'^*_j)$ passing the test of Step 2, check that $D_k(C_i) \oplus D_{k'}(C'_j) \in \Delta T^5$ and $D_{k^*}(C_i^*) \oplus D_{k'^*}(C'^*_j) \in \Delta T^5$.
  3.2 If the number of quartets $(C_i, C_i^*, C'_j, C'^*_j)$ passing Step 3.1 is greater than or equal to 3, output the guessed key quartet (k, k^*, k', k'^*) as the right key quartet of round 7. Otherwise, go to Step 3.

This attack requires two pools of $2^{84.5}$ plaintext pairs and thus the data complexity of this attack is $2^{86.5}$ related-key chosen plaintexts. This attack also requires about $2^{90.83}$ $(= 2^{86.5} \cdot 20)$ memory bytes since the memory complexity of this attack is dominated by Step 1.

From the two pools of $2^{84.5}$ plaintext pairs we can make 2^{169} plaintext quartets. Step 2 requires $2^{84.5}$ searches of $2^{84.5}$ ciphertext pairs. This procedure can be done efficiently by sorting the ciphertext pairs, (C'_j, C'^*_j)'s by C'_j's. In Step 2, by assuming that the intermediate encryption values are distributed uniformly over all possible values we get $C_i \oplus C'_j \in \Delta O$ and $C_i^* \oplus C'^*_j \in \Delta O$ with probability 2^{-176} $(= 2^{-11 \cdot 8 \cdot 2})$ for a wrong quartet $(C_i, C_i^*, C'_j, C'^*_j)$. This probability follows from the fact that all elements of ΔO have a identically fixed value in 11 bytes. However, the difference set ΔO is induced by the difference set ΔT and the probability that each ciphertext quartet passes the test of Step 2 is same as that of our 7-round related-key rectangle distinguisher. Hence, the expected number of ciphertext quartets passing the test of Step 2 is about $2^{169} \cdot (2^{-166.97} + 2^{-176}) \approx 2^2$. Using this expected number we can estimate the time complexity of Step 3, i.e., Step 3 requires about 2^{63} $(= 2^{63} \cdot 2^2 \cdot 2^2 \cdot \frac{1}{8} \cdot \frac{1}{2})$ 8-round AES encryptions on average. Hence, the time complexity of this attack is dominated by Step 1 and thus this attack requires about $2^{86.5}$ 8-round AES encryptions.

For each wrong key quartet and each ciphertext quartet, the probability of passing the test of Step 3.1 is at most 2^{-6}. Note that the largest number in DC table of S-box used in AES is 4. This probability may occur when two of k, k^*, k', k'^* are correct and each of the rest two of them is correct except for one byte. In this case the probability that Step 3 outputs the guessed wrong key quartet is at most $(2^{-6})^3$. Since the number of these kinds of wrong key quartets is at most $2 \cdot 5 \cdot (2^8 - 1)$, the probability that Step 3 outputs such a wrong key quartet is at most 0.01. In this manner we can check all cases for wrong key quartets. For each of all other cases the probability that Step 3 outputs a wrong key quartet is much less than 0.01 and thus the probability that this attack outputs a wrong key quartet is approximately 0.01. Since the probability that for the right key quartet there exist at least 3 quartets passing the test of Step 3.1 is about 0.77 ($\approx \sum_{i=3}^{2^{169}} \binom{2^{169}}{i} (2^{-166.97})^i (1-2^{-166.97})^{2^{169}-i}$), the success rate of this attack is about 0.76 ($\approx 0.77 \cdot (1-0.01)$).

5 Conclusion

In this paper we proposed a new notion of related-key rectangle attack using 4 related keys and showed that it could break SHACAL-1 with 512-bit keys up to 70 rounds out of 80 rounds and AES with 192-bit keys up to 8 rounds out of 12 rounds. It is worthwhile to apply this attack to other block ciphers and to study simple key scheduling algorithms which may be resistant to this kind of attack.

References

1. E. Biham and A. Shamir, *Differential Cryptanalysis of DES-like Cryptosystems*, Proceedings of CRYPTO 1990, LNCS 537, pp. 2-21, Springer, 1990.
2. E. Biham, *New Types of Cryptanalytic Attacks Using Related Keys*, Proceedings of EUROCRYPT 1993, pp. 398-409, LNCS 765, 1993.
3. E. Biham, A. Biryukov and A. Shamir, *Cryptanalysis of skipjack reduced to 31 rounds using impossible differentials*, Proceedings of EUROCRYPT 1999, LNCS 1592, pp. 12-23, Springer, 1999.
4. E. Biham, O. Dunkelman and N. Keller, *The Rectangle Attack - Rectangling the Serpent*, Proceedings of EUROCRYPT 2001, LNCS 2045, pp. 340-357, Springer, 2001.
5. E. Biham, O. Dunkelman and N. Keller, *Rectangle Attacks on 49-Round SHACAL-1*, Proceedings of Fast Software Encryption 2003, LNCS 2887, pp. 22-35, Springer, 2003.
6. M. Blunden and A. Escott, *Related Key Attacks on Reduced Round KASUMI*, Proceedings of Fast Software Encryption 2001, LNCS 2355, pp. 277-285, Springer, 2002.
7. J. Daemen and V. Rijmen, *The Design of Rijndael*, Springer, 2002
8. N. Ferguson, J. Kelsey, B. Schneier, M. Stay, D. Wagner and D. Whiting, *Improved Cryptanalysis of Rijndael*, Proceedings of Fast Software Encryption 2000, LNCS 1978, pp. 213-230, Springer, 2001.
9. H. Handschuh and D. Naccache, *SHACAL*, Proceedings of NESSIE first workshop, Leuven, 2000.
10. P. Hawkes, *Differential-Linear Weak-Key Classes of IDEA*, Proceedings of EUROCRYPT 1998, LNCS 1403, pp. 112-126, Springer, 1998.
11. G. Jakimoski and Y. Desmedt, *Related-Key Differential Cryptanalysis of 192-bit Key AES Variants*, Proceedings of Selected Areas in Cryptography 2003, LNCS 3006, pp. 208-221, Springer, 2004.
12. J. Kelsey, B. Schneier and D. Wagner, *Key Schedule Cryptanalysis of IDEA, G-DES, GOST, SAFER, and Triple-DES*, Proceedings of CRYPTO 1996, LNCS 1109, pp. 237-251, Springer, 1996.

13. J. Kelsey, B. Schneir and D. Wagner, *Related-Key Cryptanalysis of 3-WAY, Biham-DES, CAST, DES-X, NewDES, RC2, and TEA*, Proceedings of International Conference on Information and Communications Seucrity 1997, LNCS 1334, pp. 233-246, Springer, 1997.
14. J. Kelsey, T. Kohno and B. Schneier, *Amplified Boomerang Attacks Against Reduced-Round MARS and Serpent*, Proceedings of Fast Software Encryption 2001, LNCS 1978, pp. 75-93, Springer, 2002.
15. J. Kim, G. Kim, S. Hong, S. Lee and D. Hong, *The Related-Key Rectangle Attack-Application to SHACAL-1*, Proceedings of International Conference on Information Security and Privacy 2004, LNCS 3108, pp. 123-136, Springer, 2004.
16. J. Kim, G. Kim, S. Lee, J. Lim and J. Song, *Related-Key Attacks on Reduced Rounds of SHACAL-2*, Proceedings of INDOCRYPT 2004, To appear.
17. J. Kim, D. Moon, W. Lee, S. Hong, S. Lee and S. Jung, *Amplified Boomerang Attack against Reduced-Round SHACAL*, Proceedings of ASIACRYPT 2002, LNCS 2501, pp. 243-253, Springer, 2002.
18. L.R. Knudsen, *Trucated and Higher Order Differentials*, Proceedings of Fast Software Encryption 1996, LNCS 1039, pp. 196-211, Springer, 1996.
19. U.S. Department of Commerce. *FIPS 180-1*: Secure Hash Standard, Federal Information Processing Standards Publication, N.I.S.T., April 1995.
20. S.K. Langford and M.E. Hellman, *Differential-Linear Cryptanalysis*, Proceedings of CRYPTO 1994, LNCS 839, pp. 17-25, Springer, 1994.
21. S. Lucks, *Attacking seven rounds of Rijndael under 192-bit and 256-bit keys*, Proceedings of AES3, NIST.
22. Y. Ko, S. Hong, W. Lee, S. Lee and J. Kang, *Related Key Differential Attacks on 26 Rounds of XTEA and Full Rounds of GOST*, Proceedings of Fast Software Encryption 2004, LNCS 3017, pp. 299-316, Springer, 2004.
23. D. Wagner, *The Boomerang Attack*, Proceedings of Fast Software Encryption 1999, LNCS 1636, pp. 156-170, Springer, 1999.

A The First Related-Key Differential Characteristic and the Associated Key Differences of SHACAL-1

Table 2. Key Differences Used in the First Related-Key Differential Characteristic

i	ΔK_i^*	i	ΔK_i^*	i	ΔK_i^*	i	ΔK_i^*	i	ΔK_i^*	i	ΔK_i^*	i	ΔK_i^*
0	e_{31}	10	0	20	0	30	0	40	e_3	50	$e_{3,7}$	60	$e_{3,7}$
1	e_{31}	11	0	21	0	31	e_0	41	e_4	51	e_5	61	$e_{2,4,7,9,10}$
2	e_{31}	12	0	22	0	32	e_1	42	0	52	e_7	62	$e_{3,7,11}$
3	e_{31}	13	0	23	0	33	0	43	$e_{1,3,4}$	53	e_8	63	$e_{2,3,4,9}$
4	0	14	0	24	0	34	e_1	44	e_5	54	0	64	$e_{3,5,11}$
5	e_{31}	15	e_{31}	25	0	35	e_2	45	$e_{2,3}$	55	$e_{3,5,7,8}$	65	$e_{3,12}$
6	0	16	0	26	0	36	0	46	e_5	56	e_9	66	$e_{3,5}$
7	e_{31}	17	0	27	0	37	$e_{2,3}$	47	$e_{1,2,6}$	57	$e_{2,5,6}$	67	$e_{3,5,6,9,11,12}$
8	0	18	0	28	0	38	e_3	48	e_{31}	58	e_9	68	e_{13}
9	0	19	0	29	e_0	39	e_1	49	$e_{3,5,6}$	59	$e_{2,3,5,6,10}$	69	$e_{3,5,9,10}$

Table 3. The First Related-Key Differential Characteristic

Round (i)	ΔA_i	ΔB_i	ΔC_i	ΔD_i	ΔE_i	ΔK_i^*	Prob.
0	0	$e_{8,22,1}$	$e_{1,15}$	e_{10}	$e_{5,31}$	e_{31}	
1	e_5	0	$e_{6,20,31}$	$e_{1,15}$	e_{10}	e_{31}	2^{-2}
2	0	e_5	0	$e_{6,20,31}$	$e_{1,15}$	e_{31}	2^{-5}
3	$e_{1,15}$	0	e_3	0	$e_{6,20,31}$	e_{31}	2^{-6}
4	0	$e_{1,15}$	0	e_3	0	0	2^{-3}
5	0	0	$e_{13,31}$	0	e_3	e_{31}	2^{-3}
6	e_3	0	0	$e_{13,31}$	0	0	2^{-3}
7	e_8	e_3	0	0	$e_{13,31}$	e_{31}	2^{-3}
8	0	e_8	e_1	0	0	0	2^{-2}
9	0	0	e_6	e_1	0	0	2^{-2}
10	0	0	0	e_6	e_1	0	2^{-2}
11	e_1	0	0	0	e_6	0	2^{-2}
12	0	e_1	0	0	0	0	2^{-1}
13	0	0	e_{31}	0	0	0	2^{-1}
14	0	0	0	e_{31}	0	0	2^{-1}
15	0	0	0	0	e_{31}	e_{31}	2^{-1}
16	0	0	0	0	0	0	1
⋮	⋮	⋮	⋮	⋮	⋮	⋮	⋮
28	0	0	0	0	0	0	1
29	0	0	0	0	0	e_0	1
30	e_0	0	0	0	0	0	2^{-1}
31	e_5	e_0	0	0	0	e_0	2^{-1}
32	e_{10}	e_5	e_{30}	0	0	e_1	2^{-2}
33	$e_{1,5,15,30}$	e_{10}	e_3	e_{30}	0		2^{-4}

B The Second Related-Key Differential Characteristic and the Associated Key Differences of SHACAL-1

Table 4. Key Differences Used in the Second Related-Key Differential Characteristic

i	$\Delta K'_i$	i	$\Delta K'_i$	i	$\Delta K'_i$	i	$\Delta K'_i$	i	$\Delta K'_i$	i	$\Delta K'_i$	i	$\Delta K'_i$
0	0	10	0	20	e_{31}	30	0	40	0	50	0	60	e_1
1	e_{31}	11	e_{30}	21	e_{30}	31	e_{31}	41	e_{31}	51	0	61	e_2
2	e_{31}	12	0	22	e_{31}	32	0	42	0	52	0	62	0
3	e_{30}	13	e_{30}	23	$e_{30,31}$	33	e_{31}	43	0	53	0	63	$e_{2,3}$
4	0	14	e_{31}	24	e_{31}	34	0	44	0	54	0	64	e_3
5	$e_{29,30,31}$	15	$e_{30,31}$	25	e_{30}	35	0	45	0	55	e_0	65	e_1
6	e_{31}	16	e_{31}	26	e_{31}	36	0	46	0	56	0	66	e_3
7	0	17	$e_{30,31}$	27	e_{31}	37	0	47	0	57	e_0	67	e_4
8	e_{31}	18	e_{31}	28	e_{31}	38	0	48	0	58	e_1	68	0
9	e_{29}	19	$e_{30,31}$	29	e_{31}	39	0	49	0	59	0	69	$e_{1,3,4}$

Table 5. The Second Related-Key Differential Characteristic

Round (i)	ΔA_i	ΔB_i	ΔC_i	ΔD_i	ΔE_i	$\Delta K'_i$	Prob.
33	$e_{1,8}$	0	$e_{3,6,31}$	$e_{1,3,31}$	$e_{3,13,31}$	e_{31}	
34	$e_{1,3}$	$e_{1,8}$	0	$e_{3,6,31}$	$e_{1,3,31}$	0	2^{-4}
35	0	$e_{1,3}$	$e_{6,31}$	0	$e_{3,6,31}$	0	2^{-4}
36	e_1	0	$e_{1,31}$	$e_{6,31}$	0	0	2^{-3}
37	e_1	e_1	0	$e_{1,31}$	$e_{6,31}$	0	2^{-2}
38	0	e_1	e_{31}	0	$e_{1,31}$	0	2^{-1}
39	0	0	e_{31}	e_{31}	0	0	2^{-1}
40	0	0	0	e_{31}	e_{31}	0	1
41	0	0	0	0	e_{31}	e_{31}	2^{-1}
42	0	0	0	0	0	0	1
⋮	⋮	⋮	⋮	⋮	⋮	⋮	⋮
54	0	0	0	0	0	0	1
55	0	0	0	0	0	e_0	1
56	e_0	0	0	0	0	0	2^{-1}
57	e_5	e_0	0	0	0	e_0	2^{-1}
58	e_{10}	e_5	e_{30}	0	0	e_1	2^{-3}
59	$e_{1,15}$	e_{10}	e_3	e_{30}	0		2^{-4}

How to Maximize Software Performance of Symmetric Primitives on Pentium III and 4 Processors

Mitsuru Matsui and Sayaka Fukuda

Information Technology R&D Center,
Mitsubishi Electric Corporation,
5-1-1 Ofuna Kamakura Kanagawa, Japan
{matsui, sayaka}@iss.isl.melco.co.jp

Abstract. This paper discusses the state-of-the-art software optimization methodology for symmetric cryptographic primitives on Pentium III and 4 processors. We aim at maximizing speed by considering the internal pipeline architecture of these processors. This is the first paper studying an optimization of ciphers on Prescott, a new core of Pentium 4. Our AES program with 128-bit key achieves 251 cycles/block on Pentium 4, which is, to our best knowledge, the fastest implementation of AES on Pentium 4. We also optimize SNOW2.0 keystream generator. Our program of SNOW2.0 for Pentium III runs at the rate of 2.75 μops/cycle, which seems the most efficient code ever made for a real-world cipher primitive. For FOX128 block cipher, we propose a technique for speeding-up by interleaving two independent blocks using a register group separation. Finally we consider fast implementation of SHA512 and Whirlpool, two hash functions with a genuine 64-bit architecture. It will be shown that new SIMD instruction sets introduced in Pentium 4 excellently contribute to fast hashing of SHA512.

1 Introduction

Recent microprocessors, especially Intel processors, have long pipeline stages to raise clock frequency, which, on the other side, often leads to new performance penalty factors. Now it is not rare that a program runs slower on a newer processor with a higher clock frequency. It seems that the clock-raising of modern processors is approaching to its margin. That is, for maximizing performance of a software program on a processor, it is becoming increasingly important for programmers to understand its hardware architecture and programming techniques specific to the processor.

This paper deals with Intel Pentium III and 4 processors, which are most widely used in modern PCs, and studies methodology for optimizing speed of recently proposed symmetric ciphers and hash functions on these processors. Intel recently shipped a new Pentium 4 (Prescott) with an architecture different from the previous Pentium 4 (Willamette, Northwood) under the same name.

H. Gilbert and H. Handschuh (Eds.): FSE 2005, LNCS 3557, pp. 398–412, 2005.

Since we often have to discuss these different cores separately, we call the old cores Pentium 4-N, and the new core Pentium 4-P to distinguish them.

First, in section 2, we refer to Gladman's code of Serpent block cipher [9] to see to what extent an architecture of microprocessors affects performance of the same code. It will be seen that even if everything is on the first level cache, the number of execution cycles of a given code significantly varies on a processor with a different version. Then we briefly summarize structural characteristics of Pentium III and 4. It should be noted that Pentium 4 has successfully raised its clock frequency by increasing the number of pipeline stages, but in return, SIMD instructions work only in a longer latency on this processor.

In section 3, we show how to measure a speed of a target assembly code on these processors. We have adopted a common method for the measurement; i.e. we count the number of clock cycles of a target subroutine using an internal timer of the processor. However in repeating our measurement experiments, we have found that an accurate measurement of execution cycles is not a simple issue as it looks on Pentium 4 particularly with Hyperthread Technology. Since this is a separate matter of interest but a less cryptographic topic, we will give a further observation in an appendix.

In subsequent sections, we specifically discuss software optimization techniques for symmetric cryptographic primitives. Our first target algorithm is AES [6]. The structure of AES is very suitable for 32-bit processors, but if we aim at ultimate performance, a dependency chain and a register starvation are likely a bottleneck of the speed. We carefully selected and arranged registers and instructions, and as a result, our optimized code with 128-bit key runs in 251 cycles/block on Pentium 4, which is, to our best knowledge, the fastest implementation of AES on Pentium 4.

The next algorithm is stream cipher SNOW2.0 [4]. This algorithm has two highly independent functions inside, and hence is suitable for superscalar processors. We derive a possible minimum number of μops on Pentium III and 4, and show that this number can be achieved in practice. Our program generates a key stream very efficiently at the rate of 2.75 μops/cycle on Pentium III, which is very close to the structural limit of Pentium III and 4 (3 μops/cycle), and is, as far as we know, the most efficient code designed for a read-world cipher primitive.

We also give the first performance analysis of FOX128 block cipher [11]. Since this cipher has an 8-byte$\times$8-byte matrix inside, we should use the 64-bit MMX registers, but due to a long dependency chain, a straightforward program runs inefficiently. However, fortunately this algorithm does not require many registers, and we can improve performance by assigning two independent register sets to two independent blocks respectively and interleaving the two codes in an internal block loop. It will be seen that this technique excellently improves the speed of FOX128.

Finally we deal with hash functions SHA512 [7] and Whirlpool [3]. These hash functions have a genuine 64-bit structure, and use of 64-bit MMX instructions is essential. Nakajima et al. [14] studied performance analysis of these hash functions on Pentium III, but due to missing "64-bit add" instructions, SHA512 had a heavy performance penalty on Pentium III. This paper first gives detailed performance analysis of SHA512 and Whirlpool on Pentium 4. We show that a two-block parallel implementation (in the sense of [14]) using the 128-bit XMM registers significantly boosts its hashing speed.

All the results shown in this paper were obtained using the following PCs.

Table 1. Our reference machines and environments

Processor	Pentium III	Pentium 4	Pentium 4
Core	Coppermine	Northwood	Prescott
Clock	800MHz	2.0GHz	2.8GHz
Hyperthread	no	no	yes
Memory	256MB	1GB	512MB
OS	Windows 2000	Windows XP Professional	Windows XP Professional
Compiler	Microsoft Visual Studio .NET 2003/Macro Assembler Version 7		

2 Pentium III and 4 Processors

2.1 Pentium III and 4 at a Glance

Table 2 shows our performance measurement results of Gladman's implementation [9] of Serpent block cipher [1] written in an assembly language. In [9] we can find two assembly language source codes: one is coded using 32-bit x86 instructions only (Program 1) and the other encrypts two blocks in parallel using 64-bit MMX SIMD instructions, where the first block is put on the upper 32-bit half of the MMX registers and the second block on the lower half (Program 2). This parallel encryption technique works well because Serpent was designed so that the entire algorithm could be efficiently implemented using 32-bit logical and shift operations only. This implementation technique can be used for encrypting not only two independent message streams but also one single stream with a non-feedback mode of operation such as a counter mode. In addition, we modified Program 2 to enable us to encrypt four blocks in parallel using 128-bit XMM SIMD instructions (Program 3); this translation is very straightforward.

Table 2. Encryption speed of Gladman's Serpent codes (cycles/block)

	Pentium III	Pentium 4-N	Pentium 4-P
Program 1 (32-bit code)	773	1267	689
Program 2 (64-bit code)	570	1052	1119
Program 3 (128-bit code)	-	656	681

Program 1 is very slow on Pentium 4-N; this is probably because 32-bit shift instructions have long latencies (4 or 5 cycles) on this processor, which was later improved on the new Prescott core (more than one shift in one cycle). Program 2 runs faster on Pentium III but not twice as fast, because we need four instructions to do a rotate shift on the MMX registers. The reason why Program 2 is again so slow on Pentium 4 is totally different; the MMX units of Pentium 4 work only in half speed. On the other side, Program 3 is fast as expected due to the SIMD computation, as compared with Program 2. (Program 3 does not work on Pentium III because Pentium III does not have 128-bit XMM shift instructions).

Note that these programs are not optimized for Pentium 4, and hence this table should not be seen as a maximal performance figure of Serpent. It was intended to show a typical example where the same code runs in a totally different efficiency on a processor with a different version. Table 2 clearly shows that a selection of a processor and a careful optimization on the processor are critically important for maximizing performance.

2.2 Pentium III and 4 a Bit More

We here sketch structural characteristics of Pentium III and 4 for later sections. For more details about the internal architecture and optimization tips of Pentium III and 4-N, see an excellent article written by Agner Fog [8], which tells us much more than any published documents about these processors.

[Pentium III] One of the biggest stall factors of Pentium III comes from the decoding stage, where a sequence of x86 instructions is broken down into RISC-style micro operations (μops). This break-down rule is quite complex, and a programmer must carefully arrange the order of instructions in order not to suffer a stall in this stage.

The executing stage has five independent pipes p0–p4, where p0 and p1 handle arithmetic and logical μops, p2 is used for reading from memory, and p3/p4 are used for writing to memory. This means that to aim at 3 μops/cycle, which is the maximal execution rate of Pentium III, we have to assign at least one μop out of three μops to memory read/write.

[Pentium 4 Northwood] In Northwood, instructions are cached after decoding. This means that the decoding stage is no longer a bottleneck of speed, assuming that the size of a critical loop is sufficiently small. An important feature of Northwood is that execution units for simple 32-bit μops run in double speed, but those for 64-bit/128-bit SIMD μops work only in half speed.

A special penalty of Northwood comes from 32-bit shift instructions, which have long latencies, typically 4 or 5 cycles. Also reading from memory to the MMX/XMM registers is very slow, taking approximately 8 cycles, according to [8]. The maximal execution rate of this processor remains 3 μops/cycle.

[Pentium 4 Prescott] This new core of Pentium 4 has not been well documented. The speed of a 32-bit shift instruction is greatly improved; more than one shift can be issued in a single cycle (but not exactly two shifts in our experiments, unlike what Intel's manual says). This is a good news.

However, many μops of Prescott have longer latencies than those of Northwood due to a deeper pipeline of this processor. The latency of a 32-bit load and an xor, for instance, is 4 and 1, respectively (2 and 0.5 for Northwood), which can be a new performance constraint.

Table 3 summarizes major differences of Pentium III and 4. The sixth and seventh rows show a latency of a sequence of two instructions, whose results were obtained by our own experiments. This type of sequence often appears on a dependency chain of block cipher codes.

Table 3. Pentium III vs. Pentium 4

	Pentium III	Pentium 4-N	Pentium 4-P
Pipeline Stages	10	20	32
L1 data cache	16KB	8KB	16KB
32-bit load latency/throughput	3/1	2/1	4/1
32-bit xor latency/throughput	1/0.5	0.5/0.5	1/0.5
32-bit shift latency/throughput	1/1	4/1	>0.5/1
`mov ebx,TABLE[eax] / mov eax,ebx`	4 cycles	3 cycles	5 cycles
`movq mm0,TABLE[eax] / movd eax,mm0`	4 cycles	13 cycles	18 cycles

3 How to Measure Execution Cycles

A common method for measuring a speed of a piece of code is to insert the code to be measured between two CPUID-RDTSC sequences, where CPUID flushes the pipeline and RDTSC reads processor's internal clock value as follows:

```
xor     eax,eax
cpuid
rdtsc
mov     CLK1,eax
xor     eax,eax
cpuid

FUNCTION(..., int block)

xor     eax,eax
cpuid
rdtsc
mov     CLK2,eax
xor     eax,eax
cpuid
```

Code 1. Measurement of FUNCTION

```
xor     eax,eax
cpuid
rdtsc
mov     CLK3,eax
xor     eax,eax
cpuid

; nothing here

xor     eax,eax
cpuid
rdtsc
mov     CLK4,eax
xor     eax,eax
cpuid
```

Code 2. Measurement of overhead

Clearly CLK2-CLK1 shows clock cycles from line 4 to line 11, but this value contains an overhead of measurement itself, which corresponds to CLK4-CLK3.

Hence we define the speed of FUNCTION as ((CLK2-CLK1)-(CLK4-CLK3))/block (cycles/block). In our reference PCs, the overhead CLK4-CLK3 is 214, 632 and 847 cycles for Pentium III, 4-N and 4-P, respectively.

In practice, the measured number of cycles varies due to various reasons. We hence ran the code above many times and adopted an average value, not a minimal value, as the speed of FUNCTION. For more details about the measurement issue on Pentium 4, see appendix.

4 AES

The first example of our implementation is AES. For the description and notations of the AES algorithm, refer to [6]. The fastest AES codes currently known on Pentium III and Pentium 4-N were designed by Lipmaa [12][13]. However no information about his implementation details has been published. Our implementation below is hence independent of his works.

AES is a typical cipher from an implementation viewpoint in the sense that we have to make use of data registers also as address registers alternatively on its critical path, which means that a dependency chain is likely a performance bottleneck. A common x86 code of one round of AES consists of (1) a four-time repetition of the following sequence (with different input registers), which corresponds to Subbytes+Shiftrows+Mixcolumns, and (2) four xors, which correspond to AddRoundKey. Note that, while the final round of AES is different from other rounds, it can be implemented using the same sequence below with another tables, which will be referred as table5 to table8. We hence need a total of 8KB memory for the lookup tables.

```
movzx    esi,al                    ; lowest byte of input eax
mov/xor  register1,table1[esi*4]   ; first table lookup (4 byte)
movzx    esi,ah                    ; second byte of input eax
mov/xor  register2,table2[esi*4]   ; second table lookup (4 byte)
shr      eax,16                    ; move higher 16 bits to lower side
movzx    esi,al                    ; third byte of input eax
mov/xor  register3,table3[esi*4]   ; third table lookup (4 byte)
movzx    esi,ah                    ; highest byte of input eax
mov/xor  register4,table4[esi*4]   ; fourth table lookup (4 byte)
```

Code 3. An example of 1/4 component of Subbytes+Shiftrows+Mixcolumns (mov for the first time and xor for the second to the forth times)

In an actual assembly program, how to minimize the latency of one round sequence is not trivial due to a "register starvation". Since we need four one-byte components of registers1 to register4 in the next round, these four registers should be eax, ebx, ecx and edx, but this is impossible without saving/restoring at least one input register in each round, which requires additional instructions. Assigning 64-bit MMX registers to registers1 to 4 also requires additional instructions for copying them to eax, ebx, ecx and edx for the next round, since we can not direct extract a byte of an MMX register to an x86 register.

[Pentium III] Our implementation on Pentium III uses four specially arranged lookup tables. These tables have an 8-bit input and a 64-bit output, where `table1` to `table4` (in **Code 3**) are put on the lower 32-bit half of each entry of our new tables, and `table5` to `table8` for the final round are put on the higher 32-bit half of the entry. We also assign two x86 registers and two MMX registers to `register1` to `register4` for all rounds except the final round, and assign four MMX registers to all of `register1` to `register4` in the final round. `movq/pxor` instructions are used to access MMX registers.

This lookup table structure contributes to reducing code size and hence increasing decoding efficiency. This structure also works very well in the final round, because the output of the final round no longer has to be copied to x86 registers and can be treated as full 64-bit data, as shown below. Our implementation of AES with 128-bit key on this strategy runs at the speed of 232 cycles/block in our measurement policy. When the block loop overhead (see appendix) is taken into consideration, this is almost the same performance as Lipmaa's best known result.

```
punpckhdq  mm0,mm1                    ; two upper 32-bit -> 64-bit
punpckhdq  mm2,mm3                    ; two upper 32-bit -> 64-bit
pxor       mm0, Final_Subkey1         ; AddRoundKey (8 bytes)
pxor       mm2, Final_Subkey2         ; AddRoundKey (8 bytes)
movq       [memory+0], mm0            ; store ciphertext (8 bytes)
movq       [memory+8], mm2            ; store ciphertext (8 bytes)
```

Code 4. Our AES code after the final round

[Pentium 4] Since MMX memory instructions have a very long latency on Pentium 4 (for both Northwood and Prescott), we have to write a code using x86 registers and instructions only. In addition, using a high 8-bit partial register, such as `ah`, leads to a special penalty on Pentium 4, while no penalty takes place in using a low 8-bit partial register. Specifically, `movzx esi,ah` is decomposed into two μops on Pentium 4 unlike Pentium III. **Code 1** uses this type of instruction twice, but one of them can be avoided by changing the last two lines as follows:

```
shr       eax,8                      ; only one uop (upper 24 bits = 0)
mov/xor   register4, table4[eax*4]   ; fourth table lookup
```

Code 5. Modification of Code 1 for Pentium 4

For Northwood, which has only 8KB L1 data cache, we reduced the size of our lookup tables to 6KB by removing `table5` and `table6` of the final round without increasing the number of instructions. This is possible by using a `movzx` instruction as shown in Table 4 (note that Pentium is a little-endian processor). As a result, our code runs at the speed of 251 cycles/block on Pentium 4 Northwood. This is, as far as we know, the fastest implementation of AES on Pentium 4. On the other hand, our implementation on Prescott is unfortunately slower than that on Northwood. We feel that this is due to a high latency of load instructions (4 cycles for Prescott and 2 cycles for Northwood).

Table 5 summarizes our performance results. Our codes have the following interface and we assume that the subkey has been given in the third argument. We set `block` to 128 or 256, which was fastest in our environments. We did not use any static memory except read-only lookup tables.

```
FUNCTION( uchar *plaintext, uchar *ciphertext, uint *subkey, int block )
```

Table 4. Reduction of lookup tables of the final round

	Operation	Instruction	Table Size
`table5`	`x` $\rightarrow$ (0‖0‖0‖`S[x]`)	`movzx Register,BYTE PTR table8+3[esi*4]`	0
`table6`	`x` $\rightarrow$ (0‖0‖`S[x]`‖0)	`movzx Register,WORD PTR Table8+2[esi*4]`	0
`table7`	`x` $\rightarrow$ (0‖`S[x]`‖0‖0)	`mov Register,table7[esi*4]`	1KB
`table8`	`x` $\rightarrow$ (`S[x]`‖0‖0‖0)	`mov Register,table8[esi*4]`	1KB

Table 5. Our implementation results of AES

	Pentium III	Pentium 4-N	Pentium 4-P
μops/block	596	654	654
cycles/block	232	251	284
cycles/byte	14.5	15.7	17.8
μops/cycles	2.57	2.61	2.30

5 SNOW2.0

Our next example of implementation is stream cipher SNOW2.0, which was designed by Ekdahl and Johansson and presented at SAC2002 [4]. SNOW2.0 was intended to overcome a slight weakness of its earlier version, which was initially submitted to the NESSIE project [15]. SNOW2.0 is based on a firm theoretical background and is very fast. It is now under discussion for an inclusion in the next version of the ISO/IEC 18033 standard. Our paper gives the first detailed performance analysis of SNOW2.0 in an assembly language.

Figure 1 illustrates the keystream generation algorithm of SNOW2.0, which consists of sixteen 32-bit registers s_i with feedback mechanism with two 32-bit memories $R1$ and $R2$. α and α^{-1} are multiplicative constants over $GF(2^{32})$, and S is an AES-like 4-byte$\times$4-byte matrix multiplication. Clearly SNOW2.0 strongly targets at 32-bit processors from the implementation point of view.

According to the authors' document, α and α^{-1} were chosen so that the multiplications with these values over $GF(2^{32})$ could work efficiently using two pre-calculated tables `MUL_a` and `MUL_ainv` as follows:

$$\texttt{s} * \alpha = \texttt{(s << 8) xor MUL\_a[s >> 24]}$$
$$\texttt{s} * \alpha^{-1} = \texttt{(s >> 8) xor MUL\_ainv[s \& 0xff]}$$

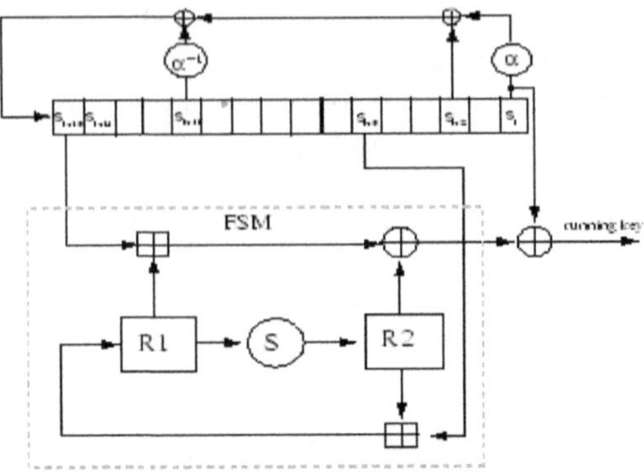

Fig. 1. SNOW 2.0

The straightforward implementation of the first operation (multiplication with α) requires four instructions (five μops), but we have found that it can be done with fewer instructions as shown below by preparing a new table `MUL_a2` such that `MUL_a2[x] = MUL_a[x] ^ (a & 0xff)`. Since we need 2KB for `MUL_a2` and `MUL_ainv` and 4KB for S, the total size of the lookup tables is 6KB, which fits in L1 data cache of both Pentium III and Pentium 4.

```
rol    eax,8                      movzx  esi,al
movzx  esi,al                     shr    eax,8
xor    eax,MUL_a2[esi*4]          xor    eax,MUL_ainv[esi*4]
```

Code 6. Multiplication with α (left) and α^{-1} (right)

The structure of SNOW2.0 is very suitable for superscalar processors, since the LFSR part (the upper half of Figure 1) and the FSM part (the lower half) can be carried out mostly independently. For fast implementation of SNOW2.0, we should treat sixteen consecutive LFSR clocks as "one round" as suggested by the designers, which enables us to skip copy operations on the sixteen 32-bit registers. We implemented the keystream generation algorithm SNOW2.0 in an assembly language for Pentium III and Pentium 4, respecting the subroutine interface given by designers' C codes at [5]. We simply added an additional variable "block" in the second argument, so that the routine can generate block*64 keystream bytes at one subroutine call as follows (the state information on s_i, $R1$ and $R2$ are passed as static variables):

```
FUNCTION( uint *keystream_block, int block )
```

Our code requires 34 and 33 μops in one LFSR clock, i.e. in every four-byte keystream generation, on Pentium III and Pentium 4, respectively. We think that this already reaches the theoretical minimum number of μops. Table 6 shows a detailed breakdown of the μops of our code. We read s_t twice for reducing the latency of the LFSR part, which is the reason why the number of μops of "read from LFSR" is 5 (not 4 as naturally expected from Figure 6), and the number of μops of "s * α" is 3 (not 4).

Table 7 gives performance of our assembly codes. Our code runs at the speed of 203 cycles/block on Pentium 4 Northwood, which is 30% faster than designers' optimized C code. The remarkable aspect of our code is its high parallelism. For example, our program on Pentium III works at the rate of 2.75 μops/cycle, which is, as far as we know, the most efficient code that was achieved in a real cryptographic primitive. This result also shows that SNOW2.0 is essentially faster than RC4. If we take a close look at the structure of RC4, we will see that RC4 requires at least 10 μops/byte including three reads and three writes. Hence even if we assume that an RC4 code works in 2.80 μops/cycle it takes at least 3.6 cycles/byte (much more in practice), while our SNOW2.0 code is running in 3.1–3.4 cycles/byte.

Table 6. μops breakdown in one LFSR clock

	S	s * α	s * α^{-1}	read from LFSR	write to LFSR/keystream	xor/add	Total μops
Pentium III	12	3	4	5	4	6	34
Pentium 4	13	3	4	5	2	6	33

Table 7. SNOW2.0 key generation speed

	Pentium III	Pentium 4-N	Pentium 4-P
μops/block	550	534	534
cycles/block	200	203	215
cycles/byte	3.13	3.17	3.36
μops/cycles	2.75	2.63	2.48

6 FOX128

FOX is a family of block ciphers, which was recently proposed by Junod and Vaudenay [11]. Here we treat a "generic version" of 128-bit block cipher FOX128 with 16 rounds. The left part of Figure 2 illustrates the round function of FOX128, and the right part gives the details of the $f64$ function in the round function. The $f64$ function consists of a sequence of (1) key xor, (2) eight parallel *sbox* lookups, (3) 8-byte × 8-byte matrix $mu8$, (4) key xor again, and (5) eight parallel *sbox* lookups again. This is essentially a 64-bit structure, suitable for use of the MMX instructions on Pentium III and 4.

The straightforward implementation of the $f64$ function requires eight 2KB tables for the first sbox layer and $mu8$, and additional one to four 1KB tables for the second sbox layer. If we take a close look at the $mu8$ matrix, it is easily seen that we can reduce one 2KB table in the first layer at the cost of three or

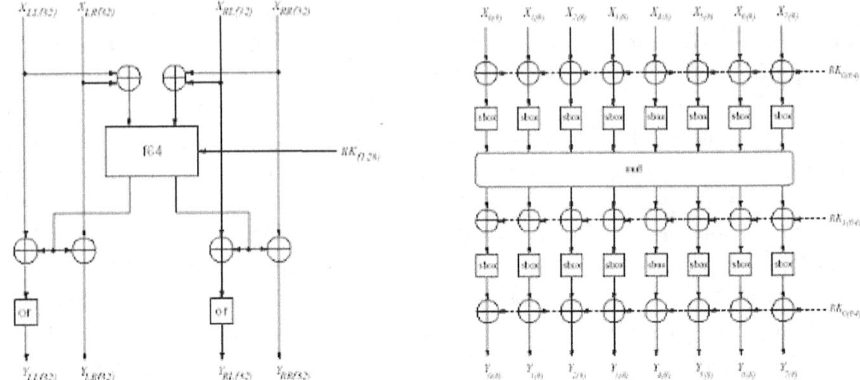

Fig. 2. FOX128

four additional MMX μops. Our first implementation fully uses MMX registers and MMX instructions in the main stream and in the $f64$ function, reducing the table size to a total of 15KB (14KB/1KB for the first/second layer) so that the entire tables are covered within 16KB. This runs at the speed of 692 cycles/block on Pentium III, which is approximately 20% faster than designers' optimized C implementation. However, this program becomes very slow in Pentium 4 because the $f64$ function has a long dependency chain and moreover Northwood suffers a lot of cache miss penalties.

On the other hand, our implementation is free from a "register starvation"; that is, four out of the eight MMX registers are enough to implement the entire cipher, which means that half of the MMX registers can remain free. This leads us to a possibility of another parallel implementation technique "register group separation". Specifically, we assign four MMX registers to one message stream and the remaining four to another message stream, and interleave the two independent codes inside a block loop. This technique is expected to contribute to an efficient use of superscalar pipelines and improve an overall performance accordingly. Our code remarkably reduces the number of execution cycles on Pentium 4. In particular, the improvement on Northwood is prominent; although the

Table 8. Our implementation results of FOX128

	Pentium III		Pentium 4-N		Pentium 4-P	
	(I)	(II)	(I)	(II)	(I)	(II)
μops/block	1269	1388	1395	1505	1395	1505
cycles/block	692	622	1986	1187	1481	981
cycles/byte	43.3	38.9	124.1	74.2	92.6	61.3
μops/cycle	1.83	2.23	0.70	1.27	0.94	1.53

Northwood core is still paying the penalty of cache misses, its performance is now very close to that on Prescott.

Table 8 summarizes performance measurement results of our FOX128 codes, where (I) and (II) show the straightforward method and the two-block parallel method, respectively. We adopted the same subroutine interface and coding policy as that of the AES block cipher. In general, it is difficult to apply the register group separation technique to codes using x86 registers only, but new registers such as MMX and XMM have opened up a new possibility of this parallel computation technique.

7 SHA512 vs. Whirlpool

We here discuss two genuine 64-bit hash functions SHA512 [7] and Whirlpool [3], both of which are now under consideration for an inclusion in the next version of the ISO/IEC 18033 standard. These algorithms well suit for 64-bit processors and it is expected that the 64-bit MMX instructions can be efficiently used for gaining performance. Nakajima et al. [14] discussed speed of these hash functions on Pentium III, and reported that Whirlpool is slightly faster than SHA512.

SHA512 suffered heavy penalty cycles on Pentium III because Pentium III does not have an instruction for 64-bit addition, which is an essential operation of this algorithm. Pentium 4 solves this problem, but a high latency of MMX memory instructions can be a possible penalty factor. On the other side, we can hash two independent messages in parallel using 128-bit XMM instructions, which is expected to boost the hashing speed.

The structure of Whirlpool is similar to AES. It uses an 8-byte $\times$ 8-byte matrix (a 4-byte $\times$ 4-byte matrix for AES), and hence a straightforward implementation requires eight 2KB tables. Our coding method is basically the same

Table 9. Our implementation results of SHA512

1block = 128bytes	Pentium III	Pentium 4-N		Pentium 4-P	
	single	single	double	single	double
μops/block	13924	8710	4363	8710	4363
cycles/block	5148	4666	2826	5294	3111
cycles/byte	40.2	36.5	22.1	41.4	24.3
μops/cycles	2.70	1.87	1.54	1.65	1.40

Table 10. Our implementation results of Whirlpool

1block = 64bytes	Pentium III	Pentium 4-N	Pentium 4-P
μops/block	5206	5526	5526
cycles/block	2061	3024	2319
cycles/byte	32.2	47.3	36.2
μops/cycles	2.53	1.83	2.38

as [14]; that is, we have only four tables and generate other data when necessary using the `pshufw` (word shuffling) instruction.

Table 9 and Table 10 show our performance figures of SHA512 and Whirlpool, respectively. We also made our own programs for Pentium III, where Whirlpool runs faster than [14]. This is due to a better instruction scheduling. "single" and "double" denote straightforward single message hashing using 64-bit MMX instructions and double message hashing using 128-bit XMM instructions, respectively. In the single message hashing, Whirlpool is still faster than SHA512 on Pentium 4 Prescott, (Northwood is slow simply due to cache miss penalties), but the effect of double message hashing is evident; SHA512 then becomes more than 30% faster than Whirlpool.

8 Concluding Remarks

This paper discussed various implementation trade-offs of cryptographic primitives on Pentium III and 4 processors, introducing parallel encryption techniques. The clock-raising of modern processors is approaching to its margin and it seems that a next generation of processors goes toward independent multiple cores, rather than a deeper pipeline and a higher superscalability. Hence we believe that parallel encryption/hashing techniques will be increasingly important in a very near future.

References

1. R. Anderson, E. Biham, L. Knudsen: "Serpent: A proposal for the Advanced Encryption Standard", http://www.ftp.cl.cam.ac.uk/ftp/users/rja14/serpent.pdf
2. K. Aoki, H. Lipmaa: "Fast Implementations of AES Candidates", Proceedings of The Third AES Candidate Conference, 2000. Available at http://www.tcs.hut.fi/~helger/papers/al00/fastaes.pdf
3. P. Barreto: "The Whirlpool Hash Function", http://planeta.terra.com.br/informatica/paulobarreto/WhirlpoolPage.html
4. P. Ekdahl, T. Johansson: "A new version of the stream cipher SNOW", Proceedings of 9th Annual Workshop on Selected Areas in Cryptography SAC2002, Lecture Notes in Computer Science, Vol.2595, pp 47-61, Springer-Verlag, 2002.
5. P. Ekdahl: SNOW Homepage, http://www.it.lth.se/cryptology/snow/
6. Federal Information Processing Standards Publication 197, "Advanced Encryption Standard (AES)", NIST, 2001.
7. Federal Information Processing Standards Publication 180-2, "Secure Hash Standard", NIST, 2002.
8. A. Fog: "How To Optimize for Pentium Family Processors", Available at http://www.agner.org/assem/
9. B. Gladman: "Serpent Performance", Available at http://fp.gladman.plus.com/cryptography_technology/serpent/
10. IA-32 Intel Architecture Optimization Reference Manual, Order Number 248966-011, http://developer.intel.ru/download/design/Pentium4/manuals/24896611.pdf

11. P. Junod, S. Vaudenay: "FOX : a new Family of Block Ciphers", Preproceedings of 11th Annual Workshop on Selected Areas in Cryptography SAC2004, pp131-146, 2004.
12. H. Lipmaa: "Fast Software Implementations of SC2000", Proceedings of Information Security Conference ISC2002, Lecture Notes in Computer Science, Vol.2433, pp.63-74, Springer-Verlag, 2002.
13. H. Lipmaa: "AES / Rijndael: speed", http://www.tcs.hut.fi/~helger/aes/rijndael.html
14. J. Nakajima, M. Matsui: "Performance Analysis and Parallel Implementation of Dedicated Hash Functions on Pentium III", IEICE Trans. Fundamentals, Vol.E86-A, No.1, pp.54-63, 2003.
15. New European Schemes for Signatures, Integrity, and Encryption (NESSIE), https://www.cosic.esat.kuleuven.ac.be/nessie/

Appendix: On Measuring Execution Cycles

Our measurement method shown in section 3 agrees with [14], but not with [2]. Aoki et al. [2] regarded execution time for decrementing the block counter and branching conditionally inside `FUNCTION` also as an overhead. Specifically, they subtracted the number of execution cycles of the following "Null function" from that of `FUNCTION`, and defined its result as performance of the target primitive.

```
        /* push all used registers */
        cmp   dword ptr [block], 0
        jz    L1
        align 16
L0:
        dec   dword ptr [block]
        jnz   L0
L1:
        /* pop these registers once more */
```

We do not adopt this method because our definition is more practical and visible for users (application programmers) and moreover it is difficult to measure the overhead of the loop processing accurately, due to the nature of superscalar and out-of-order architecture of the processors. It should be noted that in Pentium 4 micro-operations in a small loop are likely rearranged on the trace cache so that the number of branches can be reduced [8].

Also, it is common to count execution cycles many times and regard the minimum value as a "real" cycle count in practice. This is based on the assumption that an interruption by an operation system always increases execution cycles. But this does not always hold for Pentium 4 with Hyperthread Technology (HT), which enables a single processor to run two multi-threaded codes simultaneously. Our experiments show that **Code 2**, for example, runs in 632 cycles on Northwood without HT (or with disabled HT) in almost all cases, and takes more cycles in some rare cases; however on Northwood with HT, **Code 2** runs in 636 cycles in almost all cases and takes more or **less** cycles in some rare cases. We saw it run even in 600 cycles!

Another implicit assumption is that we should always obtain a constant cycle count if no interruption takes place during the measurement. To make sure this, we again measured the speed of **Code 2** under DOS with disabling interruptions for more than 20 processors with different stepping/revision numbers. As a result, we found that only one type of processor (Prescott Stepping 3 Revision 0) did not run in a constant time. We do not know reason of this instability.

This suggests that if we measure a target code many and many times, we might finally obtain an exceptionally fast result, but clearly this does not make sense in practice. We hence decided to take the most frequent value or an average value in measuring a speed of a code. Our experiments show that the number of cycles obtained by Aoki et al's method [2] is smaller than ours by typically 6, 14 and 7 cycles/block for Pentium III, 4-N and 4-P, respectively. Hence to compare our results with Aoki's or Lipmaa's, simply subtract these numbers from ours.

A Side-Channel Analysis Resistant Description of the AES S-Box*

Elisabeth Oswald[1], Stefan Mangard[1], Norbert Pramstaller[1], and Vincent Rijmen[1,2]

[1] Institute for Applied Information Processing and Communciations (IAIK), TU Graz, Inffeldgasse 16a, A–8010 Graz, Austria

[2] Cryptomathic A/S Jægergårdsgade 118, DK-8000 Århus C, Denmark

{elisabeth.oswald, stefan.mangard, norbert.pramstaller, vincent.rijmen}@iaik.tugraz.at

Abstract. So far, efficient algorithmic countermeasures to secure the AES algorithm against (first-order) differential side-channel attacks have been very expensive to implement. In this article, we introduce a new masking countermeasure which is not only secure against first-order side-channel attacks, but which also leads to relatively small implementations compared to other masking schemes implemented in dedicated hardware.

Our approach is based on shifting the computation of the finite field inversion in the AES S-box down to $GF(4)$. In this field, the inversion is a linear operation and therefore it is easy to mask.

Summarizing, the new masking scheme combines the concepts of multiplicative and additive masking in such a way that security against first-order side-channel attacks is maintained, and that small implementations in dedicated hardware can be achieved.

Keywords: AES, side-channel analysis, masking schemes.

1 Introduction

Securing small hardware implementations of block ciphers against differential side-channel attacks [8] has proven to be a challenging task. Hardware countermeasures, which are based on special leakage-resistant logic styles, typically lead to a significant increase of area and power consumption [14]. Algorithmic countermeasures also lead to a significant increase of area, if implemented in hardware. Nevertheless, algorithmic countermeasures can be tailored towards a particular algorithm, and hence, they can be optimized to a certain extent.

* The work described in this paper has been supported in part by the European Commission through the IST Programme under Contract IST-2002-507932 ECRYPT and by the FWF "Investigations of Simple and Differential Power Analysis" project (P16110).

H. Gilbert and H. Handschuh (Eds.): FSE 2005, LNCS 3557, pp. 413–423, 2005.

In case of the AES algorithm, several algorithmic countermeasures have been proposed [2], [6], and [13]. They are all based on masking, *i.e.*, the addition of a random value (the mask) to the intermediate AES values. However two of them, [2] and [13], are both susceptible to a certain type of (first-order) differential side-channel attack, the zero-value attack. The latter one has turned out to be vulnerable even to standard differential side-channel attacks as well [1]. The countermeasure presented in [6] is not suitable for hardware implementations. The weakness of these three countermeasures is the way in which they secure the intermediate values occurring in the AES **SubBytes** operation. The **SubBytes** operation is the non-linear component within AES, which makes it particularly difficult to mask.

In this article, we propose a secure masking scheme for the AES algorithm, which is particularly suited for implementation in dedicated hardware. In order to achieve security, we use a combination of additive and multiplicative masking. The most tricky part when masking AES is to mask its non-linear operation, which is the finite field inversion (short: inversion) in the S-box, *i.e.*, the **SubBytes** operation. All other operations are linear and can be masked in a straightforward manner as it is for example shown in [2]. Hence, this article focuses on the inversion operation in the S-box only.

The masking scheme for the inversion presented in this article is based on composite field arithmetic, which has already been previously used for efficient S-box implementations in hardware [15]. However, while in [15] the inversion is performed in $GF(16)$, we shift the inversion down to $GF(4)$ in this article. The motivation for this is the fact that the inversion in $GF(4)$ is a *linear* operation, which can be masked easily.

Because of that we can build a masking scheme with very nice properties. The approach presented in this article for example has a much smaller area-time product than [2]. It also has the advantage of being secure against *all* first-order differential side-channel attacks. In addition, it can be implemented in software as well.

The remainder of this article is organized as follows. In Section 2 we motivate our research by discussing zero-value attacks on multiplicative masking schemes. Our analysis shows the need for masking schemes which are secure against zero-value attacks. Such a new secure scheme is introduced in Section 3. Arguments for the security of our scheme are provided in Section 4. The efficiency in hardware compared to other masking schemes is discussed in Section 5. We conclude our research in Section 6.

2 Discussion of Multiplicative Masking Schemes

The masking schemes proposed in [2] and [13] are susceptible to so-called zero-value attacks. In this section, we analyze the effectiveness of zero-value attacks against these masking schemes.

In AES, an **AddRoundKey** operation is performed prior to the first encryption round and thus, prior to the first time when the inversion needs to be computed. If a key byte k equals a data byte d, then the result of **AddRoundKey**, which is $x = d + k$, equals zero. This observation can readily be used in an attack which is referred to as **zero-value attack** and was introduced in [6].

Let t denote a power measurement (trace) and let the set of all traces t be denoted by T. Suppose a number of AES encryptions is executed and their power consumption is measured. Assume that the input texts are known. For all 256 possible key-bytes k', we do the following. We define a set M_1 which contains those measurements with $k' = d$ right before the **SubBytes** transformation. We also define a set M_2 which contains the measurements with $k' \neq d$ right before the **SubBytes** transformation.

$$M_1 = \{t \in T : k' = d\} \tag{1}$$

$$M_2 = \{t \in T : k' \neq d\} \tag{2}$$

If $k = k'$, then the difference-of-means trace $M_d = \overline{M_1} - \overline{M_2}$ shows a considerable peak at the point in time when the masked **SubBytes** operation has been performed. This is due to the fact that set M_1 contains the measurements in which the 0-value is manipulated in the inversion. If $k \neq k'$, then the definition of the sets is meaningless. Hence, no difference between the sets can be observed.

The difficulty in this scenario is that one needs enough traces in M_1 to reduce the variance, *i.e.* to get rid of noise. In [9] it has been estimated that around 64 times more measurements are needed in a zero-value attack than in a standard differential side-channel attack. This number indicates that zero-value attacks still pose a serious practical threat and must be avoided.

3 Combined Masking in Tower Fields

In order to thwart zero-value attacks, we have developed a new scheme which works with combinations of additive and multiplicative masks. Throughout the whole cipher, including the **SubBytes** computation, the data is concealed by an additive mask.

Before going into the details of our new scheme, we review some necessary facts about the efficient implementation of **SubBytes** first.

3.1 Inversion in • • (256)

Our **SubBytes** design follows the architecture we have proposed in [15] (we call this approach **S-IAIK** from now on) . This architecture is based on composite field arithmetic [5], and has very low area requirements. Thus, it is ideally suited for small hardware implementations. In this approach, each element of $GF(256)$ is represented as a linear polynomial $a_h x + a_l$ over $GF(16)$.

The inversion of such a polynomial can be computed using operations in $GF(16)$ only:

$$(a_h x + a_l)^{-1} = a'_h x + a'_l \tag{3}$$
$$a'_h = a_h \times d' \tag{4}$$
$$a'_l = (a_h + a_l) \times d' \tag{5}$$
$$d = (a_h^2 \times p_0) + (a_h \times a_l) + a_l^2 \tag{6}$$
$$d' = d^{-1} \tag{7}$$

The element p_0 is defined in accordance with the field polynomial which is used to define the quadratic extension of $GF(16)$, see [15].

In the following subsections we present the mathematical formulae for our masking scheme.

3.2 Masked Inversion in • • (256)

In our masking scheme for the inversion, which we call **Masked SubBytes IAIK** (short: **MS-IAIK**) from now on, all intermediate values as well as the input and the output are masked additively. In order to calculate the inversion of a masked input value, we first map the value as well as the mask to the composite field representation as defined in [15]. This mapping is a linear operation and therefore it is easy to mask. After the mapping, the value that needs to be inverted is represented by $(a_h + m_h)x + (a_l + m_l)$. Note that both elements in the composite field representation are masked additively.

Our goal is that all input and output values in the computation of the inverse are masked. Hence, we have to modify (3)-(7), by introducing functions f_{a_h}, f_{a_l}, f_d and $f_{d'}$, as follows:

$$((a_h + m_h)x + (a_l + m_l))^{-1} = (a'_h + m'_h)x + (a'_l + m'_l) \tag{8}$$

$$\begin{aligned} a'_h + m'_h &= f_{a_h}((a_h + m_h), (d' + m'_d), m_h, m'_h, m'_d) \\ &= a_h \times d' + m'_h \end{aligned} \tag{9}$$

$$\begin{aligned} a'_l + m'_l &= f_{a_l}((a'_h + m'_h), (a_l + m_l), (d' + m'_d), m_l, m'_h, m'_l, m'_d) \\ &= (a_h + a_l) \times d' + m'_l \end{aligned} \tag{10}$$

$$\begin{aligned} d + m_d &= f_d((a_h + m_h), (a_l + m_l), p_0, m_h, m_l, m_d) \\ &= a_h^2 \times p_0 + a_h \times a_l + a_l^2 + m_d \end{aligned} \tag{11}$$

$$\begin{aligned} d' + m'_d &= f_{d'}(d + m_d, m_d, m'_d) \\ &= d^{-1} + m'_d \end{aligned} \tag{12}$$

The function f_{a_h}, f_{a_l}, f_d and $f_{d'}$ are functions on $GF(16)$.

3.3 Derivation of the Functions $\bullet_{a_h}\bullet\bullet_{a_l}\bullet\bullet_d$ and $\bullet_{d'}$

This section shows how to transform (4)-(7) into (9)-(12).

Transforming Equation 4 into Equation 9. Suppose that we calculate (4) with masked input values, *i.e.*, with $a_h + m_h$ instead of a_h and with $d' + m'_d$ instead of d':

$$(a_h + m_h) \times (d' + m'_d) = a_h \times d' + m_h \times d' + a_h \times m'_d + m_h \times m'_d. \quad (13)$$

Comparing the result of this calculation to (9) shows that the desired and masked result, $a_h \times d' + m'_h$, is only part of the result of (13). All the terms that occur in addition due to the masks, have to be removed. These terms can be easily removed by adding the terms $(d' + m'_d) \times m_h$, $(a_h + m_h) \times m'_d$, $m_h \times m'_d$ and m'_h. This is done by the function f_{a_h}, which takes five elements of $GF(16)$ as input, and produces an element of $GF(16)$ as output.

$$f_{a_h}(r, s, t, u, v) = r \times s + s \times t + r \times v + t \times v + u \quad (14)$$

If we choose $r = (a_h + m_h)$, $s = (d' + m'_d)$, $t = m_h$, $u = m'_h$ and $v = m'_d$ and compute $f_{a_h}((a_h + m_h), (d' + m'_d), m_h, m'_h, m'_d)$, we get the desired result $a_h \times d' + m'_h$ (see (9)).

One has to take care when adding correction terms that no intermediate values are correlated with values, which an attacker can predict. It needs to be pointed out that the formulae, which we derive in this section, do not lead to a secure implementation when directly implemented. The secure implementation of these formulae requires the addition of an independent value to the first intermediate value that is computed. This becomes clear from the security proof given in Section 4.

Another aspect, which we do not treat in this article, is the discussion of the particular choice of the masks m'_h, m'_l, m_d and m'_d. In our implementation in dedicated hardware, see [11] for details, we decided to re-use masks as often as possible. For example, in our implementation of (9) we set $v = m'_d = m_l$ and $u = m'_h = m_h$. Consequently, in our implementation we calculate the function f_{a_h} as shown in (15).

$$\begin{aligned} f_{a_h} = & \underbrace{(a_h + m_h) \times (d' + m_l)}_{dm_4} \\ & + \underbrace{(d' + m_l) \times m_h}_{c_7} + \underbrace{(a_h + m_h) \times m_l}_{c_1} \\ & + \underbrace{m_h}_{c_6} + \underbrace{m_h \times m_l}_{c_5} \end{aligned} \quad (15)$$

The term which is labelled as dm_4 refers to the masked data. The terms which are labelled as c_1 to c_7 in this equation are the so-called correction terms which are applied by the function f_{a_h}. It can be seen in the subsequent paragraphs that we can re-use several of these correction terms. This significantly reduces the area required for our implementation.

At first sight our numbering scheme for the masked-data terms and the correction terms might look erratic. However, the indices of the dm_i and c_j indicate when a certain value would be calculated during an implementation. For instance, the masked data is labelled by dm_4 in this formula, because it would be calculated only later. Equations (9)-(12) show, that the result of (12) is needed for (9) and (10). Therefore, (9) would be calculated later.

The reason why we make a difference in labelling masked-data terms and correction terms is that it makes it easier to see how many additional operations are introduced by the masking scheme. All terms labelled by dm_i have to be calculated in the original S-box design (**S-IAIK**, [15]) as well. However, all terms labelled by c_j are the corrections that we have to apply. Thus, these are the additional operations, which are introduced by the masking.

In the same style as for (4), we subsequently transform (5) and (6).

Transforming Equation 5 into Equation 10. In order to transform (5) into (10) we define a function f_{a_l} that applies the appropriate correction terms.

The function f_{a_l} takes seven elements of $GF(16)$ as input and gives one element of $GF(16)$ as output.

$$f_{a_l}(r,s,t,u,v,w,x) = r + s \times t + t \times u + s \times x + v + w + u \times x \qquad (16)$$

If we choose $r = a'_h + m'_h$, $s = a_l + m_l$, $t = d' + m'_d$, $u = m_l$, $v = m'_h$, $w = m'_l$ and $x = m'_d$ we indeed get (10).

In our implementation, we set $u = w$ (*i.e.* $m'_l = m_l$) and $x = m'_d = m_h$. Hence, in our implementation, we calculate f_{a_l} as is shown in (17).

$$\begin{aligned} f_{a_l} = {} & (a_h \times d' + m_h) + \underbrace{(a_l + m_l) \times (d' + m_h)}_{dm_5} \\ & + \underbrace{(d' + m_h) \times m_l}_{c_8} + \underbrace{(a_l + m_l) \times m_h}_{c_2} \\ & + \underbrace{m_l}_{c_9} + \underbrace{m_h}_{c_6} + \underbrace{m_l \times m_h}_{c_5} \end{aligned} \qquad (17)$$

As in the previous paragraphs, the terms that are labelled by c_i are correction terms.

Transforming Equation 6 into Equation 11. In order to transform (6) into (11) we define a function f_d that applies the appropriate correction terms (as demonstrated in the previous paragraphs).

The function takes six elements of $GF(16)$ as inputs and gives an element of $GF(16)$ as result.

$$f_d(r,s,t,u,v,w) = r^2 \times t + r \times s + s^2 + r \times v + s \times u + u^2 \times t + v^2 + u \times w + u \qquad (18)$$

If we choose $r = a_h + m_h$ and $s = a_l + m_l$, $t = p_0$, $u = m_h$, $v = m_l$ and $w = m_d$ then we get (11).

In our implementation we set $w = m_l$. Consequently, we calculate f_d in our implementation as shown in (19).

$$\begin{aligned} f_d = & \underbrace{(a_h + m_h)^2 \times p_0}_{dm_1} + \underbrace{(a_h + m_h) \times (a_l + m_l)}_{dm_2} + \underbrace{(a_l + m_l)^2}_{dm_3} \\ & + \underbrace{(a_h + m_h) \times m_l}_{c_1} + \underbrace{(a_l + m_l) \times m_h}_{c_2} \\ & + \underbrace{\overbrace{m_h^2}^{c_3'} \times p_0}_{c_3} + \underbrace{m_l^2}_{c_4} \\ & + \underbrace{m_h \times m_l}_{c_5} + \underbrace{m_h}_{c_6} \end{aligned} \tag{19}$$

As in the previous paragraphs, the terms that are labelled by c_j or c_j' are correction terms.

Transforming Equation 7 into Equation 12. Calculating the inverse in $GF(16)$ can be reduced to calculating the inverse in $GF(4)$ by representing $GF(16)$ as quadratic extension of $GF(4)$.

In short, an element of $GF(4) \times GF(4)$ is a linear polynomial with coefficients in $GF(4)$, *i.e.*, $a = (a_h x + a_l)$, with a_h and $a_l \in GF(4)$. The same formulae as given in (17) – (19) can be used to calculate the masked inverse in $GF(4) \times GF(4)$. In $GF(4)$, the inversion operation is equivalent to squaring: $x^{-1} = x^2 \, \forall x \in GF(4)$. Hence, in $GF(4)$ we have that $(x+m)^{-1} = (x+m)^2 = x^2 + m^2$; the inversion operation preserves the masking in this field.

4 Security of Our Masking Scheme

In this section, we show that all the operations discussed in Section 3, are secure. We follow the security notion that has been introduced in [4] and strengthened by [3]:

Definition 1. An algorithm is said to be secure if for all adversaries A and all realizable distributions M_1 and M_2, M_1 equals M_2.

This definition is equivalent to the *perfect masking* condition given in [3] for standard differential SCA. Counteracting higher-order differential SCA is not within the scope of this article.

In the following paragraphs we will show that all data-dependent intermediate values that occur in (17) – (19) fulfill Definition 1. These values are masked data $a+m$, masked multiplications $(a+m_a)\times(b+m_b)$, multiplications of masked values with masks $(a+m_a) \times m_b$, and masked squarings $(a+m_a)^2$ and $(a+m_a)^2 \times p$.

Definition 1 does imply that regardless of the hypotheses, which an attacker can make, the distributions, which are derived by using these hypotheses, are identical. Consequently, we must proof that every operation that is performed in our masking scheme, leads to an output whose distribution does not depend (in a statistical sense) on the input.

Our proof is divided into two parts. In the first part, which consists of the Lemmas 1 to 4, we show that the data-dependent intermediate values are all secure. In the second part, which consists of Lemma 5, we show that also the summation of the intermediate results can be done securely.

We re-use the Lemmas 1 and 2 of [3]:

Lemma 1. Let $a \in GF(2^n)$ be arbitrary. Let $m \in GF(2^n)$ be uniformly distributed in $GF(2^n)$ and independent of a. Then, $a + m$ is uniformly distributed regardless of a. Therefore, the distribution of $a + m$ is independent of a.

Lemma 2. Let $a, b \in GF(2^n)$ be arbitrary. Let $m_a, m_b \in GF(2^n)$ be independently and uniformly distributed in $GF(2^n)$. Then the probability distribution of $(a + m_a) \times (b + m_b)$ is

$$\Pr((a + m_a) \times (b + m_b) = i) = \begin{cases} \frac{2^{n+1}-1}{2^{2n}} & \text{, if } i = 0, \textit{ i.e.}, \text{ if } m_a = a \text{ or } m_b = b \\ \frac{2^n-1}{2^{2n}} & \text{, if } i \neq 0. \end{cases}$$

Therefore, the distribution of $(a + m_a) \times (b + m_b)$ is independent of a and b.

These two lemmas cover almost all data-dependent operations in our masking scheme. The operation $(a + m_a) \times m_b$ is covered by Lemma 3.

Lemma 3. Let $a \in GF(2^n)$ be arbitrary. Let $m_a, m_b \in GF(2^n)$ be independently and uniformly distributed in $GF(2^n)$. Then the distribution of $(a+m_a) \times m_b$ is

$$\Pr((a + m_a) \times m_b = i) = \begin{cases} \frac{2^{n+1}-1}{2^{2n}} & \text{if } i = 0, \textit{ i.e.}, \text{ if } m_a = a \text{ or } m_b = 0 \\ \frac{2^n-1}{2^{2n}} & \text{if } i \neq 0. \end{cases}$$

Therefore, the distribution of $(a + m_a) \times m_b$ is independent of a.

Lemma 3 is a special case ($b = 0$) of Lemma 2. The proof is therefore omitted.

The two remaining operations that occur in our masking scheme, $(a + m_a)^2$ and $(a + m_a)^2 \times p$ are covered by Lemma 4.

Lemma 4. Let $a \in GF(2^n)$ be arbitrary and $p \in GF(2^n)$ a constant. Let $m_a \in GF(2^n)$ be independently and uniformly distributed in $GF(2^n)$.

Then, the distribution of $(a + m_a)^2$ and $(a + m_a)^2 \times p$ is independent of a.

Proof. According to Lemma 1, $a + m_a$ is uniformly distributed in $GF(2^n)$. This is straightforward because for an arbitrary but fixed a, $a+m_a$ is a permutation of $GF(2^n)$. Hence, $(a+m_a)^2$ gives all quadratic residues of $GF(2^n)$, regardless of a. This implies that the distribution of $(a+m_a)^2$ is independent of a. Consequently, also the distribution of $(a + m_a)^2 \times p$ (p is a constant) is independent of a.

The Lemmas 1 to 4 show that all major operations of our masking scheme are secure. However, more intermediate results occur in the masking scheme because we add the major operations, and thus, produce implicitly more intermediate results than are directly visible from the formulae.

Lemma 5 shows that these intermediate results can be added in a secure way.

Lemma 5. Let $a_i \in GF(2^n)$ be arbitrary and $M \in GF(2^n)$ be independent of all a_i and uniformly distributed in $GF(2^n)$.

Then the distribution of $\sum_i a_i + M$ is (pairwise) independent of a_i.

Proof. The proof for this lemma follows directly from Lemma 1.

Lemma 5 shows that for secure implementations, the order in which the terms of a sum are added, is important! In particular, every summation of variables must start with the addition of an independent mask M.

5 Comparison of Masking Schemes

A high-level comparison of the three masking schemes, **S-Akkar**, **S-Blömer** and **MS-IAIK** shows that in terms of area, our scheme leads to the smallest implementation. Table 1 lists the number of high-level operations (multiplication, multiplication with a constant and square) in $GF(16)$ of each of the three schemes.

Table 1. High-level comparison of masking schemes

	Mult	MultConst	Square
S-Akkar	18	6	4
S-Blömer	12	1	2
MS-IAIK	9	2	2

We have not included a count of the $GF(4)$ operations for **S-Blömer** and **MS-IAIK** because they do not contribute significantly to the area. We have also not included an XOR count in $GF(16)$ because the number of XORs is highly dependent on the amount of fresh masks which are available. In the following, we discuss hardware implementations of **S-Akkar** and **S-Blömer** in more detail.

5.1 S-Akkar

S-Akkar makes use of multiplications in finite fields. In particular, 4 multiplications, 1 inversion and 2 XORs in addition to the original inversion have to be computed. All operations are performed in the finite field with 256 elements. For a fair comparison, we assume that all multipiers are based on the same, optimized multipliers which we consider for the implementation of **S-IAIK**. Because **S-IAIK** uses composite field arithmetic, three $GF(16)$ multipliers and one $GF(16)$ constant-coefficient multiplier had to be combined according to [10] in order to build a $GF(2)[x]/(x^8 + x^4 + x^3 + x + 1)$ multiplier. Counting only the largest component of this circuit, which are the $GF(16)$ multipliers, we see that this implementation requires $4 \times 3 = 12$ $GF(16)$ multipliers. Hence, the implementation of **S-Akkar** is much bigger than an implementation of our masking scheme.

5.2 S-Blömer

S-Blömer suggests a comparable technique for counteracting (single-order) differential side-channel attacks as we do. In this article, an implementation strategy for a dedicated hardware implementation is outlined. In this strategy, the architecture of the **SubBytes** operation is based on [12]. In [12], the authors have used the Itho-Tsujii algorithm [7] for computing the multiplicative inverse over a finite field. The Itho-Tsujii inversion algorithm leads to the same inversion formulae as used in [15].

The major difference between [15] and [12] is that in [15], field polynomials have been chosen that lead to particularly efficient finite field arithmetic. In [12], other field polynomials have been chosen that lead to a less efficient finite field arithmetic. Thus, **S-Blömer** suffers from this drawback.

S-Blömer leads to a larger **SubBytes** implementation than our proposal **MS-IAIK**. This is based on the fact that in **S-Blömer** high-level operations such as the finite field multiplication and the finite field squaring are masked. As a consequence, correction terms are computed more often than necessary. In particular, correction terms which involve multiplications cannot be re-used. This, in combination with the less efficient finite field arithmetic, leads to an increase in area.

S-Blömer requires the computation of three masked multiplications in $GF(16)$. One masked multiplication requires 4 ordinary $GF(16)$ multiplications. Hence, **S-Blömer** requires 12 $GF(16)$ multipliers according to [12].

6 Conclusions

In this article, we have presented a new secure and efficient scheme for masking the intermediate value of an AES **SubBytes** implementation. To motivate our research, we have discussed zero-value attacks on multiplicative masking schemes first. Zero-value attacks pose a serious practical threat. Therefore we have introduced a new masking scheme which does not succumb to these attacks. We have given arguments for the security of our scheme. In addition, we have compared the number of operations needed in our scheme with other masking schemes; our scheme requires the least amount of operations.

References

1. M.-L. Akkar, R. Bevan, and L. Goubin. Two Power Analysis Attacks against One-Mask Methods. In Bimal K. Roy and Willi Meier, editors, *Fast Software Encryption – FSE 2004*, volume 3017 of *Lecture Notes in Computer Science*, pages 332-347. Springer, 2004.
2. Mehdi-Laurent Akkar and Christophe Giraud. An Implementation of DES and AES, Secure against Some Attacks. In Çetin Kaya Koç, David Naccache, and Christof Paar, editors, *Cryptographic Hardware and Embedded Systems – CHES 2001*, volume 2162 of *Lecture Notes in Computer Science*, pages 309–318. Springer, 2001.

3. Johannes Blömer, Jorge Guajardo Merchan, and Volker Krummel. Provably Secure Masking of AES. Cryptology ePrint Archive (http://eprint.iacr.org/), Report 2004/101, 2004.
4. Suresh Chari, Charanjit S. Jutla, Josyula R. Rao, and Pankaj Rohatgi. Towards Sound Approaches to Counteract Power-Analysis Attacks. In Michael J. Wiener, editor, *Advances in Cryptology - CRYPTO '99*, volume 1666 of *Lecture Notes in Computer Science*, pages 398–412. Springer, 1999.
5. John Horton Conway. On Numbers and Games. 2nd edition, AK Peters, 2001
6. Jovan D. Golić and Christophe Tymen. Multiplicative Masking and Power Analysis of AES. In Burton S. Kaliski Jr., Çetin Kaya Koç, and Christof Paar, editors, *Cryptographic Hardware and Embedded Systems - CHES 2002*, volume 2535 of *Lecture Notes in Computer Science*, pages 198–212. Springer, 2003.
7. Toshiya Itoh and Shigeo Tsujiis. A Fast Algorithm for Computing Multiplicative Inverses in $GF(2^m)$ Using Normal Bases. *Information and Computation*, 78(3):171–177, September 1988.
8. Paul C. Kocher, Joshua Jaffe, and Benjamin Jun. Differential Power Analysis. In Michael Wiener, editor, *Advances in Cryptology - CRYPTO '99*, volume 1666 of *Lecture Notes in Computer Science*, pages 388–397. Springer, 1999.
9. Elisabeth Oswald, Stefan Mangard, and Norbert Pramstaller. Secure and Efficient Masking of AES – A Mission Impossible? Cryptology ePrint Archive (http://eprint.iacr.org/), Report 2004/134, 2004.
10. Christof Paar. *Efficient VLSI Architectures for Bit-Parallel Computation in Galois Fields.* PhD thesis, Institute for Experimental Mathematics, University of Essen, 1994.
11. Norbert Pramstaller, Frank K. Gürkaynak, Simon Haene, Hubert Kaeslin, Norbert Felber, and Wolfgang Fichtner. Towards an AES Crypto-chip Resistant to Differential Power Analysis. In *Proccedings 30th European Solid-State Circuits Conference - ESSCIRC 2004, Leuven, Belgium, Proceedings - to appear*, 2004.
12. Akashi Satoh, Sumio Morioka, Kohji Takano, and Seiji Munetoh. A Compact Rijndael Hardware Architecture with S-Box Optimization. In Colin Boyd, editor, *Advances in Cryptology - ASIACRYPT 2001*, pages 239–254. Springer, 2001.
13. Elena Trichina, Domenico De Seta, and Lucia Germani. Simplified Adaptive Multiplicative Masking for AES. In Burton S. Kaliski Jr., Çetin Kaya Koç, and Christof Paar, editors, *Cryptographic Hardware and Embedded Systems - CHES 2002*, pages 187–197. Springer, 2003.
14. Kris Tiri and Ingrid Verbauwhede. Securing Encryption Algorithms against DPA at the Logic Level: Next Generation Smart Card Technology. In Burton S. Kaliski Jr., Çetin Kaya Koç, and Christof Paar, editors, *Cryptographic Hardware and Embedded Systems - CHES 2002*, pages 125–136. Springer, 2003.
15. Johannes Wolkerstorfer, Elisabeth Oswald, and Mario Lamberger. An ASIC implementation of the AES SBoxes. In Bart Preneel, editor, *Topics in Cryptology - CT-RSA 2002*, volume 2271 of *Lecture Notes in Computer Science*, pages 67–78. Springer, 2002.

DPA Attacks and S-Boxes

Emmanuel Prouff

Oberthur Card Systems,
25 rue Auguste Blanche, 92800 Puteaux, France
e.prouff@oberthurcs.com

Abstract. For the power consumption model called *Hamming weight model*, we rewrite DPA attacks in terms of correlation coefficients between two Boolean functions. We exhibit properties of S-boxes (also called (n, m)-functions) relied on DPA attacks. We show that these properties are opposite to the non-linearity criterion and to the propagation criterion. To quantify the resistance of an S-box to DPA attacks, we introduce the notion of *transparency order of an S-box* and we study this new criterion with respect to the non-linearity and to the propagation criterion.

1 Introduction

Block cipher algorithms embedded in cryptographic devices are sensitive to two main kinds of attacks, which are usually investigated in parallel. The first kind relies on the properties of the cryptographic primitives involved in the cryptosystem. The second kind is based on the analysis of the hardware's leakages.

The most well-known attacks against block ciphers algorithms are the *known-plaintext attacks* called *differential cryptanalysis* [2, 13] and *linear cryptanalysis* [21]. Most block cipher algorithms (such as DES or AES) use vectorial functions, also called *S-boxes*, as cryptographic primitives. To protect such cryptosystems against linear and differential attacks, S-boxes are designed to fulfil some cryptographic criteria (balancedness, high nonlinearity or high algebraic degree).

Since electronic components are not usually perfectly tamper-proof, one can obtain sensitive information from side channels such as the timing of operations or the power consumption. In 1996, Kocher successfully used this approach to exhibit a first *side-channel attack* effective enough to recover secret keys in numerous cryptosystems [16]. Since Kocher's original paper, a large number of very efficient attacks has been reported on a wide variety of cryptographic implementations (see for instance [4,5,8,11,22,25]). Among these attacks, the *Differential Power Analysis* (DPA) is one of the most powerfull against iterated block ciphers. DPA are usually used to attack on either the first or the last round but it can sometimes be applied to attack on intern rounds of the block ciphers. It requires the knowledge of either the plaintext or the ciphertext. It relies on a statistical analysis of a large number of samples where the same key operates

H. Gilbert and H. Handschuh (Eds.): FSE 2005, LNCS 3557, pp. 424–441, 2005.

on different data. For this strategy of attacks, S-boxes involved in the cryptosystems are usually considered by cryptanalysts and also by cryptographers as oracles providing the output corresponding to a given data. So, to withstand DPA attacks, countermeasures are added at the implementation level to make the signals needed for these attacks useless.

The efficiency of DPA attacks is much greater than the efficiency of differential or linear cryptanalysis [1]. Moreover, in the area of embedded cryptography, because of the life expectancy of the device, known-plaintext attacks requiring a large number of pairs plaintext/ciphertext or requiring a large number of encryptions are unpracticable. The difference between the efficiencies of the two categories of attacks is not taken into account to design block ciphers for smart cards. Indeed, nearly all the algorithms embedded in smart cards have been designed to resist at high level to linear, differential and high-order differential attacks, whereas nothing has been done to make them inherently resistant to DPA attacks. Countermeasures against DPA attacks are generally added to the algorithms when implemented on devices. Following this addition, the performances and the code sizes of the resulting embedded algorithms are approximately multiplied by two. This increase is damaging in the area of embedded cryptography where the computation power and the memory capability are limited. The design of DPA-resistant algorithms would make the addition of countermeasures innecessary. Such a design could be done by selecting pertinent S-boxes.

For a very particular power consumption model, Guilley *et al.* studied in [9] the *single-bit* DPA attack in terms of *correlation coefficients* between two Boolean signals, the first one depending on linear combinations of output-bits of S-boxes and the second one depending on consumption. The authors pointed out that the better shielded against linear cryptanalysis an S-box is, the more vulnerable it is to side-channel attacks such as DPA. In this paper, we extend the study of Guilley *et al.* for *multi-bit* DPA attacks and for the power consumption model called *Hamming weight model.* We exhibit the properties of S-boxes related to DPA attacks. We argue that these new properties and the classical cryptographic criteria (such as the high non-linearity or the satisfaction of propagation criteria at high level) cannot be satisfied simultaneously. Since a highly non-linear S-box cannot withstand DPA attacks in an optimal way, we point out that a trade-off between the classical cryptographic criteria and resistance to DPA attacks has to be found. We introduce a new cryptographic criteria, that we call *transparency order of an S-box*, to quantify the resistance of an S-box to DPA attacks. We exhibit lower and upper bounds on it and we study their tightness. We prove in particular that bent functions (and more generally functions satisfying $PC(l)$ for a high level l) cannot by definition resist DPA attacks. To ensure the resistance of an algorithm to these attacks, we argue that the new criterion must be satisfied

[1] For example, a DPA of a software DES without any countermeasure requires between 50 and 200 plaintext/ciphertext pairs, whereas the best non-side-channel attack against DES requires under 64 terabytes of plaintexts and ciphertexts encrypted under a single key.

at a certain level and that this level depends on the amount of noise inside the device and/or the number of encryptions that a cryptanalyst can do with the same key.

This paper is organized as follows. In Sect. 2, we recall the basic facts about the main cryptographic properties of S-boxes. In Sect. 3, we give the formal definition of an iterated block cipher and we recall the theory behind DPA attacks. To establish the relationship between these attacks and the cryptographic properties of S-boxes, we rewrite in Sect. 4 the DPA attacks in terms of *correlation coefficients.* After arguing that the efficiency of (*single-bit* or *multi-bit*) DPA attacks relies on the behavior of the so-called *differential trace*, we analyze it in Sect. 5. We use this analysis in Sect. 6 to investigate how S-boxes can withstand DPA attacks. In Sect. 7, we introduce and we briefly study the notion of *transparency order of a function*, whose aim is to quantify the resistance of an S-box to DPA attacks.

2 Notation and Preliminaries

In this paper, we distinguish the additions of integers in $\mathbb{R}$, denoted by $+$, and the additions mod 2, denoted by $\oplus$. For simplicity and because there will be no ambiguity, we denote by $+$ the addition of vectors of $\mathbb{F}_2^n$ (words) with $n > 1$.

We call (n, m)-function any mapping F from $\mathbb{F}_2^n$ into $\mathbb{F}_2^m$. If m equals 1, then the function is called Boolean. If F is an affine (n, m)-function, then we call *direction of* F, the linear (n, m)-function L such that there is a vector $B \in \mathbb{F}_2^m$ for which $F(x) = L(x) + B$, $x \in \mathbb{F}_2^n$.

For every vector $a \in \mathbb{F}_2^n$, $n \in \mathbb{N}$, we denote by $H(a)$ the Hamming weight of a. We denote the all-zero vector (resp. the all-one vector) on $\mathbb{F}_2^m$, by 0_m (resp. by 1_m). The set $\{x \in \mathbb{F}_2^n / F(x) \neq 0_m\}$ is called *support* of F: it is denoted by *Supp* F. An (n, m)-function F is said to be *balanced* if every element $y \in \mathbb{F}_2^m$ admits the same number 2^{n-m} of pre-images by F.

To every (n, m)-function F, we associate the m-tuple $(f_1, \cdots, f_m)$ of Boolean functions on $\mathbb{F}_2^n$, called *the coordinate functions of* F, such that we have $F(x) = (f_1(x), \cdots, f_m(x))$ for every $x \in \mathbb{F}_2^n$. The usual scalar product is denoted by "$\cdot$". We recall that it is defined for every pair of vectors $a = (a_1, \cdots, a_m)$ and $b = (b_1, \cdots, b_m)$ by $a \cdot b = \bigoplus_{i=0}^m a_i b_i$.

To make the study of the properties of F easier, we introduce the *sign function* of F, that is the function $(x, v) \mapsto (-1)^{v \cdot F(x)}$ (if F is Boolean, the sign function is the function $x \mapsto (-1)^{F(x)}$). For every (n, m)-function F and for every vector $v \in \mathbb{F}_2^m$, we have:

$$v \cdot F = \frac{1}{2} - \frac{1}{2}(-1)^{v \cdot F} \ . \tag{1}$$

The Fourier transform of the sign function of an (n, m)-function F (that we call *Walsh transform* of F) is the function W_F defined on $\mathbb{F}_2^n \times \mathbb{F}_2^m$ by the formula:

$$W_F(u, v) = \sum_{x \in \mathbb{F}_2^n} (-1)^{v \cdot F(x) + u \cdot x} \ . \tag{2}$$

As we recall in the following proposition, the balancedness of a function can be characterized through its Walsh transform's coefficients.

Proposition 1. *A (n, m)-function F is balanced if and only if $W_F(0, v)$ equals zero for every vector $v \in \mathbb{F}_2^{m*}$.*

Let n be a positive integer and let f and g be two Boolean functions defined on $\mathbb{F}_2^n$, the *correlation coefficient* of f and g, denoted by Cor (f, g), is defined by:

$$Cor(f, g) = \sum_{x \in \mathbb{F}_2^n} (-1)^{f(x)+g(x)} \quad . \tag{3}$$

If the output-bits of two Boolean functions are statistically independent, then their correlation coefficient equals zero.

The nonlinearity of a function F is one of the parameters which quantify the level of *confusion* brought in the system by the function (another such parameter is the degree). The nonlinearity of a vectorial function F is defined as the minimum Hamming distance between the nonzero linear combinations of the coordinate functions of F and the set of all Boolean affine functions (that is functions $x \mapsto a \cdot x \oplus b$, $a, b \in \mathbb{F}_2^n$). Cryptographic functions used in block ciphers must have high nonlinearities to prevent linear attacks (see [21]).

For every (n, m)-function F, the nonlinearity N_F and the Walsh transform W_F satisfy the relation $N_F = 2^{n-1} - \frac{1}{2} \max_{u \in \mathbb{F}_2^n, v \in \mathbb{F}_2^{m*}} |W_F(u, v)|$. The nonlinearity N_F of every (n, m)-function F is upper bounded by $2^{n-1} - 2^{n/2-1}$. If n is even and $m \leq \frac{n}{2}$, then this bound is tight. The functions achieving it are called *bent.*

Another useful tool for quantifying the cryptographic resistance of functions is the notion of *derivative.* The derivative of F with respect to a vector $a \in \mathbb{F}_2^n$ is the (n, m)-function $D_aF : x \mapsto F(x) + F(x + a)$. The notion of derivative is related to differential and higher-order differential attacks [2, 13, 19]. A vector $a \in \mathbb{F}_2^n$ such that D_aF is a constant function is called *linear structure* of F. The space $\{a \in \mathbb{F}_2^n;\ D_aF = \text{cst}\}$ is called *linear space* of F and it is denoted by ε_F. As argued by Evertse in [7], the linear spaces of functions used as cryptographic primitives in iterated block ciphers have be reduced to the null vector in order to protect the systems against differential attacks.

Remark 1. Notice that for every (n, m)-function F and for every pair $(a, v) \in \mathbb{F}_2^n \times \mathbb{F}_2^m$, the correlation coefficient between Boolean functions $x \mapsto v \cdot F(x)$ and $x \mapsto v \cdot F(x + a)$ equals $W_{D_aF}(0, v)$. ◇

The *Strict Avalanche Criterion* (*SAC*) was introduced by Webster and Tavares in [32] and this concept was generalized into the *Propagation Criterion* (*PC*) by Preneel [30]. These properties describe the behavior of a function whenever some input coordinates are complemented. They must be satisfied at high levels, in particular by functions involved in block ciphers. A function F *satisfies* $PC(l)$ if the function D_aF is balanced for every vector a of weight at most l. In [31], Rothaus showed that a function is bent if and only if it satisfies $PC(n)$.

In the next section, our aim is to highlight the role that (n, m)-functions play in DPA attacks on block ciphers.

3 DPA Attacks on Iterated Block Ciphers

3.1 Introduction to Iterated Block Ciphers

To define an iterated block cipher in a formal way, we usually consider a family $(F_K)_{K\in\mathcal{K}}$ of (n,n)-functions indexed by a value $K \in \mathcal{K}$, where $\mathcal{K}$ is called the *round key space*. The *encryption function* of the iterated block cipher with block size n, with R rounds and with round functions F_K is defined by:

$$X^{(i)} = F_{K_i}\left(X^{(i-1)}\right) \text{ for } 1 \le i \le R, \tag{4}$$

where $X^{(0)}$ is the plaintext and $X^{(R)}$ is the ciphertext.

The vector $(K_1, \ldots, K_R)$ is called the *key* and its coordinates are the *round keys*.

As recalled in Sect. 2, balancedness is a fundamental property which has to be satisfied by every designed round function F_K, $K \in \mathcal{K}$. A classical way to define the balanced functions F_K is to design or select the coordinates functions of each F_K being pairwisely independent. We assume in this paper that the coordinate functions of every round function F_K are pairwisely independent.

3.2 Introduction to Differential Power Analysis

Differential Power Analysis uses the fact that computers and microchips leak information about the operations they process. Specific methods for analyzing the power consumption measurements to find secret keys from *tamper-resistant* devices have been studied in [3, 17, 25]. In what follows, we use notations introduced in [17]. Moreover, we assume that the set $\mathcal{K}$ equals $\mathbb{F}_2^r$, where r is a positive integer.

Let $(F_K)_{K\in\mathbb{F}_2^r}$ be a family of (n,n)-functions used as round functions in an iterated block cipher embedded in a smart card, the power consumption of the smart card after one round of the encryption of a message $X \in \mathbb{F}_2^n$ using a round key $\dot{K} \in \mathbb{F}_2^r$ is usually (*cf.* [3, 17]) denoted by $C_{\dot{K}}(X)$. Function $C_{\dot{K}}$ is called *power consumption function related to* $\dot{K}$ or *power consumption function* if there is no ambiguity on $\dot{K}$.

To describe the DPA attacks, one usually introduces a Boolean function D called *selection function* and defined for every 3-tuple $(X, K, j) \in \mathbb{F}_2^n \times \mathbb{F}_2^r \times \{1, \cdots, n\}$ as the value of the jth bit of $F_K(X)$.

A DPA attack is done by computing a so-called *differential trace* whose values are related to the selection function and to the power consumption function. In what follows, we recall the definition of the differential trace.

Definition 1. *[17] Let $(F_K)_{K\in\mathbb{F}_2^r}$ be a family of permutations on $\mathbb{F}_2^n$ and let D be a selection function related to this family. Let $(X_i)_{i\le N}$ be a family of N distinct vectors of $\mathbb{F}_2^n$ (randomly chosen if $N < 2^n$). Then, for every pair $(K, \dot{K}) \in {\mathbb{F}_2^r}^2$ and for every integer $j \le n$, the* differential trace of K with respect to the 3-tuple $(\dot{K}, N, j)$ *is denoted by $\Delta_{K,\dot{K}}(N, j)$ and defined by:*

$$\Delta_{K,\dot{K}}(N,j) = \frac{\sum_{i=1}^{N} D(X_i,K,j)C_{\dot{K}}(X_i)}{\sum_{i=1}^{N} D(X_i,K,j)} - \frac{\sum_{i=1}^{N}(1-D(X_i,K,j))C_{\dot{K}}(X_i)}{\sum_{i=1}^{N}(1-D(X_i,K,j))} , \quad (5)$$

where $C_{\dot{K}}$ is the power consumption function related to $\dot{K}$.

For large values N, the value $\Delta_{K,\dot{K}}(N,j)$ approximately equals $\Delta_{K,\dot{K}}(2^n,j)$. To simplify notations, we denote $\Delta_{K,\dot{K}}(2^n,j)$ by $\Delta_{K,\dot{K}}(j)$.

In Relation (5), information about the secret parameter $\dot{K}$ is given by the power consumption function $C_{\dot{K}}$. Each value $C_{\dot{K}}(X)$ can be viewed as the energy to flip bits from a previous state to state $F_{\dot{K}}(X)$. To better understand the kind of information this function can give about the round key $\dot{K}$, a theoretical model for the power consumption of devices must be introduced.

In this paper, we use the *Hamming distance model* introduced in [3] as a generalization of the *Hamming weight model* (*cf.* [1]). In the Hamming distance model, it is assumed that switching a bit from 0 to 1 requires the same amount of energy as switching it from 1 to 0. The average power consumption to switch a bit from 0 to 1 is denoted by c and for every pair $(X,K) \in \mathbb{F}_2^n \times \mathbb{F}_2^r$, one denotes by $\alpha(X,K) \in \mathbb{F}_2^n$ the value of the data which is replaced by $F_K(X)$ on the device. We call *state function* function α. For every pair $(X,K) \in \mathbb{F}_2^n \times \mathbb{F}_2^r$, we assume throughout this paper that the power consumption $C_K(X)$ satisfies the relation $C_K(X) = c \times H(\alpha(X,K) + F_K(X)) + w$, where w denotes a noise.

Remark 2. Due to Relation (1), we have:

$$C_K(X) = \frac{nc}{2} - \frac{c}{2} \times \sum_{\substack{u \in \mathbb{F}_2^n \\ H(u)=1}} (-1)^{u \cdot (\alpha(X,K)+F_K(X))} + w \ . \quad (6)$$

◇

In the following section, we describe DPA attacks more formally and we rewrite the differential trace in terms of correlation coefficients for balanced S-boxes and for constant noise w.

4 DPA Attacks and Correlations

4.1 Single-Bit DPA Attacks

One denotes by $\dot{K}$ the first round key used in an iterated block cipher encrypting messages X. We assume in this section that a cryptanalyst wants to retrieve $\dot{K}$ and that he has measured nearly all the values $C_{\dot{K}}(X)$, X ranges over $\mathbb{F}_2^n$.

In the rest of the paper, since we only consider the restriction $\alpha(\cdot,\dot{K})$ of the state function α, we denote by α the function $X \mapsto \alpha(X,\dot{K})$ to simplify notations.

Let $(\dot{K},N,j) \in \mathbb{F}_2^r \times \{1,\cdots,2^n\} \times \{1,\cdots,n\}$ be a fixed 3-tuple. In a DPA attack, coefficients $\Delta_{K,\dot{K}}(N,j)$ are computed for different round keys $K \in \mathbb{F}_2^r$

until one value is significantely greater than the others. Let us denote by K_t the corresponding key. The core of the attack is the following: if $\Delta_{K_t,\dot{K}}(N,j)$ is significantely greater than the other values $\Delta_{K,\dot{K}}(N,j)$, $K \in \mathbb{F}_2^r$, then equality $K_t = \dot{K}$ holds with high probability. Since such an attack uses one single bit (of index j) of the outputs $F_{\dot{K}}(X)$, it is usually called *single-bit DPA attack*.

Currently, the main cryptographic properties of S-boxes (nonlinearity, resiliency, balancedness and propagation criteria) are characterized through the Walsh transform. Therefore, to reveal the properties of balanced S-boxes that are related to DPA attacks, we start rewriting the differential trace of a vector in terms of correlation coefficients.

Lemma 1. *Let $(F_K)_{K\in\mathbb{F}_2^r}$ be a family of (n,n)-functions. Let α denote the state function of a cryptographic system implementing F_K, $K \in \mathbb{F}_2^r$, as round functions and let c denote the average power consumption to switch a bit in the system. If all the functions F_K are balanced, then for every pair $(K,\dot{K}) \in \mathbb{F}_2^{r\,2}$ and for every positive integer $j \leq n$, we have :*

$$\Delta_{K,\dot{K}}(j) = \frac{c}{2^n} \sum_{\substack{u\in\mathbb{F}_2^n \\ H(u)=1}} \mathrm{Cor}\left(v \cdot F_K, u \cdot (F_{\dot{K}} + \alpha)\right) \ , \tag{7}$$

where $v = (v_1, \cdots, v_n) \in \mathbb{F}_2^n$ is such that $v_j = 1$ and $v_i = 0$ if $i \neq j$.

Proof. By definition of v, we have $D(X,K,j) = v \cdot F_K(X)$, which implies equalities $\sum_{X\in\mathbb{F}_2^n} D(X,K,j) = \#Supp(v \cdot F_K)$ and $\sum_{X\in\mathbb{F}_2^n}(1 - D(X,K,j)) = 2^n - \#Supp(v \cdot F_K)$. Because we assume that every F_K is balanced, it follows that cardinality of $Supp(v \cdot F_K)$ equals 2^{n-1} for every pair $(v,K) \in \mathbb{F}_2^n \times \mathbb{F}_2^r$, $v \neq 0$. Thus, Relation (5) applied for $N = 2^n$ implies the equality $\Delta_{K,\dot{K}}(j) = \frac{-1}{2^{n-1}}\left(\sum_X (1 - 2(v \cdot F_K(X)))\, C_{\dot{K}}(X)\right)$. Using Relation (1), we obtain $\Delta_{K,\dot{K}}(j) = \frac{-1}{2^{n-1}} \sum_X (-1)^{v\cdot F_K(X)} C_{\dot{K}}(X)$. This equality and Relation (6) imply

$$\begin{aligned} \Delta_{K,\dot{K}}(j) = \frac{-nc - 2w}{2^n} & \sum_{X\in\mathbb{F}_2^n} (-1)^{v\cdot F_K(X)} \\ & + \frac{c}{2^n} \sum_{\substack{u\in\mathbb{F}_2^n \\ H(u)=1}} \sum_{X\in\mathbb{F}_2^n} (-1)^{v\cdot F_K(X) + u\cdot(\alpha(X) + F_{\dot{K}}(X))} \ , \end{aligned} \tag{8}$$

where we recall that w denotes a constant noise. Due to the balancedness of F_K and Proposition 1, the first summation in Relation (8) is null for every non-zero vector v and for every $K \in \mathbb{F}_2^r$. Because the second summation in Relation (8) equals $\frac{c}{2^n} \sum_{\substack{u\in\mathbb{F}_2^n \\ H(u)=1}} \mathrm{Cor}\left(v \cdot F_K, u \cdot (F_{\dot{K}} + \alpha)\right)$, Relations (8) and (7) are equivalent. ◇

More generally a DPA attack can be done by studying correlations between a non-zero linear combination $v \cdot F_K$ and all the coordinate functions of $F_{\dot{K}}$, when

K ranges over $\mathbb{F}_2^r$.. To take this remark into account, we extend Definition 1 by assuming that the differential trace of a vector K is defined with respect to a pair $(\dot{K}, v)$ by:

$$\Delta_{K,\dot{K}}(v) = \frac{c}{2^n} \sum_{\substack{u \in \mathbb{F}_2^n \\ H(u)=1}} \mathrm{Cor}\left(v \cdot F_K, u \cdot (F_{\dot{K}} + \alpha)\right) . \tag{9}$$

In our model, a single-bit DPA attack on the first round of a block cipher is led by designing, for a vector $v \in \mathbb{F}_2^{n*}$, the set of round keys K such that $|\Delta_{K,\dot{K}}(v)|$ is maximal.

4.2 Multi-bit DPA Attacks

Single-bit DPA attacks were generalized in multi-bit DPA attacks in [3,23,24,26, 29]. Among these generalizations, the multi-bit DPA attack proposed by Brier *et al.* in [3] is the most efficient. It is led by searching for high correlations between functions $X \mapsto H(F_K(X))$, $K \in \mathbb{F}_2^r$, and the power consumption function $X \mapsto C_{\dot{K}}(X)$, where $\dot{K}$ is the expected round key. One can prove as in Lemma 1 (and for the same assumptions on $(F_K)_{K\in\mathbb{F}_2^r}$) that multi-bit DPA attack is done by selecting round keys K which maximize the value $\delta_{\dot{K}}(K)$ defined for every pair $(K, \dot{K}) \in \mathbb{F}_2^{r\,2}$ by:

$$\delta_{\dot{K}}(K) = \Big| \sum_{v \in \mathbb{F}_2^n,\ H(v)=1} \Delta_{K,\dot{K}}(v) \Big| . \tag{10}$$

To better understand how the candidate round keys are selected, we study the differential trace in the next section.

5 Analysis of the Differential Trace

Values $\Delta_{K,\dot{K}}(v)$ (and hence $\delta_{\dot{K}}(K)$) are strongly related to the assumptions which are made on the state function α. Indeed, as noticed in [3, 5, 8], if α is supposed to be unknown and dependent on $F_{\dot{K}}$, then the values taken by $(K, v) \mapsto \Delta_{K,\dot{K}}(v)$ cannot be used to get information about the round key $\dot{K}$. Consequently, it is usually assumed either that functions α and F_K are independent for every round key $K \in \mathbb{F}_2^r$, or that α is constant.

5.1 Functions F_K and Function α Are Independent

To prevent statistical attacks, round functions $(F_K)_{K\in\mathbb{F}_2^r}$ of iterated block ciphers are currently designed such that the coordinates of vectors $Y = F_K(X)$, $X \in \mathbb{F}_2^n$, are statistically independent. Moreover, to withstand differential and statistical attacks, the functions in $(F_K)_{K\in\mathbb{F}_2^r}$ are defined to be as uncorrelated as possible. Then, for every pair of distinct elements $(K, \dot{K}) \in \mathbb{F}_2^{r\,2}$ and for every pair $(u, v) \in \mathbb{F}_2^{n*} \times \mathbb{F}_2^{n*}$, $u \neq v$, one can realistically assume in a cryptographic area that $\mathrm{Cor}(v \cdot F_K, u \cdot F_{\dot{K}})$ equals zero (let us notice that in the particular case $u = v$, it cannot be usually assumed that functions $u \cdot F_K$ and $u \cdot F_{\dot{K}}$ are uncorrelated). This assumption is related to the *hypothesis of wrong-key randomization* [10,18]. Under this assumption, we argue in the following proposition that the differential trace has a very simple behavior.

Proposition 2. *Let $(F_K)_{K\in\mathbb{F}_2^r}$ be a family of (n,n)-functions. Let α denote the state function of a cryptographic system implementing functions F_K, $K \in \mathbb{F}_2^r$, as round functions and let c denote the average power consumption to switch a bit in the system. If for every pair $(K,\dot{K}) \in \mathbb{F}_2^{r\,2}$ and for every pair of distinct vectors $(u,v) \in \mathbb{F}_2^{n\,2}$ s.t. $H(u)=1$, functions $u \cdot F_K$ and $v \cdot F_{\dot{K}}$ are independent and if α is independent of the round functions F_K for every $K \in \mathbb{F}_2^r$, then for every 3-tuple $(v,K,\dot{K}) \in \mathbb{F}_2^{n*} \times \mathbb{F}_2^{r\,2}$, coefficient $\Delta_{K,\dot{K}}(v)$ equals $\frac{c(-1)^{v\cdot(F_K+F_{\dot{K}})}}{2^n} W_\alpha(0,v)$ if $v \cdot (F_K + F_{\dot{K}})$ is constant and equals 0 otherwise.*

Proof. Because the functions α and F_K are independent for every $K \in \mathbb{F}_2^r$, the correlation coefficient $\mathrm{Cor}(v \cdot F_K \oplus u \cdot F_{\dot{K}}, u \cdot \alpha)$, $(u,v) \in \mathbb{F}_2^{n\,2}$, $H(u)=1$, equals zero if $v \cdot F_K \oplus u \cdot F_{\dot{K}}$ is not constant and equals $\pm W_{u\cdot\alpha}(0)$ if $v \cdot F_K \oplus u \cdot F_{\dot{K}}$ is constant. We assumed that Boolean functions $v \cdot F_K$ and $u \cdot F_{\dot{K}}$ are independent for every pair $(K,\dot{K}) \in \mathbb{F}_2^{r\,2}$ and every pair of distinct vectors (u,v) such that $H(u)=1$. One deduces that if $v \cdot F_K \oplus u \cdot F_{\dot{K}}$ is constant, then u equals v (and $H(v)=1$). ◇

5.2 Study of $\Delta_{K,\dot{K}}$ When α Is Constant

It is realistic to assume that during the execution of an algorithm embedded in smart cards, state function α is constant. This can be assigned to the so-called *pre-charged logic* where the bus is cleared between each significant transferred value or when the previous operation concerning the bus is an opcode loading (*cf.* [5]). As explained in [3], another possible reason is that complex architectures implement separated busses for data and addresses, that may prohibit certain transitions.

Proposition 2 was established after assuming in particular that functions $v \cdot F_K \oplus u \cdot F_{\dot{K}}$ and $u \cdot \alpha$ are independent for every pair of distinct nonzero vectors $(u,v) \in \mathbb{F}_2^{n\,2}$, $H(u)=1$, and every pair $(K,\dot{K})$. When α is assumed to be constant, this assumption cannot be satisfied. However when α is constant, Relation (9) can be rewritten $\Delta_{K,\dot{K}}(v) = \frac{c}{2^n}\sum_{\substack{u\in\mathbb{F}_2^n \\ H(u)=1}} (-1)^{u\cdot\beta}\mathrm{Cor}\left(v \cdot F_K, u \cdot F_{\dot{K}}\right)$, after denoting by β the constant value of α. Thus, one straightforwardly deduces the following proposition:

Proposition 3. *Let $(F_K)_{K\in\mathbb{F}_2^r}$ be a family of (n,n)-functions. Let α denote the state function of a cryptographic device implementing functions F_K, $K \in \mathbb{F}_2^r$, as round functions and let c denote the average power consumption to switch a bit in the system. Let us assume that functions $v \cdot F_K$ and $u \cdot F_{\dot{K}}$ are independent for every pair $(K,\dot{K}) \in \mathbb{F}_2^{r\,2}$ and for every pair of distinct elements $(u,v) \in \mathbb{F}_2^{n\,2}$, $H(u)=1$. If α is constant, equal to $\beta \in \mathbb{F}_2^n$, then for every 3-tuple $(v,K,\dot{K}) \in \mathbb{F}_2^{n*} \times \mathbb{F}_2^{r\,2}$, the differential trace of K with respect to $(\dot{K},v)$ satisfies:*

$$\Delta_{K,\dot{K}}(v) = \frac{c \times (-1)^{v \cdot \beta}}{2^n} \mathrm{Cor}\,(v \cdot F_K, v \cdot F_{\dot{K}}) \ . \qquad (11)$$

5.3 Efficiency of the Discrimination of Round Keys in DPA Attacks

Usually, DPA attacks do not permitt to obtain the expected key $\dot{K}$ immediately but allow to isolate it in a subset of $\mathbb{F}_2^r$. For single-bit DPA attacks (resp. multi-bit DPA attacks), the elements of this subset correspond to *ghost peaks* in the distribution of the values of the function $K \in \mathbb{F}_2^r - \{\dot{K}\} \mapsto |\Delta_{K,\dot{K}}(v)|$ (resp. $K \in \mathbb{F}_2^r - \{\dot{K}\} \mapsto |\delta_{\dot{K}}(K)|$). Clearly, the greater the number of ghost peaks, the smaller the efficiency of the attack. Indeed, wrong guesses have to be tested again.

Under assumptions done in Propositions 2 and 3, the set of round keys selected in a single-bit DPA attack with respect to a pair $(v, \dot{K})$ contains the set $\{K \in \mathbb{F}_2^r |\ v \cdot (F_K + F_{\dot{K}}) = \mathrm{cst}\}$. Indeed, when the state function is constant or independent of functions F_K, then the value $|\Delta_{K,\dot{K}}(v)|$ is maximal for every K belonging to $\{K \in \mathbb{F}_2^r |\ v \cdot (F_K + F_{\dot{K}}) = \mathrm{cst}\}$. For multi-bit DPA attacks on a device with random (or null) state function, the set of selected round keys admits the set $\{K \in \mathbb{F}_2^r |\ F_K + F_{\dot{K}} \in \{0_n, 1_n\}\}$ as a subset.

6 Resistance of S-Boxes to DPA Attacks When Round Keys Are Introduced by Addition

In many iterated block ciphers such as DES [27] or AES [28], the round key is introduced *by addition*. In this case, we have $r = n$ and, for every round key $K \in \mathbb{F}_2^n$, the round function F_K is the function $X \mapsto F(X + K)$, where F is a robust cryptographic permutation on $\mathbb{F}_2^n$. In such a system we call *S-box* the function F.

In the rest of the paper, we assume that the round keys are introduced by addition. Under this assumption, Propositions 2 and 3 imply the following corollary:

Corollary 1. *Let F be an (n, n)-function whose coordinate functions are pairwisely independent and let α be the state function of a cryptographic device in which F is embedded as an S-box. If α is independent of F or constant, then the number of round keys selected after a single-bit DPA attack with respect to the vector $v \in \mathbb{F}_2^n$ (resp. after a multi-bit DPA attack) is greater than or equal to $\#\varepsilon_{v \cdot F}$ (resp. $\#\varepsilon_F$).*

One cannot withstand multi-bit DPA attacks by increasing the size of the linear space ε_F of F, since the elements of $\dot{K} + \varepsilon_F$ act in a very similar way in the cryptosystem. Indeed, by definition of ε_F, for every element K in $\dot{K} + \varepsilon_F$, there exists a constant $C \in \mathbb{F}_2^n$ such that $X \mapsto F(X + \dot{K})$ and $X \mapsto F(X + K) + C$ are equal.

We showed in Sect. 4.2 that only vectors K such that $\delta_{\dot{K}}(K)$ is maximal have to be stored as good candidate round keys. In practice, because of the imperfections of the measurements (and also because the values of $K \mapsto \Delta_{K,\dot{K}}(N,j)$, $N \leq 2^n$ and $j \leq n$ fixed, are not computed for $N = 2^n$ but for large $N \ll 2^n$), every tested vector such that $\delta_{\dot{K}}(K)$ is significantely high, is stored as a good candidate key (even if $\delta_{\dot{K}}(K)$ is not the maximal value achieved). For this reason, it is difficult to mount an efficient DPA attack when the amplitude of the peaks in the distribution of the values $\delta_{\dot{K}}(K)$, $K \in \mathbb{F}_2^n$, are not high enough (*cf.* [5,6]). Indeed, let us denote by σ the assumed margin of error on the computation of values $\delta_{\dot{K}}(K)$. We argued in Sect. 4 that under some realistic assumptions, the value $\delta_{\dot{K}}(K)$ is always maximal for $K = \dot{K}$. Thus, if the average value

$$D(\dot{K}) = \frac{1}{2^n - 1} \sum_{K \in \mathbb{F}_2^n - \{\dot{K}\}} \left(\delta_{\dot{K}}(\dot{K}) - \delta_{\dot{K}}(K)\right) \tag{12}$$

is smaller than σ, then the peaks in the distribution of values $\delta_{\dot{K}}(K)$ could not be identified by an attacker because of the imperfections of the measurements. Reciprocally, if Difference (12) is significantly higher than σ, then the peak corresponding to $\delta_{\dot{K}}(\dot{K})$ will clearly appear in the distribution of values $\delta_{\dot{K}}(K)$ when K ranges over $\mathbb{F}_2^n$.

Let us develop the computation of $D(\dot{K})$ for α independent of F and for α constant.

Lemma 2. *Let F be a (n,n)-function whose coordinate functions are pairwisely independent and let α be the state function of a cryptographic system implementing F as an S-box.*
If α is independent of F, then for every element $\dot{K} \in \mathbb{F}_2^n$ we have :

$$D(\dot{K}) = \frac{c}{2^n} \Big| \sum_{\substack{v \in \mathbb{F}_2^n \\ H(v)=1}} W_\alpha(0,v) \Big| - \frac{1}{2^n - 1} \sum_{K \in \mathbb{F}_2^n - \{\dot{K}\}} \delta_{\dot{K}}(K) \ . \tag{13}$$

If α is constant and equals $\beta \in \mathbb{F}_2^n$, then for every element $\dot{K} \in \mathbb{F}_2^n$ we have :

$$D(\dot{K}) = c|n - 2H(\beta)| - \frac{c}{2^{2n} - 2^n} \sum_{a \in \mathbb{F}_2^{n*}} \Big| \sum_{\substack{v \in \mathbb{F}_2^n \\ H(v)=1}} (-1)^{v \cdot \beta} W_{D_a F}(0,v) \Big| \ . \tag{14}$$

Proof. Due to Proposition 2, if α and F are independent, then the summations $\sum_{v \in \mathbb{F}_2^n,\ H(v)=1} \Delta_{\dot{K},\dot{K}}(v)$ and $\frac{c}{2^n} \sum_{v \in \mathbb{F}_2^n,\ H(v)=1} W_\alpha(0,v)$ are equivalent: one straightforwardly deduces Relation (13). If the function α equals the constant value β, coefficient $W_\alpha(0,v)$ in Relation (13) equals $(-1)^{v \cdot \beta} \times 2^n$. Moreover, due to Remark 1 and Relation (11), one has

$$\Delta_{K,\dot{K}}(v) = \frac{c}{2^n} \times (-1)^{v \cdot \beta} W_{D_{K+\dot{K}}F}(0,v) \ . \tag{15}$$

From Relations (1), (10), (13) and (15), one deduces Relation (14). ◇

Remark 3.
1. More generally, one can rewritte Relation (14) for (n, m)-functions as:

$$D(\dot{K}) = c|m - 2H(\beta)| - \frac{c}{2^{2n} - 2^n} \sum_{a \in \mathbb{F}_2^{n*}} | \sum_{\substack{v \in \mathbb{F}_2^n \\ H(v)=1}} (-1)^{v \cdot \beta} W_{D_a F}(0, v)| \ . \quad (16)$$

Moreover, due to Relation (1), summation $\sum_{\substack{v \in \mathbb{F}_2^n \\ H(v)=1}} (-1)^{v \cdot \beta} W_{D_a F}(0, v)$ is also equal to $[n2^n - 2\sum_{X \in \mathbb{F}_2^n} H(\beta + D_a F(X))]$. 2. For every vector $t \in \mathbb{F}_2^n$, let τ_t denotes the function $X \in \mathbb{F}_2^n \mapsto X + t$. Since $W_{D_{K+\dot{K}} F}(0, v)$ equals the function $X \mapsto \mathrm{Cor}(v \cdot F, v \cdot F \circ \tau_{K+\dot{K}})$, Relation (15) relates the differential trace function to the *cross-correlation function* of the coordinate functions of F viewed as binary sequences (see for instance [12] for more details about the cross-correlation function of binary sequences). ◇

As we recalled in Sect. 5.2, it is realistic to assume that during the execution of an algorithm running in a smart card environment, state function α is constant. For such a case, we introduce a new notion, that we call transparency order of a function, to quantify the resistance of an S-box to (single bit or multi-bit) DPA attacks.

7 Transparency Order of S-Boxes

7.1 Definition

Let us assume that the state function is constant. Usually, one cannot presuppose the constant value taken by α, which depends on the implementation. Thus, to thwart DPA attacks on one round of an iterated block cipher, the $D(\dot{K})$ values have to be small enough not only for any round key $\dot{K}$ but also for every possible value β. This remark leads us to introduce a new criterion on S-boxes. In order to be as general as possible, we introduce the notion for (n, m)-functions and not only for permutations on $\mathbb{F}_2^n$.

Definition 2. *Let n and m be two positive integers and let F be an (n, m)-function. The transparency order of F, denoted by $\mathcal{T}_F$, is defined by:*

$$\mathcal{T}_F = \max_{\beta \in \mathbb{F}_2^m} (|m - 2H(\beta)| - \frac{1}{2^{2n} - 2^n} \sum_{a \in \mathbb{F}_2^{n*}} | \sum_{\substack{v \in \mathbb{F}_2^m \\ H(v)=1}} (-1)^{v \cdot \beta} W_{D_a F}(0, v)|). \quad (17)$$

The smaller the transparency order of an S-box, the higher its resistance to DPA attacks. Indeed, to make the peak corresponding to $\delta_{\dot{K}}(\dot{K})$ undistinguishable from noise of measurements, value $\delta_{\dot{K}}(\dot{K})$ must be approximately equal to the average amplitude $\delta_{\dot{K}}(K)$ when K ranges over $\mathbb{F}_2^n$. Thus, the greatest transparency order that an S-box can achieve without compromising its resistance to DPA attacks depends on the quality of the measurements an attacker can achieve [2].

7.2 Study of Transparency Order of S-Boxes

In order to determine what a reasonably high transparency order is, there is a need for an upper bound on the transparency order of (n, m)-functions. In what follows, we introduce an upper bound and a lower bound on the transparency order of a function. We show that these bounds can be achieved.

Theorem 1. *Let n and m be two positive integers, transparency order $\mathcal{T}_F$ of every (n, m)-function F satisfies the following relation:*

$$0 \leq \mathcal{T}_F \leq m \ . \tag{18}$$

If every coordinate function of F is bent, then $\mathcal{T}_F = m$. Moreover, $\mathcal{T}_F$ is null if and only if F is an affine function, whose direction L satisfies $Im(L) \subseteq \{0_m, 1_m\}$.

Remark 4. Since n-variables bent functions only exist for n even, the tightness of the upper bound in Relation (18) is still an open problem for n odd. ◇

Being unbalanced, bent functions are never used as cryptographic primitives. However, due to their properties recalled in Sect. 2 (optimal non-linearity and only balanced non-zero derivatives), they resist in an optimal way to linear and differential cryptanalysis. By showing that bent functions are the weakest possible functions from DPA attacks viewpoint, Theorem 1 establishes that it is impossible to design a function that can resist in an optimal way to linear, differential and DPA attacks. In the following proposition, we show more generally that the functions satisfying $PC(l)$ for a large (but not necessarily optimal) order l do not have a good transparency order.

Proposition 4. *Let m and n be two positive integers such that $m \leq n$. Let F be a (n, m)-function. Let $l \leq n$ be a positive integer. If F satisfies the $PC(l)$ criteria, then the transparency order of F satisfies:*

$$\mathcal{T}_F \geq m\left(1 - \frac{2^n - \sum_{j=0}^{l}\binom{n}{j}}{2^n - 1}\right) \ . \tag{19}$$

Proof. Because function F satisfies $PC(l)$, then function D_aF is balanced for every vector a s.t. $H(a) \leq l$. Due to Proposition 1, one deduces that for every vector a such that $H(a) \leq l$ and for every non-zero vector $v \in \mathbb{F}_2^m$, we have $W_{D_aF}(0, v) = 0$. Thus, if F satisfies $PC(l)$, then for every vector $\beta \in \mathbb{F}_2^m$, we have

$$\sum_{a \in \mathbb{F}_2^{n*}} \mid \sum_{\substack{v \in \mathbb{F}_2^m \\ H(v)=1}} (-1)^{v \cdot \beta} W_{D_aF}(0, v)| = \sum_{\substack{a \in \mathbb{F}_2^{n*} \\ H(a)>l}} \mid \sum_{\substack{v \in \mathbb{F}_2^m \\ H(v)=1}} (-1)^{v \cdot \beta} W_{D_aF}(0, v)| \ .$$

[2] By adding Hardware's countermeasures to the device, it is possible to ensure a minimal margin of error for any measurement of the power consumption.

The cardinality of the set $\{a \in \mathbb{F}_2^n,\ H(a) > l\}$ is $2^n - \sum_{j=0}^{l} \binom{n}{j}$. Moreover, since every value $W_{D_aF}(0,v)$ is lower than or equal to 2^n, then the inequality $|\sum_{\substack{v\in\mathbb{F}_2^m\\ H(v)=1}} (-1)^{v\cdot\beta} W_{D_aF}(0,v)| \leq m2^n$ is satisfied for every $\beta \in \mathbb{F}_2^m$. One deduces the following relation for every vector $\beta \in \mathbb{F}_2^m$:

$$\sum_{a\in\mathbb{F}_2^{n*}} |\sum_{\substack{v\in\mathbb{F}_2^m\\ H(v)=1}} (-1)^{v\cdot\beta} W_{D_aF}(0,v)| \leq m2^n \left(2^n - \sum_{j=0}^{l} \binom{n}{j}\right) . \quad (20)$$

From Relations (17) and (20) and the fact that $\max_\beta |m - 2H(\beta)|$ is maximal for $\beta \in \{0_m, 1_m\}$, one deduces Inequality (19). ◇

In the next proposition, we investigate the transparency order of affine (n,m)-functions. In particular, we argue that the transparency of an affine function is related to the *weight enumerators* of the *cosets* of $\mathrm{Im}(L)$, where $\mathrm{Im}(L)$ is seen as a *binary linear code.*

Proposition 5. *Let n and m be two positive integers. Let F be an affine (n,m)-function admitting L for direction, then its transparency order satisfies the following relation:*

$$\mathcal{T}_F = \max_{\beta\in\mathbb{F}_2^m} \left(\frac{2^n}{2^n-1}|m - 2H(\beta)| - \frac{1}{2^n-1} \sum_{j=0}^{m} |m-2j| \mathcal{N}_{j,\beta} \right) , \quad (21)$$

where $\mathcal{N}_{j,\beta}$ denotes the cardinality of the set $\{a \in \mathbb{F}_2^n;\ H(L(a)+\beta) = j\}$. Moreover, if F is balanced, then its transparency order satisfies:

$$\mathcal{T}_F = \begin{cases} \frac{2^n}{2^n-1}\left(m - \frac{m}{2^m}\binom{m}{\frac{m}{2}}\right) & \text{if } m \text{ is even} \\ \frac{2^n}{2^n-1}\left(m - \frac{2m}{2^m}\binom{m-1}{\frac{m-1}{2}}\right) & \text{if } m \text{ is odd} \end{cases} . \quad (22)$$

Remark 5.
1. In Proposition 5, the set $\beta + \mathrm{Im}\ (L)$ can be viewed as a coset of a linear code. Let C denotes this code. If β belongs to C, then $\beta + \mathrm{Im}(L) = \mathrm{Im}(L)$ and values $\mathcal{N}_{j,\beta}$, $j \leq m$, are the coefficients of the *weight enumerator* of C (see for instance [20] for more details about weight enumerators of codes).
2. We recall that for m even and due to *Stirling's formula*, we have $\binom{m}{m/2} \simeq 2^m/\sqrt{\frac{m}{2}\pi}$ for large values of m. Thus for large values m and for balanced affine (n,m)-functions F, the transparency order of F equals approximately $\frac{2^n}{2^n-1}(m - \sqrt{\frac{2m}{\pi}})$ if m is even and to $\frac{2^n}{2^n-1}\left(m - \frac{m}{\sqrt{\frac{(m-1)\pi}{2}}}\right)$ if m is odd. ◇

Due to Proposition 5 and to Remark 5, the transparency order of balanced affine functions is not close to 0 for high values m. Moreover, Relation (21) relates the problem of the construction of affine functions with small transparency order to the problem of defining linear codes whose elements have a Hamming weight either close to 0 or close to m.

8 Conclusion

The study of DPA attacks in terms of correlation coefficients enables us better to understand these attacks. It allows us to characterize the properties of S-boxes related to DPA attacks. To quantify the information leakage of devices involving S-boxes, we introduced the notion of transparency order. We established a spectral characterization of the transparency order of S-boxes and we exhibit its upper and lower bounds. We proved that the lower bound is achieved by particular affine functions and we proved that the transparancy order of bent functions achieves the upper bound. The construction of highly-nonlinear S-boxes with small transparency order (close to 0) is an open problem. The definition of such S-boxes would allow the design of specific block cipher algorithms for smart cards which are less resistant to linear or differential attacks but are inherently resistant to DPA attacks. To make up for this security loss, such algorithms can be implemented in smart cards without the high penalties due to DPA-countermeasures.

References

1. M.-L. Akkar, R. Bévan, P. Dischamp, and D. Moyart. Power Analysis, What is Now Possible. In T. Okamoto, editor, *ASIACRYPT 2000*, volume 1976 of *LNCS*, pages 489–502. Springer, 2000.
2. E. Biham and A. Shamir. Differential cryptanalysis of DES-like cryptosystems. *Journal of Cryptology*, 4(1):3–72, 1991.
3. E. Brier, C. Clavier, and F. Olivier. Correlation Power Analysis with a Leakage Model. In M. Joye and J.-J. Quisquater, editors, *CHES 2004*, volume 3156 of *LNCS*, pages 16–29. Springer, 2004.
4. S. Chari, C. Jutla, J. Rao, and P. Rohatgi. Towards Sound Approaches to Counteract Power-Analysis Attacks. In M. Wiener, editor, *CRYPTO '99*, volume 1666 of *LNCS*, pages 398–412. Springer, 1999.
5. C. Clavier, J.-S. Coron, and N. Dabbous. Differential power analysis in the presence of hardware countermeasures. In Ç. Koç and C. Paar, editors, *CHES 2000*, volume 1965 of *LNCS*, pages 252–263. Springer, 2000.
6. J.-S. Coron, P. Kocher, and D. Naccache. Statistics and secret leakage. In Y. Frankel, editor, *Financial Cryptography - FC 2000*, volume 1962 of *LNCS*. Springer, 2000.
7. J. Evertse. Linear structures in blockciphers. In D. Chaum and W. Price, editors, *EUROCRYPT '87*, volume 304 of *LNCS*, pages 249–266. Springer, 1987.
8. L. Goubin and J. Patarin. DES and Differential Power Analysis – The Duplication Method. In Ç. Koç and C. Paar, editors, *CHES '99*, volume 1717 of *LNCS*, pages 158–172. Springer, 1999.
9. S. Guilley, P. Hoogvorst, and R. Pascalet. Differential power analysis model and some results. In J.-J. Quisquater, P. Paradinas, Y. Deswarte, and A. E. Kalam, editors, *Smart Card Research and Advanced Applications VI - CARDIS 2004*, pages 127–142. Kluwer Academic Publishers, 2004.
10. C. Harpes. Cryptanalysis of iterated block ciphers. In *ETH Series in Information Processing*, volume 7. Hartung-Gorre Verlag, 1996.

11. A. A. Hasan. Power analysis attacks and algorithmic approaches to their countermeasures for Koblitz cryptosystems. In Ç. Koç and C. Paar, editors, *CHES 2000*, volume 1965 of *LNCS*, pages 93–108. Springer, 2000.
12. T. Helleseth and P. V. Kumar. Sequences with low correlation. In *Handbook of coding theory, Vol. II*, pages 1765–1853. North-Holland, 1998.
13. L. Knudsen. Truncated and Higher Order Differentials. In B. Preneel, editor, *Fast Software Encryption - FSE '94*, volume 1008 of *LNCS*, pages 196–211. Springer, 1994.
14. P. Kocher. Timing attacks on implementations of Diffie-Hellman, RSA, DSS, and other systems. In N. Koblitz, editor, *CRYPTO '96*, volume 1109 of *LNCS*, pages 104–113. Springer, 1996.
15. P. Kocher, J. Jaffe, and B. Jun. Differential Power Analysis. In M. Wiener, editor, *CRYPTO '99*, volume 1666 of *LNCS*, pages 388–397. Springer, 1999.
16. Z. Kukorelly. On the validity of certain hypotheses used in linear cryptanalysis. In *ETH Series in Information Processing*, volume 13. Hartung-Gorre Verlag, 1999.
17. X. Lai. Higher order derivatives and differential cryptanalysis. In *Symposium on Communication, Coding and Cryptography*, 1994. en l'honneur de J.L. Massey à l'occasion de son 60ème anniversaire.
18. F. J. MacWilliams and N. J. A. Sloane. *The theory of error-correcting codes.* North-Holland Publishing Co., 1977. North-Holland Mathematical Library, Vol. 16.
19. M. Matsui. Linear cryptanalysis method for DES cipher. In T. Helleseth, editor, *EUROCRYPT '93*, volume 765 of *LNCS*, pages 386–397. Springer, 1993.
20. R. Mayer Sommer. Smartly Analyzing the Simplicity and the Power of Simple Power Analysis on Smartcards. In Ç. Koç and C. Paar, editors, *CHES 2000*, volume 1965 of *LNCS*, pages 78–92. Springer, 2000.
21. T. Messerges. *Power Analysis Attacks and Countermeasures for Cryptographic Algorithms.* PhD thesis, University of Illinois, 2000.
22. T. Messerges, E. Dabbish, and R. Sloan. Investigations of Power Analysis Attacks on Smartcards. In *the USENIX Workshop on Smartcard Technology (Smartcard '99)*, pages 151–161, 1999.
23. T. Messerges, E. Dabbish, and R. Sloan. Power Analysis Attacks of Modular Exponentiation in Smartcard. In Ç. Koç and C. Paar, editors, *CHES '99*, volume 1717 of *LNCS*, pages 144–157. Springer, 1999.
24. T. Messerges, E. Dabbish, and R. Sloan. Examining Smart-Card Security under the Threat of Power Analysis Attacks. *IEEE Transactions on Computers*, 51(5), May 2002.
25. National Bureau of Standards. *FIPS PUB 46: The Data Encryption Standard*, January 1977.
26. National Institute of Standards and Technology. *FIPS PUB 197: Advanced Encryption Standard*, 2001.
27. E. Oswald. *On Side-Channel Attacks and the Application of Algorithmic Countermeasures.* PhD thesis, Institute for Applied Information Processing and Communications - Graz University of Technology, May 2003.
28. B. Preneel, R. Govaerts, and J. Vandewalle. Boolean functions satisfying higher order propagation criteria. In F. Pichler, editor, *EUROCRYPT '85*, volume 219 of *LNCS*, pages 141–152. Springer, 1985.
29. O. S. Rothaus. On bent functions. In *Journal of Combinatorial Theory*, volume 20a, pages 300–305. Academic Press, 1976.
30. A. Webster and S. Tavares. On the design of S-boxes. In H. Wiliams, editor, *CRYPTO '85*, volume 218 of *LNCS*, pages 523–534. Springer, 1985.

A Proofs of Theorem 1 and of Proposition 5

A.1 Proof of Theorem 1

Proof. The value of $|m - 2H(\beta)|$ is upper bounded by m and equals m for $\beta = 0_m, 1_m$. On the other hand, values taken by the summation in Relation (17) belong to $[0; m]$. One straightforwardly deduces Inequality (18).
$\mathcal{T}_F$ equals m if and only if $\beta \in \{0_m, 1_m\}$. In this case, the value of the summation $\sum_{a\in\mathbb{F}_2^{n*}} |\sum_{\substack{v\in\mathbb{F}_2^m\\ H(v)=1}} (-1)^{v\cdot\beta} W_{D_aF}(0,v)|$ is null if and only if the summation $\sum_{\substack{v\in\mathbb{F}_2^m\\ H(v)=1}} W_{D_aF}(0,v)$ is null for every non-zero vector a. On the other hand, if every coordinate function of F is bent, then for every $a \in \mathbb{F}_2^n$ and every $v \in \mathbb{F}_2^m$ such that $H(v) = 1$, the function $D_a(v \cdot F)$ is balanced and (due to Proposition 1) satisfies $W_{D_aF}(0,v) = 0$. One concludes that such functions F, $\mathcal{T}_F$ is maximal and equals m.
Now, we show that if $\mathcal{T}_F$ is null, then F is an affine function, whose direction L satisfies $\mathrm{Im}(L) \subseteq \{0_m, 1_m\}$. By definition, $\mathcal{T}_F$ is greater than or equal to each value

$$|m - 2H(\beta)| - \frac{1}{2^n(2^n-1)} \sum_{a\in\mathbb{F}_2^{n*}} |\sum_{v\in\mathbb{F}_2^m, H(v)=1} (-1)^{v\cdot\beta} W_{D_aF}(0,v)| \ ,$$

$\beta \in \mathbb{F}_2^m$, which implies (for $\beta \in \{0_m, 1_m\}$):

$$m - \frac{1}{2^n(2^n-1)} \sum_{a\in\mathbb{F}_2^{n*}} |\sum_{v\in\mathbb{F}_2^m,\ H(v)=1} W_{D_aF}(0,v)| \leq \mathcal{T}_F \ . \tag{23}$$

The left-hand side of Relation (23) being always positive or null, if $\mathcal{T}_F$ equals 0, then $m - \frac{1}{2^n(2^n-1)} \sum_{a\in\mathbb{F}_2^{n*}} |\sum_{v\in\mathbb{F}_2^m,\ H(v)=1} W_{D_aF}(0,v)|$ must equal 0, which is equivalent to:

$$\sum_{a\in\mathbb{F}_2^{n*}} |\sum_{v\in\mathbb{F}_2^m,\ H(v)=1} W_{D_aF}(0,v)| = m2^n(2^n-1) \ . \tag{24}$$

Relation (24) is satisfied if and only if $|W_{D_aF}(0,v)|$ equals 2^n for every pair $(a,v) \in \mathbb{F}_2^{n*} \times \mathbb{F}_2^m$, $H(v) = 1$, which implies that F is affine. Let L denote the direction of F, then Relation (24) is equivalent to

$$\sum_{a\in\mathbb{F}_2^{n*}} |\sum_{v\in\mathbb{F}_2^m,\ H(v)=1} (-1)^{v\cdot L(a)}| = m(2^n-1) \ ,$$

and the equality holds if and only if $\sum_{v\in\mathbb{F}_2^m,\ H(v)=1} (-1)^{v\cdot L(a)}$ (that is the value $m - 2H(L(a))$) equals $\pm m$ i.e. if and only if $L(a)$ equals 0_m or 1_m. One deduces that if $\mathcal{T}_F$ equals 0, then F is an affine function whose direction L satisfies $\mathrm{Im}(L) \subseteq \{0_m, 1_m\}$. Let us prove now that this necessary condition is a sufficient one.

Let F be an affine function whose direction L satisfies $\mathrm{Im}(L) \subseteq \{0_m, 1_m\}$. Then summation $|\sum_{v\in\mathbb{F}_2^n, H(v)=1} W_{D_aF}(0,v)|$ equals $2^n|m-2H(\beta)|$ if $L(a)=0_m$ and equals $2^n|m-2H(\beta+1_m)|$ if $L(a)=1_m$. Since one has $|m-2H(\beta+1_m)| = |m-2H(\beta)|$, one deduces the equality

$$\sum_{a\in\mathbb{F}_2^{n*}} |\sum_{v\in\mathbb{F}_2^n, H(v)=1} W_{D_aF}(0,v)| = 2^n(2^n-1)|m-2H(\beta)| \ ,$$

and hence, that $\mathcal{T}_F$ is null. ◇

A.2 Proof of Proposition 5

Before providing proof of Proposition 5, let us first introduce the following technical lemma:

Lemma 3. *For every positive integer m, the following relation is satisfied:*

$$\sum_{j=0}^{m} |m-2j| \binom{m}{j} = \begin{cases} m\binom{m}{\frac{m}{2}} & \text{if } m \text{ is even} \\ 2m\binom{m-1}{\frac{m-1}{2}} & \text{if } m \text{ is odd} \end{cases} . \quad (25)$$

Using Lemma 3, a proof of Proposition 5 is:

Proof. Function L being the direction of F, for every pair $(a,v) \in \mathbb{F}_2^n \times \mathbb{F}_2^m$, coefficient $W_{D_aF}(0,v)$ equals $2^n(-1)^{v\cdot L(a)}$. Thus, for every $\beta \in \mathbb{F}_2^m$, summation $\sum_{\substack{v\in\mathbb{F}_2^m \\ H(v)=1}} (-1)^{v\cdot\beta} W_{D_aF}(0,v)$ equals $2^n\left(m-2H\left(\beta+L(a)\right)\right)$. Hence, from Relation (17) one deduces:

$$\mathcal{T}_F = \max_{\beta\in\mathbb{F}_2^m} \left(\frac{2^n}{2^n-1}|m-2H(\beta)| - \frac{1}{2^n-1}\sum_{a\in\mathbb{F}_2^n} |m-2H\left(\beta+L(a)\right)| \right) . \quad (26)$$

Because summation $\sum_{a\in\mathbb{F}_2^n} |m-2H\left(\beta+L(a)\right)|$ can be rewritten on the form $\sum_{j=0}^{m}\sum_{a\in\mathbb{F}_2^n, H(\beta+L(a))=j} |m-2j|$, Relation (21) is satisfied.
If F is balanced, then $\mathrm{Im}(L) = \beta + \mathrm{Im}(L) = \mathbb{F}_2^m$ and $\mathcal{N}_{j,\beta}$ equals $2^{n-m} \times \binom{m}{j}$ for every vector β and every integer $j \leq m$. In this case, summation $\sum_{a\in\mathbb{F}_2^n} |m-2H\left(\beta+L(a)\right)|$ equals $2^{n-m}\sum_{j=0}^{m}|m-2j|\binom{m}{j}$. By applying Lemma 3, one deduces that for every balanced affine (n,m)-function, Relations (22) and (26) are equivalent. ◇

Author Index

Lecture Notes in Computer Science

For information about Vols. 1–3463

please contact your bookseller or Springer

Vol. 3573: S. Etalle (Ed.), Logic Based Program Synthesis and Transformation. VIII, 278 pages. 2005.

Vol. 3569: F. Bacchus, T. Walsh (Eds.), Theory and Applications of Satisfiability Testing. XII, 492 pages. 2005.

Vol. 3562: J. Mira, J.R. Álvarez (Eds.), Artificial Intelligence and Knowledge Engineering Applications: A Bioinspired Approach, Part II. XXIV, 636 pages. 2005.

Vol. 3561: J. Mira, J.R. Álvarez (Eds.), Mechanisms, Symbols, and Models Underlying Cognition, Part I. XXIV, 532 pages. 2005.

Vol. 3560: V.K. Prasanna, S. Iyengar, P.G. Spirakis, M. Welsh (Eds.), Distributed Computing in Sensor Systems. XV, 423 pages. 2005.

Vol. 3559: P. Auer, R. Meir (Eds.), Learning Theory. XI, 692 pages. 2005. (Subseries LNAI).

Vol. 3557: H. Gilbert, H. Handschuh (Eds.), Fast Software Encryption. XI, 443 pages. 2005.

Vol. 3556: H. Baumeister, M. Marchesi, M. Holcombe (Eds.), Extreme Programming and Agile Processes in Software Engineering. XIV, 332 pages. 2005.

Vol. 3555: T. Vardanega, A. Wellings (Eds.), Reliable Software Technology – Ada-Europe 2005. XV, 273 pages. 2005.

Vol. 3552: H. de Meer, N. Bhatti (Eds.), Quality of Service – IWQoS 2005. XV, 400 pages. 2005.

Vol. 3551: T. Härder, W. Lehner (Eds.), Data Management in a Connected World. XIX, 371 pages. 2005.

Vol. 3548: K. Julisch, C. Kruegel (Eds.), Intrusion and Malware Detection and Vulnerability Assessment. X, 241 pages. 2005.

Vol. 3547: F. Bomarius, S. Komi-Sirviö (Eds.), Product Focused Software Process Improvement. XIII, 588 pages. 2005.

Vol. 3543: L. Kutvonen, N. Alonistioti (Eds.), Distributed Applications and Interoperable Systems. XI, 235 pages. 2005.

Vol. 3541: N.C. Oza, R. Polikar, J. Kittler, F. Roli (Eds.), Multiple Classifier Systems. XII, 430 pages. 2005.

Vol. 3540: H. Kalviainen, J. Parkkinen, A. Kaarna (Eds.), Image Analysis. XXII, 1270 pages. 2005.

Vol. 3537: A. Apostolico, M. Crochemore, K. Park (Eds.), Combinatorial Pattern Matching. XI, 444 pages. 2005.

Vol. 3536: G. Ciardo, P. Darondeau (Eds.), Applications and Theory of Petri Nets 2005. XI, 470 pages. 2005.

Vol. 3535: M. Steffen, G. Zavattaro (Eds.), Formal Methods for Open Object-Based Distributed Systems. X, 323 pages. 2005.

Vol. 3533: M. Ali, F. Esposito (Eds.), Innovations in Applied Artificial Intelligence. XX, 858 pages. 2005. (Subseries LNAI).

Vol. 3532: A. Gómez-Pérez, J. Euzenat (Eds.), The Semantic Web: Research and Applications. XV, 728 pages. 2005.

Vol. 3531: J. Ioannidis, A. Keromytis, M. Yung (Eds.), Applied Cryptography and Network Security. XI, 530 pages. 2005.

Vol. 3530: A. Prinz, R. Reed, J. Reed (Eds.), SDL 2005: Model Driven Systems Design. XI, 361 pages. 2005.

Vol. 3528: P.S. Szczepaniak, J. Kacprzyk, A. Niewiadomski (Eds.), Advances in Web Intelligence. XVII, 513 pages. 2005. (Subseries LNAI).

Vol. 3527: R. Morrison, F. Oquendo (Eds.), Software Architecture. XII, 263 pages. 2005.

Vol. 3526: S.B. Cooper, B. Löwe, L. Torenvliet (Eds.), New Computational Paradigms. XVII, 574 pages. 2005.

Vol. 3525: A.E. Abdallah, C.B. Jones, J.W. Sanders (Eds.), Communicating Sequential Processes. XIV, 321 pages. 2005.

Vol. 3524: R. Barták, M. Milano (Eds.), Integration of AI and OR Techniques in Constraint Programming for Combinatorial Optimization Problems. XI, 320 pages. 2005.

Vol. 3523: J.S. Marques, N. Pérez de la Blanca, P. Pina (Eds.), Pattern Recognition and Image Analysis, Part II. XXVI, 733 pages. 2005.

Vol. 3522: J.S. Marques, N. Pérez de la Blanca, P. Pina (Eds.), Pattern Recognition and Image Analysis, Part I. XXVI, 703 pages. 2005.

Vol. 3521: N. Megiddo, Y. Xu, B. Zhu (Eds.), Algorithmic Applications in Management. XIII, 484 pages. 2005.

Vol. 3520: O. Pastor, J. Falcão e Cunha (Eds.), Advanced Information Systems Engineering. XVI, 584 pages. 2005.

Vol. 3519: H. Li, P. J. Olver, G. Sommer (Eds.), Computer Algebra and Geometric Algebra with Applications. IX, 449 pages. 2005.

Vol. 3518: T.B. Ho, D. Cheung, H. Liu (Eds.), Advances in Knowledge Discovery and Data Mining. XXI, 864 pages. 2005. (Subseries LNAI).

Vol. 3517: H.S. Baird, D.P. Lopresti (Eds.), Human Interactive Proofs. IX, 143 pages. 2005.

Vol. 3516: V.S. Sunderam, G.D.v. Albada, P.M.A. Sloot, J.J. Dongarra (Eds.), Computational Science – ICCS 2005, Part III. LXIII, 1143 pages. 2005.

Vol. 3515: V.S. Sunderam, G.D.v. Albada, P.M.A. Sloot, J.J. Dongarra (Eds.), Computational Science – ICCS 2005, Part II. LXIII, 1101 pages. 2005.

Vol. 3514: V.S. Sunderam, G.D.v. Albada, P.M.A. Sloot, J.J. Dongarra (Eds.), Computational Science – ICCS 2005, Part I. LXIII, 1089 pages. 2005.

Vol. 3513: A. Montoyo, R. Muñoz, E. Métais (Eds.), Natural Language Processing and Information Systems. XII, 408 pages. 2005.

Vol. 3512: J. Cabestany, A. Prieto, F. Sandoval (Eds.), Computational Intelligence and Bioinspired Systems. XXV, 1260 pages. 2005.

Vol. 3510: T. Braun, G. Carle, Y. Koucheryavy, V. Tsaoussidis (Eds.), Wired/Wireless Internet Communications. XIV, 366 pages. 2005.

Vol. 3509: M. Jünger, V. Kaibel (Eds.), Integer Programming and Combinatorial Optimization. XI, 484 pages. 2005.

Vol. 3508: P. Bresciani, P. Giorgini, B. Henderson-Sellers, G. Low, M. Winikoff (Eds.), Agent-Oriented Information Systems II. X, 227 pages. 2005. (Subseries LNAI).

Vol. 3507: F. Crestani, I. Ruthven (Eds.), Information Context: Nature, Impact, and Role. XIII, 253 pages. 2005.

Vol. 3506: C. Park, S. Chee (Eds.), Information Security and Cryptology – ICISC 2004. XIV, 490 pages. 2005.

Vol. 3505: V. Gorodetsky, J. Liu, V. A. Skormin (Eds.), Autonomous Intelligent Systems: Agents and Data Mining. XIII, 303 pages. 2005. (Subseries LNAI).

Vol. 3504: A.F. Frangi, P.I. Radeva, A. Santos, M. Hernandez (Eds.), Functional Imaging and Modeling of the Heart. XV, 489 pages. 2005.

Vol. 3503: S.E. Nikoletseas (Ed.), Experimental and Efficient Algorithms. XV, 624 pages. 2005.

Vol. 3502: F. Khendek, R. Dssouli (Eds.), Testing of Communicating Systems. X, 381 pages. 2005.

Vol. 3501: B. Kégl, G. Lapalme (Eds.), Advances in Artificial Intelligence. XV, 458 pages. 2005. (Subseries LNAI).

Vol. 3500: S. Miyano, J. Mesirov, S. Kasif, S. Istrail, P. Pevzner, M. Waterman (Eds.), Research in Computational Molecular Biology. XVII, 632 pages. 2005. (Subseries LNBI).

Vol. 3499: A. Pelc, M. Raynal (Eds.), Structural Information and Communication Complexity. X, 323 pages. 2005.

Vol. 3498: J. Wang, X. Liao, Z. Yi (Eds.), Advances in Neural Networks – ISNN 2005, Part III. L, 1077 pages. 2005.

Vol. 3497: J. Wang, X. Liao, Z. Yi (Eds.), Advances in Neural Networks – ISNN 2005, Part II. L, 947 pages. 2005.

Vol. 3496: J. Wang, X. Liao, Z. Yi (Eds.), Advances in Neural Networks – ISNN 2005, Part II. L, 1055 pages. 2005.

Vol. 3495: P. Kantor, G. Muresan, F. Roberts, D.D. Zeng, F.-Y. Wang, H. Chen, R.C. Merkle (Eds.), Intelligence and Security Informatics. XVIII, 674 pages. 2005.

Vol. 3494: R. Cramer (Ed.), Advances in Cryptology – EUROCRYPT 2005. XIV, 576 pages. 2005.

Vol. 3493: N. Fuhr, M. Lalmas, S. Malik, Z. Szlávik (Eds.), Advances in XML Information Retrieval. XI, 438 pages. 2005.

Vol. 3492: P. Blache, E. Stabler, J. Busquets, R. Moot (Eds.), Logical Aspects of Computational Linguistics. X, 363 pages. 2005. (Subseries LNAI).

Vol. 3489: G.T. Heineman, I. Crnkovic, H.W. Schmidt, J.A. Stafford, C. Szyperski, K. Wallnau (Eds.), Component-Based Software Engineering. XI, 358 pages. 2005.

Vol. 3488: M.-S. Hacid, N.V. Murray, Z.W. Raś, S. Tsumoto (Eds.), Foundations of Intelligent Systems. XIII, 700 pages. 2005. (Subseries LNAI).

Vol. 3486: T. Helleseth, D. Sarwate, H.-Y. Song, K. Yang (Eds.), Sequences and Their Applications - SETA 2004. XII, 451 pages. 2005.

Vol. 3483: O. Gervasi, M.L. Gavrilova, V. Kumar, A. Laganà, H.P. Lee, Y. Mun, D. Taniar, C.J.K. Tan (Eds.), Computational Science and Its Applications – ICCSA 2005, Part IV. XXVII, 1362 pages. 2005.

Vol. 3482: O. Gervasi, M.L. Gavrilova, V. Kumar, A. Laganà, H.P. Lee, Y. Mun, D. Taniar, C.J.K. Tan (Eds.), Computational Science and Its Applications – ICCSA 2005, Part III. LXVI, 1340 pages. 2005.

Vol. 3481: O. Gervasi, M.L. Gavrilova, V. Kumar, A. Laganà, H.P. Lee, Y. Mun, D. Taniar, C.J.K. Tan (Eds.), Computational Science and Its Applications – ICCSA 2005, Part II. LXIV, 1316 pages. 2005.

Vol. 3480: O. Gervasi, M.L. Gavrilova, V. Kumar, A. Laganà, H.P. Lee, Y. Mun, D. Taniar, C.J.K. Tan (Eds.), Computational Science and Its Applications – ICCSA 2005, Part I. LXV, 1234 pages. 2005.

Vol. 3479: T. Strang, C. Linnhoff-Popien (Eds.), Location- and Context-Awareness. XII, 378 pages. 2005.

Vol. 3478: C. Jermann, A. Neumaier, D. Sam (Eds.), Global Optimization and Constraint Satisfaction. XIII, 193 pages. 2005.

Vol. 3477: P. Herrmann, V. Issarny, S. Shiu (Eds.), Trust Management. XII, 426 pages. 2005.

Vol. 3476: J. Leite, A. Omicini, P. Torroni, P. Yolum (Eds.), Declarative Agent Languages and Technologies. XII, 289 pages. 2005. (Subseries LNAI).

Vol. 3475: N. Guelfi (Ed.), Rapid Integration of Software Engineering Techniques. X, 145 pages. 2005.

Vol. 3474: C. Grelck, F. Huch, G.J. Michaelson, P. Trinder (Eds.), Implementation and Application of Functional Languages. X, 227 pages. 2005.

Vol. 3472: M. Broy, B. Jonsson, J.-P. Katoen, M. Leucker, A. Pretschner (Eds.), Modell-Based Testing of Reactive Systems. VIII, 659 pages. 2005.

Vol. 3468: H.W. Gellersen, R. Want, A. Schmidt (Eds.), Pervasive Computing. XIII, 347 pages. 2005.

Vol. 3467: J. Giesl (Ed.), Term Rewriting and Applications. XIII, 517 pages. 2005.

Vol. 3466: S. Leue, T.J. Systä (Eds.), Scenarios: Models, Transformations and Tools. XII, 279 pages. 2005.

Vol. 3465: M. Bernardo, A. Bogliolo (Eds.), Formal Methods for Mobile Computing. VII, 271 pages. 2005.

Vol. 3464: S.A. Brueckner, G.D.M. Serugendo, A. Karageorgos, R. Nagpal (Eds.), Engineering Self-Organising Systems. XIII, 299 pages. 2005. (Subseries LNAI).

Printed in the United States
By Bookmasters

Printed in the United States
By Bookmasters